DIGITAL
AND
MICROPROCESSOR
TECHNOLOGY

DIGITAL AND MICROPROCESSOR TECHNOLOGY

Second Edition

PATRICK J. O'CONNOR

DeVry Institute of Technology
Chicago, Illinois

PRENTICE HALL
Englewood Cliffs, New Jersey 07632

Library of Congress Cataloging-in-Publication Data

O'Connor, Patrick J.,
    Digital and microprocessor technology / Patrick J. O'Connor.—
2nd ed.
      p. cm.
    Includes index.
    ISBN 0-13-212754-7
    1. Electronic digital computers—Circuits.   2. Logic circuits.
3. Switching circuits. I. Title.
TK7888.4.026 1988
621.395—dc19
                                      88-12439
                                         CIP

Editorial/production supervision and
  interior design: *Anne Kenney*
Cover design: *Joel Mitnick Design, Inc.*
Manufacturing buyer: *Robert Anderson*

© 1989, 1983 by Prentice-Hall, Inc.
A Division of Simon & Schuster
Englewood Cliffs, New Jersey 07632

Printed in the United States of America

ISBN 0-13-212754-7

Prentice-Hall International (UK) Limited, *London*
Prentice-Hall of Australia Pty. Limited, *Sydney*
Prentice-Hall Canada Inc., *Toronto*
Prentice-Hall Hispanoamericana, S.A., *Mexico*
Prentice-Hall of India Private Limited, *New Delhi*
Prentice-Hall of Japan, Inc., *Tokyo*
Simon & Schuster Asia Pte. Ltd., *Singapore*
Editoria Prentice-Hall do Brasil, Ltda., *Rio de Janeiro*

To Leah, Risha, and Renata

CONTENTS

PART II
Synchronous Circuits
160

PART III
Microprocessors and Microcomputers
259

PREFACE

What is important in computers and digital electronics today? A few years ago it was punchcards and self-modifying program code. Today, it is custom large-scale integration and structured programming. Tomorrow, it may be optical switching elements and non-Von Neumann processing. We always have to think of tomorrow and to try to anticipate what may be the next generation of technology to "take hold."

The first edition of *Digital and Microprocessor Technology* was begun almost six years ago. That's a long time in digital electronics. It takes several years between the time the typing starts and the time the book finally rests in the reader's hands. That is time for technology to change, so the writer's task is that of a crystal-ball gazer as well as a presenter of facts. In some sections of the first edition, the guesses were good, and we anticipated trends well enough that these sections did not have to be changed. In other sections, the guesses fell short, or strayed from the course of actual progress in technology. As a result, many changes were needed in order to bring the second edition of this book up to date.

One of the changes we successfully anticipated was a greater interest in digital electronics on the part of students beginning their course of studies in electronics. More students than ever before are studying digital electronics as one of their first courses in electronics. Thus, our assumption that we should start from "ground zero" seems to have been justified. One modification we have felt it necessary to make is to postpone the study of the electrical composition of gates until a later time in the first digital course; this gives the student more time to learn the names and functions of passive and active electrical devices, before applying them to digital switching applications.

It has also become apparent that the troubleshooting and repair of digital systems, while requiring a good grasp of the function of individual parts, also needs a "systems" approach to be successful. In the beginning of the book, this is not obvious for the simple reason that there are no systems, really, just individual building blocks that can be *made* into systems in the later chapters.

As the density of very-large-scale integration increases, the memory devices chosen as state-of-the-art examples repeatedly become obsolete. Understanding a large system by first looking at a small example continues to be a useful learning tool, even though the first example chosen may be technologically "dated." To present "modern" examples of memory-devices to the student, the IC's chosen for discussion must be constantly replaced by newer and larger-scale memory chips. It is unlikely that any example chosen as "typical" or "outstanding" will be either typical or outstanding by the time the book goes to press. *No word gets old faster than the word "modern."* This is especially true in the world of digital electronics packing-density, and the devices most responsive to new economies of scale are memory devices. Also, some of the "blue sky" technologies that were anticipated as a "next generation" in memory have fallen by the wayside while others, like bubble memory, have been adopted for use in certain restricted applications. Today's "blue sky" ideas with potential for becoming the "next generation" of tomorrow are not the ones discussed in the past, and we have tried to respond in the "speculative" part of this chapter.

Rather short shrift was given to 16-bit processors in the first edition. At that time, most personal computers and microprocessor-based controllers used by industry employed 8-bit machines. Now the personal computer industry has moved to standardize on IBM-PC compatibility and use PC-compatible, or upward-compatible, processors. The selection of what the student must know about 16-bit processors is fairly clear, and this additional information has been incorporated into the chapters introducing microprocessors and their instruction sets.

The choice of Microsoft BASIC for the examples of programming in a high-level language has held up remarkably well. What is called MS-BASIC in today's parlance is a bit different from the various dialects of BASIC that Microsoft developed for Apple, Radio Shack TRS-DOS, and the Commodore PET, but the changes needed to bring the examples in line with "compatibles" in use today have been quite minimal.

Finally, the telecommunications/data communications chapter has a somewhat different outlook. There is just *too much* happening in digital telecommunications for even an attempt at an overview. Instead, we have concentrated on adding 'meat' to a few areas that are clearly important to the digital/computer technician, such as the RS-232 interface, and added discussion of fields that apply this technology, which may provide employment to the digital technician in the future. The student who wishes to delve more deeply into the data/voice telecommunications field may build on this chapter (which is admittedly just a brief survey) through additional reading (i.e., *Voice/Data Telecommunications*, by Gurrie and O'Connor Prentice Hall, Englewood Cliffs, New Jersey, 1985).

For the teacher, we have added a set of chapter objectives to the beginning of each chapter, a set of drill problems to the end of each chapter (totalling 745 additional problems) and more than seventy examples of worked-out problems throughout the book.

We have tried to adapt this introduction to digital electronics to the changing needs of today's students. We hope that you like the results.

Patrick J. O'Connor
Chicago, Illinois

DIGITAL
AND
MICROPROCESSOR
TECHNOLOGY

PART I
Asynchronous Circuits

1
SWITCH CIRCUITS AND LOGIC

CHAPTER OBJECTIVES

By the time you finish this chapter, you will be able to:

1. Identify the TRUE and FALSE logical conditions of a system or device.
2. Relate the TRUE and FALSE conditions by the use of the numerical values 1 and 0.
3. Relate the TRUE and FALSE conditions of a *normally open* or a *normally closed* switch to its operational state.
4. Recognize the *Boolean logic expressions* for AND, OR, NOT, and *noninverted* switch diagrams.
5. Identify the two major types of solid-state switching elements in common use, and relate their actions to those of a manual pushbutton.
6. Relate the terms HIGH and LOW to the TRUE and FALSE or 1 and 0 states described previously.
7. Identify the INPUT(s) and OUTPUT(s) of a logical system.

1

All digital logic circuits are switching circuits. Whatever else they may contain, each circuit begins with a circuit component that does the same job as a switch or pushbutton. At the beginning, we are going to look at circuits made out of pushbuttons, and later, we'll find out what types of switches are really used in logic circuits. Let's begin with some basic ideas.

1-1 LOGIC STATES 1 AND 0

There are several schematic symbols for switches. The one you're probably most familiar with looks like that shown in Figure 1-1. Since there's not much point to having a switch unless there's something to switch on and off, we included a light at one end of the circuit and a power supply (V^+) at the other. The first piece of information that's new in this picture is the way we indicate whether the light is off or on. A light that's ON is marked with a 1 or with the word TRUE. When the light is OFF, we've marked it with a 0 or the word FALSE. The use of the words TRUE and FALSE makes sense if you remember that the thing at the end of the wire is called a light bulb. When the bulb is operating normally, it's lighted. That's the TRUE condition for an operating light bulb. (Otherwise, we'd have to call it a *dark* bulb!) When the bulb is not operating normally, we indicate that with the word FALSE. Remember that anything, when it's operating normally, is TRUE, and when it's not operating normally, it's FALSE.

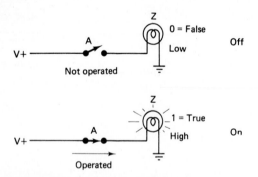

FIGURE 1-1 Output 0 and 1 conditions.

The numbers 1 and 0 are another way to indicate TRUE and FALSE. For the light that is lighted, a 1 indicates that there's something there. For the light that's not lighted, the 0 indicates that there's nothing there. On the diagram, there's a third way of indicating what's happening to the light bulb. The word "LOW" by the bulb that's OFF and "HIGH" by the bulb

that's ON indicates what *voltage* is applied. A LOW voltage (in this case, nothing) is applied to the bulb when the switch is open. The HIGH voltage is the voltage of the power supply, connected to the light bulb when the switch is closed.

Now that we've seen the meaning of FALSE, LOW, and 0 and TRUE, HIGH, and 1, let's look at another type of switch (Figure 1-2).

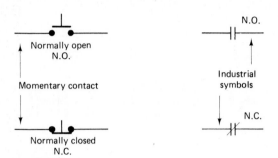

FIGURE 1-2 Momentary-contact switch symbols.

1-2 SWITCH CIRCUIT REPRESENTATION FOR 1 AND 0 AS INPUTS AND OUTPUTS

The two switches in the picture are called **normally open** (N.O.) and **normally closed** (N.C.) types. Another name for switches of this type is *momentary contact*, or *pushbuttons*. The N.O. (normally open) pushbutton is open until you push the button, then it closes and completes the connection. The N.C. (normally closed) pushbutton is closed until you push the button and break the connection. We've also included the industrial standard symbol for these types of switch contacts, which isn't as easy to recognize, but is easier to draw. The circuit with a momentary-contact pushbutton instead of the switch in Figure 1-1 looks like the one shown in Figure 1-3.

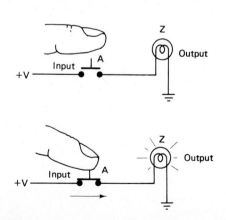

FIGURE 1-3 Input 0 and 1 conditions (normally open).

Now, we have a circuit with an input and an output. The **input** operation, A, is whatever you're doing to the pushbutton, and the **output** operation, Z, is whatever the light bulb is doing. How is the pushbutton being operated? Remember the definition we had for 1 and 0; if the pushbutton is pushed, it's being operated, so it's a 1 (TRUE); if the pushbutton is not being pushed, its a 0 (FALSE). For the input of the pushbutton circuit, a 1 is a pushed pushbutton, and a 0 is one that's been released.

Now, let's look at the output. We already know that a lighted light bulb is a 1, and a dark one is a 0. How does the input affect the output? If what you're doing at A is a 1 (pushing the button), the output at Z is also a 1 (the light is lit). If what you're doing at A is a 0 (not pushing the button), the output at Z (the state of the bulb) is also a 0 (dark). Notice that whatever you do at A, the **logic state** of Z (the 1 or 0) is the same. We could say that "Z is a 1 if A is a 1" and "Z is a 0 if A is a 0," or we could say the same thing with the expression

$$Z = A \qquad \text{(Boolean expression)}$$

The **Boolean expression** is named after George Cayley Boole, who first used this kind of representation to show the relation between a cause and an effect. Actually, Boole did a great deal more, but we'll see that later.

EXAMPLE 1-1

A simple telegraph (shown in the diagram) uses a key, a battery, wires, and a sounder to transmit information.

(a) Which parts are the input, output, and power supply?

(b) If the key is normally open, as shown, in which state is it a logic 1, and in which state is it a 0?

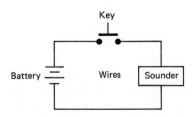

Solution: (a) The **key** is the input, since it *controls* the action of the rest of the circuit. The **sounder** is the output, since it *responds* to the input received by the circuit. Since the key does not create the energy that operates the sounder,

but controls energy from the battery, the **battery** is the circuit's power supply.

(b) When we push on the pushbutton, that is a logic 1. In this diagram, as in schematic diagrams generally, the key is shown in an inactive condition. Pushing on the key will cause the open contacts to close—the active condition—and produce a sound at the sounder. A 0 is already shown in the diagram. The pushbutton is not being used—unpushed—and no sound is being produced at the sounder.

The circuit in Figure 1-3 shows everything you can do with one (N.O.) switch. Let's see how many different ways we can use two switches:

1. We can attach two switches together in a **series** connection.
2. We can use a **parallel** connection.

Figure 1-4 shows both of these.

1-3 SWITCH CIRCUIT REPRESENTATION OF AND LOGIC AS A SERIES CIRCUIT OF NORMALLY OPEN SWITCHES

The *series* circuit in Figure 1-4 is identified as an **AND circuit.** Both switch A AND switch B must be operated before the light will go on. We say that a 1 must be input to A AND B to get a 1 at the output, Z. There is a Boolean expression for this relationship written to the right of the series diagram. We'll find out later just why the "dot" is used to represent the word AND in this expression.

1-4 SWITCH CIRCUIT REPRESENTATION OF OR LOGIC AS A PARALLEL CIRCUIT OF NORMALLY OPEN SWITCHES

The *parallel* circuit in Figure 1-4 is identified as an **OR circuit.** Either switch A OR switch B (or both) must be operated before the light will go on. We say that a 1 must be input to A OR B to get a 1 at the output, Z. There is a Boolean expression for this relationship written to the right of the parallel diagram. We'll find out later just why the "plus" is used to represent the word OR in the Boolean expression.

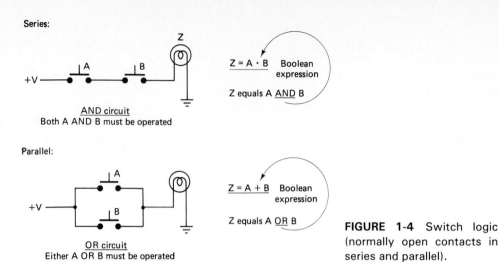

Series:

AND circuit
Both A AND B must be operated

$Z = A \cdot B$ Boolean expression

Z equals A AND B

Parallel:

OR circuit
Either A OR B must be operated

$Z = A + B$ Boolean expression

Z equals A OR B

FIGURE 1-4 Switch logic (normally open contacts in series and parallel).

1-5 SWITCH CIRCUIT REPRESENTATION OF NOT LOGIC AS A SWITCH WITH NORMALLY CLOSED CONTACTS

There's one more thing that can be done using the pushbuttons. Up to now, we've only seen what normally open contacts do. Let's take a look at the *normally closed* pushbutton. If we attach it like the switch in Figure 1-3, what we get is shown in Figure 1-5.

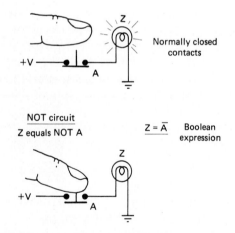

Normally closed contacts

NOT circuit
Z equals NOT A

$Z = \bar{A}$ Boolean expression

FIGURE 1-5 NOT gate (normally closed contacts).

The input A has a different effect on the light Z than it did before. Because the switch *breaks* the circuit when it is pushed, the light goes out only when the button is pushed. Now, if we recall that a *pushed* pushbutton is a 1, and a *lighted* light bulb is a 1, we can see that a 1 input does NOT produce a 1 output. In fact, the A input and the Z output are exactly the *opposite*. When

the switch is a 1 (pushed), the light is a 0 (dark); when the switch is a 0 (released), the light is a 1 (lit). Whatever the A input is, the Z output is NOT. If the switch is ON (pushed), the light is NOT ON. If the switch is OFF (released), the light is NOT OFF. The way this circuit behaves leads us to call it a **NOT circuit.**

Earlier, we promised that we'd see what the switches used in logic circuits were really like. After all, you can't make a computer (with tens of thousands of switches) out of pushbuttons. First, you'd need tens of thousands of people to push the buttons! And second, to operate at anything like the speed of real computers, the people would need to have awfully fast hands! There ought to be better things for all those people with fast hands to do, so let's look at another solution to the "fast-pushbuttons-without-fingers" problem.

1-6 BIPOLAR TRANSISTORS USED AS SWITCHES

Figure 1-6 illustrates the use of a **bipolar transistor** as a switch. Instead of a finger pressing on a button, the thing that operates this "switch" is a current coming in the wire marked *base*. When this happens, another (usually larger) current begins to flow in the wire marked *collector*. You can see that this current is going to light the bulb. Without the base current, there's no collector current. You can think of the transistor as a pushbutton, like the one shown below the transistor in the diagram. Instead of a push, the button is activated by a current. This means that there'll be no moving parts involved in switching the light on and off. An *NPN transistor* like the

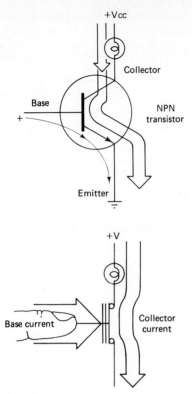

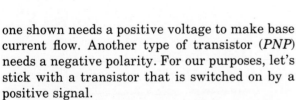

FIGURE 1-6 Bipolar transistor used as a switch.

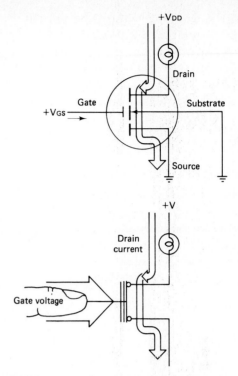

FIGURE 1-7 Field effect transistor used as a switch.

one shown needs a positive voltage to make base current flow. Another type of transistor (*PNP*) needs a negative polarity. For our purposes, let's stick with a transistor that is switched on by a positive signal.

Transistors can switch millions of times faster than a pushbutton. Mechanical switches wear out, and fail sometimes because things get caught between the moving parts. The transistor, having no moving parts, is not only faster, but is more reliable than a pushbutton. We'll see types of logic circuitry later that put together bipolar transistors exactly as we put together switches in the AND or OR circuits.

1-7 FIELD-EFFECT TRANSISTORS USED AS SWITCHES

Figure 1-7 shows a **field-effect transistor** (FET) used to replace a pushbutton in the same way as

Figure 1-6 used a bipolar transistor. Like bipolar transistors, there are two types of FETs, one of which is operated by a positive polarity (the N-channel type) and the other by a negative polarity (P-channel). Again, for convenience, we'll stick with the type that's operated with a positive polarity. In Figure 1-7, the N-channel FET is turned on by a positive voltage on its *gate* wire. The FET (more accurately called a **MOSFET**) is conducting as long as its gate is more positive than the wire marked *substrate*, and stops conducting if the gate is not positive (as measured from the substrate).

The MOSFET input does not require current, so the FET uses a lot less power than the bipolar transistor. On the other hand, the FET is slower than the bipolar transistor, so the designer has to trade-off power economy versus speed in deciding whether to use MOS or bipolar switching devices.

QUESTIONS

1-1. Identify whether the items that follow are 0 or 1.
   (a) A faucet with water running
   (b) An empty coffee cup

   (c) A furnace with no fuel
   (d) A motor with its armature spinning
   (e) A doghouse with a dog in it
   (f) A full coffee cup

(g) A gas stove with the burners lit

(h) A television with no picture

(i) A recording tape with music on it

(j) An automobile with its engine running

1-2. Is the switch shown below always closed, closed only when operated, or closed except when operated?

1-3. Indicate whether the statements below are represented by a logic 1 (HIGH) or a logic 0 (LOW) state.

(a) All lizards can fly.

(b) Grass is a mineral.

(c) Frogs have four legs.

(d) A normally open switch opens when operated.

(e) A normally open switch closes when operated.

(f) A momentary-contact switch returns to its normal state when released.

(g) If you push on a normally closed push-button, the circuit it is operating will stop.

1-4. Identify which of the items below are inputs and which are outputs.

(a) A key put into the automobile ignition switch to start its engine

(b) The engine of an automobile starting up

(c) A hamburger cooked by a charcoal fire

(d) Someone starting a charcoal fire to cook hamburgers

DRILL PROBLEMS

Fill in the blanks.

1-1. TRUE is represented by _____. (1 or 0)

1-2. FALSE is represented by _____. (1 or 0)

1-3. HIGH (voltage level) is the same as _____. (1 or 0)

1-4. LOW (voltage level) is the same as _____. (1 or 0)

1-5. A is a normally _____ switch symbol. (open, closed)

1-6. B is a normally _____ switch symbol. (open, closed)

1-7. C is a normally _____ switch symbol. (open, closed)

C —| |—

1-8. D is a normally _____ switch symbol. (open, closed)

D —|⊢

1-9. A doorbell is usually made with a normally _____ switch.

1-10. Boolean arithmetic is named after _____. (full name)

1-11. In the circuit below, the switches are in _____. (series, parallel)

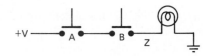

1-12. The switches above represent _____ (AND, OR, NOT) logic.

1-13. Write the Boolean expression for the circuit above. _____

1-14. In the diagram above, a X symbol is an _____. (input, output)

1-15. In the diagram above, the lamp is an _____. (input, output)

1-16. Draw the circuit used to explain OR logic.

1-17. In Boolean arithmetic, a "+" sign is used for _____.

1-18. In Boolean arithmetic, a "·" is used for _____.

1-19. The part of a bipolar transistor that causes the collector current to "switch on" is called the _____

1-20. Name the three parts of a bipolar transistor. _____, _____, _____

1-21. A bipolar transistor that is switched on by a positive current is called a(n) _____ type. (NPN, PNP)

1-22. A transcriptor _____ (does, does not) have moving parts.

1-23. A(n) _____ (NPN, PNP) is switched on by a negative current.

1-24. Which requires more power? _____ (MOSFET, bipolar)

1-25. Which is faster? _____ (MOSFET, bipolar)

1-26. Name the four parts of a MOSFET. _____, _____, _____, _____

1-27. The part of the FET that causes the drain current to "switch on" is called the _____.

1-28. FET stands for _____ _____ _____.

1-29. The P-channel FET is operated by a _____ (positive, negative) input voltage.

1-30. The N-channel FET is operated by a _____ (positive, negative) input voltage.

2
INTRODUCTION TO GATES

_____CHAPTER OBJECTIVES_____

By the time you finish this chapter, you will be able to:

1. Construct a truth table for a logical device with any number of inputs, using the *binary numbering system*.
2. Recognize the symbolic logic gate diagrams for the basic three positive-logic gates AND, OR, and NOT.
3. Identify the function of switch diagram schematics for the basic AND, OR, and NOT gates.
4. See why AND is called logic multiplication, and OR is called logic addition.
5. Understand the meaning of the NOT gate logic identities:

$$\overline{1} = 0 \qquad \text{and} \qquad \overline{0} = 1$$

6. Understand how the two major types of solid-state switching elements may be used to produce NOT logic.
7. See the reason why a NOT gate is called an *inverter*.
8. Understand how to read a *timing diagram* of a logic gate, and see what its relationship is to the gate's truth table.
9. See that a *timing diagram* is related to the display shown on a measuring instrument called an *oscilloscope* more closely than it is to a gate's truth table.

2-1 THE TRUTH TABLE, GATE SYMBOL, AND BOOLEAN EXPRESSION

We're going back to the switching circuit made with two normally open switches in series (we'll replace them with transistors later) to examine in more detail how it works. We'll see several new concepts while we analyze this circuit, and we'll apply these concepts to the other circuits we talked about, as well.

The first new concept we'll look at answers the question: How do all *possible* inputs affect the output? To refresh your memory, look at the AND circuit in Figure 2-1. You can see that there's only one way for an electric current to reach the light and turn it on. Both pushbuttons have to be pushed. Remember that a pushed pushbutton is represented by a 1, and one that hasn't been pushed is represented by a 0. Next to the AND circuit in Figure 2-1 is a table called a **truth table.** It shows the state of the inputs, A and B, and the output, Z, on four lines. Each line represents what is happening at one moment in time. Let's look at the line marked with a small 0 under the heading "A   B   Z." As we read across this line, we see that A and B are both zero at this time. That means that neither button has been pressed. The value under "Z" on this line is also a zero, which indicates that the light is off. This line, reading from left to right, tells us that at a time when neither pushbutton is being operated, the light is off. That is exactly the way things are represented in the schematic diagram of the switch circuit. For some reason, schematic diagrams are always shown with buttons unpressed (probably because it's so much easier than drawing fingers on each pushbutton in the schematic!).

Next, we can look at what happens when someone *does* push on a button. Reading the line on the truth table marked with a small 1, we see what happens when someone pushes button B. Reading across this line, we see that A is unpressed, B is pressed, and the light, Z, is still off. On the next two lines, we can see what happens if we press A alone, and if we press both A and B. After the line marked with a small "3," there are no more lines. It turns out that the four lines already represent all possible ways the two switches can be operated. If you're not sure about this, consider that we've pressed buttons in the following order: line 0—none; line 1—B only; line 2—A only; line 3—both A and B. If you think about it, and try to come up with more combinations of pressed and unpressed buttons, you'll keep coming up with combinations that are already on the table. In a little while, we'll show you why this is true, and how to figure out what combinations are possible with *any* number of switches.

One other concept that we're including with Figure 2-1 is the **symbol** (also called a symbolic logic diagram). The symbol for an AND circuit is not a schematic. It doesn't show *how* the circuit is connected to make the output go on when inputs A and B are operated. It only tells you that, whatever is inside, its output *will* go on only when A AND B are operated. This symbol is used for anything that has the same relationship between the inputs and the output, regardless of what type of energy is used for the input, or what type of switching devices are used inside. In fact, instead of using the name AND *circuit*, we refer to the symbol as an AND *gate* (neatly solving the problem of what to call the AND if it doesn't use electrical energy).

Notice that the truth table is like the symbol. It doesn't matter what the AND is made of, only how it *works.*

EXAMPLE 2-1

A jury trial requires that all 12 jurors vote a defendant "guilty" to convict, and that all 12 jurors vote a defendant "innocent" to acquit. If an electric polling process gives each juror two buttons (momentary-contact, normally open switches), one marked "guilty," and the other marked "innocent," how should the buttons be connected to the other jurors' buttons so that a lamp marked "convicted" lights if all 12 jurors vote "guilty" and another lamp, marked "acquitted," lights if all 12 jurors vote "innocent"?

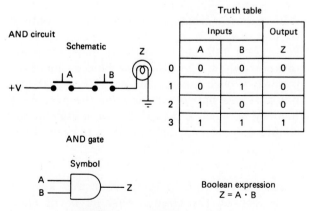

Truth table

	Inputs		Output
	A	B	Z
0	0	0	0
1	0	1	0
2	1	0	0
3	1	1	1

Boolean expression
$$Z = A \cdot B$$

FIGURE 2-1   AND logic circuit.

Solution: All 12 inputs must be "guilty" to convict; thus they are AND logic. Normally open switches are connected in *series* to get AND logic. Therefore, a 12-input AND circuit for the "convicted" lamp would look like this:

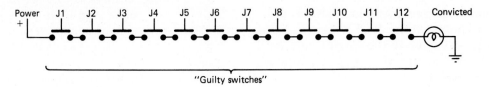

The "innocent" switches would be wired to the "acquitted" lamp in the same way:

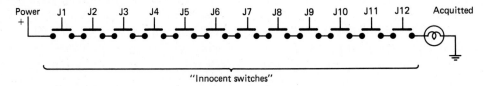

What happens when we start with the parallel circuit instead of the series circuit? It was called an OR circuit, so its truth table is an OR truth table, and its symbol is the symbol of an OR gate (Figure 2-2).

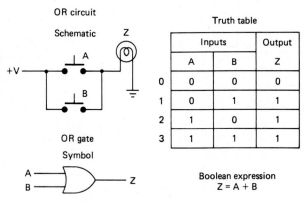

FIGURE 2-2 OR logic circuit.

The truth table shows what happens at the output of the OR circuit for every possible combination of 0s and 1s at the two inputs. You'll notice that the same pattern of inputs is used on this table that we used on the AND. We'll see later why this particular pattern is used. For now, the truth tables are matched, so we can compare the behavior of the OR output with the AND.

As you read across the lines, you'll notice that the "1," "2," and "3" lines turn on the output. The "0" line does not. This matches up with the fact that on lines "1" and "2," there is an active input at A OR B. On line "3," both A and B are active. In the circuit, you can see that the light should be on as long as *any* switch is operated. On line "0," the circuit is operated just the way it looks on the diagram. With neither A nor B pressed, there is no way the light can light up.

The symbolic diagram for the OR gate is shown beneath the schematic diagram. As with the AND, the OR symbol merely tells what the logic of the circuit is, not how the circuit is connected.

EXAMPLE 2-2

A **keypad** is a group of pushbuttons (momentary-contact, normally open switches), each used to generate a different binary code. It is important to know when a button—any button—has been pushed. If each pushbutton on a 12-key keypad has an extra pair of contacts for the purpose of identifying when a "keypress" has occurred, how should they be attached to a lamp to show that a key on the keypad has been pressed?

Solution: To light a lamp when any key has been pressed, the circuit must have OR logic. Normally open switches are connected together in *parallel* connection to get OR logic, so the keypress circuit should look like this:

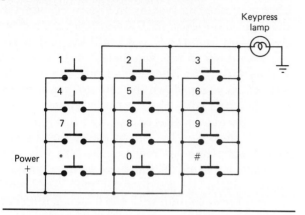

2-2 TRUTH TABLES FOR *AND* AND *OR* SWITCH CIRCUITS

We promised to explain how to set up a truth table for any number of inputs, and to explain why that particular pattern of input lines used in Figures 2 1 and 2-2 was the one we used. We're going to get there, but we'll start in a rather strange place—with a license plate.

Imagine that you're living in a state where the autos have licenses that look like Figure 2-3. Each license plate has six places on it for a digit (we'll call them six *registers*). Each digit can be one of 10 different *values* (0, 1, 2, 3, 4, 5, 6, 7, 8, or 9). Now, you ask yourself: "How many different plates can they make like this before they can't find any more numbers to use?" One answer is to try *counting*. If you start counting at 000-000, and keep counting all the way up to 999-999, there's no possible way we could miss a six-digit number. There aren't any six-digit numbers larger than 999,999, and any six-digit number you make up will turn out to be one that you counted between 000,000 and 999,999. If you count every plate, there will be 1 million of them (not 999,999, because there's a plate with a 000-000 on it).

$$1 \text{ million} = 10^6$$

Notice that there were 6 registers on the plate, each of which could have 10 different values. Both the number 6 and the number 10 are part of the 10^6 possible plates. The 10 (number of values) is the number on the bottom (it's called the **base**). The 6 (number of registers) is the number on the top (it's called the **exponent**).

What does all this have to do with truth tables? Simply that to list all the possible ways in which the inputs A and B can be arranged, we're dealing with a problem very similar to the license-plate problem. There are only two registers (A and B), and each register can have only two different values (OFF = 0 and ON = 1). The total number of plates was

$$(\text{values})^{(\text{registers})} = (10)^6 = 1 \text{ million}$$

Let's see how the same method of solution fits the problem of finding the number of lines on a truth table.

$$(\text{values})^{(\text{registers})} = (2)^2 = 4$$

So for a circuit with two pushbuttons, each with two positions (pushed or unpushed), there should be four lines on the truth table. We cheated. We knew the answer in advance. There *were* four lines on the truth table.

Let's try the same thing for a circuit we haven't already worked with (Figure 2-4). This circuit has three input registers (A, B, and C),

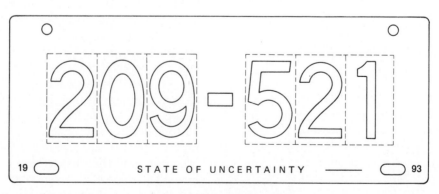

How many total plates are possible?

FIGURE 2-3

each of which can have two values (0 or 1). According to our previous reasoning,

$$(\text{values})^{(\text{registers})} = (2)^3 = 8$$

there should be eight lines on the truth table.

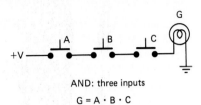

AND: three inputs

$$G = A \cdot B \cdot C$$

FIGURE 2-4 Three-input AND gate.

Next, we must answer the question: How do we decide what to put on each line? Numbering the license plates (without missing any numbers) was just a matter of counting. Just count all the numbers, we said, and you can't miss any. Counting all the numbers on the truth table is the same thing, except that we can't count numbers that have digits bigger than 1. This means that we can't count in the decimal system with values of 0, 1, 2, 3, 4, 5, 6, 7, 8, and 9 in one place. We have to count in the **binary numbering system,** which has only the values 0 and 1. Our truth table, with the first eight binary numbers, will look like Figure 2-5. Notice that the first eight numbers are zero through seven. As we did with the license plates, we start counting at zero in the truth table. If you check the truth tables in Figures 2-1 and 2-2, you'll see that the top line is zero on those tables, also. In fact, we'll *always* start counting with zero in digital logic. It may seem strange at first, but it turns out to be very important in a lot of places.

Most people aren't comfortable counting in binary. In fact, most people don't know how to count in binary (and manage to live very nicely

in spite of it!). Here's a shortcut that lets you count in binary without knowing much about the binary numbering system: In Figure 2-5, start at the right-hand column (C) of numbers. Go down the column from the C and you'll see that, starting with the 0 on top, you have 0, 1, 0, 1, 0, 1, 0, 1, alternating. Go over to the next column to the left (B). As you go down the column, you'll notice that, starting with 0 again, there are *two* 0s, *two* 1s, *two* 0s, *two* 1s, alternating. The A column has *four* 0s, then *four* 1s. You probably see the pattern. Every time you move to the left, you have alternating 0s and 1s, but twice as many of each as you did in the previous column. You'd start the rightmost column with one 0, one 1, alternating, until you had as many as you needed. Moving to the left, the next column would have two 0s, two 1s, alternating, until you reached the bottom of the table. The next column over would have four 0s, four 1s, alternating, until you reached the bottom of the table. If there were another column, you'd start it with eight 0s, then eight 1s, alternating eight of one with eight of the other until you reached the bottom of the table.

At this point we have shown how to generate truth tables but not *why* we generate truth tables. Why do we do all this? What is the point of having the truth table for a logic circuit?

Think of the truth table as a list of steps in an experiment. We want to identify how a certain logic circuit behaves. What we do is to try out different inputs to see what they do at the output. Once we have a complete list of things that make the output work, we know something important about the circuit. We know how to describe the circuit, because we know what it takes to turn its output ON or OFF.

Imagine that you have a "black box." It's a logic circuit made of several pushbuttons and a light bulb, with a power source attached. You know that by operating the pushbuttons, you can make the light go on, but you have no idea how the pushbuttons are wired together inside the box. You want to figure out how the pushbuttons are connected inside the box. To do this, you need to find out just which combinations of button-states light the bulb. You could start pushing buttons in different patterns by hit or miss. You'd try a pattern of buttons, note down what result it produced, and then ask yourself: "Now . . . what haven't I tried yet?" If you kept this up until you couldn't think of anything you hadn't tried, you *might* have a complete table. Then again, you might have forgotten something, or tried something twice, without realizing it. To test the cir-

	Inputs			Output
	A	B	C	G
0	0	0	0	0
1	0	0	1	0
2	0	1	0	0
3	0	1	1	0
4	1	0	0	0
5	1	0	1	0
6	1	1	0	0
7	1	1	1	1

FIGURE 2-5 Truth table for AND gate.

cuit, you should really have a *system*. Your system should let you know in advance exactly how many different "things" there are to try. This will help you be sure that you didn't forget anything. Your system should also tell you what's the next thing to try after you log into your notebook the results from the previous thing that you did. This will assure you that you didn't do any step twice, or forget a step. The system we're talking about is the truth table. What we just described (figuring the number of lines on the truth table, and counting in binary until the table is full) is a systematic approach to testing out the circuit. Whatever circuit we have to evaluate, if we use this approach, we're bound to find all the ways to turn the output ON or OFF.

You can use your truth table for the black box to find out what's in the box, by matching it up with truth tables of known logic gates with the same number of inputs. Suppose that the circuit you had in the black box produces a truth table that matches an AND gate. You can say that there's an AND gate in the box no matter how it's wired. If the truth tables match, the circuits are *interchangeable*. That means that you can swap the circuit inside the box with the matching gate circuit and, from the outside, you can't tell the difference.

This is called a *proof by perfect induction*. You prove that the circuit in the box is an AND gate when you show that their truth tables match everywhere.

EXAMPLE 2-3

Suppose that you wish to make up a truth table for the 12-pushbutton keypad in Example 2-2. How many lines would be required for a complete truth table? You do *not* have to draw up this table, only tell how many lines it *would have* if you *did* make a truth table for this circuit.

Solution: There are 12 inputs (pushbuttons). Each can be two ways (pushed or unpushed). Consequently, the table has

$$2^{12} = 4096 \text{ lines (!)}$$

(You can see why you wouldn't want to draw up the actual table.)

2-3 *AND* GATE, SHOWING LOGIC MULTIPLICATION

You may wonder why a dot was used to represent the word AND. It would seem that the phrase "A AND B" should be written "A + B" instead of "A · B," because in everyday language we use "and" and "plus" to mean the same thing most of the time. In *logic*, things are different. An AND circuit is AND because inputs A AND B must both be pressed if you want the lamp to light.

Figure 2-6 shows a two-input AND logic circuit. There's a truth table for the circuit at the top of the diagram, and below it, a table showing what happens when we *multiply* two one-digit binary numbers. You can see that the two tables match exactly.

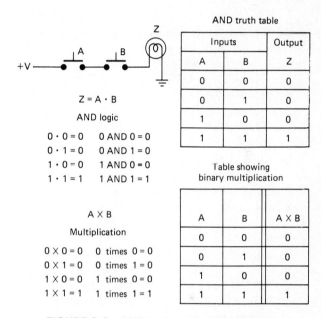

AND truth table

Inputs		Output
A	B	Z
0	0	0
0	1	0
1	0	0
1	1	1

Z = A · B

AND logic

0 · 0 = 0   0 AND 0 = 0
0 · 1 = 0   0 AND 1 = 0
1 · 0 = 0   1 AND 0 = 0
1 · 1 = 1   1 AND 1 = 1

A X B

Multiplication

0 X 0 = 0   0 times 0 = 0
0 X 1 = 0   0 times 1 = 0
1 X 0 = 0   1 times 0 = 0
1 X 1 = 1   1 times 1 = 1

Table showing binary multiplication

A	B	A X B
0	0	0
0	1	0
1	0	0
1	1	1

FIGURE 2-6 AND logic and multiplication.

The dot is often used in algebra as a multiplication symbol rather than an x, because x may be one of the letters being used to represent a number you're multiplying. Imagine trying to write "x times y" by using "xxy." You would have no way of knowing that one of the x symbols meant "multiply." In algebra, as well as other places, it's clear that a *special* symbol is needed for multiplication that won't be confused with anything else. We used the x on the bottom of Figure 2-6 to stand for *algebra multiplication*, so that we could tell the number multiplication from the logic AND, but you can see how AND logic matches multiplication for one-digit binary numbers. This is why the symbol for multiplication is used to represent AND. In logic, the AND function is sometimes called *logic multiplication*. You must remember, however, that this "multiplication" works only when you stick to one-digit numbers. Because AND (logic multiplication) isn't really the same as number multiplication, other symbols are also used to represent the AND function. The dot is the most popular one, however, so it's the one we'll use from now on.

2-4 OR GATE, SHOWING LOGIC ADDITION

We look next at why the + sign is used in place of the word "OR." Reading "A + B" as "A OR B" is particularly hard, because in everyday language we often read "A + B" by saying "A and B." This is *not so* in logic.

In Figure 2-7, you can see the comparison of a truth table for an OR circuit with an *addition* table for one-digit binary numbers. Compare the table for "A plus B" with the truth table for "A AND B" of Figure 2-6. They don't match. It's clear why the "+" isn't used for AND. Now compare the table for "A OR B" with the "A plus B" table. They don't match everywhere, either, but they match in a lot more places than the AND gate did. In fact, "A OR B" is the same as "A plus B" everywhere except for the last line. On the last line of the truth table, Z represents whether the light is OFF (0) or ON (1). There's no way for Z to be a 2, so it's impossible for the OR gate (which has only a single output) to match the binary addition table. This means that OR logic won't work the same way as *number addition* everywhere, even if you confine yourself to one-digit binary numbers. This will turn out to be important later, when we look at Boolean algebra.

In spite of the failure to match everywhere, the + symbol is often used for OR, and the OR logic function is sometimes called *logic addition*. There are other symbols for OR, but they're not used as often as the +, so we'll use the + for OR from now on.

One more thing; in Figures 2-6 and 2-7, besides the circuit diagrams and truth tables, we've included "number" and "word" descriptions of the AND and OR gates' action. In Figure 2-6, the first line under "AND logic" reads

$$0 \cdot 0 = 0 \qquad 0 \text{ AND } 0 = 0$$

You can read this as saying that when an input that is a 0 is ANDed (in series, in this circuit) with another input 0, the output is a 0. The "0 AND 0" or "0 OR 0" of Figures 2-6 and 2-7 show that the two inputs are combined by being ANDed or ORed in the circuit wiring. *Never* mistake "0 AND 0" for "0 plus 0" in logic. Reading it aloud, "0 AND 1 = 0" *sounds* very wrong (because of how we are used to using "and" in everyday language), but AND has a different logic meaning from everyday language, so "0 AND 1 = 0" is *right*. You just have to remember that logical AND really works more like multiplication than addition. This is why it helps to use a multiplication symbol in place of the word "AND" in Boolean expressions.

2-5 NOT GATE, SHOWING THAT $\overline{0} = 1$, $\overline{1} = 0$, AND $\overline{\overline{A}} = A$

The NOT circuit of Figure 2-8 is the same circuit that we used in Figure 1-5. We have added the truth table and symbolic logic diagram to the circuit as well as "word" description of the NOT gate. Notice that if we use the words "OFF" and

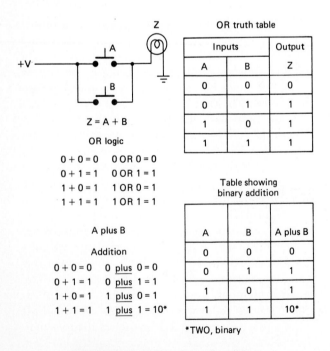

FIGURE 2-7   OR logic and addition.

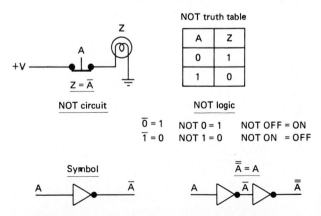

FIGURE 2-8   NOT logic and double negation.

"ON" for "0" and "1," we can see that a NOT gate's output is easy to understand. When we say "NOT 1 = 0" that's hard to follow, but when we say "NOT ON = OFF," it seems almost *too* simple. Let's look at the Boolean arithmetic of the NOT gate. The important relationships are:

$$\overline{1} = 0 \quad \text{and} \quad \overline{0} = 1$$

which can be read

$$\text{NOT } 1 = 0 \quad \text{and} \quad \text{NOT } 0 = 1$$

which might be hard to remember, but

$$\text{NOT ON} = \text{OFF} \quad \text{and} \quad \text{NOT OFF} = \text{ON}$$

says the same thing.

The bar over the letter or number in a NOT expression takes the place of writing the word "NOT" in front of it. An example of a way to make a NOT circuit with no moving parts is a circuit that uses a bipolar transistor as a switch (Figure 2-9).

In Figure 2-9(a), the input to the transistor (base current) is OFF, and the transistor does not conduct. Since the transistor does not conduct, the current flows through the lamp instead. (The lamp is not as good a conductor as a *conducting* transistor, but it is a lot better conductor than a *nonconducting* transistor.) If the lamp, which is lighted, is the output, it is NOT OFF at the time when the input is OFF.

In Figure 2-9(b), the input to the transistor is ON, and the transistor conducts. Since the transistor is a better conductor than the lamp, the current flows through the transistor instead of the lamp, and the lamp is NOT ON. You can see that the input (base current) and the output (lamp current) are opposites. When the input is OFF, the output is NOT OFF, and when the input is ON, the output is NOT ON.

In Figure 2-9(c), we have followed several cycles of voltage input and output to this circuit. The wave diagrams, called **timing diagrams,** show what happens at the input and output at the same time. You can see that when the input goes ON–OFF–ON, the output goes OFF–ON–OFF at the same time. Since the waveform diagram Z is *inverted* (upside down) in comparison to the waveform at A, the NOT circuit is also called an **inverter.** Although "inverter" is a longer and clumsier word than "NOT," we'll be using both names from now on, because the name "inverter" is used more often than "NOT" in the technical literature.

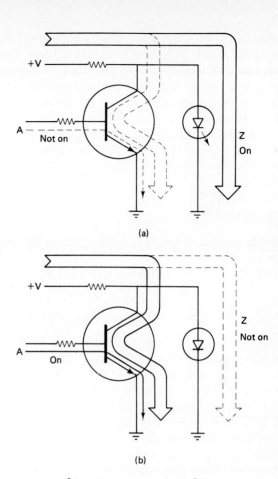

FIGURE 2-9 Transistor NOT gate.

One last idea. If the inverter turns a waveform upside down, what would be the effect of inverting a signal twice? When you turn anything upside down twice, it's right-side up again. A circuit that inverts *twice* acts as though *no change* took place in its input. We can say that

$$Z = \overline{\overline{A}} \quad \text{(inverted twice)}$$

is the same as

$$Z = A$$

or we could say that

$$\overline{\overline{A}} = A$$

Since a circuit like Figure 2-9 takes a transistor and several resistors, and two of them take twice as many parts, if the entire circuit does nothing to a signal, you might want to replace the whole thing with a *wire* (which also does nothing to the signal, but is a lot cheaper than two inverters).

Later, identities like this will lead us into the topic of Boolean *simplification*, which we'll use to replace complicated logic circuits with simpler ones.

EXAMPLE 2-4

Using switch symbols (—● ●—) for normally open switches, draw the schematic diagram for a circuit whose Boolean expression is

$$Z = A + B + C$$

(Output Z should be represented by a lamp.)

Solution: The "+" sign indicates Boolean addition, which is OR logic. There are three inputs, A, B, and C, indicated in the expression, and an output, Z. We already know that the output should be indicated by a lamp, and we can control it with three inputs in the form of pushbutton switches. To get the OR logic we need from normally open pushbuttons, we should place three switches in *parallel* connection, and label each one with A, B, or C:

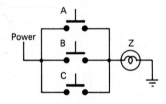

2-6 TIMING DIAGRAMS FOR *AND* AND *OR* SWITCH CIRCUITS

In Section 2-5 we saw that a **timing diagram** is another way to show how a circuit's inputs affect its outputs. The timing diagram for the NOT gate provided an explanation for calling it an *inverter*, but that's not the only reason for having timing diagrams.

In a way, the timing diagram is a truth table laid on its side, with HIGH and LOW represented by voltages on a graph of voltage versus time, instead of by 1's and 0's in a grid. An example of a timing diagram could be provided by using the three-input AND gate in Figure 2-4. First, flip its truth table (Figure 2-5) onto its side, as in Figure 2-10:

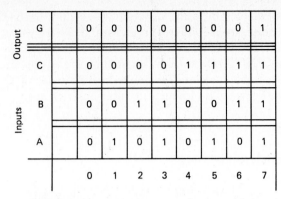

	0	1	2	3	4	5	6	7
Output **G**	0	0	0	0	0	0	0	1
C	0	0	0	0	1	1	1	1
B	0	0	1	1	0	0	1	1
A	0	1	0	1	0	1	0	1

FIGURE 2-10 Rotated truth table for a Three-input AND gate (see fig. 2-5)

Now, we have the basis for a timing diagram of the three-input AND gate. All we have to do is transform all the 1's into HIGH voltages and all the 0's into LOW voltages. If we leave off the exact amount of voltage on the diagram, just noting that the 0's are places where we *don't have* a voltage, and 1's are places where we *do have* a voltage, then we have this diagram:

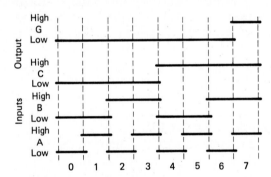

FIGURE 2-11 Truth table converted to a timing diagram (3-input AND)

The truth table has become a timing diagram. Each of the numbered lines on Figure 2-5 represented inputs and outputs happening at the same time. Line 3, for example, showed what happened at the output when the inputs were CBA = 011. In Figure 2-11, it is the *vertical columns* (still numbered 0 through 7) that represent events happening at the inputs and outputs simultaneously.

There *is* an important difference, however, between a timing diagram and a truth table. The timing diagram for a gate doesn't *have* to match the truth table for the same gate. A truth table for a three-input gate like the one above only has eight lines, and never has more, because it represents a sequence of unrelated *test procedures* and there are only eight ways to test a three-input gate. Once the results of a particular test are shown on a truth table, there is no reason to show the same test (with the same results) again. It's also important to show *all* the possible inputs and all the possible outputs in a truth table. A timing diagram, however, is something entirely different. In Figure 2-11, the vertical columns numbered from 0 to 7 are called **timing states,** and they represent a series of events happening one after another, in *real time.* Unlike Figure 2-11, in real systems, the numbers on these timing states don't have to bear any relation to the 1's and 0's at the gate-inputs. For instance, in a real digital system, the AND circuit's C, B and A inputs could be 000, 011 and 111, over and over, with 011 being present at time 1, time 4 and time 7. In this case, the AND gate's timing diagram shows *only* these states; the other five possible states never appear. Instead of some abstract test procedure, this diagram (Figure 2-12, below) looks like it does because the circuit is *actually being run that way.*

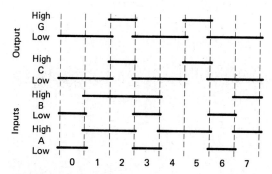

FIGURE 2-12 Timing diagram for an AND gate given inputs 1, 4 and 7, only

You might well ask, "What's the point? If the timing diagram isn't a truth table, exactly what *is* it?"

Electrical test equipment for testing and troubleshooting real digital circuits (ones that use solid-state switches, not manually-operated pushbuttons) often includes an **oscilloscope,** which is a sort of glorified, graph-plotting, high-speed voltmeter. It plots its voltage measurements as a graph of voltage versus time, and does it on a screen that resembles a television set or computer video screen. For sophisticated digital troubleshooting, it is often necessary to visualize the voltage patterns (waveforms) at several places in a circuit simultaneously, to see if the signals are related to one another in the proper way indicated by the digital system's logic. Multi-channel oscilloscopes fulfill this troubleshooting need, and are commonly used to track down digital system faults, but somewhere, to troubleshoot the circuit, you must have a picture of the way the oscilloscope traces should look when the circuit is working correctly. If you know the circuit—and know it 'cold'—you'll have this picture in your head, but if you have to look it up, a copy of the oscilloscope waveform on paper will be very important in finding where the system is malfunctioning. Your paper copy of the 'good system' waveforms is a **timing diagram**. Figure 2-12, for instance, must represent the AND circuit's input and output states, as they come along in a real circuit (which is only operated by three input patterns repeatedly), while a truth table only tries to represent what an AND gate might do, in general, in any circuit.

QUESTIONS

2-1. Select the Boolean expression that matches the circuit shown.

(a)

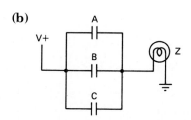

(b)

(c)

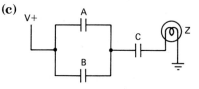

(1) $Z = A + (B \cdot C)$ (4) $Z = A + B + C$
(2) $Z = (A \cdot B) + C$ (5) $Z = A \cdot B \cdot C$
(3) $Z = (A + B) \cdot C$

2-2. Select the expression for the circuit shown.

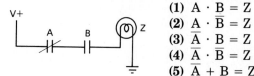

(1) $A \cdot B = Z$
(2) $A \cdot \overline{B} = Z$
(3) $\overline{A} \cdot B = Z$
(4) $\overline{A} \cdot \overline{B} = Z$
(5) $\overline{A} + B = Z$

2-3. Identify the logic of the following switch diagram.

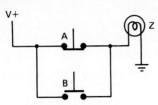

(1) $Z = A + \overline{B}$
(2) $Z = \overline{A} \cdot B$
(3) $Z = \overline{A} + B$
(4) $Z = A \cdot \overline{B}$
(5) $Z = \overline{A} + \overline{B}$

2-4. Is the Boolean expression $Z = A \cdot B$ correct for the circuit of the following figure?

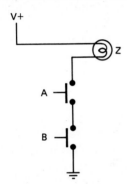

2-5. Complete the truth table for the AND circuit shown here.

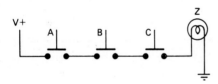

A	B	C	Z
0	0	0	
0	0	1	
0	1	0	
0	1	1	
1	0	0	
1	0	1	
1	1	0	
1	1	1	

2-6. Complete the truth table for the OR circuit.

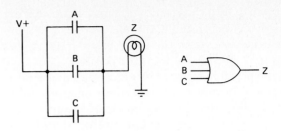

A	B	C	Z
0	0	0	
0	0	1	
0	1	0	
0	1	1	
1	0	0	
1	0	1	
1	1	0	
1	1	1	

2-7. A license plate has a two-letter alphabetic prefix followed by a four-digit decimal number.
(a) How many plates like this are possible?

SN 7400

(b) Would more license plates be possible if all six places were alphabetic instead of two alphabetic and four decimal?

2-8. Draw up the input portion of a truth table for four inputs called A, B, C, and D (complete the example).

	D	C	B	A
first lines	0	0	0	0
	0	0	0	1
to be completed	:	:	:	:
last lines	1	1	1	0
	1	1	1	1

2-9. Match the name with the logic function.
(a) $A + B$ (1) AND
(b) $A \cdot B$ (2) OR
(c) Logic addition   (3) NOT

(d) Logic multiplication

(e) Complement

2-10. Complete the Boolean expression with "0" or "1."

  (a) $1 \cdot 1 =$

  (b) $0 \cdot 0 =$

  (c) $1 + 1 =$

  (d) $1 + 0 =$

  (e) $\overline{1} =$

2-11. Which switching circuit will invert, the one on the left below or on the right?

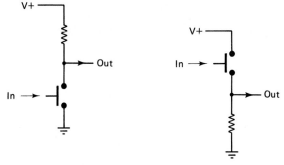

2-12. Complete the timing diagram for the AND gate begun in the following figure.

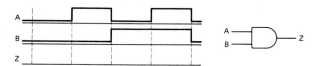

2-13. Convert the truth table of an **OR** gate (Figure 2-2) into a timing diagram, filling waveforms in on the following figure.

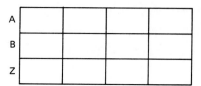

2-14. Suppose that the three-input **AND** gate in Figure 2-4 is operated in a circuit where its inputs are always 010, 101, or 111, repeating in this order. Complete the following timing diagram for five states of this circuit's operation, starting with the number 010.

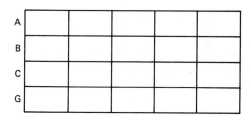

DRILL PROBLEMS

2-1. Draw the truth table for the Boolean expression

$$X = D \cdot E$$

2-2. How many different combinations of D and E must be shown on this truth table?

2-3. Draw the symbol for an AND gate with two inputs.

2-4. In the symbol of Problem 2-3, which leads represent the inputs?

2-5. Draw the truth table for the Boolean expression

$$X = D + E$$

2-6. How many combinations of D and E produce a TRUE output for X?

2-7. Draw the symbol for an OR gate with two inputs.

2-8. In the symbol of Problem 2-7, which lead represents the output?

2-9. In the expression 10^6, which number is the base?

2-10. In the expression 10^6, which number is the exponent?

2-11. When counting the lines on a truth table, you should start counting with the number _____. (one, zero)

2-12. In binary, what number comes after 1010?

2-13. In binary, what number comes after 0011?

2-14. How many lines are there in a truth table for three inputs?

2-15. How many lines are there in a truth table for four inputs?

2-16. Explain the difference between algebra multiplication and logic multiplication.

2-17. The term "logic multiplication" is often used for _____. (AND, OR)

2-18. Which symbol $(+, -, \cdot, \div)$ is used for the AND function?

2-19. Which symbol $(+, -, \cdot, \div)$ is used for the OR function?

2-20. The term "logic addition" is often used for _____. (AND, OR)

2-21. Explain why the phrase "0 and 1 = 0" sounds wrong even though it is right in Boolean logic.

2-22. Draw the symbol for the NOT gate.

2-23. In the symbol of Problem 22, which lead is the input?

2-24. In the symbol of Problem 22, which lead is the output?

2-25. "OFF" is the same as logic ———. (0, 1)

2-26. "OFF" is the same as logic ———. (TRUE, FALSE)

2-27. Write the Boolean expression that says "NOT OFF = ON."

2-28. Write the Boolean expression that says "NOT ON = OFF."

2-29. An **inverter** is the same as a(n) ——— (AND, OR, NOT) circuit.

2-30. What is the effect of inverting a signal twice?

3
LOGIC FUNCTIONS DERIVED FROM BASIC GATES

By the time you finish this chapter, you will be able to:

1. Describe the behavior of the remaining four basic logic gates, NAND, NOR, exclusive-OR, and exclusive-NOR.
2. Identify the symbolic logic gate diagrams for the above-mentioned positive-logic gates.
3. Identify the switch diagrams (using normally open and normally closed pushbutton switches) for the above-named gates.
4. Identify how the aforementioned gates may be implemented via positive or negative logic, and how this is described by DeMorgan's theorem.
5. Describe how to make all other gate types from combinations of NAND and NOR gates.
6. Describe the *designation number* of a two-input logic gate.
7. Describe how two-input logic gates may be used to *enable* or *disable* the passage of a signal (for AND, OR, NAND, and NOR gates).
8. Understand how the Boolean identities

$$A \cdot 0 = 0 \qquad A \cdot 1 = A$$

and

$$A + 0 = A \qquad A + 1 = 1$$

relate to the enabled or disabled states of AND and OR gates.
9. Understand how XOR and XNOR gates do not enable or disable signals, but instead become *controlled inverters*.
10. Understand the NAND and NOR identities corresponding to item 8 for AND and OR.
11. See, from the *timing diagrams*, what is meant by *enabled* and *disabled* (or *inhibited*) conditions.

In Chapter 2 we introduced three basic gates, AND, OR, and NOT. In this chapter we're going to see four more gates, NAND, NOR, exclusive-OR, and exclusive-NOR, which can be made by combining the basic gates.

3-1 THE NAND GATE

In previous series and parallel circuits, we've put normally open switches in series (**AND gate**) and parallel (**OR gate**), but we've never combined normally closed switches. Figure 3-1 shows a circuit made of normally closed switches in parallel. At first, it looks like an OR circuit, because the switch contacts are in parallel. If current passes through the upper OR lower contacts, the light will light. The logic of the circuit doesn't depend on how the contacts are wired, though; what's really important is how you must operate the contacts to turn on the output. Looking closely at the circuit, we see that the contacts are already closed, and current passes through both top and bottom contacts. The output is already ON. How do we turn it OFF? If we try pressing a pushbutton (which *opens* the contacts now), can we turn off the light by pressing A OR B? The answer is "no." Current must be stopped at both A AND B. If just one switch passes current, the light will still be ON, so we must press *both* buttons to turn the light OFF.

When we pressed both switches to do something, we called the logic of the operation, AND logic. In the past, we were always asking the question: What turns the output ON? Now, we ask: What turns the output OFF? Since we are answering a different question, we can't call this AND logic, too. The circuit that turns OFF when an AND turns ON is called a NAND. The name is a contraction of "NOT" and "AND," and it's used for this circuit because we could get the same output from a circuit made of an AND gate with two inputs and a NOT gate, connected as shown in the symbolic diagram. To make this easier to see, we've included an extra column in the truth table for our NAND, showing what an AND would do with the same inputs. If we invert each output of the AND gate, turning 1s to 0s and 0s to 1s, the outputs match the NAND gate. Note that the symbol for the NAND combines the AND and NOT symbols, but that's left of the NOT symbol is the "meatball" (the little circle).

The Boolean expression for NAND logic combines the "dot" of AND logic with the "bar" of NOT logic, as you might have expected. The expression

$$Z = \overline{A \cdot B}$$

shows that the output of an AND gate ($Z = A \cdot B$) has been inverted. Since only one thing (the output of the gate) has been inverted, there's only one bar covering the expression.

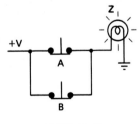

NAND circuit

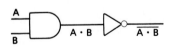

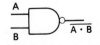

A	B	NAND Z	AND A · B
0	0	1	0
0	1	1	0
1	0	1	0
1	1	0	1

$Z = \overline{A \cdot B}$ Boolean expression

FIGURE 3-1   NAND gate.

3-2 THE NOR GATE

Figure 3-2 shows the results of attaching two normally closed switches in series. Although the switch contacts are connected so that current must pass through A AND B, the circuit is not an AND circuit.

Since the normally closed contacts are conducting only when they are NOT pushed, the output in Figure 3-2 is ON. To turn OFF the output, it isn't necessary to push both A AND B. In fact, the output will go off as soon as you push either A or B. Because the circuit turns OFF when you push A OR B, this circuit is called a **NOR**. Since the output is just the opposite of an OR gate's, its name is a contraction of "NOT" and "OR." On the truth table, we've included the output of an OR gate alongside the NOR output for comparison. You can see that if the OR is inverted, you'll get the NOR output. The symbolic diagram shows a two-input OR and an inverter (NOT) combined to form the symbol of the NOR. The Boolean expression, as you might suspect, contains both the "+" of OR logic, and the "bar" of NOT logic:

$$Z = \overline{A + B}$$

You'll notice that the expression has one bar when the symbolic diagram has one "meatball." We'll see later that this fits into a pattern.

Summarizing what we've found out so far, the series and parallel circuits made with normally closed switches have given us two new types of logic gates. They are called NOR and NAND, and follow the rule that:

NAND = AND with outputs inverted

NOR = OR with outputs inverted

3-3 DEMORGAN'S THEOREM: INVERTED INPUTS VERSUS OUTPUTS

In Figure 3-1 we saw that a parallel circuit with two normally closed switches was like an AND gate whose output is inverted. In fact, the actual circuit does not have an inverted output. It has two inverted inputs. Think about this: The normally closed switch is already inverting logic. When you activate the input (push it), the output stops being active (stops conducting). Two of these inverters have been put in parallel, so that current may flow through the contacts of one OR the other.

In Figure 3-3(a), we're taking apart the NAND gate symbol introduced in Figure 3-1. The truth table with Figure 3-3(a) shows the output of the AND part (called Z) and the output of the inverter (called Y). Since Z is the input of the inverter, Y is "NOT Z."

Figure 3-3(b)–(d) show the truth tables for symbolic diagrams of an OR with inverted inputs, an OR with an inverted output (a NOR gate), and an AND with inverted inputs. To see how we got the truth tables for circuits whose inputs are inverted, we've included the inverted value of each input on the same line with the inputs. If you OR together the $\overline{A}$ and $\overline{B}$ numbers, you get the output listed on each line of the truth table for an OR with inverted inputs. Similarly, in the AND with inverted inputs, if you AND together the numbers in the $\overline{A}$ and $\overline{B}$ columns, you will get the values that appear in the V column.

Examine the outputs V, W, X, and Y. You'll see that Y matches X, and W matches V. According to what we said about proof by perfect induction, the truth tables with the same inputs and

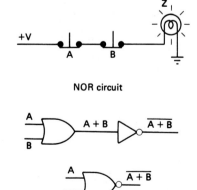

NOR circuit

FIGURE 3-2 NOR gate.

A	B	NOR Z	OR $\overline{A + B}$
0	0	1	0
0	1	0	1
1	0	0	1
1	1	0	1

$Z = \overline{A + B}$ Boolean expression

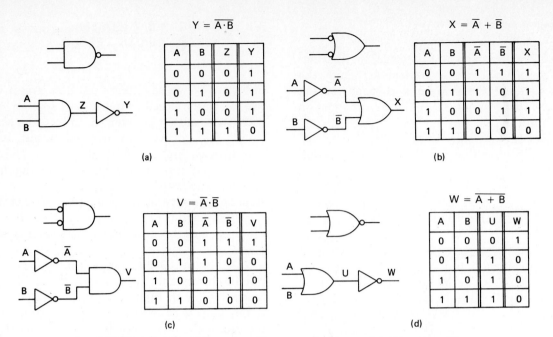

FIGURE 3-3 (a) AND gate with inverted output; (b) OR gate with inverted inputs; (c) AND gate with inverted inputs; (d) OR gate with inverted output.

outputs on every line represent circuits that are *interchangeable*, and can be *substituted* for one another. The Boolean expressions that show this interchangeability are:

$$\overline{A \cdot B} = \overline{A} + \overline{B} \quad (Y = X)$$

and

$$\overline{A + B} = \overline{A} \cdot \overline{B} \quad (W = V)$$

This relationship is called **DeMorgan's theorem**. (It will be extended later to cover a wide variety of things.)

The "new" gates in Figure 3-3(b) and (c) have Boolean expressions which reflect something of the structure of the symbolic logic diagram. For instance, Figure 3-3(b) is an OR gate with two inverted inputs. Its Boolean expression is

$$X = \overline{A} + \overline{B}$$

Each input has its own bar, and in the diagram, each input has its own inverter. Since there are two inverters in the diagram, there are two bars in the expression. In Figure 3-3(c), the same thing is true. The other symbolic diagrams have just one inverter apiece, and their Boolean expressions have just one bar apiece.

Although circuits Y and X have different symbolic diagrams, they have the same truth ta-

bles, and are both called NAND gates. Circuits W and V are both called NOR gates.

A final thought about the DeMorgan identities

$$\overline{A \cdot B} = \overline{A} + \overline{B} \quad \text{and} \quad \overline{A + B} = \overline{A} \cdot \overline{B}$$

To remember how these identities work, we like to use the phrase: "Break the bar and change the sign." What that means is that if you break the bar over an expression with a single bar over it (e.g., $\overline{A \cdot B}$) so that there's a bar fragment above each letter (e.g., $\overline{A} \cdot \overline{B}$), and then change the sign from + to ·, or vice versa (e.g., $\overline{A} + \overline{B}$), you get the equivalent expression for a circuit with inverted inputs, instead of one inverted output. To go in the opposite direction, you could use the phrase "Join the bar and change the sign" to get an inverted output expression from a circuit with inverted inputs.

3-4 NAND AND NOR IMPLEMENTATION OF OTHER GATES

NAND and NOR gates are not as fundamental as AND, OR, and NOT, but are actually more useful. With a combination of AND, OR, and NOT we can make all other logic circuits. We could build a NAND gate, for instance, out of an AND

and a NOT, or out of an OR and two NOTs. To do this, we have to use two different types of gates. NAND and NOR gates are **universal gates**. We could make an AND out of NAND gates, a NOT out of NANDs, or an OR out of NANDs. Since NANDs can be used to make AND, OR, and NOT gates, and we already know that everything else can be built as a combination of those, it's possible to make any logic circuit entirely of NAND gates. The same thing is true for NOR gates. With enough NOR gates, you could make any other type of logic gate (Figure 3-4 and 3-5).

Let's take a look at **NAND implementation**. This is the name given to the process of designing logic entirely of NAND gates. An AND can be made by using two NANDS, with one NAND inverting the output of the other two-input NAND. Here's what that does: Since the NAND is an AND with an inverted output to begin with, the NAND with its output inverted is an AND with its output inverted *twice*. To show just what this does, Figure 3-4(b) shows the first NAND broken down into an AND and a NOT, and the second NAND (used as a NOT) is drawn as another NOT. If we call the output of the AND part inside the NAND gate "W," as shown, then after the first inversion, it becomes $\overline{W}$, and after the second inversion, it is $\overline{\overline{W}}$.

We already know that something inverted twice is the same as if it had never been inverted at all. If the output of the two NANDs is W (because W inverted twice is W), then the two NANDs work exactly like an AND.

How many NANDs do we need to put together to make an OR? Figure 3-4(c) shows a circuit containing three NANDs. We've "pulled apart" the NAND in front, and reduced the two NAND inverters at the back to inverters only (since all they're being used for is NOT gates), to show how this works. Remember how a NOR gate can be made by using inverted inputs to AND logic? DeMorgan's theorem says that you can use either an OR with an inverted output, or an AND with inverted inputs. We "DeMorganized" the AND and the two inverters into a NOR made of an OR with one inverter at its output. What's left is an OR gate with its output inverted twice. Since you already know about twice-inverted logic, you can see that this is an OR gate, after we do all these substitutions.

In Figure 3-5, we've decided to let the symbolic diagrams, speak for themselves. The **NOR implementation** shown for the NOT, OR, and AND gates is done using almost exactly the same rules as we used in the NAND implementation. We decided to show the step-by-step breakdown of each diagram, adding a few words next to each step to show the rule used to get there, instead of a detailed word description here.

A word about symbolic-logic diagrams:

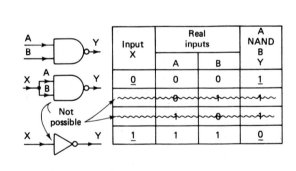

Input X	Real inputs		A NAND B Y
	A	B	
0	0	0	1
~~0~~	~~0~~		
		~~0~~	
1	1	1	0

(a) NAND inverter

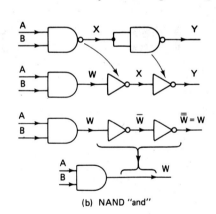

(b) NAND "and"

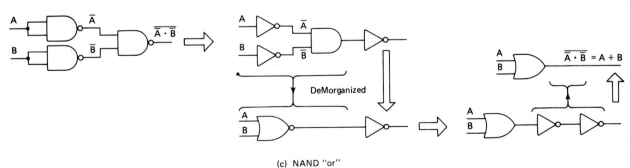

(c) NAND "or"

FIGURE 3-4 NAND implementation of other gates: (a) NAND inverted; (b) NAND "and"; (c) NAND "or."

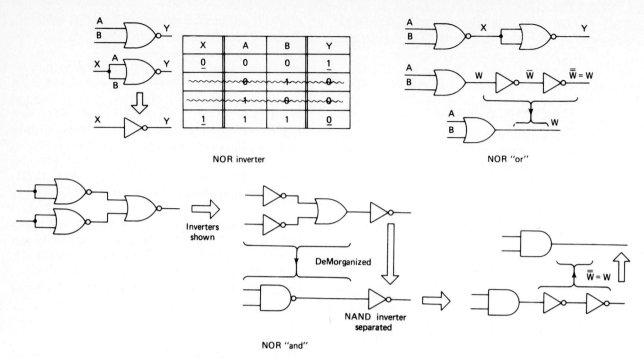

X	A	B	Y
0	0	0	1
	0	1	0
	1	0	0
1	1	1	0

NOR inverter

NOR "or"

Inverters shown

DeMorganized

NAND inverter separated

$\overline{\overline{W}} = W$

NOR "and"

FIGURE 3-5 NOR implementation of other gates.

We've used symbolic diagrams wherever we wanted to represent the logic apart from any real-world circuit, and we've used switch contact diagrams wherever we wanted to use a real circuit to show how a particular gate works. These aren't the only possible real circuits for AND, OR, NOT, NAND, and NOR—or even the best ones. For the most part, real solid-state integrated circuits are treated like the logic symbols you see in the symbolic diagrams. The output of one gate can be used to drive the input of another (not, as with switches, a mechanical input and an electrical output). The switch diagrams have their limitations. Remember, we use them as convenient examples of the way a circuit for each type of logic *might* be constructed. Real logic circuitry is assembled by using the output of one gate to drive the inputs of others. This is hard to show with mechanical switches (and obsolete!), so we've avoided "combination" circuits where *relays* might have to be used to let a signal from one switch "push" another.

3-5 THE EXCLUSIVE-OR GATE (EQUAL/ UNEQUAL COMPARE)

We've seen four types of gates with two inputs: the NAND, NOR, AND, and OR gates. We already know that the two inputs can be operated in four different ways. That was how we knew

how many lines belonged in a truth table. Now, we ask ourselves how many different types of gates, in total, are possible with two inputs. Could these four possibly be all there are?

The answer is "no." To see how many possible two-input gates there could be, we need to know how many outputs each gate has, and how many possible combinations of 1s and 0s are possible with this many outputs. It's the outputs that define what the gate is. If we look at the truth table for an AND gate, for instance, and read the outputs from top to bottom, we get 0001. This is called the **designation number** of the AND gate. It is also a binary 1, so we might say that the designation number of the AND gate is 1. If we look at the OR gate's truth table, reading from top to bottom, the outputs form the number 0111. This is a binary "seven," and we can say that the designation number for the OR gate is "7." If the designation number of the AND gate is "1," and the OR gate has a designation number of "7," there must be at least seven different types of two-input gates. Let's look at Table 3-1, which shows all possible designation numbers for two-input gates, and what each gate's logic is. You'll see, in the table, three kinds of things. First, there are gates in which both inputs are useful. The NAND, NOR, AND, and OR are among these. Some of the others combine one inverted input with one that is not inverted. Second, there are gates in this table for which only one input is use-

TABLE 3-1

Designation numbers for two-input gates

0000 = always off	
0001 = A AND B	(A · B)
0010 = A AND (NOT B)	
0011 = A	
0100 = (NOT A) AND B	
0101 = B	
0110 = ?	
0111 = A OR B	(A + B)
1000 = A NOR B	$(\overline{A + B})$
1001 = ?	
1010 = NOT B	
1011 = A OR (NOT B)	
1100 = NOT A	
1101 = (NOT A) OR B	
1110 = A NAND B	$(\overline{A \cdot B})$
1111 = always on	

ful, marked A, $\overline{A}$, B, and $\overline{B}$. We see our two-input NAND and NOR inverters in this part of the table. Third, there are gates which have no useful input. These are always ON and always OFF. They are no use at all!

There are also two designation numbers marked with a "?", on lines 0110 and 1001. These are the **exclusive-OR** and **exclusive-NOR gates**. If you look at the truth table in Figure 3-6, you will be able to see that, for the exclusive OR (XOR) gate, the output is ON only if A and B are *different*. This gives the XOR one of its other names: the *inequality gate*. The XOR gets its name from the fact that its output is ON when A OR B is ON, but not for A AND B. It's an OR gate but not an AND gate.

The exclusive NOR (XNOR) gets its name from the fact that its outputs are exactly the opposite of the XORs on every line of the truth table. You could make an XNOR by inverting the output of an XOR. This is why the XNOR symbolic-logic diagram looks like an XOR with a "meatball" at the front.

The XOR and XNOR are useful as "comparator" gates. We've already mentioned that the XOR can be called an inequality gate, because its output is ON only when A differs from B. The XNOR gate's output is ON only when A matches B, so it is called an *equality gate*.

3-6 ENABLING AND INHIBITING SIGNALS GOING THROUGH GATES

When we discussed the XOR and XNOR gates, we developed a table of outputs for every possible two-input gate. It turned out that some of these weren't very useful. Some only had one usable input, and some had none. It turns out that these

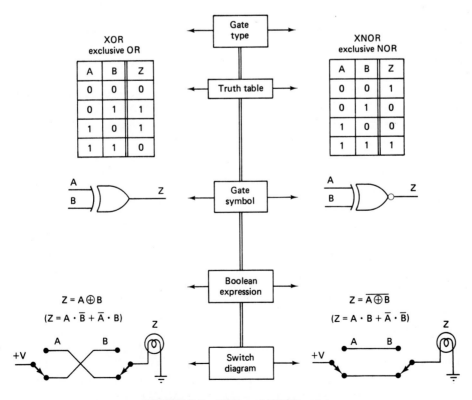

FIGURE 3-6   XOR and XNOR gates.

"useless" truth tables could be gotten from perfectly normal AND, OR, NAND, and NOR gates, depending on how we use the inputs. We'll find out, in fact, that these input conditions give us one of the most useful features of these two-input gates.

3-6.1 AND Gates

Figure 3-7 shows a table we saw previously (in Figures 2-6 and 2-7), and some Boolean identities we can discover by looking at the Boolean arithmetic. For the AND logic table, notice that the Z output is 0 on lines where B is a 0, and the Z output is the same as A on lines where B is a 1. We can say that the B input is **enabling** and **disabling** the A input. While B is a 0, the A input has no control over the Z output. Z is always a 0. When B becomes a 1, the A input *can* control the output. In fact, the output *matches* A exactly. We can think of the signal coming into the gate at A, and getting out at Z, as though it were passing through a switch. When B is a 1, and the A signal "gets through" the switch, it is enabled. When B is a 0, and the A signal is "blocked," it is disabled.

From this discovery, we find that the enabling and disabling of one input by another in an AND gate can be expressed by the Boolean **identity**

$$A \cdot 0 = 0 \qquad A \cdot 1 = A$$
$$\text{(disabled)} \qquad \text{(enabled)}$$

Notice that the AND identities above are exactly what you would expect from algebra (number multiplication). When you multiply something by 0, you always get a 0, and when you multiply "thing times 1," you get "thing." This is what you'd expect when the AND table and the multiplication table match everywhere, but it's nice to know that it's working with the letters of algebra as well as the numbers 0 and 1.

3-6.2 OR Gates

Figure 3-7 also contains a table listing the Boolean arithmetic identities for OR logic. Looking at these, we can see that on the top two lines, the signal at A can control Z, but on the bottom two lines, A has no effect on Z, and the output is "stuck" at 1. The OR gate is enabled and disabled. When the B input is 0 (on the top two lines) the A signal is able to "get through" to the output (enabled). When the B input is 1, the A signal is *unable* to get through to the output (*disabled*). A disabled signal is sometimes called an *inhibited signal*.

The identities for the OR gate

$$A + 0 = A \qquad A + 1 = 1$$
$$\text{(enabled)} \qquad \text{(disabled)}$$

don't look anything like algebra (number addition). That shouldn't surprise you; OR logic didn't exactly match addition, and we expected that

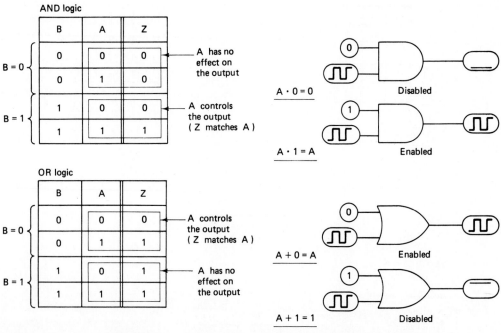

FIGURE 3-7 Enabling and disabling gates (AND or OR).

logic addition wouldn't always match number addition.

The identities for an OR gate indicate that when the OR gate is enabled (by a 0 at B), its output can switch on and off as A does. When the OR gate is disabled (by a 1 at B), its output is **frozen** at 1. Remember that the AND gate was enabled by a 1, and **frozen low** when it was disabled by a 0, and now, the OR gate is enabled by a 0, and **frozen high** when it is disabled by a 1.

EXAMPLE 3-1

Using industrial switch symbols for normally open (—| |—) and normally closed (—|/|—) momentary-contact switches, draw the switch-diagram schematics for:

(a) A · 1
(b) A · 0
(c) A + 1
(d) A + 0

(Use a lamp to indicate the output in each case.) In two of these circuits, the switch cannot control the lamp. Which two circuits? What is this called?

Solution: In switch logic, a 1 is a connection that is always connected, and a 0 is a connection that is never connected. This can be accomplished with a jumper wire between two points (a 1) or an open circuit between the two points (a 0). Connecting the four items listed above, using series connection of normally open switches for AND and parallel connection of normally open switches for OR, we get:

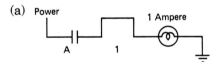

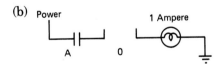

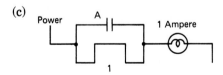

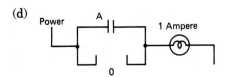

The switch (A) cannot control the lamp in circuits b and c. In circuit b the switch cannot turn on the lamp; it is always off. In circuit c the switch cannot turn off the lamp; it is always on. Both circuits are **disabled.**

3-6.3 The Commutative Law for AND and OR

These two gates, the AND and OR, may be used as switches to pass or block signals transferring from one place to another in a logic circuit. We can think of the remaining terminal (B) in the gate as a *control* terminal. Actually, the names "A" and "B" are entirely arbitrary; the signal could be put into B and controlled by A. It really doesn't matter which terminal we call A or B; the logic of AND and OR circuits works exactly the same either way. This is called the **commutative law,** and here's how it works for AND gates. You draw an AND gate, calling the top wire "A" and the bottom wire "B," or you build a series circuit, calling the first normally open switch "A" and the second "B." Now, for comparison, you draw an AND gate with "B" on top and "A" on the bottom, or a series circuit with the first switch called "B" and the second called "A." Now, you put the two circuits into "black boxes" where you can't see the circuit inside, only the inputs (marked "A" and "B") and the output. Imagine trying to tell which box has the "A" on top and which has the "A" on the bottom. Since the output will only go ON when you push *both* switches, you can't tell the two boxes apart. Imagine the same experiment with OR logic. Since *any* switch you push will turn ON the output, it really doesn't matter which you call "A" or "B." This commutative law is summed up in the expression

$$A \cdot B = B \cdot A \qquad A + B = B + A$$

Obviously, in the gate symbols, it doesn't matter who's on top and who's on the bottom ("A" and "B," of course!) as long as things come out the same.

3-6.4 NAND and NOR Gates

The NAND and NOR gates are also enabled and disabled by the same signals as AND and OR gates. To enable a NAND, you do exactly what you would have done to enable an AND. The signal at A can control the output when B is a 1. There *is* a difference, though. When output Z is controlled by input A, this time the output doesn't match the input; it is inverted. Since NAND is just like AND with an inverter added, this should be no big surprise.

To enable a NOR gate, you do exactly what you would have done to an OR, and the enabled signal comes out the output of the NOR as an inverted version of the input.

As you might expect, when the NAND and NOR gates are disabled, their outputs are frozen, but not at the levels of a disabled AND and OR. The AND would have been frozen LOW, so the NAND is frozen HIGH. (A NAND is an inverted AND, so that's "logical.") Since an OR would have been frozen HIGH, its NOR counterpart is frozen LOW. To summarize:

$$A \text{ NAND } 0 = 1 \quad A \text{ NAND } 1 = \overline{A}$$
$$\quad \text{(disabled)} \qquad\qquad \text{(enabled)}$$

and

$$A \text{ NOR } 0 = \overline{A} \quad A \text{ NOR } 1 = 0$$
$$\quad \text{(enabled)} \qquad\qquad \text{(disabled)}$$

3-7 THE EXCLUSIVE-OR (XOR), SHOWING THAT IT IS A CONTROLLED INVERTER INSTEAD OF BEING ENABLED AND DISABLED

Finally, we come to the XOR and XNOR gates. We've seen that AND, OR, NAND, and NOR

gates can be used as one-input gates to switch a signal "through" or "block" it, using the other input as a **control line.** In fact, this switching process is called **gating;** but what do the **XOR** and **XNOR** gates do with a signal put into A, when we control B? Do the **XOR** and **XNOR** gates enable and disable?

We look at Figure 3-8, and find that the lines on the XOR truth table with B = 0 have output Q matching input A. Does this mean that XOR is enabled, and will disable when B = 1? The lines on the truth table with B = 1 have Q controlled by A also, but this doesn't mean that B has no effect; the Q outputs this time are inverted A. The XOR is at no time ever disabled, but switches from a **buffer** (gate whose output is the same as its input) into an inverter, at the control of the B input. On the symbolic-logic diagram in Figure 3-8, we've included a timing diagram at the input and output of the XOR gate, to show how this works.

The XNOR gate is exactly the opposite of the XOR, as you might expect. A 0 causes the XNOR to be in the inverting condition, and the 1 at input B causes the XNOR to pass the signal at A to output R without inversion. This is also shown in Figure 3-8.

A **controlled inverter,** made from either the XOR or XNOR, is a useful device. Just why we might want to pass a signal from one place in a logic circuit to another buffered, and then suddenly change it to inverted, hasn't been explained yet, but we'll find a use for it soon.

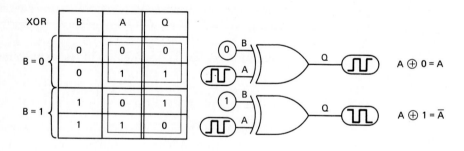

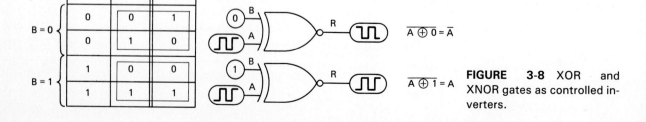

FIGURE 3-8 XOR and XNOR gates as controlled inverters.

3-8 TIMING DIAGRAMS FOR *NAND, NOR,* AND *XOR* CIRCUITS

Have you wondered why we call the logic circuits we have discussed "logic *gates*"? We've seen that AND, OR, NAND, and NOR circuits (with two inputs) act to pass or block a signal entering one input, depending on what the other input is doing. This is like a gate you open to let someone pass through, or close to block them from getting through. When the signal which is carrying some information at A controls the output, we say that the B input has **enabled** signal A to reach the output. In digital electronics, we mean the same thing when we say that signal A is **gated** through to the output.

3-8.1 AND and OR Gates

Look at the timing diagrams (Figure 3-9) for the A input and Z output of the AND gates we saw in Figure 3-7. In the timing diagrams for the AND gate, the signal turns on and off at Z as the control line (input B) is turned on and off. This is how gating, or enabling, the AND gate looks on the timing diagram of a good AND gate. If the control line (B) does not make the waveform at Z go on and off this way, we can tell that we have a bad AND gate. If you are troubleshooting a logic circuit and don't have an oscilloscope that can show three signals (A, B, and Z) simultaneously, you can still troubleshoot, or check the effectiveness of the gate, using a two-trace (dual-trace) oscilloscope, provided that you know *exactly* what

condition the control (B) terminal is in. Just view input A and output Z. As you change the state of the (unscoped) B input, the wave at Z should switch on and off.

Now, let's look at timing diagrams (Figure 3-10) for the OR gates, also shown in Figure 3-7. The OR gate is different from the AND gate, but again, the signal can be gated on and off by control line B. In this case, however, the signal appears at the output when B is off. Using an oscilloscope to troubleshoot an OR gate can be done as easily as testing an AND gate. A signal injector puts a wave into one of the OR gate's inputs, and the other input is used as a control terminal. If the output signal waveform switches on and off as the control line is varied, and matches the timing diagrams above, it is a good OR gate.

It's possible that you may have a problem here. In the lower diagram [Figure 3-10(b)] we said that the output had *no signal*, yet its level is HIGH (on) all the time. How can we call that *disabled*?

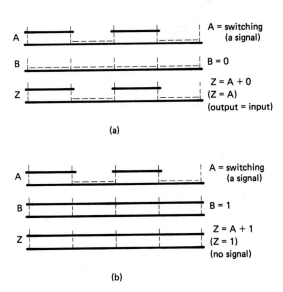

FIGURE 3-10 (a) OR gate, enabled; (b) OR gate, disabled.

The important thing, in this case, is whether the "information" at A is able to affect output Z. Suppose that you are tapping out a message in Morse code on a telegraph key attached to input A. The HIGH parts of the waveform put into A represent the dots and dashes of a message. The *message* is information that "gets through" in Figures 3-9 and 3-10, where the output has the same waveform as the input you put in. Someone at that end of the gate can see what information you wanted to send. In the other cases, the output

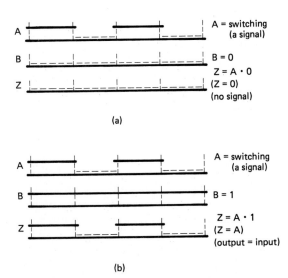

FIGURE 3-9 (a) AND gate, disabled; (b) AND gate, enabled.

waveform is a flat, straight line. In one case, the straight line is LOW (voltage OFF) and in the second, the straight line is HIGH (voltage ON), but *neither one contains any information.* Nobody looking at the output can see what information your Morse-code message contains! A voltage that is HIGH (on) all the time is as useless as one that is LOW (off) all the time.

This is what we mean by a **signal.** The "information" or "message" that you were tapping in at input A is a signal. If the signal doesn't reach the output from the input, we say that the gate is **disabled,** whether it is stuck in a 1 or a 0 state. Being stuck isn't a signal, and the message you tapped in isn't getting through the gate.

3-8.2 NAND and NOR Gates

For NAND and NOR circuits, these gates are enabled and disabled, also. The timing diagrams of the NAND gate are shown in Figure 3-11, and the diagrams for the NOR gate are shown in Figure 3-12. Why do we still call the outputs of Figures 3-11b and 3-12a "enabled" when they are clearly upside down? The reasoning here depends on the answer to the question "From what outputs can I still get the information?" In Figures 3-11a and 3-12b, there is nothing we could do to

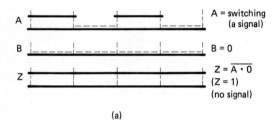

(a)

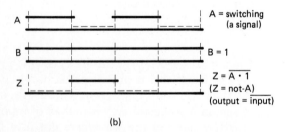

(b)

FIGURE 3-11 (a) NAND gate, disabled; (b) NAND gate, enabled.

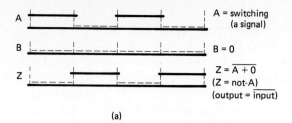

(a)

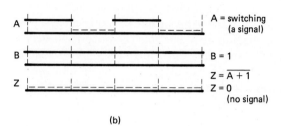

(b)

FIGURE 3-12 (a) NOR gate, enabled; (b) NOR gate, disabled.

find out what dots and dashes our Morse-code message contained (from our earlier example); the information isn't getting through. In Figure 3-11(b) and 3-12(a), on the other hand, all we have to do to interpret what is going into the input is to take the (inverted) output and invert it again. The signal is *enabled, although it is inverted.*

3-8.3 XOR and XNOR Gates

In this chapter we have also seen two other gates, the exclusive-OR and exclusive-NOR gates. According to what we can see in Figure 3-8, these gates cannot be disabled. They are always enabled, either in the straightforward sense, when the output is the same as the input, or in the inverted sense, when the output is an inverted mirror image of the input. Since there is no way to disable an XOR or XNOR gate, we can't really use them for *gating,* in the sense that we've spoken about earlier.

Rather than showing how the timing diagrams for the XOR and XNOR gates look, they will be left for you to work out in two of the questions at the end of this chapter. It should be fairly easy to work them out from the diagrams in Figure 3-8, since you've just seen the same thing done for Figure 3-7.

3-1. Identify the logic of the switch diagrams shown.

(a)

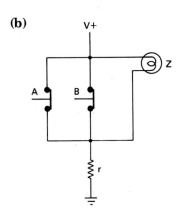

(b)

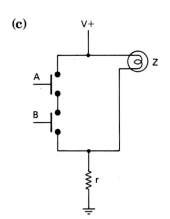

(c)

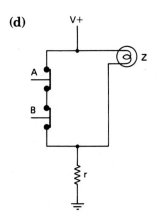

(d)

(Assume that r represents a resistor that passes enough current to light Z.) Answer parts (a) to (d) with these choices; (1) AND; (2) OR; (3) NAND; (4) NOR.

3-2. Match the gate symbols of the figures on the left with their DeMorganized equivalent among those on the right.

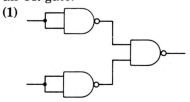

(a)   **(1)**

(2)

(b)   **(3)**

(4)

3-3. Find a Boolean expression for each gate symbol shown.

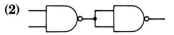

(a) A   Z   **(1)** $Z = 0$

(b) A   Z   **(2)** $Z = 1$

(c) A   Z   **(3)** $Z = A$

(d) A   Z   **(4)** $Z = \overline{A}$

3-4. Identify which NAND implementation is an OR gate.

(1)

(2)

(3)

3-5. What is the four-digit binary designation number for an OR gate with two inputs?

3-6. What is the four-digit binary designation number for:
(a) A two-input AND gate?
(b) A two-input NAND gate?
(c) An XOR gate?

3-7. What (0 or 1?) enables an:
  (a) AND gate?
  (b) OR gate?
  (c) NAND gate?
  (d) NOR gate?

3-8. When disabled, the output of a gate is frozen HIGH or LOW. Which way (0 or 1?) is the disabled output frozen for the:
  (a) AND gate?
  (b) OR gate?
  (c) NAND gate?
  (d) NOR gate?

3-9. Is it possible to enable and disable the XOR and XNOR logic gates by using one input as a control lead and the other input for a signal?

3-10. Identify the results of the logic expression X + 1 = ?.
  (1) X
  (2) 0
  (3) $\overline{X}$
  (4) 1

3-11. From the diagrams provided in Figure 3-8, can you make two timing diagrams for the XOR gate, showing the conditions

$$Z = A \oplus 0 \quad \text{and} \quad Z = A \oplus 1?$$

(Assume that there is an input B and that its input is permanently LOW in the first expression and permanently HIGH in the second expression.)

3-12. From the diagrams provided in Figure 3-8, can you make two timing diagrams for the XNOR gate, showing the conditions

$$Z = \overline{A \oplus 0} \quad \text{and} \quad Z = \overline{A \oplus 1}?$$

(Assume that there is an input B and that its input is permanently LOW in the first expression and permanently HIGH in the second expression.)

In the diagrams that follow, we have redrawn some familiar two-input switching logic circuits, but replaced switch B with a short circuit or an open circuit. Indicate whether the remaining input signal at A is enabled or disabled to the output (can control the output or not) for each case.

Example: (Two normally open switches, A and B, are in series, leading power to a light bulb; B has been replaced by a short circuit.) Since the signal at A always gets through part B to the light, A controls the

light at all times, and B has enabled input A to control the light.

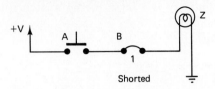
Shorted

3-13. What is the condition of each of the following switch circuits?
  (a)

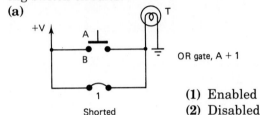
Shorted
OR gate, A + 1
  (1) Enabled
  (2) Disabled

  (b)

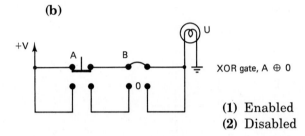
XOR gate, A ⊕ 0
  (1) Enabled
  (2) Disabled

  (c)

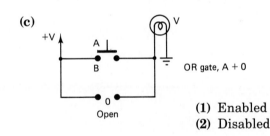
Open
OR gate, A + 0
  (1) Enabled
  (2) Disabled

  (d)

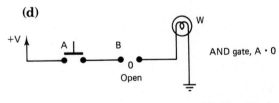
Open
AND gate, A · 0
  (1) Enabled
  (2) Disabled

3-14. From the NAND gate timing diagrams in Figure 3-11, make a diagram like the ones in Figures 3-8 and 3-9 (including the truth tables) for an enabled and disabled NAND.

3-15. From the NOR gate timing diagrams in Figure 3-12, make a diagram like the ones in Figures 3-8 and 3-9 (including the truth tables) for an enabled and disabled NOR.

3-1. Draw the truth table for the Boolean expression

$$X = \overline{D \cdot E}$$

3-2. How many different combinations of D and E must be shown on this truth table?

3-3. How many combinations of D and E produce a TRUE output for X?

3-4. Draw the symbol for a NAND gate with two inputs.

3-5. NAND is a contraction of the words _____ and _____.

3-6. Draw the truth table for the Boolean expression

$$X = \overline{D + E}$$

3-7. How many combinations of D and E produce a TRUE output for X?

3-8. Draw the symbol for a NOR gate with two inputs.

3-9. NOR is a contraction of the words _____ and _____.

3-10. An AND gate with an inverted output is a _____. (AND, OR, NAND, NOR)

3-11. An OR gate with an inverted output is a _____. (AND, OR, NAND, NOR)

3-12. Write the two expressions used to describe DeMorgan's theorem.

3-13. Write the English phrase used to summarize both expressions.

3-14. The expression $\overline{A \cdot B}$ indicates _____ (one, two) inverters on the _____. (inputs, outputs)

3-15. The expression $\overline{A} \cdot \overline{B}$ indicates _____

(one, two) inverters on the _____. (inputs, outputs)

3-16. Draw the logic symbol diagram for $\overline{A \cdot B}$.

3-17. Draw the logic symbol diagram for $\overline{A} \cdot \overline{B}$.

3-18. Which two types of gates are called "universal gates"?

3-19. Draw the truth table for the Boolean expression

$$X = D \oplus E \quad \text{(exclusive OR)}$$

3-20. Draw the symbol for an XOR gate with two inputs.

3-21. Draw the truth table for the Boolean expression

$$X = \overline{D \oplus E} \quad \text{(exclusive NOR)}$$

3-22. How many combinations of D and E produce a TRUE output for X?

3-23. Draw the symbol for an XNOR gate with two inputs.

3-24. How many NAND gates does it take to implement a NOT gate?

3-25. How many NAND gates does it take to implement an OR gate?

3-26. How many NOR gates does it take to implement an AND gate?

3-27. How many NOR gates does it take to implement an OR gate?

3-28. What is the designation number of a two-input NAND gate?

3-29. What is the designation number of a two-input NOR gate?

3-30. What is the designation number of a two-input XOR gate?

4
COMBINATIONAL LOGIC

By the time you finish this chapter, you will be able to:

1. Obtain a (sum-of-products) expression, in Boolean algebra, from a truth table.

2. Use DeMorgan's theorem to simplify Boolean expressions gotten by solving for the "output = 0" states of the truth table, instead of the "output = 1" states (*maxterms* instead of *minterms*).

3. Use a table of basic Boolean identities to simplify logic expressions.

4. Understand the *associative*, *distributive*, and *commutative* principles as applied to Boolean algebra (logic) expressions.

5. Understand how to use the distributive principle in reverse, to simplify logic expressions by *factoring*.

6. Understand how to use the commutative principle to *rearrange terms* or *rearrange factors* in a Boolean expression.

7. Use the principle of *redundancy* to simplify Boolean expressions.

8. Use the principle of *double inversion* to simplify the Boolean expression of systems with multiple inversion of signals.

9. Understand how to convert a Boolean expression into a symbolic logic diagram, which is really a wiring diagram for the logic circuit.

10. Understand that in simplifying a Boolean expression, we are making the symbolic logic diagram, and thus the actual parts and wiring of the circuit simpler, and that the circuit is thus smaller, easier to assemble, and cheaper than its nonsimplified equivalent.

4-1 EQUATION SIMPLIFICATION USING IDENTITIES

In Chapter 2 we saw that a truth table was used to describe a logic circuit. With the truth table of a circuit, it is possible to figure out what the circuit looks like (at the logic-gate level, or using switches). This is possible because there is a way to read a truth table that results in a Boolean expression. You'll recall that the Boolean expression is another way to describe the logic circuit. With the Boolean expression, drawing a schematic for a circuit is just a matter of drawing AND gates for the · in the expression, and drawing OR gates for the +, and so on.

A	B	Z
0	0	0
0	1	1
1	0	1
1	1	0

How do we get the Boolean expression from the truth table? Here is the truth table for a circuit (we know that this is the truth table for an **XOR** gate). Reading the truth table's top line, we can see that when A and B are both LOW, the output is LOW. Let's express that another way: In terms of what we have on this line, all the items on this line are things that are *happening at the same time*. To use this information in a Boolean way, we also have to know what we have, and what we do not have. On the top line, we do NOT have A, we do NOT have B, and we do NOT have an output Z. This can be written in terms of the inputs we have, and the output we have as a result:

$$\overline{A} \cdot \overline{B} = \overline{Z}$$

The Boolean expression says that we do NOT have A (not-A) and we do NOT have B (not-B). Since these conditions both exist at the same time, we say "not-A AND not-B." Since this input produces a "not-Z" condition, we say that "not-A and not-B *causes* not-Z." The relationship between *cause* and *effect* (the input is the cause, and the output is the effect) in Boolean is always represented with an "=" sign. We did not have to write the Boolean expression as it appears above. It could have been

$$\overline{Z} = \overline{A} \cdot \overline{B}$$

and it would still say the same thing. For some reason, although truth tables usually have the causes (inputs) on the left and the effects (outputs) on the right, most of the time the Boolean expression has it the other way (with the inputs on the right, as in the second expression).

On the second line of the truth table, there are two new factors to consider. One input is a 0 and the other is a 1. Also, the output is now a 1. How do we handle these 1s in the Boolean expression? Since the 0s told us what we did not have, the 1s obviously tell us what we do have. In this case, we do NOT have an A (not-A), we do have a B (B), and we do have an output Z. Reading this line gives us

$$Z = \overline{A} \cdot B$$

"Z is caused by not-A AND B." In other words, you get Z ON when you have A NOT ON and B ON. Just write everything on the line of the truth table that's TRUE by its letter, and everything that's FALSE by NOT (letter), and for circuits with more than one input, connect the inputs with ANDs, since they *all* have to be present in their stated conditions *at the same time* to produce the output that you see on that line.

Let's see how this works for a circuit with more than two inputs. Suppose that there is a circuit whose truth table has a line like this:

A	B	C	X
0	1	1	1

You'd read the line (since X is ON) based on what turns X ON. Your expression would start with "X =." Then there is a listing of the input variables (A, B, and C). A is NOT ON, and B and C are ON, at the same time. This translates as

$$\overline{A} \cdot B \cdot C$$

(Notice that everything on the line that was a number zero has a bar over its letter, and everything else is just the letter.) The whole Boolean expression is

$$X = \overline{A} \cdot B \cdot C$$

That seems simple enough. Put the output first, instead of last, and indicate all the letters for the output and inputs, putting bars over the letters whose value is 0 on this line, and stringing all the inputs together with AND signs.

EXAMPLE 4-1

If one line of a truth table has inputs and output as shown, what is the Boolean expression for that line?

D	C	B	A	g
1	0	0	0	1

Solution: The inputs and output must be active HIGH or active LOW, according to the 1s and 0s in the table, so we write

D	C	B	A	g
1	0	0	0	1

$$D \quad \overline{C} \quad \overline{B} \quad \overline{A} \quad g$$

Since the inputs must all be present at the same time to deliver the output shown, we AND the inputs together:

D	C	B	A	g
1	0	0	0	1

$$D \cdot \overline{C} \cdot \overline{B} \cdot \overline{A} \quad g$$

The output is active (HIGH) when the inputs are in the state indicated, so we complete the expression:

D	C	B	A	g
1	0	0	0	1

$$D \cdot \overline{C} \cdot \overline{B} \cdot \overline{A} = g$$

And that's it. All that's left is to rearrange the input and output sides of the Boolean expression, if we feel like it:

$$g = D \cdot \overline{C} \cdot \overline{B} \cdot \overline{A}$$

It's not quite as simple as that. Truth tables don't usually have just one line. The truth table for the XOR gate with two inputs had four lines, and the table with inputs A, B, and C would have eight lines if completely written out. We know how to get an expression from one line of a truth table, but not how to get an expression for the whole table. Without using the whole table, we have only a fraction of the information we need to develop the logic circuit, since many different circuits may have the same top line on their truth tables (AND, OR, and XOR do, for example). This is true for other lines as well as the first. To get a complete Boolean expression for the circuit, we need to use the complete truth table. For example:

A	B	Z	
0	0	0	$Z = \overline{A} \cdot \overline{B}$
0	1	1	$Z = \overline{A} \cdot B$
1	0	1	$Z = A \cdot \overline{B}$
1	1	0	$\overline{Z} = A \cdot B$

Since there are four lines on the XOR truth table, there are four expressions written, one for each line. Two of these expressions are for Z. The other two are for $\overline{Z}$. The expressions for Z are expressions that tell you what it takes to turn the output ON. The expressions for not-Z tell you what it takes to turn the output OFF. The expressions that describe how to turn the output ON are called **minterms**. Those that show what conditions at the inputs turn the output OFF are called **maxterms**.

Most people who design logic circuits are interested in what it takes to turn the output ON. All they need to write a complete expression is to list all the minterms. The question is: How do we put the minterms together?

In the XOR truth table, there are two minterms. That means there are two different ways to turn on the output. One is to apply the inputs indicated in the second line of the table ($\overline{A} \cdot B$). The other is to apply the inputs shown on the third line ($A \cdot \overline{B}$). Since either line 2 OR line 3's input conditions will turn ON the output, the most suitable way to show that either condition will work is to OR the two minterms together:

$$Z = (\overline{A} \cdot B) + (A \cdot \overline{B})$$

The expression above is a complete description of all the ways to turn the output of this gate ON. From this, we can get a general rule about how to write a Boolean expression from the truth table of a logic circuit.

Step 1. List the minterms from the truth table as shown here:

A	B	Z	
0	0	0	maxterm
→ 0	1	1	$Z = \overline{A} \cdot B$
→ 1	0	1	$Z = A \cdot \overline{B}$
1	1	0	maxterm

Step 2. OR together all minterms:

$$Z = (\overline{A} \cdot B) + (A \cdot \overline{B})$$

The expression above (although we got it from the truth table) can be found in Figure 3-6.

Let's try the same thing with another familiar truth table. The table below is the truth table for an OR gate with two inputs. This table has one maxterm (on the first line) and three minterms (on the remaining lines).

A	B	Z
0	0	0
0	1	1
1	0	1
1	1	1

The minterm expression (for Z = ...) is expressed with three **terms** (things separated by + signs):

$$Z = (\overline{A} \cdot B) + (A \cdot \overline{B}) + (A \cdot B)$$

Each term represents a line on the truth table whose output is a logic 1. This is a fairly awkward expression, especially when we remember that the expression for an OR gate is supposed to be

$$Z = A + B$$

In this case it would actually be easier to solve the truth table for the one maxterm. There is only one maxterm, and the maxterm expression (for $\overline{Z}$ = ...) would be much shorter:

$$\overline{Z} = \overline{A} \cdot \overline{B} \quad \text{from the top line of the truth table}$$

The only problem with the minterm expression (three terms) and the maxterm expression is that neither one of them is right. Neither of the expressions we have looks anything like

$$Z = A + B$$

Why?

It turns out that *both* of the expressions from the truth table are correct. They're well disguised, but they can both be converted into the "A OR B" form using Boolean identities and doing a little substitution. This is the same process you use in ordinary number algebra to "cook down" an equation into a simpler form. Knowing

a few rules, it's possible to swap a clumsy expression with a simpler one, until the final result is as simple as it possibly can be. In this case, one of the tricks that we already know is **De-Morganizing**. If we start with the maxterm expression

$$\overline{Z} = \overline{A} \cdot \overline{B}$$

we have an expression that tells us how to turn the output of the gate OFF. That's not what we want. We want an expression (for Z = ...) that tells us how to turn the output of the gate ON. Since that's just the opposite of OFF, we can convert the expression for ($\overline{Z}$ = ...) into an expression for (Z = ...) by inverting the maxterm expression. If we invert what's on the left-hand side of the = sign (to get Z = ...), we must also invert what's on the right-hand side of the = sign (since that's an expression for $\overline{Z}$):

$$\overline{\overline{Z}} = \overline{\overline{A} \cdot \overline{B}}$$

On the left-hand side of the expression, the result of inverting Z twice is Z. On the right, to see what the result is, we DeMorganize the expression under the long bar: "Break the bar and change the sign":

$$Z = \overline{\overline{A}} + \overline{\overline{B}}$$

The letters A and B now have two bars each, meaning that each input is inverted twice. That's the same as not being inverted at all, so the double bar over A can be dropped, and so can the double bar over B:

$$Z = A + B$$

What's left is the standard expression for an OR gate.

If we want to convert the minterm expression for Z into its simplest form, we have a much more formidable task, since the minterm expression is so large. Before we try, it's necessary to have the tools to do the job. These tools are the Boolean identities listed in Table 4-1. We'll go through the list and see what each identity says before we try to use them to simplify the expression for (Z = ...).

Rule 0 is the **double-inversion rule**, which actually says that any *even* number of inversions will cancel out.

Rules 1, 2, 3, and 4 are familiar rules about the effects of 1s and 0s on enabling and disabling AND and OR gates.

TABLE 4-1

Basic Boolean identities

0. $\overline{\overline{M}} = M$
1. $M \cdot 0 = 0$
2. $M \cdot 1 = M$
3. $M + 0 = M$
4. $M + 1 = 1$
5. $M \cdot \overline{M} = 0$
6. $M + \overline{M} = 1$
7. $M \cdot N = N \cdot M$
8. $M + N = N + M$
9. $M \cdot M = M$
10. $M + M = M$
11. $L \cdot (M \cdot N) = (L \cdot M) \cdot N = L \cdot M \cdot N$
12. $L + (M + N) = (L + M) + N = L + M + N$
13. $L \cdot (M + N) = (L \cdot M) + (L \cdot N)$
14. $L + (M \cdot N) = (L + M) \cdot (L + N)$
15. $M \cdot (M \cdot N) = M$
16. $M \cdot (M + N) = M$
17. $M + (\overline{M} \cdot N) = M + N$
18. $(\overline{M} \cdot N) + (M \cdot \overline{N}) + (M \cdot N) = M + N$

Rules 5 and 6 are identities that tell you what happens when you AND or OR something with its opposite. In rule 5, the expression indicates that M is ON AND OFF at the same time. Since this can never happen, the expression can never be true. It isn't, so the output is always a 0. In rule 6, the expression indicates that M is ON OR OFF at some time. This is always true. M only has two states; it can be ON or it can be OFF. Since one OR the other is always TRUE, the output is always a 1.

Rules 7 and 8 are the commutative law, which was discussed in chapter 3. It says that regardless of which input of a two-input AND or OR gate you choose to call M (calling the remaining one N) the gate behaves the same way. In other words, in an expression where things are ANDed together, you can shuffle the **factors** (the name for things ANDed together) around in any order, and the gate behaves the same way no matter what order the factors are in. This applies to gates with any number of inputs, not just two-input gates. Where there are things ORed together (we call these things **terms**), the same law applies. You can shuffle the terms around in any order, for an OR gate with as many inputs as you want, and the gate has the same action, and the same truth table, regardless of the position of each input.

Rules 9 and 10 are related to something called **redundancy**. Rule 9, for instance, says that when A AND A are TRUE, A is TRUE. Stated in words, this is nonsense. Nobody talks that way! The whole point of the rule is that if you AND together a bunch of things, and one of the inputs appears *twice*, that's one time too many. Only *one* of each thing is needed. Rule 10 says the same thing for OR logic. If you have an OR gate with a lot of different things ORed together, and one of the inputs appears in two places (or even three or four) you can drop the extras. Only one of each input is needed. Redundancy just means that you have more of "them" than you need.

EXAMPLE 4-2

Simplify the Boolean expression

$$X = AB + BC + AB + AC$$

to an expression containing only three terms.

Solution: According to identity 10 in Table 4-1, if a term in an expression is the same as another term, you need write that term only once. If we identify the terms in this expression as shown here:

$$X = \underset{1}{AB} + \underset{2}{BC} + \underset{3}{AB} + \underset{4}{AC}$$

we can see that terms 1 and 3 are the same. We can either drop term 1 or term 3, whichever we please. If we drop term 3, the expression becomes

$$X = AB + BC + AC$$

Rules 11 and 12 are called the **associative law**. Rule 11 says that for AND logic, if you have three things ANDed together, it really doesn't matter if two of them were ANDed together first, and another thing was ANDed later. No matter what order things were ANDed first, second, third, and so forth, if they are all ANDed together, the result is one big AND gate. The logic of the gate would be the same if all the inputs were ANDed together at the same time (see Figure 4-1). Rule 12 says that OR logic follows the same associative law that AND logic does. Anything ORed together in any order becomes a big OR gate.

Rules 13 and 14 are called the **distributive principle** or **distributive law**. Rule 13 states for AND logic that something ANDed with an expression having several terms will end up *distributed* over each of the terms by the AND process. In other words, in the example, the L was ANDed with the expression "(M + N)." In the result, the M term and the N term each had an L of their own, ANDed with the term. The L had been distributed to the two terms (which is why we call it the distributive rule). Rule 14 shows that the same rule applies to OR logic. (In alge-

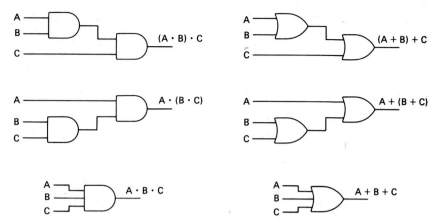

FIGURE 4-1 Associative principle.

bra, where "·" means "multiply" and "+" means "add," this rule does not work for "+," but it does in Boolean algebra—further evidence that the "+" of Boolean does not always work the same way as the "+" of arithmetic.)

In the example, the variable L is being ORed with the expression "(M · N)." The result shows that an L has been distributed to each of the two factors found inside the parentheses. The M factor and the N factor have each been provided with an L of their own, ORed with the factor. Both of these rules can be used in reverse, mainly, to get an L that shows up in two or more places, as on the right of the examples, and reduce the expression to just one L (any other letter will do, we just chose L for this example). When we substitute the shorter left-hand-side version of the expression (which has L in just one place) for the longer right-hand-side version, the simplification is called **factoring** if the L is ANDed with a bunch of terms, as in rule 13. I suppose that the simplification we get by going from an expression like that on the right in rule 14 to the shorter form on the left in rule 14 could be called *terming*, but I've never heard anyone use the word for extracting a common term found in a bunch of factors. In rule 13, though, where we extract a common factor found in a bunch of terms, the extraction of the factor is definitely called "factoring."

Rules 15 and 16 both show simplification rules that permit you to drop a letter out of an expression in certain cases. In rule 15, one of the two terms contains the other term as a part of itself. The longer term is dropped, and the shorter term remains. In rule 16, one of the factors contains another factor as a part of itself. The longer factor is dropped, and the shorter factor remains. From this, we get a general rule: "If an expression made of *terms* (or *factors*) has a *term* (*factor*) which *completely contains* another shorter *term* (*factor*) as part of itself, the longer *term* (*factor*) is dropped and the shorter one remains."

The reason this rule works (to give one example) in rule 15 might be more understandable if the *words* instead of the Boolean symbols are examined. Rule 15 reads: "If M is ON, OR if M AND N are ON, you get an output from the OR that exactly matches M." Reading between the lines, this says that N does *nothing*. Exploring the wording of rule 15, we see that we're really saying: "If M is ON (regardless of whether N is HIGH or LOW), OR if M is ON AND N is ON, the output should go ON." Since the output will be on for *either* condition, the second condition (when M is ON AND N is ON) is really a part of the first condition (when M is ON, regardless of N's condition). Since the first condition turns the output ON at *all* the times the second condition would, and more besides, the second condition isn't needed. It turns out that the shorter term actually covers *more* situations than the longer one (and the longer one becomes redundant, so it is dropped).

EXAMPLE 4-3

Simplify the Boolean expression

$$X = AB + B + BC + BCD$$

Solution: According to identity 15 in Table 4-1, if an expression contains several terms, and one of the terms *completely contains* another shorter term, the longer term can be dropped and the shorter term remains. In the case above, we will identify the terms as shown:

$$X = \underset{1}{AB} + \underset{2}{B} + \underset{3}{BC} + \underset{4}{BCD}$$

Term 2 is B, and terms 1, 3, and 4 all contain the letter B. Can terms 1, 3, and 4 all be dropped because term 2 is part of them? The answer is yes. This leaves

$$X = B$$

Rule 17 is an expression that says the output will be ON if M is ON (regardless of the condition of N), or if N is ON (and M is OFF). This ends up the same as "M OR N" because, if M is ON by itself, the output will be ON, and if N is ON by itself, the output will be ON. When both M and N are ON, M is ON, and according to the first term in the expression, as long as M is on, it doesn't matter what N is doing, the output will be ON. So the "exclusion" that appears in the second term of the expression really affects nothing in the truth table. For us, this means that if a larger term in an expression contains the **complement** (inverted form) of a smaller term, the complement part can be dropped out of the larger term. Note that unlike rule 15, we don't drop out the whole larger term, but just the part of it that is the complement of another term.

You may have noticed a pattern in rules 7 and 8, 9 and 10, 11 and 12, 13 and 14, and 15 and 16, taken as pairs. Each rule in the pair could be transformed into the other rule by just switching + for ·, and vice versa. You could do the same thing for this rule 17, and you would get

$$M \cdot (\overline{M} + N) = M \cdot N$$

(Is this a true statement? Can you prove it?)

Rule 18 is simply the truth table of an OR gate converted into a minterm expression. We know that it's true, although we haven't seen it worked out. Perhaps now, with all these other rules, we can find some ways to check out the truthfulness of our expression for the OR gate.

We'll start out with the expression we got from the OR truth table, where we called the inputs A and B instead of M and N:

$$Z = (\overline{A} \cdot B) + (A \cdot \overline{B}) + (A \cdot B)$$

As a starting point, let's take the fact that the last two terms in the expression contain A as a common factor. We can factor the A out using the identity:

$$L \cdot (M + N) = (L \cdot M) + (L \cdot N)$$
Rule 13

In this case, we'll rewrite rule 13 as

$$A \cdot (\overline{B} + B) = (A \cdot \overline{B}) + (A \cdot B)$$

On the right-hand side of this expression we see the last two minterms from the OR truth table. On the left is a simplified version having just one A distributed over the (not-B) and B terms. Our

expression for the OR is now somewhat simpler. It has one less letter and one less sign, and it looks like

$$Z = (\overline{A} \cdot B) + (A \cdot (\overline{B} + B))$$

If we look at this expression closely, we can find a place to use another identity, namely rule 6:

$$M + \overline{M} = 1$$

which we can rewrite as

$$B + \overline{B} = 1$$

Since we have this at the end of the expression for the OR, we can replace the "not-B OR B" with a 1, like this:

$$Z = (\overline{A} \cdot B) + (A \cdot 1)$$

Now the expression for the OR gate is even shorter and simpler, but within this expression, we can find yet another place to use an identity, rule 2, which says:

$$M \cdot 1 = M$$

which we rewrite for our problem as

$$A \cdot 1 = A$$

We can now replace the "A AND 1" part at the end of the expression with just an A, like so:

$$Z = (\overline{A} \cdot B) + A$$

Swap the two terms around using rule 8 (the commutative law for OR logic):

$$Z = A + (\overline{A} \cdot B)$$

and compare the results with rule 17:

$$M + (\overline{M} \cdot N) = M + N$$

which we can rewrite with As and Bs as

$$A + (\overline{A} \cdot B) = A + B$$

If we substitute the right-hand form of the expression above for the left, we have the final simplification:

$$Z = A + B$$

which is what we were supposed to get all along. (I didn't say it would be simple!)

It took six steps to simplify the OR expression from the minterms on the truth table to the final, simplest form. In doing this, we used at least six different identities in the steps. To see what each identity can do to simplify Boolean expressions, let's look at some simpler examples used with each type of identity.

4-2 THE COMMUTATIVE LAW FOR *AND* AND *OR* LOGIC

The commutative law is seldom used to *simplify* Boolean expressions; instead, it's usually used to *rearrange* their terms or factors when trying to spot relationships that could be simplified with other rules. For example, the terms

$$Z = A + BC + BD + A + B$$

can be reshuffled using the commutative law as

$$Z = A + A + B + BC + BD$$

With this rearrangement, it is much easier to see the "A + A" at the beginning of the expression and that "B + BC + BD" is at the end of the expression. Deciding what to do with these is another matter, but the commutative law has made it possible to rearrange these terms so that sections of the expression bear a strong resemblance to some of the identities in the list.

Another example shows the use of the commutative law on a series of factors in an expression:

$$Z = A \cdot (B + C) \cdot (D + C) \cdot A$$

This expression has four factors, which can be rearranged as

$$Z = A \cdot A \cdot (B + C) \cdot (D + C)$$

This groups together the two As (redundant!) and the factors which both contain a C term (which can be "termed" out of both factors using rule 14). It doesn't really change anything, it just makes certain relationships easier to see.

4-3 THE EFFECT OF REDUNDANCY ON *AND* AND *OR* LOGIC

The observation of redundancy in a Boolean expression is at times the easiest, and at times

the most difficult of simplification steps. Easy examples are provided in the two commutative law examples above. The first,

$$Z = A + A + B + BC + BD$$

uses the "implied" AND operation (BC is really $B \cdot C$, etc.) What's easy to see, however, is the "A + A" part of the expression. This is clearly redundant. Only one A is needed; accordingly, "A + A = A" yields

$$Z = A + B + BC + BD$$

which may still be simplified further, but not on the basis of redundancy. The second commutative law example was

$$Z = A \cdot A \cdot (B + C) \cdot (D + C)$$

in which "A · A = A" may be used to get

$$Z = A \cdot (B + C) \cdot (D + C)$$

4-4 THE ASSOCIATIVE LAW FOR *AND* AND *OR* LOGIC

This is another rule more useful for shuffling things around than actual replacement of complicated stuff with simple stuff. It can be used in the example below:

$$X = A \cdot (B + D) \cdot (A \cdot D)$$

to rearrange the expression into

$$X = A \cdot A \cdot D \cdot (B + D)$$

which contains a "redundant A" and a factor "(B + D)" that contains another factor within itself, "D."

4-5 THE EFFECT OF DOUBLE INVERSION ON LOGIC

In simplifying Boolean expressions, there are often times when the rule "$\overline{\overline{A}} = A$" can be used. These occasions usually arise after DeMorganizing some expression or finding its complement. Example:

$$\overline{Z} = \overline{A} \cdot \overline{B} \qquad (1)$$

is the maxterm expression for an OR gate. To find

out what the expression for Z looks like (we have the expression for not-Z), we've got to invert both sides of the Boolean expression

$$\overline{\overline{Z}} = \overline{\overline{A} \cdot \overline{B}} \qquad (2)$$

using the double-inversion rule on the Z to get the expression

$$Z = \overline{\overline{A} \cdot \overline{B}} \qquad (3)$$

At this point, we have a problem. Is the bar over the A and the B a double bar (which can be dropped)? Answer: No. One bar covers both letters, while the short bars cover one letter apiece. This is not one of those places where the double-inversion rule applies. It can only work (in Boolean expression terms) where both bars cover *exactly the same space*. The long bar covers the whole expression (which represents the output), whereas the short bars cover one letter apiece (which represent each of the inputs). As we know from an earlier chapter, inverting an output does not cancel an inverter at an input. What is called for here is a DeMorganizing operation to "break the bar" into short fragments that cover one letter apiece. This will make the expression more manageable because each letter will now have a double inversion, and the bars can be dropped. First, we'll DeMorganize:

$$Z = \overline{\overline{A}} + \overline{\overline{B}} \leftarrow \begin{array}{l}\text{break the bar and change} \\ \text{the sign under the break}\end{array} \qquad (4)$$

Then we'll drop the double bars over each input (A, B) that has been inverted twice.

$$Z = A + B \qquad (5)$$

In this case, we can see that we needed to use the double-inversion rule two times in simplifying the expression. The expression is simpler, not because there are any fewer terms, but because the number of inverters needed to implement the circuit of expression (3) is three, whereas the final form of the expression (5) needs none. The number of two-input gates [an AND in expression (3) and an OR in expression (5)] is the same, so the total number of gates, and the total number of *types of gates*, is reduced.

4-6 GATE REPRESENTATION OF BOOLEAN EXPRESSIONS

After looking at the last example, you might be inclined to ask "How does he know that?" about the number of gates. Here's how it's done:

We'll use the example of expression (3):

$$Z = \overline{\overline{A} \cdot \overline{B}}$$

In this expression we'll work from the inside of the expression outward as we build a symbolic diagram. The innermost part of the expression is the AND logic of

$$A \cdot B$$

which we'll build using an AND gate attached to inputs A and B [Figure 4-2(a)].

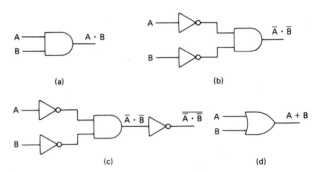

(a)

(b)

(c)

(d)

FIGURE 4-2 DeMorganizing an AND gate: (a) AND gate; (b) with inverted input; (c) with inverted inputs and output; (d) simplification using DeMorgan's theorem.

As we work our way out of the expression, we find next that the inputs A and B have short bars over each letter. This shows that the inputs each have an inverter [Figure 4-2(b)]:

$$\overline{A} \cdot \overline{B}$$

At the final, outermost part of the expression, we find a long bar covering the whole expression for Z. The long bar indicates that whatever circuit was already constructed, its logic has been *totally* inverted (i.e., inverted after everything else has been already put together). This requires an inverter at the output of the symbolic diagram, because when we see the *whole expression* on the right side of the = sign inverted, we are actually seeing Z, and we are saying that the circuit which makes Z is inverted before its output arrives at Z [Figure 4-2(c)]:

$$\overline{\overline{A} \cdot \overline{B}}$$

For comparison, the simplified expression for this gate is represented by an OR symbol in Figure 4-2(d).

Example 4-4

Using logic symbols for AND, OR, and NOT gates, and for gates with active-HIGH and active-LOW inputs, draw schematics for the following Boolean expressions:

(a) $Z = \overline{A \cdot B}$
(b) $Z = \overline{\overline{A} \cdot \overline{B}}$
(c) $Z = \overline{A + B}$
(d) $Z = \overline{\overline{A} + \overline{B}}$

Solution: Observe that the expressions with a single bar covering the entire output expression produce symbolic diagrams with a single inverter "meatball" at the device's output. On the other hand, note that the expressions with *two* bars, one over the letter for each input, produce symbolic diagrams with *two* inverter meatballs, one at each input.

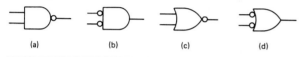

(a)   (b)   (c)   (d)

4-7 SIMPLIFICATION OF BOOLEAN EXPRESSIONS, SHOWING CONSEQUENT SIMPLIFICATION OF THE GATE DIAGRAM

We're going to work out an example at this point to show the idea of Boolean simplification and the simplification that results in the circuit when the expression is simplified. Here is a little background on the problem we're going to solve:

A professor of digital electronics has a beautiful daughter named Alice. The professor and his wife, Betty, are taking Alice and her two boy friends, Chuck and Dave, out to the summer cottage for a fishing weekend. The professor wants to be sure that Alice is chaperoned at all times, and wishes to design a logic circuit that will set off an alarm if Alice is alone with either one or both of the boys while the professor is out fishing in the boat.

To solve this, we'll need some definitions at the start:

logic 0 = person outside the cottage
logic 1 = person inside the cottage
alarm ON = logic 1
alarm OFF = logic 0

For the sake of simplicity, we'll indicate who's where with a truth table showing all the possible combinations of Alice (A), Betty (B), Chuck (C), and Dave (D), indicating whether each person is

outside the cottage (0) or inside the cottage (1). The truth table will have to show all the possible combinations of people in each place. Then we'll have to look at each line of the truth table and decide whether the professor should be "alarmed."

TRUTH TABLE

Line	A	B	C	D	Alarm
0	0	0	0	0	
1	0	0	0	1	
2	0	0	1	0	
3	0	0	1	1	
4	0	1	0	0	
5	0	1	0	1	
6	0	1	1	0	
7	0	1	1	1	
8	1	0	0	0	
9	1	0	0	1	
10	1	0	1	0	
11	1	0	1	1	
12	1	1	0	0	
13	1	1	0	1	
14	1	1	1	0	
15	1	1	1	1	

Let's go through this line by line.

Line 0 = Everybody's outside (no problem) Alarm = 0

Line 1 = Dave's inside (nothing to worry about) Alarm = 0

Line 2 = Chuck's inside (no alarm here) Alarm = 0

Line 3 = Chuck and Dave inside (???) Alarm = 0 (for Alice)

Line 4 = Betty's inside (A, C, and D outside alone) Alarm = 1

Line 5 = Betty and Dave inside (A and C outside) Alarm = 1

Line 6 = Betty and Chuck inside (A and D outside) Alarm = 1

Line 7 = Betty, Chuck, and Dave inside (Alice is alone outside, so everything's OK) Alarm = 0

Line 8 = Alice is alone inside (no problem) Alarm = 0

Line 9 = Alice is alone inside with Dave (pulling down the window shades) Alarm = 1 (!)

Line 10 = Alice is alone inside with Chuck (!!) Alarm = 1

Line 11 = Alice is inside with Chuck and Dave (!!!!) Alarm = 1

Line 12 = Alice is inside with Betty (OK) Alarm = 0

Line 13 = Alice with Dave, Betty chaper-
oning   Alarm = 0
Line 14 = Alice with Chuck, Betty chaper-
oning   Alarm = 0
Line 15 = Everybody's inside   Alarm = 0

Looking through this list, we find 6 minterms and 10 maxterms. Since there are fewer minterms, we'll write the expression for these:

Line 4: (0100) expression = $\overline{A} \cdot B \cdot \overline{C} \cdot \overline{D}$
Line 5: (0101) expression = $\overline{A} \cdot B \cdot \overline{C} \cdot D$
Line 6: (0110) expression = $\overline{A} \cdot B \cdot C \cdot \overline{D}$
Line 9: (1001) expression = $A \cdot \overline{B} \cdot \overline{C} \cdot D$
Line 10: (1010) expression = $A \cdot \overline{B} \cdot C \cdot \overline{D}$
Line 11: (1011) expression = $A \cdot \overline{B} \cdot C \cdot D$

The expression for "Alarm =" (X =) should be the OR of all these lines, so the expression for the "Alice alarm" is

$$X = \overline{A} \cdot B \cdot \overline{C} \cdot \overline{D} + \overline{A} \cdot B \cdot \overline{C} \cdot D$$
$$+ \overline{A} \cdot B \cdot C \cdot \overline{D} + A \cdot \overline{B} \cdot \overline{C} \cdot D$$
$$+ A \cdot \overline{B} \cdot C \cdot \overline{D} + A \cdot \overline{B} \cdot C \cdot D$$

We now have a six-input OR gate receiving inputs from six four-input AND gates, each of which has an average of two inverters at its in-puts—12 inverters, total. The circuit for this mess is shown in Figure 4-3.

To actually build one of these things is going to require a six-input OR gate (we can get this on one integrated circuit that contains a two four-input ORs, by using the associative law to put together a six-input gate). Then we'll need six four-input ANDs (this'll take three dual, four-in-put AND circuit packages). Then, we'll need 12 NOT gates (which come six to a package, making two six-packs). This is a total of six integrated cir-cuit packages, plus gobs of wiring connections to goof up, not counting the alarm and power sup-ply—mind you, this is just the logic!

Now, in the interest of sanity, and because it'll save us money, let's see what we can do to simplify this thing.

We start with the expression

$$X = \overline{A} \cdot B \cdot \overline{C} \cdot \overline{D} + \overline{A} \cdot B \cdot \overline{C} \cdot D$$
$$+ \overline{A} \cdot B \cdot C \cdot \overline{D} + A \cdot \overline{B} \cdot \overline{C} \cdot D$$
$$+ A \cdot \overline{B} \cdot C \cdot \overline{D} + A \cdot \overline{B} \cdot C \cdot D$$

Notice that the first three terms contain not-A and that the last three terms contain A. We'll ap-ply the distributive principle to these, factoring

out the not-A and A from terms that contain them:

$$X = \overline{A} \cdot (B \cdot \overline{C} \cdot \overline{D} + B \cdot \overline{C} \cdot D + B \cdot C \cdot \overline{D})$$
$$+ A \cdot (\overline{B} \cdot \overline{C} \cdot D + \overline{B} \cdot C \cdot \overline{D} + \overline{B} \cdot C \cdot D)$$

We've already reduced 24 inputs to 20 (we can see that by counting the letters on the right of the = sign in each case).

Now, if we look closely, we'll see that the three BCD terms inside the upper parentheses all contain B, and in the lower parentheses, the three BCD terms all contain not-B. Factoring out the Bs and not-Bs, we get

$$X = \overline{A} \cdot B \cdot (\overline{C} \cdot \overline{D} + \overline{C} \cdot D + C \cdot \overline{D})$$
$$+ A \cdot \overline{B} \cdot (\overline{C} \cdot D + C \cdot \overline{D} + C \cdot D)$$

This brings the circuit down to 16 inputs. We can compare the items inside parentheses to the last identity (18) on our list, the one that says

$$(\overline{M} \cdot N) + (M \cdot \overline{N}) + (M \cdot N) = M + N$$

which we could rewrite as

$$(\overline{C} \cdot D) + (C \cdot \overline{D}) + (C \cdot D) = C + D$$

this is the exact expression that we find in the lower set of parentheses, which means we could reduce X to

$$X = \overline{A} \cdot B \cdot (C \cdot \overline{D} + \overline{C} \cdot D + \overline{C} \cdot \overline{D})$$
$$+ A \cdot \overline{B} \cdot (C + D)$$

But we're not yet through with identity 18. If we use

$$M = \overline{C} \quad \text{and} \quad N = \overline{D}$$

we rewrite the identity as

$$(\overline{\overline{C}} \cdot \overline{D}) + (\overline{C} \cdot \overline{\overline{D}}) + (\overline{C} \cdot \overline{D}) = \overline{C} + \overline{D}$$

and drop out the double bars to get

$$(C \cdot \overline{D}) + (\overline{C} \cdot D) + (\overline{C} \cdot \overline{D}) = \overline{C} + \overline{D}$$

It turns out that the expression for "not-C OR not-D" is exactly what's in the upper parentheses, so we replace the longer expression with the shorter one, and get

$$X = \overline{A} \cdot B \cdot (\overline{C} + \overline{D}) + A \cdot \overline{B} \cdot (C + D)$$

eight terms. Probably, we can do this in three in-

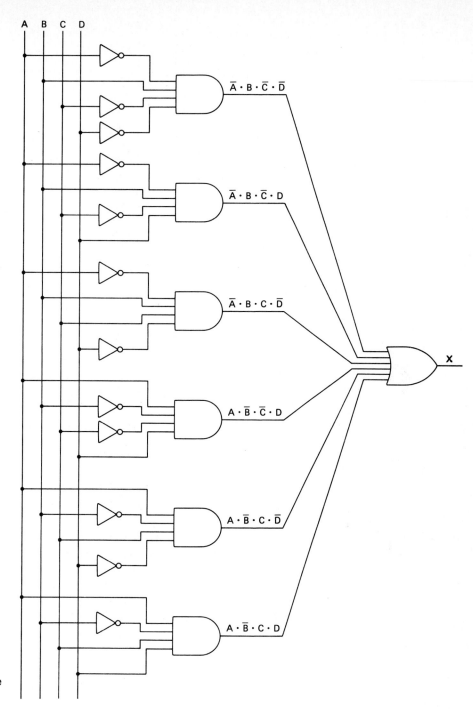

FIGURE 4-3 The "Alice problem" (not simplified).

tegrated-circuit packages, with a whole lot less wiring, as shown in the symbolic diagram (Figure 4-4).

4-8 OTHER SIMPLIFICATION RULES

The table of identities used in this chapter (Table 4-1) does not include DeMorgan's theorem, but we've used it in earlier chapters, and used it liberally in the examples in this chapter. It is indis-

pensable when you want to derive a Boolean expression from the maxterms of a truth table. After inverting the maxterm expression, some simplification can almost always be accomplished by DeMorganizing.

In the "Alice, Betty, Chuck, and Dave" problem we saw a truth table derived from first principles—deciding line by line what we wanted. When we were done, there were fewer minterms (6) than maxterms (10). We chose to simplify the minterm expression because there was less work.

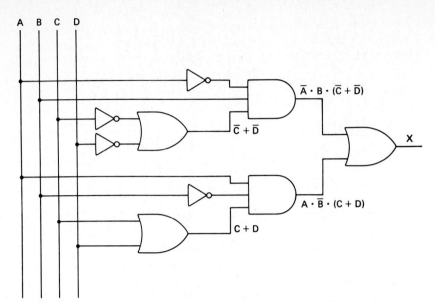

$$\overline{A} \cdot B \cdot (\overline{C} + \overline{D})$$

$$\overline{C} + \overline{D}$$

$$A \cdot \overline{B} \cdot (C + D)$$

$$C + D$$

X

FIGURE 4-4 The "Alice problem" (simplified).

If the results had come out the other way, we'd have simplified the maxterm expression, and inverted it at the end. In the final analysis, you should choose whatever approach seems to involve less work. If the expression you want to simplify has fewer terms, that will be fewer places to make a mistake. Less work, and more reliable answers, are the result.

We should point out that in the "Alice, Betty, Chuck, and Dave" problem, the simplification you saw was not the only way that expression could be simplified. It may have looked simple. Things always look simple when you're seeing someone else—who knows exactly how to get where he or she is going—do the problem solving. When you're doing real problem solving, there will be false starts, mistakes, and dead-ends in any Boolean simplification. There'll be a lot of trial and error involved in spotting what identities can be useful. It won't always be easy. You should also keep in mind the idea of checking your work when your simplification is finished. The only way you'll know if you made a mistake along the line is to work out a truth table for your final, simplified expression and see if that table matches the truth table you started with. If it doesn't, it's "back to the drawing board." You'll have to try working over the steps you used to see if any of them is in error. This is called "debugging" in the design field, and is considered a normal part of any real design project. It's okay to goof up as long as you have a way of *knowing* when you've goofed. (A final piece of advice here—most of the errors made in Boolean simplification happen when you lose a bar copying a part of an expression from one line to another. Count them to double-check yourself; it can't hurt.)

There's another way the "Alice, Betty, Chuck, and Dave" truth table can be implemented using a single chip and requiring no Boolean simplification at all, but we'll save that for you as a "surprise" in future chapter.

QUESTIONS

4-1. Write a truth table "$Z = A + \overline{B}$." Write an expression for this truth table that lists the minterms.

4-2. Write minterm and maxterm expressions for the truth table.

A	B	Y
0	0	1
0	1	1
1	0	0
1	1	0

4-3. Simplify the expression

$$X = A \cdot (B + \overline{B})$$

4-4. DeMorganize the expression

$$\overline{Z} = (C \cdot \overline{D}) + (\overline{C} \cdot D)$$

to get an expression for ($Z = \ldots$) by inverting the whole expression shown above.

4-5. A very small airport has a control tower and three air traffic controllers. Two of these people must be on duty at all times. A light atop the control tower will go ON to warn incoming air traffic if fewer than two air traffic controllers are on duty. Complete the truth table for the logic circuit that turns ON the light. Additional information:

Air traffic controllers are Al (A), Bob (B), and Cal (C).

Any ATC who's not on duty is represented as a logic 0.

Any ATC who's on duty is represented as a logic 1.

When the light is ON, L = 1.

When the light is OFF, L = 0.

A	B	C	L
0	0	0	1
0	0	1	1
.	.	.	.
.	.	.	.

4-6. List the minterms in the truth table for Question 4-5, as a Boolean expression (L = . . .).

(*Note:* Check your truth table carefully before beginning to develop this expression. If there's an error on the table, the "snowball effect" will get to you on Question 4-6, as well as Question 4-5.)

4-7. Below is the symbolic-logic diagram for Question 4-5. Develop an expression for the (L = . . .) output of this circuit from the gate logic shown. This may be helpful in checking your work in Question 4-6, since the minterms from Question 4-6 should simplify down to the same expression as you would get from this circuit.

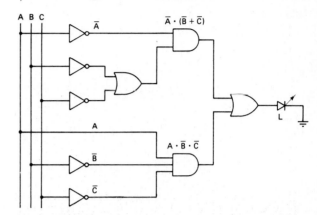

4-8. One of the rules in the list of identities (rule 17) is

$$M + (\overline{M} \cdot N) = M + N$$

Work out the truth table for "$M + (\overline{M} \cdot N)$" and "$M + N$." Are they the same?

DRILL PROBLEMS

Several lines (terms) on the truth table below are marked with numbers 4-1 to 4-5. They are Problems 4-1 to 4-5. Some of them have active-HIGH output and others are active LOW. In each case, determine the *not simplified* Boolean expression for the term represented by the line on the truth table.

		Inputs			Output
	D	C	B	A	Q
	0	0	0	0	0
	0	0	0	1	0
	0	0	1	0	0
4-1. →	0	0	1	1	1
	0	1	0	0	0
4-2. →	0	1	0	1	1
	0	1	1	0	1
	0	1	1	1	0
4-3. →	1	0	0	0	0
	1	0	0	1	1
	1	0	1	0	1
4-4. →	1	0	1	1	0
4-5. →	1	1	0	0	1
	1	1	0	1	0
	1	1	1	0	0
	1	1	1	1	1

Use DeMorgan's theorem to solve the five expressions below for Q = ?

4-6. $\overline{Q} = \overline{D} \cdot \overline{C} \cdot \overline{B} \cdot \overline{A}$ _____

4-7. $\overline{Q} = \overline{D} \cdot C \cdot \overline{B} \cdot A$ _____

4-8. $\overline{Q} = D \cdot \overline{C} \cdot \overline{B} \cdot \overline{A}$ _____

4-9. $\overline{Q} = D \cdot \overline{C} \cdot B \cdot A$ _____

4-10. $\overline{Q} = D \cdot C \cdot \overline{B} \cdot A$ _____

Draw gate diagrams for the five gates whose Boolean expressions are given here. Do *not* simplify the Boolean expressions before drawing the diagrams.

4-11. $\overline{Q} = \overline{D} \cdot \overline{C} \cdot \overline{B} \cdot \overline{A}$

4-12. $\overline{Q} = \overline{D} \cdot C \cdot \overline{B} \cdot A$

4-13. $Q = D \cdot \overline{C} \cdot \overline{B} \cdot \overline{A}$

4-14. $Q = D \cdot \overline{C} \cdot B \cdot A$

4-15. $\overline{Q} = D \cdot C \cdot \overline{B} \cdot A$

Simplify the following five Boolean expressions using factoring and other Boolean identities where applicable.

4-16. $Q = \overline{D} \cdot C \cdot \overline{B} \cdot A + \overline{D} \cdot C \cdot B \cdot \overline{A}$

4-17. $Q = \overline{D} \cdot \overline{C} \cdot B \cdot A + D \cdot \overline{C} \cdot B \cdot A$

4-18. $Q = D \cdot \overline{C} \cdot B \cdot A + D \cdot \overline{C} \cdot B \cdot \overline{A}$

4-19. $Q = \overline{D} \cdot C \cdot B \cdot A + \overline{D} \cdot \overline{C} \cdot B \cdot A$

4-20. $Q = D \cdot \overline{C} \cdot \overline{B} \cdot A + D \cdot \overline{C} \cdot \overline{B} \cdot A$

Simplify the following five Boolean expressions using Boolean identities from Table 4-1.

4-21. $Q = \overline{D} \cdot C \cdot B \cdot A + \overline{D} \cdot C$

4-22. $\overline{Q} = \overline{D \cdot C \cdot B \cdot A}$

4-23. $Q = \overline{D} \cdot C \cdot \overline{B} \cdot A + \overline{D} \cdot C \cdot B \cdot \overline{A} + \overline{D} \cdot C \cdot B \cdot A$

4-24. $Q = C \cdot B + C \cdot \overline{B} \cdot A$

4-25. $Q = D \cdot C \cdot B \cdot A + \overline{D \cdot C \cdot B \cdot A}$

Simplify the following five expressions containing Boolean variables and logic states.

4-26. $Q = \overline{D} \cdot C \cdot 0 \cdot A$

4-27. $Q = D \cdot \overline{C} \cdot B \cdot A + 1$

4-28. $Q = D \cdot \overline{C} \cdot 1 + D \cdot C \cdot 0$

4-29. $Q = D \cdot C \cdot C \cdot A$

4-30. $Q = \overline{D} \cdot C \cdot \overline{B} \cdot A + \overline{D} \cdot B \cdot C \cdot A$

5

APPLICATIONS OF GATES

By the time you finish this chapter, you will be able to:

1. Describe the binary numbering system and show how AND and NOT logic gates can be applied to ''recognize'' any one of the binary numbers (a *decoder*).

2. Describe how to construct a *decoder matrix* of AND gates using NOT gates to provide *double-rail inputs*.

3. Convert any *decimal* number into *binary* so that you can design a *decoder* for it.

4. Recognize what binary pattern is being decoded from the symbolic logic diagram or schematic of a *decoder*.

5. Describe how an *encoder* works and how to design a matrix that will encode decimal numbers into binary.

6. Understand the reason you need diodes in the wiring matrix described in item 5.

7. Describe how OR and AND gates may be used to design an *encoder* and how they may be used to replace discrete diode circuitry for this purpose.

8. Describe the function of a *multiplexer* or a *demultiplexer* in terms of the function of a rotary switch.

9. Describe how random logic (AND, OR, and NOT gates) may be used to replace the rotary-switch model of a *multiplexer* with a circuit that has no moving parts.

10. Describe how to use NAND logic to replace the random logic of item 9 with a circuit consisting of just one type of gate everywhere.

11. Describe how to use a *multiplexer* to replace a random-logic array of gates and thus reduce the package count of a logic design.

12. Understand the *timing diagram* of a *multiplexer*.

In this chapter we leave behind the questions of what logic gates are made of, how they work, and what their symbols or truth tables look like, and take up the most important of questions: What do you *do* with them? For an answer, we look at applied logic, with the thought in mind that the gates we use will be part of a system that *does* something. Now, of course, before you do something, you must be sure you know exactly what you want to do, that it is a problem that can be solved using logic gates, and that it is a problem worth solving.

We begin each design by defining what we *want*, then work out the circuit either by common sense (if it is a simple enough circuit) or using a truth table (if it looks like it's going to take *uncommon sense* to solve it!).

5-1. AND GATES (DECODERS)

To begin with a very small step, let's look at a circuit that can "recognize" the code for the binary number 5. What we want is a circuit that will light up an output lamp when its input in 8-4-2-1 binary code is the number 0101. The light will shine through a little window with a "5" on it.

We'll begin with a brief memory refresher. The binary numbering system differs from the decimal system we use. There are no numbers bigger than 1 that can be written in a single digit; numbers that have values of 2 or more must be written in multiple digits. The binary system, like the decimal, uses **positional notation**—the *place* where a number shows up is as important as *what* it is. Our example, the number 0101, has four **binary places**. The rightmost digit, as in decimal and all other number bases, is the **ones' place**. To its left (where a decimal number would have its tens' place) is the binary number's **twos' place**. Where a decimal number would have hundreds, our binary number has its **fours' place**. Finally, where thousands would be in a decimal number, this number has its **eights' place**. This makes the number

8s	4s	2s	1s
0	1	0	1

To read it, we say: "We *have* a 4 *and* a 1, but *not* an 8 or a 2. We add to get the *sum* of what we have (4 and 1). If we have 4 and 1, that's 5 (we said it would be!), but how did we know that the places had the values indicated? In the decimal system, which is based on 10, each place as we go

left from the ones' place has a **place value** 10 times as large as the place on its right.

Decimal:

$$1000s \nearrow^{\times 10} \searrow 100s \nearrow^{\times 10} \searrow 10s \nearrow^{\times 10} \searrow 1s$$

In the binary system, which is based on the number 2, each place as we go left from the ones' place has a place value two times as large as the place on its right.

$$Binary: \ 8s \nearrow^{\times 2} \searrow 4s \nearrow^{\times 2} \searrow 2s \nearrow^{\times 2} \searrow 1s$$

Logic 1 and 0 states are often used in digital circuits as the digits of binary numbers. Supposing that a cable of four wires is used to carry decimal numbers from 0 to 9 in **binary code**, a group of four wires could represent a decimal place. We'd like to be able to identify the code for the number 5 when it appears on the wires as a combination of ON and OFF states. The circuit that makes this identification will be called a **decoder**. First, we must decide which wires are 8, 4, 2, and 1. We'll identify the 8 as wire D, the 4 as wire C, the 2 as wire B, and the 1s place as wire A. In this case we could write a truth table for the light we want to turn ON at the appearance of binary five:

D	C	B	A	L
0	0	0	0	0
0	0	0	1	0
0	0	1	0	0

(L = light)

. . . etc.

but it's pretty clear that there will be only one minterm on this truth table; the term for line 5:

D	C	B	A	L
0	1	0	1	1

Writing the **minterm expression** for this set of inputs, we have only one term, which is

$$L = \overline{D} \cdot C \cdot \overline{B} \cdot A$$

which is an expression for an AND gate with four inputs, two of them inverted. This gate and its connections to the bus (the DCBA wires) is shown in Figure 5-1. The diode with an arrow coming out of its cathode is a **light-emitting diode**, an

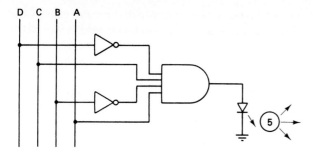

FIGURE 5-1 Decoder for binary 5 (applications of AND and NOT gates).

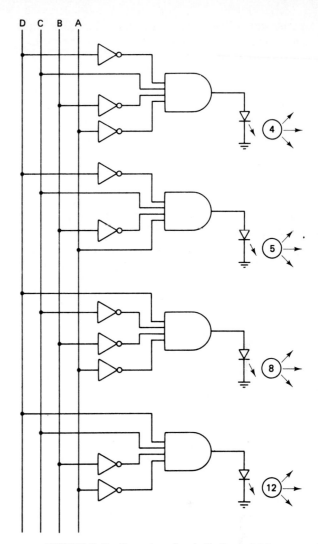

FIGURE 5-2 Decoders for 4, 5, 8, and 12.

indicator useful for direct readout of digital logic states.

At this point, you might ask yourself: "Why don't I just use L = C · A and save myself two inverters?" After all, isn't C AND A the same as 4 AND 1, and that's 5? Answer: "no." There are three other numbers that can be made from DCBA which also contain 4 and 1.

	D	C	B	A
Seven	0	1	1	1
Thirteen	1	1	0	1
Fifteen	1	1	1	1

If we build a decoder out of an AND gate with inputs 4 and 1, the little window with a 5 will light up when the code for 5 appears, but also for 7, 13, and 15. We don't want that "5" light on for those other numbers, but it will go on anyway.

This points up a mistake often made by people designing decoders: oversimplification. It's not enough to build a circuit (an AND gate) that looks for 1s in all the places where 1s are supposed to be. It's also *vitally important* to look for all the 0s where they are supposed to be at the same time as the circuit looks for 1s. This is why the expression

$$L = \overline{D} \cdot C \cdot \overline{B} \cdot A$$

stays exactly the same way we got it from the truth table.

Figure 5-2 shows the decoder for 5 that we had before, and also the decoders for 4, 8, and 12. For 4, the DCBA code is 0100. Converting 0100 to a Boolean expression we get

$$L = \overline{D} \cdot C \cdot \overline{B} \cdot \overline{A}$$

a four-input AND gate with inverters at three inputs. For 8, the DCBA code is 1000. Converted to Boolean:

$$L = D \cdot \overline{C} \cdot \overline{B} \cdot \overline{A}$$

another four-input AND with three inverters. For 12, the DCBA code is 1100. Converted to Boolean:

$$L = D \cdot C \cdot \overline{B} \cdot \overline{A}$$

a four-input AND with two inverters, at B and A.

Will each of these lights go on *once* and *only once* when a sequence of numbers is counted in binary on the wires? Yes. Since each gate represents one line in a truth table, and no two gates represent the same line, it follows that no two gates can be on at the same time, because you'll only have one number on the bus at a time. Each light is *mutually exclusive* with all of the others—when it goes on, it'll be the only light ON, because there are no other docoders in this circuit with the same code.

If you went ahead and decoded all the numbers from 0000 to 1111, that is, from 0 to 15, you'd have 16 AND gates and 32 inverters. That's a lot of gates. You can't avoid the 16 AND gates, because you've got to have an output for every light, but there is a way to cut way down on the number of inverters. It's called a **double-rail** input, and it's shown in Figure 5-3. For each input (D, C, B, and A), there's a NOT gate, and a wire running alongside the TRUE DCBA with a FALSE DCBA ($\overline{D}$, $\overline{C}$, $\overline{B}$, and $\overline{A}$). (The two side-by-side wires dedicated to each letter DCBA must have looked like railroad tracks to someone.) The availability of a wire on the bus for the FALSE as well as the TRUE of every input makes it possible to design a decoder by just attaching the inputs of a four-input AND gate to the right set of wires. The inverters needed for B in the 4, 5, 8, and 12 decoders in Figure 5-2, for instance, are replaced by just attaching an input line from the NOT-B rail to each of the four ANDs. One decoder does the job for all four gates. Since most families of logic permit driving the inputs of 10 or more gates from the output of a single gate, the FALSE or NOT rails will have no trouble supplying the entire 16 gates (only half of which will be attached to any given rail).

As an extra we added two more numbers that weren't decoded in Figure 5-2. The number 0000 = 0 is decoded at the top of the figure by attachment of the $\overline{D}$, $\overline{C}$, $\overline{B}$, and $\overline{A}$ rails to the AND decoder. The number 1001 is decoded as D, $\overline{C}$, $\overline{B}$, and A, so the "outside" AND gate inputs are attached to the TRUE rails for D and A, while the "inside" AND gate inputs are connected to the FALSE rails of B and C.

If we complete the circuit in Figure 5-3 so that there is a gate for every number from 0000 to 1111, what we have is called a **decoder matrix**. The work "matrix" describes a circuit that contains more than one gate. In this case, the matrix is called a 1-of-16 decoder matrix, or simply a 1-of-16 decoder. If the decoder has been properly designed, each code that comes in the inputs will switch on one of the outputs. A decoder matrix can be described as a circuit that has *many active inputs* and *one active output*, since for each code there's only one gate that decodes that particular pattern of 1s and 0s. If there's a gate for every possible code that enters the inputs (as there is in the 1-of-16 decoder), there will always be an output. If there are fewer than 16 decoder gates in a matrix that has four inputs, there'll be some codes that don't produce an output. When this happens, the inputs that don't produce outputs are called **don't-care** conditions. An example would be a 1-of-10 decoder matrix. This is the circuit you would use to identify what decimal number has been encoded in binary 1s and 0s. Since a decimal digit can have any value from 0 to 9, the binary code that represents it can be as small as 0 or as large as 1001 (nine). This means there must be four wires to carry the code, even though some of the four-digit binary codes will never be used. Numbers like 1100 (12) are impossible one-digit decimal numbers, so there will be six numbers that just never happen in this system. In that case, someone who's building a decimal decoder wouldn't need 16 gates, and would only put decoders for the numbers from 0 to 9 in the circuit. Now, what would happen if one of the "forbidden" codes appeared at the inputs (say, for instance, a 1100)? There would be no gate for that code, and the decoder would have no output. It would *ignore* the invalid code. This is what a don't-care input means. It's a pattern of 1s and 0s that doesn't produce an active output at any one of the gates in the matrix.

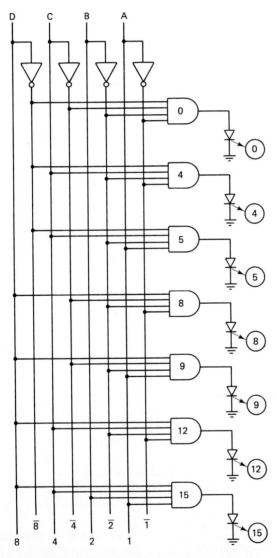

FIGURE 5-3 Decoder matrix showing double-rail inputs.

You could make a decoder matrix out of AND gates and NOT gates, but there are integrated-circuit decoders, like the TTL 74154 (a 1-of-16 decoder matrix) and 7442 (a 1-of-10 decoder matrix). The 74154 is shown in Figure 5-4. The 74154 has four inputs called D, C, B, and A. There are two more inputs, called $\overline{G0}$ and $\overline{G1}$, which are $\overline{ENABLE}$ inputs. If they are both LOW, there will be one active output for any 4-bit* code at ABCD, if they're not both LOW, all the outputs will be

*Bit is a name used by digital types for a BInary digiT.

inactive. The 16 outputs are called 0 through 15, and they are active LOW. Looking at the gate diagram for the 74154, you'll see that the gates used are NANDS rather than ANDS. This means that the output for each number will be LOW where the output for Figure 5-3 is HIGH. Appendix 4 contains a more detailed 74154 data sheet.

In Figures 5-2 and 5-3 we used decoders to light up numbers that identified what code was on the inputs of the matrix. We use decoders to *select*. Their function is to pick out one circuit or device from among a whole lot of devices, and switch that one ON, leaving all the others OFF.

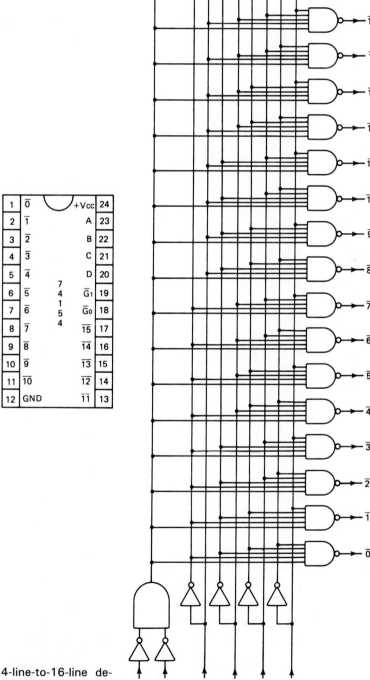

FIGURE 5-4 The 74154 4-line-to-16-line decoder. (See Appendix 4 for detailed data sheet)

A device code identifies the device. You can design a decoder to identify any binary device code as long as you know the pattern of 1s and 0s, using exactly the same techniques we applied to the 4-bit numbers. As a final example, suppose that a control cable has eight wires, called A0, A1, A2, . . . up to A7, and we have a device that we want to switch ON when the number 56 comes along. First, we have to know what 56 looks like in *binary*—which wires will be HIGH and LOW when that code comes along on the bus (the cable). Then we have to write a Boolean expression for the inputs that the code represents (the signals on the bus will be the inputs to the decoder). Finally, we have to convert the Boolean expression into a circuit made of AND and NOT gates (unless we want to trigger the gate with a LOW signal; then we'll need NAND and NOT gates).

First, we have the problem: What does 56 look like in binary? To convert a number from decimal to binary, we like the following shortcut method:

1. Write down the decimal number.
2. Divide the number in *half* (don't bother with fractions, just write down the *whole number* of times "2" goes into the number) and write the half value *above* the number.
3. Is the result "1"? If YES, go to step 4; if NO, take the half value you just got and repeat step 2 on it; keep halving until the result is "1."
4. You now have a column of numbers, with your first number at the bottom, and a 1 at the top. Next to each number, write a 0 if the number is even and a 1 if the number is odd.
5. Read the string of 1s and 0s from top to bottom; this is the binary value of the decimal number.

To see how this works, let's use the number 56. *(You read this starting at the bottom).*

2 goes into 3: 1	time	Odd	1
2 goes into 7: 3	times	Odd	1
2 goes into 14: 7	times	Odd	1
2 goes into 28: 14	times	Even	0
2 goes into 56: 28	times	Even	0
→ we start with: 56		Even	0

Now, we read the string of 1s and 0s on the right of our table from top to bottom: 111000. Is this the binary form of 56? We can check it (and also assign a wire to each bit) by writing the number, indicating each digit's place value:

A7	A6	A5	A4	A3	A2	A1	A0	wire
		1	1	1	0	0	0	(56?)
128	64	32	16	8	4	2	1	place value

From what we can see above, our number is the sum of the place values where there's a 1, and we have 1s at places 32, 16, and 8. The sum is

$$32 + 16 + 8 = 56$$

Aha! It works! The number 111000 *is* 56. This will work for any decimal number. Why it works is a more complicated problem, which we'll sidestep, but it does work, everytime. Notice that the A7 and A6 wires don't have any 1s or 0s. The usual way of handling this is to assign 0 as the value of the unused bits. This gives us the input condition for our 56 decoder with eight inputs:

A7	A6	A5	A4	A3	A2	A1	A0
0	0	1	1	1	0	0	0

Let's call the output of the decoder "D56." The Boolean expression we get from the line above is

$$D56 = \overline{A7} \cdot \overline{A6} \cdot A5 \cdot A4 \cdot A3 \cdot \overline{A2} \cdot \overline{A1} \cdot \overline{A0}$$

The gate diagram (made of ANDs and NOTs) for this expression is shown in Figure 5-5. Putting a circuit together with gates like this is called **implementation in random logic**. The decoder thus constructed is called a **random logic circuit**.

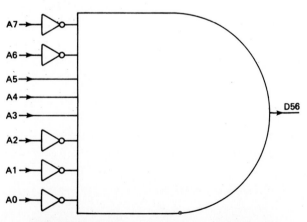

FIGURE 5-5 Binary-to-decimal decoder for the number 56.

Example 5-1

If you add three to every decimal number before converting it to binary, you get a code called **excess-3 code**. Make up a table of this code for the 10 decimal numbers, and design a decoder that converts excess-3 to decimal. It should have 10 numbered outputs, each of which becomes active high when the appropriate excess-3 code arrives at the circuit's four inputs. Call the inputs D, C, B, and A in your design, and the outputs 1, 2, 3, 4, 5, 6, 7, 8, 9, and 0.

Solution:

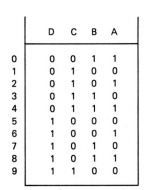

	D	C	B	A
0	0	0	1	1
1	0	1	0	0
2	0	1	0	1
3	0	1	1	0
4	0	1	1	1
5	1	0	0	0
6	1	0	0	1
7	1	0	1	0
8	1	0	1	1
9	1	1	0	0

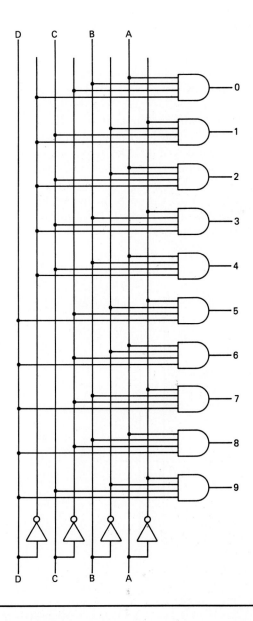

5-3 ENCODERS

5-2-1 Diode Matrix Encoders

In the preceding section we looked at one use of AND gates, the decoder. Remember that the purpose of a decoder was to take a code (a group of active signals) and produce one active output in a separate place for each code that is applied to the inputs. In this section we look at the reverse process, taking a signal from one active input and producing a code output which has many active signals.

As a small start, we'll begin with an **encoder** for the number 5, which does exactly the opposite operation done by our first decoder.

We want to construct a circuit that will create the code for the number 5, namely 0101. Whenever its input is active, this circuit's outputs at four places should be D = 0, C = 1, B = 0, and A = 1. Of course, D, C, B, and A are being used just as they were in the decoder circuit.

Figure 5-6(a) shows one way to do this. The input for the circuit is HIGH when we want a 0101 at the output. Each of the output lines D, C, B, and A is attached to a pull-down resistor. This ensures that the output will be LOW when no input is present. As soon as the input becomes HIGH, the wires from the input conduct the logic 1 to wires C and A. With C and A pulled up by the input, and the rest of the wires LOW, the outputs will be 0101, just as desired.

It's a very simple circuit. All you have to do to make any 4-bit code is attach the input where you want 1s, and leave the other wires alone. In fact, it's so simple it can't be used for two or more inputs. As soon as a second input is attached to the bus, trouble begins. This is shown in Figure 5-6(b) where a 5 and a 4 input connected to the bus are both producing 5s. This happens because no one told the wires in the 5 hookup only to conduct current coming from the 5 input. Current from the 4 input is connected to C only, but C is connected to A in the lower part of the diagram (in the 5 circuit). The wires in the 5 circuit conduct current from the 4 circuit. The connections in the 5 circuit actually form a short circuit between wires C and A. Once the 5 circuit has been hooked up, it's impossible to get a voltage on C without getting the same voltage on A.

It looks like what we need is a wire that conducts current only from its own input, and not from other places. This sounds a bit too "smart" for a wire, and it is. We need a conductor that conducts in only one direction. That's not so hard to find; it's called a diode, and the "arrow" on the diode symbol conducts positive logic in whatever direction it points. Figure 5-7 shows the new and improved model of our encoder, showing not only the inputs to create 4 and 5 on the bus, but 6, 7, 8, 9, and 10 as well. In this diagram, a *key* (a pushbutton) is used to apply HIGH positive logic signals to each input. Wherever the input has to pull up two or more lines of the output, diodes are

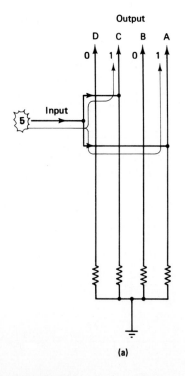

(a)

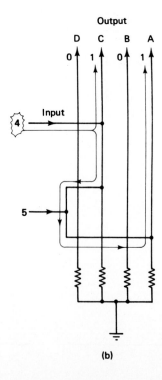

(b)

FIGURE 5-6 Decimal-to-binary encoding: (a) encoder for 5; (b) encoder for 4 and 5 showing effect of short circuits.

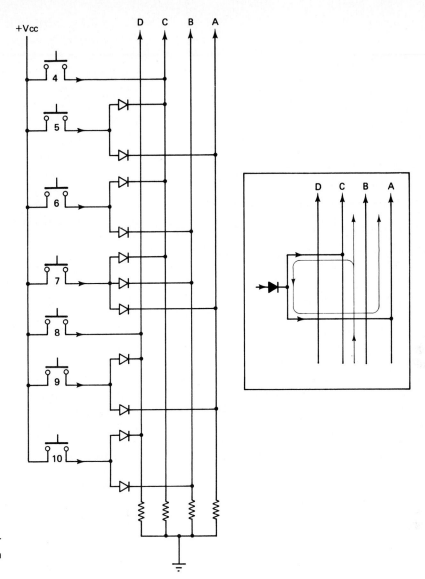

FIGURE 5-7 Diode encoder matrix for numbers 4 through 10 (active-HIGH inputs).

used to couple the input to the lines. This prevents C and A from being shorted together where the 5 input is. Both C and A can be pulled up from 5, but current into C from the 4 input, for instance, can't travel from C to A through the diodes in 5. One of the diodes is pointed the wrong direction to conduct positive logic from C to A. Notice that there *must* be one separate diode for each line to be pulled up. If we tried to use just one diode to pull up both C and A, as shown in the inset, we'd re-create exactly the same problem we were using the diodes to solve in the first place.

Another way to accomplish the same thing is shown in Figure 5-8. In this case, the inputs are active LOW (they produce a code when a LOW input is applied). As a LOW voltage is applied to each input, the diodes carry the 0 to every DCBA line that is supposed to have a 0. Pull-up resistors

provide 1s to lines that aren't getting 0s. For example, look at the 9 input. The decimal number 9 is 1001 in binary code. This means that there must be 0s in the middle two lines of the bus when the 9 input is active. Two diodes conduct a 0 to the middle two lines from the 9 input, leaving the outside lines floating. The pull-up resistors ensure that the floating lines go HIGH when not pulled down by the diodes. In this case, the idea embodied in the design of this **encoder matrix** is to conduct 0s to the places where 0s are supposed to go, and let the 1s take care of themselves.

5-2.2 Diode Matrix Implemented as Gate Logic

The circuit in Figure 5-7 is called a **diode matrix encoder**. It is the conceptually simplest way to

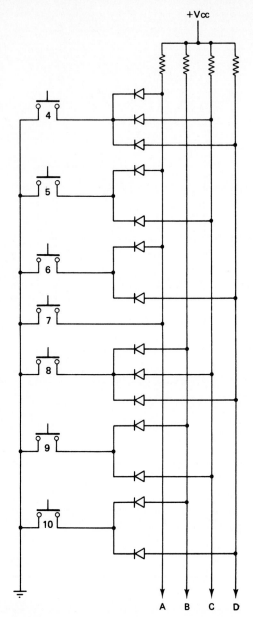

FIGURE 5-8 Diode matrix encoder for 4 through 10 (active-LOW inputs).

of encoding, although more elegant than the diode matrix, are also more complicated to understand. As an attempt, though, we'll present a **decimal-to-binary encoder** made of random logic. It turns out that the diodes comprise a number of OR gates, using real OR gate symbols to replace each group of diodes that go to a particular line. This circuit conversion is shown in Figure 5-9.

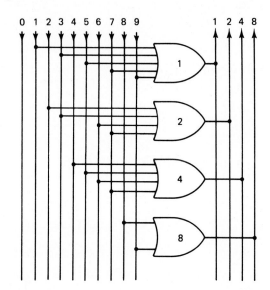

FIGURE 5-9 Encoder matrix using OR gates.

OR gates will output a 1 if a 1 appears at any input. The four OR gates used to drive the DCBA lines are always HIGH or LOW at their outputs, so pull-down resistors are not needed in this circuit. Each OR gate is attached to whatever input lines produce its particular output. The C gate, for example, drives the C bit (the 4-bit), and is attached to all the inputs for numbers that contain a 4-bit. The B gate is supposed to output a HIGH when any number that contains a 2-bit is active, so its inputs are tied to all numbers that contain 2-bits in binary code. The A and D bits are used the same way.

A gate circuit that copies the circuit in Figure 5-8 uses AND gates instead of OR gates, since the output of an AND gate will receive a 0 from any LOW input. Examining the diode matrix of Figure 5-8, you can see that the diode connections to each of the DCBA lines and their pull-up resistors form a diode AND gate. This is why the logic-gate replacement for the diodes is made of ANDS. Figure 5-10 shows the AND implementation of the diode circuit in Figure 5-8.

Encoders and decoders. Sometimes it's hard to remember the difference between a decoder and an encoder. We'll give you a couple of

do the encoding of the pushbuttons in the keyboard we drew in Figure 5-8. By the way, the purpose of the encoder in this case is to allow you to push keys on a decimal keyboard to create binary codes. Inside devices like pocket calculators, the familiar decimal numbers on the key you hit aren't there at all. Running around inside the calculator case are signals that are all 1s and 0s, no 7s or 8s. These signals travel around in the circuitry on buses, just like the DCBA bus we saw in the encoder and decoder examples. There are several other ways besides the diode matrix that can be used to encode decimal into binary, and calculators do not use diode logic, which is hard to integrate into a calculator chip. The other ways

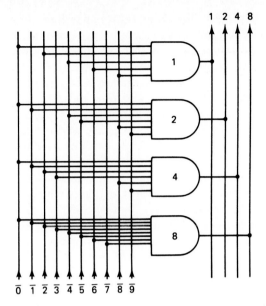

FIGURE 5-10 Encoder matrix using AND gates.

definitions here, to clarify the difference and show that the two are *opposites*, and we'll also give you a shortcut way to recall which is which:

> **Decoder** = A circuit with *many* active inputs that produce *one* active output.

> **Encoder** = A circuit with *one* active input that produces *many* active outputs.

To remember what's an encoder and what's a decoder, you need only remember what a code is. What's code, and what isn't, is in the eye of the beholder. For example, binary numbers like 111000 are code, whereas, decimal numbers like 56 are not code because decimal numbers are familiar to us, and *binary* numbers are "mysterious"—everybody doesn't know how to read them—so they are code. Going with this definition, a decoder is something that takes mysterious stuff like binary (which is "in code") and decodes it into familiar stuff like decimal. An encoder does the opposite. It takes familiar stuff like decimal numbers on a keyboard and converts them to mysterious stuff like binary code.

Just to add to the confusion, what's *called* a decoder by the manufacturers doesn't always fit the definition above. A device like a 7447 *binary-to-seven-segment decoder* could be called an encoder just as easily, because it converts 4-bit binary code into 7-bit code used to switch lighted segments on and off in a display. Once displayed, the 7-bit codes appear in the shapes of familiar decimal numbers, which anyone can read on the display. Since the seven-segment shapes are less mysterious then the 4-bit binary patterns, the 7447 device is called a decoder. Strictly a matter of taste.

Example 5-2

Design an encoder for the excess-3 code described in 5-1, using OR gates. The encoder should have four outputs, called D, C, B, and A, representing the 4-bits of excess-3 code, and 10 active-HIGH inputs numbered from 0 to 9.

Solution:

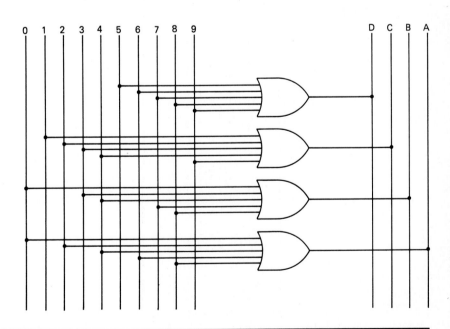

5-3 MULTIPLEXERS

A *multiplexer* is a circuit that uses AND, OR, and NOT gates to do the same job as a **rotary switch**. Let's review what a rotary switch does. Figure 5-11(a) shows how a rotary switch is constructed. It is a mechanical device with a movable rotating contact (the *wiper arm*) that brushes across several stationary contacts, connecting with each one as it is turned by a manual knob. The first application of a rotary switch illustrates how a source can decide which destination its signal will go to. The second application shows how the destination selects one of several sources it will receive its signal from. The first application, in which a source signal selects its destination, is called a **demultiplexer**. The second, in which a receiving circuit at the destination selects which one of the sources it will receive from, is called a **multiplexer**. The same rotary switch can be used for either job, since wires and mechanical contacts conduct equally well in either direction, but for a digital circuit, the two tasks are completely different, and require two different circuits. One circuit has one input and four outputs, but the other has four inputs and just one output. You can't just turn around the one circuit and use it for the other. Digital inputs can't be used for outputs, or vice versa.

The multiplexer uses logic gates to make a rotary switch with no moving parts which is controlled electronically, instead of by someone turning a knob. Its job is to select one of several inputs, passing its signal to the output.

There are three parts to the symbolic diagram of a multiplexer shown in Figure 5-12. The

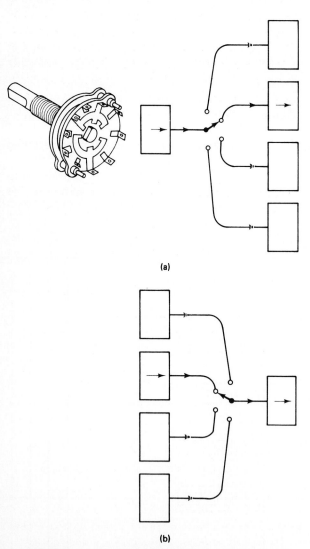

FIGURE 5-11 Rotary switch used as: (a) a demultiplexer; (b) a multiplexer.

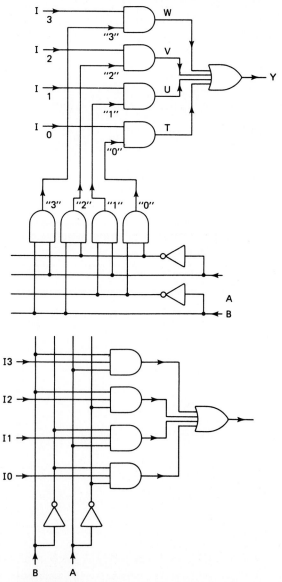

FIGURE 5-12 Four-input multiplexer made of random logic.

first is a decoder, which has two inputs called A and B. Since a 2-bit binary number like AB has four possible codes—00, 01, 10, and 11—the decoder has four outputs. To simplify things, these outputs are known by their decimal names, 0, 1, 2, and 3. The second part of the multiplexer circuit is a group of four AND gates (T, U, V, and W) used as ENABLE/DISABLE gates. Since only one of the 0123 outputs will be HIGH from the decoder, only one of these four AND gates will be enabled, and the other three will be frozen low at their outputs. The signal inputs to these four AND gates are I0, I1, I2, and I3 (the control inputs are, of course, 0, 1, 2, and 3). Suppose that the code entering A and B is 00. The active output of gate 0 will be HIGH, and all the other decoder outputs will be LOW. This will enable signal I0 to pass through gate T while the signals I1, I2, and I3 are blocked and the output of their gates is LOW. The third part of this circuit is an OR gate (Y) that puts T, U, V, and W together. Since only one of the four can ever be HIGH (the enabled gate, T), the output of the OR becomes HIGH and LOW when the I0 input becomes HIGH and LOW. Another way to look at what happens in the multiplexer when AB is 00 is to see what logic is entering and leaving each gate. Gate 0 gets 11 from the double-rail inputs (it's attached to NOT-A and NOT-B). The output of gate 0 is HIGH, and connects to gate T, whose output matches its input I0, since

$$I0 \cdot 1 = I0$$

The outputs of gates U, V, and W are LOW, since

$$I1 \cdot 0 = 0 \qquad I2 \cdot 0 = 0 \qquad I3 \cdot 0 = 0$$

so the inputs to Y are

$$T = I0 \qquad U = 0 \qquad V = 0 \qquad W = 0$$

and these are OR'd together, so that

$$Y = T + U + V + W = I0 + 0 + 0 + 0 = I0$$

because ORing something with 0 is just like adding 0 to a number.

You might have noticed that the control signals from the AND gates in the decoder were being sent to another set of AND gates in the enable/disable section. The associative law permits the use of one three-input AND instead of two two-input ANDs, so the multiplexer has been redrawn at the bottom of Figure 5-12 to show this compact version of the circuit. The logic is still

the same; if AB is 00, the logic of gates T, U, V and W will be

$$T = I0 \cdot 1 \cdot 1 = I0$$
$$U = I1 \cdot 0 \cdot 1 = 0$$
$$V = I2 \cdot 1 \cdot 0 = 0$$
$$W = I3 \cdot 0 \cdot 0 = 0$$

and the OR gate, Y, will have

$$Y = T + U + V + W = I0 + 0 + 0 + 0 = I0$$

The whole point of this is that any AB code put into the decoder's AB inputs enables one of the I inputs to "get through" to the Y output. We can picture the multiplexer as another kind of gate, whose symbolic diagram is shown in Figure 5-13. In this figure the multiplexer is a "box" with six inputs and one output. You can imagine it working like the picture below the symbol, a box with a SP4T (single-pole, 4-throw) rotary switch inside, and a "magic" inner box that turns the wiper arm of the rotary switch to whatever position the code on AB indicates. Of course, we know that the magic is just a decoder and enable gates, made of NOT and AND circuits, but the "fairytale" picture makes it easier to remember what the circuit is doing.

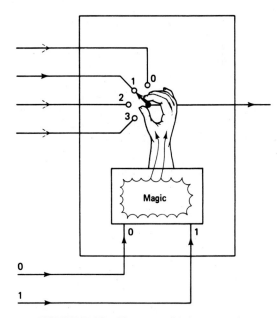

FIGURE 5-13 How a multiplexer works.

This multiplexer we've been looking at is called a *1-of-4 multiplexer*, or sometimes a *4-line-to-1-line multiplexer*. If we made a rotary switch with eight poles, or only two, they'd be called a *1-of-8* or *1-of-2 multiplexer*. Since the word "multiplexer" is such a mouthful, a lot of documentation shrinks the name down to MUX (rhymes with "ducks"). In future discussion, we'll use this contraction, MUX, as well as multiplexer, wherever it seems appropriate.

Finally, before we leave the gate diagram inside the *multiplexer* alone, and start deciding what to use the thing for, we have one more redrawing to do. Figure 5-14 shows how the MUX would look if the OR gate were DeMorganized into its ACTIVE LOW form (an AND with inverters at its inputs and outputs). The DeMorganized gate still has the same logic as the MUX always did, so this picture is also a legitimate way of representing the thing. Now, we use the old "elastic wire" trick, and "slide" the in-

verters between the ANDs and OR to the other end of each wire. This is shown just below the DeMorganized MUX. What we end up with in the third diagram is a circuit made entirely of NAND gates. This circuit is the NAND implementation version of the MUX. It's the way we would actually build a MUX out of TTL if we were designing the circuit on silicon. The reason? It's easier to make the MUX out of all one kind of gate. Your logic designer draws up the artwork for photomasking a NAND on the silicon surface, then zap, zap, zap, shoots hundreds of "pictures" of the gate on the surface of the silicon wafer. Now, all that's needed to make the NANDS into a MUX is a suitable pattern of **interconnects**, aluminum traces that connect inputs and outputs together to make

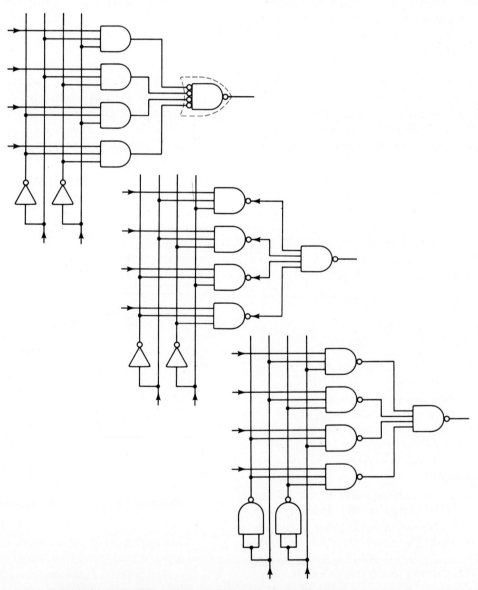

FIGURE 5-14 Four-input multiplexer showing NAND implementation. (See Appendix 4, 74153 data sheet, for more detail)

something like the circuit at the bottom of Figure 5-14.

This redrawing technique for making the MUX into all-NAND logic is useful anywhere a sum-of-products circuit has been made from a truth table. It's how the 74153 dual 1-of-4 multiplexer* is constructed in TTL, but NAND implementation is not just for designing multiplexers! It's for designing anything that comes from a truth table, so it can be made out of 7400 NAND gates (which are dirt cheap compared to custom IC circuits or random logic). It is definitely worth noting down how we did this, and practicing on a few other circuits to be sure you understand just how to do it.

We'd like to note in passing that we never showed you a truth table for the MUX. Every time we introduced a new gate before, we included its truth table along with its Boolean expression and gate symbol. In this case, there are two reasons we haven't shown the expression or table. First, the truth table is a mess! With six inputs, there are 64 different things the inputs could be doing to produce or not produce an output. Second, the Boolean expression from the truth table has only four minterms, and it hardly seems worth bothering with a truth table that has 4 minterms and 60 maxterms. The Boolean expression for the four-input MUX is

$$Y = I0 \cdot \overline{B} \cdot \overline{A} + I1 \cdot \overline{B} \cdot A$$
$$+ I2 \cdot B \cdot \overline{A} + I3 \cdot B \cdot A$$

meaning that $Y = I0$ if BA is 00, $Y = I1$ if BA is 01, $Y = I2$ if BA is 10, and $Y = I3$ if BA is 11. You can easily see that the code at BA is the binary for the number of each input (I0, I1, I2, I3) in each minterm. That's the reason the inputs were given those numbers in the first place.

5-4 USING MULTIPLEXERS TO REPLACE RANDOM LOGIC

The multiplexer is a new type of gate made from AND, OR, and NOT gates (or all NANDS) to do the job of a rotary switch. We could use the multiplexer to multiplex different data streams into a common decoder, or for bus switching in a computer, but the first application we're going to use MUX circuits for is going to be a real surprise. The MUX can be used to replace gate logic for any application. A MUX, suitably connected to the input signals, can be used to produce outputs

*A detailed data sheet for the 74153 is found in Appendix 4

for any truth table, and it can be done without deriving Boolean expressions from the truth table, or simplifying the Boolean expressions with Boolean algebra. In fact, we're looking at a "no-math" approach to circuit design. If that seems a little hard to believe, think about the circuit we developed for the "Alice, Betty, Chuck, and Dave chaperone alarm" in Chapter 4. After everything was done, we had simplified a truth table with six minterms (and which required six integrated circuits to build) down to the expression

$$alarm = A \cdot \overline{B} \cdot (C + D) + \overline{A} \cdot B \cdot (\overline{C} + \overline{D})$$

an expression that still requires three integrated circuit packages to build. With multiplexers, we can develop a circuit that needs only one integrated circuit (a 74151 1-of-8 multiplexer) and involves no Boolean conversion or substitution to design.

A 1-of-8 multiplexer is shown in Figure 5-15. It has three *select lines*: C, B, and A. These give eight combinations from 000 (zero) to 111 (seven), so the inputs of this multiplexer are I0, I1, I2, up to I7. The truth table for the "Alice problem" is also shown in the figure. Here's how we handle it. The truth table has been rearranged from its previous incarnation in Chapter 4, and is now organized with its inputs in reverse alphabetic order. The inputs CBA of the MUX are the 4s place, the 2s place, and the 1s place of binary code. For each input, there are two lines that end in the same CBA code. For the number CBA = 011, for instance, lines 3 and 11 on the truth table both end in −011. On line 3, the output is LOW, while on line 11, the output is HIGH. What is responsible for the difference? The D input, of course, is the difference between 0011 and 1011 on lines 3 and 11. To explain how this relates to the MUX, we'll point out that what we want to do is find some logic signal we can connect to I3 (in this case) so that the MUX will output exactly the same 1s and 0s as the truth table does. Well, in this case the output Y for lines 3 and 11 is LOW or HIGH exactly at the same time that D is LOW or HIGH. All we need to do is connect D to input I3, and the output will be LOW when CBA = 011 if D is LOW, but HIGH when CBA = 011 if D is HIGH. This means that for the same combination of DCBA that occurs on truth table lines 3 and 11, our MUX gives us the same output. By doing the same thing for all eight CBA combinations, the eight inputs of the multiplexer can be wired to D, NOT-D, 1, or 0, depending on what's needed. As another example, we ask ourselves what should be connected to I4 so that the output Y

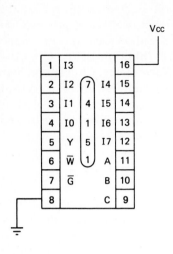

	D	C	B	A	X
0	0	0	0	0	0
1	0	0	0	1	0
2	0	0	1	0	0
3	0	0	1	1	0
4	0	1	0	0	1
5	0	1	0	1	1
6	0	1	1	0	1
7	0	1	1	1	0
8	1	0	0	0	0
9	1	0	0	1	1
10	1	0	1	0	1
11	1	0	1	1	1
12	1	1	0	0	0
13	1	1	0	1	0
14	1	1	1	0	0
15	1	1	1	1	0

FIGURE 5-15 The 74151 1-of-8 multiplexer.

matches the output of the "Alice table." The answer depends on the inputs needed to select I4. To select I4, the CBA inputs must be the binary code for a 4, CBA = 100. There are two lines on the "Alice table" that end −100 (CBA = 100). These are lines 4 and 12. On line 4, D = 0 and the output = 1. On line 12, D = 1 and the output = 0. Clearly, the output doesn't equal D in this case. Does this output equal NOT-D, or 1, or 0? It looks like NOT-D would work, so we connect input I4 of the multiplexer to the NOT-D rail.

We'll do one more line of the truth table as an example and leave the rest of the table to you to work out. (*Hint:* If you look ahead at Figure 5-16, you'll see how the connections should finally come out if you work out all eight CBA groups.) We ask what outputs from the truth table are associated with the I7 input. For I7, the CBA inputs should be CBA = 111. The two lines on the truth table that end in −111 (lines for which CBA = 111) are lines 7 and 15. The outputs for line 7 and 15 are both LOW. This means that no matter what D does, the output should be LOW when CBA = 111. This is the easiest of the three inputs we've tried. All we have to do is tie input I7 LOW. Every time CBA = 111, no matter what D is doing, the output will be LOW, and that's exactly what the truth table is doing.

Figure 5-16 shows the final results of "folding" the truth table for the "Alice problem." The eight inputs for the MUX were connected to D, NOT-D, 1, or 0 according to the pairs of outputs on the truth table for each CBA = ??? combination. This is an entirely nonmathematical method for designing a logic circuit, because the truth table is wired into the MUX instead of converted to Boolean ANDs, ORs, and NOTs. The replacement of ANDs, ORs, and NOTs (random logic) with a MUX is a "trick" worth knowing. The circuit for "Alice" in Figure 5-16 is now one package instead of three (simplified) or six (unsimplified) from Boolean techniques.

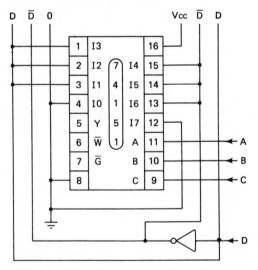

FIGURE 5-16 The "Alice problem" implemented on the 74151.

EXAMPLE 5-3

Using the method of "folded inputs," describe the connections you would make to a four-channel multiplexer (with two select lines) so that its output is HIGH whenever two out of three inputs are HIGH.

Solution: The truth table for the "2-of-3 detector" must be drawn up, and the inputs (we'll call them A, B, and C) assigned to specific inputs of the multiplexer. Here is the truth table and a diagram of the multiplexer:

A	B	C	Out
0	0	0	0
0	0	1	0
0	1	0	0
0	1	1	1
1	0	0	0
1	0	1	1
1	1	0	1
1	1	1	0

Channel 3
Channel 2
Channel 1   Y   — Output
Channel 0
          B   A

For the sake of simplicity, we'll assign A and B from the truth table to the A and B select lines. That means that we can look at adjacent pairs of lines on the truth table as though they "belong" to each of the four channels:

	A	B	C	Out
Channel 0	0	0	0	0
	0	0	1	0
Channel 1	0	1	0	0
	0	1	1	1
Channel 2	1	0	0	0
	1	0	1	1
Channel 3	1	1	0	1
	1	1	1	0

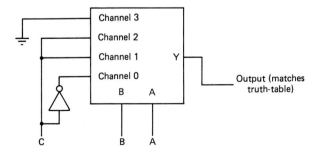

On the first two lines, we note that the output is 0 whenever A and B are selecting channel 0. We will write a logic-0 input (0 volts) to channel 0.

On the next four lines of the table, we note that the outputs are the same as the C input when AB = 01 or AB = 10. We will wire the C input to channels 1 and 2.

On the final two lines of the truth table, AB = 11 (channel 3) and the output is exactly the opposite of input C; that is, it matches $\overline{C}$. We should connect the channel 3 input to a NOT gate that inverts input C.

5-5 TIMING DIAGRAMS FOR MULTIPLEXERS

In earlier chapters we saw timing diagrams for gate circuits—which were largely extensions of the truth table of the gate in question. When we looked at enabled and disabled gates in Chapter 3, we explored a situation in which there could be only two states; a signal at the input *could* get through to the output, or else it *couldn't*. In a sense, a gate that can be enabled and disabled is a one-channel multiplexer, since its input waveform either gets through when the "channel" is enabled, or it doesn't. With multiplex circuits having several channels, the timing diagrams look like several enable gates, whose outputs are funneled into a common output.

5-5.1 Using Timing Diagrams for Troubleshooting

Figure 5-17 is a timing diagram for a four-channel MUX, perhaps one-half of a 74153, whose channel inputs are labeled 3, 2, 1, and 0. Select lines A and B determine which input is delivered to the output Y. The signals at the inputs have various frequencies, pulse widths, and duty cycles. In diagrams 5.17(a)–(f), the select lines B and A receive changing binary codes. Each of the inputs, and the output, are depicted in six schematic diagrams. In the corresponding sections of the timing diagram, we see how this looks on a standard timing-diagram format. It is easy to see, in the timing diagram, which of the input waveforms is reaching the output Y in each case. In section (a), when BA is 01, output Y matches channel 1's waveform. When BA changes to 11 in section (b), output Y matches channel 3's waveform. When BA changes to 10 in section (c), the

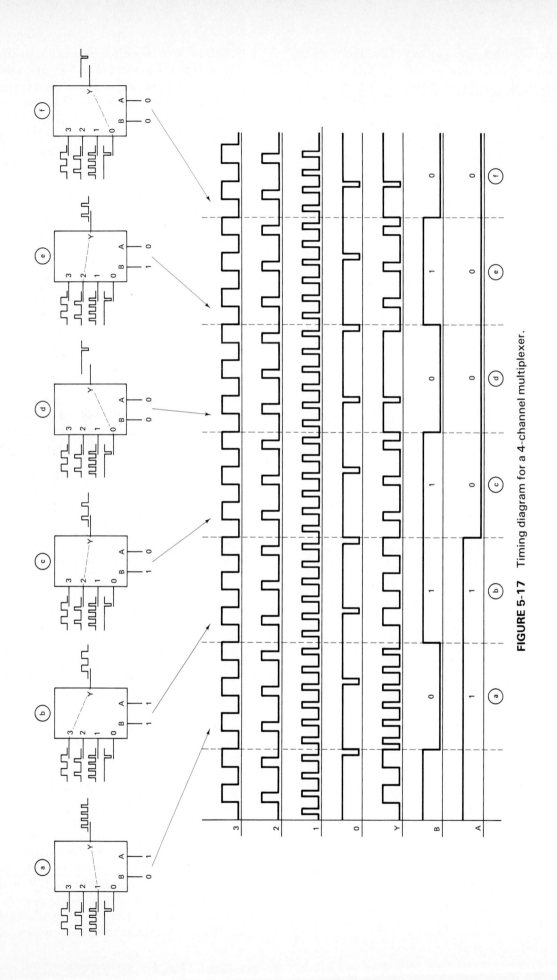

FIGURE 5-17 Timing diagram for a 4-channel multiplexer.

output matches channel 2's input, and so on to the end of the diagram. This is a characteristic timing diagram for a multiplexer, although other waveforms may clearly be substituted at the inputs. If a channel input, or the channel-selector lines, becomes defective, a test with waveforms of varying frequency, pulse width, or duty cycle at each channel should reveal this.

If, for example, setting control inputs BA to 10 does not result in the waveform injected at channel 2 appearing at the outputs, either channel 2 or the channel selector may be defective. If the waveform injected at another channel—perhaps channel 3—continues to come out the output, the selector (decoder) or its connections may be defective. If no waveform appears at the output, then channel 2's input itself may be the culprit. Either way, if the defect is inside the integrated circuit, the chip must be replaced, but it is possible that a bad connection is not leading the signal into the chip—that it is disconnected. Check this before replacing the chip. (See the Chapter 9 End-of-Chapter Lab Project for more information on how this is done.)

QUESTIONS

5-1. Complete the circuit in Figure 5-2 so that there is a decoder for all the numbers from 0000 to 1111. Note that the figure already has four of these decoders.

5-2. Complete the decoder matrix in Figure 5-3 using double-rail inputs.

5-3 Design an eight-input decoder for the decimal number 63, similar to Figure 5-5.

5-4 Design an encoder for the decimal number 15. Use active-HIGH inputs.

5-5. Why are diodes needed in the solution to Question 5-4?

5-6. Design an encoder for the decimal number 10. Use active-LOW inputs.

5-7. What is a logic probe used for?

5-8. Define "multiplexer" and "demultiplexer" and give an example of where each could be used. (Examples do not have to be digital.)

5-9. For Figure 5-12 show the four Boolean logic equations for outputs T, U, V, and W when inputs A and B are both HIGH.

5-10. Convert the following circuit into all NAND logic, using the methods shown in Figure 5-14.

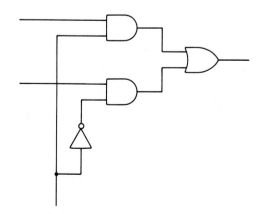

5-11. Convert the following truth table into a multiplexer circuit, using a 74153 MUX.

C	B	A	Z
0	0	0	0
0	0	1	0
0	1	0	0
0	1	1	1
1	0	0	0
1	0	1	0
1	1	0	1
1	1	1	0

5-12. Identify which of the following diagrams is an encoder and which is a decoder.

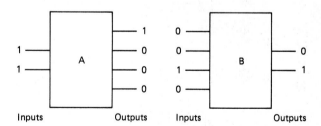

5-13. Identify which of the following diagrams is a multiplexer and which a demultiplexer.

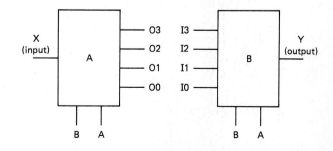

Convert from binary to decimal.

5-1. 1001

5-2. 10000000

5-3. 01010000

5-4. 1101

Convert from decimal to binary. (The size of desired binary number is indicated after each question.)

5-5. 5 (4 bits)

5-6. 120 (8 bits)

5-7. 66 (8 bits)

5-8. 200 (8 bits)

Draw a decoder circuit whose output is active when its binary inputs match each of the decimal numbers given below. (The number of input bits for each decoder circuit are indicated after each question.)

5-9. 7 (4 bits)

Fill in the blanks.

5-17. A decoder has _____ (one, many) active input(s) and _____ (one, many) active output(s).

5-18. An encoder has _____ (one, many) active input(s) and _____ (one, many) active output(s).

5-19. A multiplexer selects which of several _____ (inputs, outputs) is connected to a single _____. (input, output)

5-20. A demultiplexer selects which of several _____ (inputs, outputs) is connected to a single _____. (input, output)

For the decoder matrix shown, which output is active for the binary input given in each question?

5-21. DCBA = 1001

5-22. DCBA = 0110

5-23. DCBA = 1000

5-24. DCBA = 0011

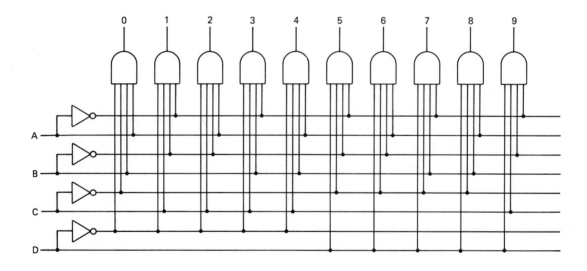

5-10. 12 (4 bits)

5-11. 27 (8 bits)

5-12. 156 (8 bits)

Draw an encoder circuit (with 4-bit outputs) of the type described in each question which will encode the decimal numbers below into binary.

5-13. 0 (active-LOW input, diode matrix)

5-14. 3 (active-HIGH input, diode matrix)

5-15. 11 (active-HIGH input, OR gates)

5-16. 13 (active-LOW input, AND gates)

5-25. What code is used by the inputs of the decoder above (based on the names of the outputs)?

(a) True binary code

(b) Twos'-complement code

(c) BCD code

(d) Ones'-complement code

(e) None of these

For the multiplexer shown below, which input is selected for the binary number given in each question?

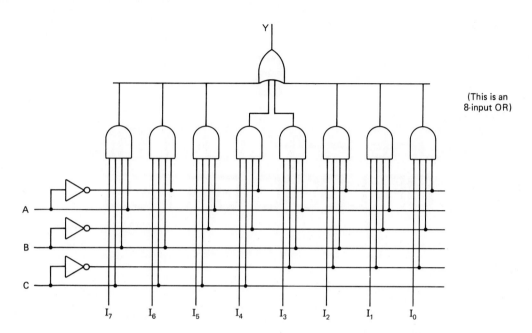

(This is an 8-input OR)

5-26. ABC = 001
5-27. ABC = 101
5-28. ABC = 110
5-29. ABC = 000
5-30. For the multiplexer above, what inputs

should be used to select channel I_1, ABC = 001 or CBA = 001?

(a) ABC = 001
(b) CBA = 001
(c) None of these

6
MORE APPLICATIONS OF LOGIC GATES

—————————CHAPTER OBJECTIVES—————————

By the time you finish this chapter, you will be able to:

1. Describe the rules for *binary addition* of two one-digit binary numbers.
2. Describe the circuit whose outputs match the results of binary addition of two one-digit binary numbers, a *half-adder*.
3. Describe the rules for *binary addition* of three one-digit binary numbers and why this is needed to perform the addition of two multidigit binary numbers of any length.
4. Describe the circuit whose outputs match the results of binary addition of three one-digit binary numbers, a *full-adder*.
5. Explain the assembly of *full-adder* and *half-adder* ''gates'' into a *parallel full adder*, used to add any number of bits of a multibit addend to an equally long augend.
6. Understand how the five items above are applied to *binary subtraction*.
7. Show how the use of a *controlled inverter* makes a circuit possible that can either *add* or *subtract*.
8. Understand the application of the XOR gate to *parity detectors* and *parity generators*.
9. Understand the application of the XOR gate to *equal/unequal* comparators.
10. See how to implement either *active-HIGH* or *active-LOW* output versions of the parity and comparator circuits.

6-1 ARITHMETIC CIRCUITRY

In Chapter 2 we found out that OR logic was called logic addition, and that the + sign was used to represent the OR operation, but the truth table for OR did not match an addition table. The OR gate didn't really add. We pointed that out, and said that the Boolean + sign would do things differently from the + sign in algebra. Of course you might ask yourself: "Is there a logic gate that does perform arithmetic addition instead of logic addition?" The answer: "Of course there is!"

6-1.1 Half-Adder

As we did before, we'll begin by defining what we want our addition circuit to do. It must be able to add two numbers according to the following rules:

Addition Rules

$$0 + 0 = 0 \quad 0 + 1 = 1 \quad 1 + 0 = 1$$

$$1 + 1 = 10$$

These are the rules of *binary addition*. Since a single wire can convey only the states 0 or 1, we have to work in the binary system of numbering. This has some advantages. Can you remember memorizing decimal addition tables and multiplication tables from $0 + 0$ to $12 + 12$ and 0×0 to 12×12? Real slave-labor, right? Now, in the binary system, the biggest table we have has only four numbers. The rules above are all the entries there are in a binary addition table. This becomes a truth table if we arrange it like this:

Addition table

A	B	A plus B
0	0	0
0	1	1
1	0	1
1	1	10

We call the numbers (inputs!) A and B the **addend** and the **augend**. A, the addend, is the number you *add to*. B, the augend, is the number you use to *augment* (make it bigger) the number that you are adding to. A plus B is the sum. Actually, if we think of this addition table as the truth table for a logic gate, we have a problem when we get to the last line on the table. How can the output of a logic gate be 10? That's both HIGH and LOW at the same time, and that wouldn't be logical! We really need a gate with two outputs, to handle the last line on this table.

We'll continue to call one output the sum, and the extra digit that is used when the result is 2, we'll call the carry. It will be clear in a little while why that name was chosen. Now, our truth table is really two truth tables, one for a sum gate whose output is listed under sum, and the other for a carry gate whose output states appear under the word carry:

Addition table

A	B	Carry	Sum
0	0	0	0
0	1	0	1
1	0	0	1
1	1	1	0

The sum gate's truth table and the carry gate's truth table are as follows:

A	B	Carry	A	B	Sum
0	0	0	0	0	0
0	1	0	0	1	1
1	0	0	1	0	1
1	1	1	1	1	0

The carry gate has the truth table of an AND gate, and the sum gate has the truth table of an exclusive-OR (XOR) gate. This is the first application in which an XOR has been useful to us to design a circuit. The two circuits can be instantly described as:

$$\text{carry} = A \cdot B \quad \text{sum} = A \oplus B$$

The whole circuit for the addition of two one-digit binary numbers is shown in Figure 6-1, and the circuit is called a **half-adder**, for reasons that will become apparent later.

This is a little circuit that adds together any two one-digit binary numbers. That isn't much. Itty-bitty kids learn in first grade to add together one-digit numbers before they go ahead to adding two- or three-digit decimals. Digital circuits start the same way. The problem of adding multiple-digit numbers can't be solved until we know how to add together one-digit numbers.

In preparation for moving on to bigger and better circuits, we'd like to point out that adding together one-digit numbers sometimes gives a two-digit result (this is true in decimal or binary). The binary carry digit represents a 2, actually, since the binary place just to the left of the 1s place is the 2s place in binary. We called it a carry because we'll be using it in the same way we use

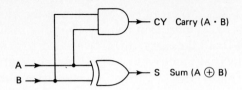 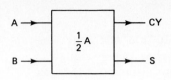

FIGURE 6-1 Half-adder circuit.

a decimal carry in doing decimal addition. For example:

(1)   (2)   (3) 1   (4)   1
    25   2   5   2 | 5   + 2   5
   +17   1 ↗ +7   1 | 7   + 1   7
   ___   ↘___   ↘___   ___
               12   2   4   2

In step (1), the sum of 25 plus 17 (decimal) is stated as a problem. In step (2), the units' column of the numbers (5 + 7) is added, giving 12. Twelve is not a one-digit number: it has a 2 in the units' place, but also a 1 in the tens' place. In step (3), the 1 is carried to the tens' place and added in with the tens, where it belongs. (Any time you get a result bigger than 10 in decimal, something—a *carry*—gets added to the tens' place.) In step (4), the tens' column (1 + 2 + 1) is added, giving 4. The result: a 4 in the tens' place and a 2 in the ones' place. It's probably been a long time since anyone bothered to tell you about an addition problem in this much detail, and you wonder why we're doing it. The answer is that you have to fix very clearly in your mind what is going on with the familiar decimal numbers you work with all the time if you've the ghost of a chance of understanding the binary addition example we're about to do below. If you get lost in what we're doing in binary, go back to the decimal example, and review what all the steps were for, checking just what was done each time, and *why*. You'll find that the same things are done for the same reason in binary. It just looks harder, because the addition process in binary isn't as familiar. Here's our binary example:

(1)   (2)   (3) 1   (4)   1
    11   1   1   1 | 1   + 1   1
   +11   1 ↗ +1   1 | 1   + 1   1
   ___   ↘___   ↘___   ___
               10   0   11 0

In step (1), the sum of 11 plus 11 (3 + 3, decimal) is stated as a problem. In step (2), the units' column of the numbers (1 + 1) is added, giving 10. Two is not a one-digit binary number: it has a 0 in the units' place, but also a 1 in the twos' place. In step (3), the 1 is carried to the twos' place and

added in with the twos, where it belongs. (Any time you get a result bigger than 2 in binary, something—a *carry*—gets added to the twos' place.) In step (4), the twos' column (1 + 1 + 1) is added, giving 11 (3). The result: a 110, which has a 1 in the fours' place, a 1 in the twos' place, and a 0 in the units' place. This is (4 + 2), the binary form of the number 6, which is certainly what we *should* get from 3 plus 3.

If that wasn't as easy to follow as 25 plus 17, we're sorry, but the same rules were used. The only awkwardness was "1 plus 1 = 10," where "10" is a 2. Another way to see what happens in the addition problem is to read "1 + 1 = 10" as "one plus one equals zero, *and carry the one*." This ensures that you remember what's the sum (0) and that the other digit (1) is to be carried to the next-higher binary place, just as the "1" from "12" was carried to the tens' place in decimal.

6-1.2 Full-Adder

In the example problem just completed, we added 3 plus 3 and got 6. When this problem was done in binary, the addend and augend were two-digit binary numbers. Since the half-adder circuit has inputs for one-digit numbers, it would seem that adding 11 plus 11 is impossible with a half-adder. Is it impossible with two half-adders, one for each digit? After all, in the four steps of the solution, we only added one column of numbers at a time, and there were only two numbers at the beginning, the addend and augend. Couldn't a couple of half-adders compute 11 + 11, one digit at a time? The answer is "no." Going from step (1) to step (2) would be no problem. A half-adder can add the 1 plus 1 and get 10. The problem arises when the carry output of the units' half-adder is carried to the twos' place. Trying to use a half-adder to go from step (3) to step (4) is hopeless. There are three numbers to add in the twos' column, and the half-adder circuit has only two inputs, A and B. The half-adder circuit can add an addend to an augend, but it cannot add an addend plus an augend plus a carry. This requires a full-adder.

The **full-adder** circuit shown in Figure 6-2 is made of two half-adders plus a gate. This is an easy way to remember why the name "half-ad-

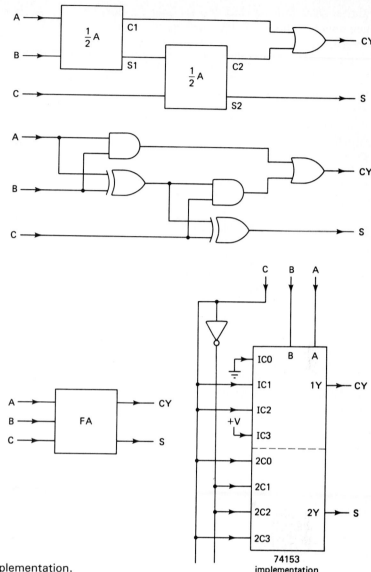

FIGURE 6-2 Full-adder circuit with 74153 implementation.

der" was given to the smaller circuit. If there are two of them in a full-adder circuit, one must be a half-adder.

The full-adder adds three numbers together instead of two like the half-adder. The three numbers are the addend (A) input, the augend (B) input, and the carry (C) input of the symbol. The outputs are still called sum and carry. Does this circuit need more outputs than the two possessed by the half-adder? If the largest number you can possibly get by adding three binary input bits is 3 (11, binary), then only two outputs are needed for this circuit.

Let's see how the two half-adders act together to make the full-adder work. A and B are added together first, giving a sum (S1) and a carry (C1). The sum is added with a third bit (C) in a second half-adder, which also has a sum (S2) and

a carry (C2) output. The S2 output is the sum of "A + B + C," and the two C1 and C2 outputs are ORed together to produce the carry (C3) output. Why? The first part makes sense; it figures that you should add the next bit to the sum of the last two bits, because the next bit is in the units' place and the sum is in the units' place. It would not make sense to add C to the carry, since the carry is in the twos' place. But why OR the carry outputs C1 and C2, instead of adding them? The answer, it happens, is that adding is what the OR gate does. When there are only three numbers to add together, the result cannot be 4. If the result cannot be 4, there'll never be a 1 at both C1 and C2 at the same time. C1 and C2 are both in the twos' place, and if they were both ON, that would mean we'd have 2 plus 2. We already know that we can't get "2 + 2" from adding three binary digits, so it follows that the C1 and C2 outputs are never both on. In that case, an OR gate is as good as an adder to add C1 to C2. We already know that an OR gate is the same as an adder except for the last line of its truth table, when both inputs are HIGH. In this case, both inputs are never HIGH, so the last line of the truth table is never used. As a result, the OR gate produces exactly the same outputs as an adder for the inputs it is given.

Full-adder circuit

A	B	C	Carry	Sum
0	0	0	0	0
0	0	1	0	1
0	1	0	0	1
0	1	1	1	0
1	0	0	0	.1
1	0	1	1	0
1	1	0	1	0
1	1	1	1	1

Now that we've explored the inner workings of the full-adder circuit, let's spend a moment looking at its truth table. The truth table above shows the eight possible combinations of A, B, and C. Each line on the truth table is treated as "add A plus B plus C" for purposes of generating an output. Sure, we know that "0 + 1 + 1" is the same as "1 + 1 + 0," but it's necessary to make the circuit work that way, too. At the bottom of Figure 6-2 is a gate diagram showing how the full-adder is implemented in XOR, AND, and OR gates. We know another way to make a circuit from a truth table: implement it at the inputs of a multiplexer. The circuit for the full-adder using a 74153 dual 1-of-4 MUX is shown below the gate diagram for comparison. The random-logic (gate) circuit would require three 14-pin IC gate packages (7486 quad-XOR, 7408 quad-AND, and 7432 quad-OR), but the MUX version would be a single 16-pin package (the 74153).

6-1.3 Multibit Full-Adder (Parallel)

We know from our development of the full-adder gate that it's impossible to add 3 plus 3 in binary without one. We also know that a half-adder can add the digits in the units' place, but a full-adder is needed in the twos' place. The circuit in Figure 6-3 combines a half-adder (for the units' place)

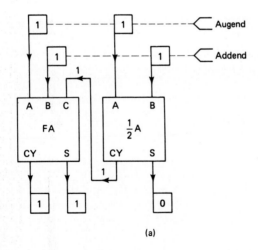

(a)

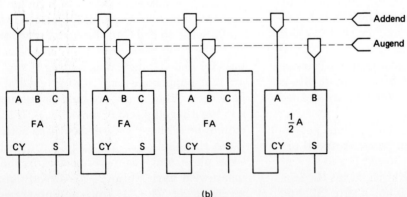

(b)

FIGURE 6-3 (a) Two-bit and (b) 4-bit parallel full-adders.

and a full-adder (for the twos' place) into a circuit that can add 3 plus 3, or any other two-digit binary addend and augend. The use of the two gates is fairly easy to understand, in light of the steps used in solving the addition problem itself. The units' digits of the addend and augend go into the half-adder at A and B. Its sum is the units' place of the parallel full-adder's output and its carry is connected to the next gate down the line. This carry wire "carries the one" to the twos' place full-adder, where it will be added in with the twos of the addend and augend. The outputs of the full-adder are the twos' place and fours' place of the parallel full-adder's output.

Inputs and outputs for the "3 plus 3" problem are shown in Figure 6-3. You can see how the half-adder responds to the two 1s at its inputs by producing a sum of 0 and a carry of 1. The carry is then taken to the full-adder, where its 1 is added with the other two 1s from the addend and augend. The sum is 3, which is binary 11, so the outputs of the full-adder are both 1. Reading the number from left to right, 110 is decimal 6, the answer for "3 plus 3."

Figure 6-3(b) shows how more full-adders are added to make a parallel full-adder for four-digit numbers. This circuit could be extended to binary numbers of any size by just adding an additional full-adder gate for each bit, and connecting its C input to the carry output of the previous full-adder. Circuits like this with 8 and 16 bits are used in microprocessors as part of their *arithmetic-logic unit* (ALU).

6-1.4 Half-Subtracter

In discussing arithmetic circuits, the ALU of a microcomputer was mentioned as a place where parallel-adder circuitry might be found. There are actually only *two* arithmetic operations which are "stock" on most microprocessors, add and subtract. The process of binary subtraction is, like the process of binary addition, a bit clumsier than the decimal operation for the same task, because we're not as familiar with the number system, but follows the same rules used to subtract decimal numbers.

Subtraction Rules

$$1 - 1 = 0 \quad 1 - 0 = 1 \quad 0 - 0 = 0$$

$$0 - 1 = 1 \quad \text{(borrow a 1)}$$

Notice the "borrow." In decimal arithmetic, when we subtract a column of two numbers, and the bottom number is larger than the top, we borrow a "1" from the next decimal place over. That

turns the top digit, X, into $1X$, which is 10 plus the digit. Since $1X$ is always larger than any digit being subtracted, no matter what X is, the subtraction can proceed, with the problem of "Where does it come from?" being pushed down the line to the next decimal digit.

Now, we try the same thing with a binary number. If we set up a subtraction table like the addition table we had for binaries, we have one line on the table for which we must borrow from down the line in order to complete the subtraction:

Subtraction table

A	B	Borrow	Difference
0	0	0	0
0	1	1	1
1	0	0	1
1	1	0	0

The two inputs to a **subtracter** gate are A, called the **minuend**, which is di*minu*ed by having something subtracted from it, and B, the **subtrahend**, which is subtracted from the minuend to make it smaller. Just think of A as the top number in the subtraction, and B as the bottom number. The two outputs from the subtracter are called the borrow and the difference. The **difference** is the number left over when the subtraction is complete, whether we had to borrow to do it or not. The **borrow** indicates whether we will have to borrow from the next place over to complete the subtraction.

In binary arithmetic, when we subtract a column of two numbers, and the bottom number is larger than the top, we borrow a "1" from the next binary place over. That turns the top digit X, into $1X$, which is 2 plus the digit. Since $1X$ is always larger than any digit being subtracted, no matter what X is, the subtraction can proceed, with the problem of "Where does it come from?" being pushed down the line to the next binary digit.

To do this, a subtracter gate would have to have two outputs, meaning it's really two gates. Their truth tables:

A	B	Borrow	A	B	Difference
0	0	0	0	0	0
0	1	1	0	1	1
1	0	0	1	0	1
1	1	0	1	1	0

indicate that the right-hand gate (difference) is an exclusive-OR (XOR), and the left-hand gate (borrow) is some kind of AND, since it has just one minterm. Evaluation of the Boolean expression from the minterm gives

$$\text{BORROW} = \overline{A} \cdot B \qquad \text{DIFFERENCE} = A \oplus B$$

The schematic for the circuit that contains the borrow and difference circuits is shown in Figure 6-4.

The circuit in Figure 6-4 is called a **half-subtracter**. Comparing it with the half-adder introduced earlier, we see that the only difference between a subtracter and an adder is the one inverter between input A (the minuend) and the AND gate.

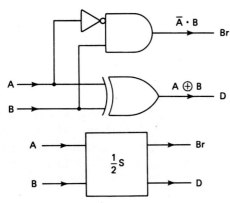

FIGURE 6-4 Half-subtracter circuit.

6-1.5 Full-Subtracter

The half-subtracter can subtract two inputs, a minuend and a subtrahend. When binary numbers larger than one digit are subtracted, we run into the same problem we had with multidigit addition: how to do the borrow from the next place to the left. We'll see what multiple-digit binary subtraction involves by looking first at decimal subtraction:

(1) 32
     -15
     ———

(2) 3 12
     1 -5
     ——————
        7 (b - 1)

(3) 3 12
     -1
     1 5
     ——————
        7

(4) 3 12
     -1
     -1 5
     ——————
     1   7

In step (1), the subtraction problem is defined as $32 - 15$. In step (2), we subtract the numbers in the units' column (2 and 5) and find that we have to *borrow* a "1" to make the "2" into a "12." 12 minus 5 is 7, but we have to remember to *borrow* a "1" from the next column when we get there. (Human beings can see all the digits and know enough to change the 3 into a 2 before subtracting this column, but digital circuits have no such "look-ahead" capability, and they have to "save" the borrow status until it's time to subtract the tens' place.) In step (3) we have moved the borrow $(b - 1)$ to the tens' column. We must remember to take one 10 away from the tens' column, because we added that 10 to the ones' column when we made a "12" there. Finally, in step (4) we subtract the subtrahend digit and the borrow digit from the "3," leaving "1". The difference calculated is "$32 - 15 = 17$" when we're finished.

This may not be exactly the way you do subtraction; we tried to do it as a computer has to. Inputs of digital logic gates can't look ahead to see if there's something to borrow from, they have to just go ahead as though they could borrow, and then subtract "1" when they get to the next column. It's a little bit like addition, but with a negative carry.

Now, let's look at a binary subtraction problem using the same approach:

(1) 110
     -101
     ————

(2) 11   0
     10   -1
     ————————
            1 (b- 1)

(3) 1   1   0
       -1
     1 -0   1
     ——————————————
        0 (b- 0) 1

(4) 1   1 0
       -0   1
     -1   0 1
     ——————————————————
        0 (b- 0) 0   1

In step (1), the problem is defined as subtraction of $6 - 5$. Step (2) subtracts the subtrahend from the minuend in the units' position. 0 minus 1 is "1, and borrow a 1." Step (3) subtracts the minuend minus the units' place borrow minus the subtrahend in the twos' place. 1 minus 1 minus 0 is "0 and no borrow." In step (4), the fours' column subtracts the minuend minus the twos' place borrow minus the subtrahend. 1 minus 0 minus 1 is "0, and no borrow." At the end of all this, 6

(110) minus 5 (101) is 1 (001), which is certainly no surprise.

Observing the four steps above, even if you don't understand them entirely, it should be easy to see that a half-subtracter won't work past the units' position [step (1)], and some kind of full-subtracter is needed to do the twos' and fours' place.

The **full-subtracter** circuit two half-subtracters (surprised?). They are connected to an OR gate (what else?) as shown in Figure 6-5. The truth table below defines what a full subtracter has to do for every input, but a simple way to define the function of the gate is: A minus B minus Br, and the result of the subtraction is a difference output and a borrow output. The *difference* is the result of the subtraction, and the borrow indicates whether it was necessary to borrow a 1 from down the line to get the difference.

Full-subtracter truth table

A	B	Br	Borrow	Difference
0	0	0	0	0
0	0	1	1	1
0	1	0	1	1
0	1	1	1	0
1	0	0	0	1
1	0	1	0	0
1	1	0	0	0
1	1	1	1	1

Four of the eight lines in the truth table require a borrow. In each case where there is a borrow, remember that the A number becomes larger by 2. The difference is the result of what you'd get if you did "A + 2 minus B minus Br." For example, the fourth line on the table is "0 − 1 − 1 = 0 (diff) and borrow a 1." Think of the borrow

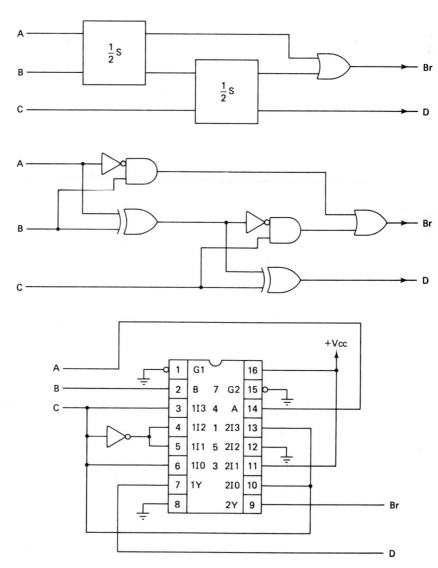

FIGURE 6-5 Full-subtracter circuit with 74153 implementation.

as changing the original statement of the problem to: "2 − 1 − 1 = 0 (diff)." This certainly makes sense. Each difference is either the result of the subtraction "A − (stuff)" (no borrow) or "A + 2 − (stuff)" (if borrow).

The full-subtracter, like the full-adder, can be implemented in a single 74153 MUX chip. This is shown at the bottom of Figure 6-5.

A 3-bit *parallel full-subtracter* to do the "6 minus 5" problem in the example is shown in Figure 6-6. You can readily see that the borrows are carried to the next binary place like the carry was in the parallel full-adder. The *parallel full-subtracter* is actually connected exactly like a parallel full-adder, if you just change the word "adder" to "subtracter" everywhere, and rename the inputs and outputs (the addend becomes the minuend, the augend becomes the subtrahend, the sum becomes the difference, and the carry becomes the borrow). To make a 4-bit parallel full-subtracter, convert the 4-bit parallel full-adder in Figure 6-3 according to the rules above.

6-1.6 Adder/Subtracter

You've probably begun to suspect this from the discussion just concluded; a subtracter can be made out of an adder. To make a half-adder into a half-subtracter, all that must be done is to invert the signal on the A line. To make a full-adder into a full-subtracter, invert the A inputs into each half-adder in the full-adder. To make a parallel full-adder into a parallel full-subtracter, invert the signal going into each and every half-adder's A input. Now, how can a signal path that didn't invert be suddenly turned into an in-

verter? The answer is to use a XOR as a controlled inverter. The XOR gate can be used to pass a signal or invert the signal by using one of its inputs as *the* input, and the other as a control line. This is shown in Figure 6-7 as a half-adder/subtracter. When the control line is HIGH, the I gate is an inverter. When the control line is LOW, the I gate acts as a buffer. Remembering the way the half-adder and half-subtracter work you can see that the circuit in Figure 6-7 is an adder when the control line is LOW and a subtracter when the control line is HIGH. Putting two of these together with an OR gate gives a **full-adder/subtracter**, and using a half-adder/subtracter and several full-adder/subtracters, a *parallel full-adder/subtracter* can be made (Figure 6-8).

One final thought on subtraction: What happens when you subtract and get a negative number? When you're working with a parallel full-subtracter, and the gates follow the rules in their truth tables, small negative numbers (such as you get by subtracting 6 minus 7) come out looking like very large positive numbers. Try this with a 4-bit parallel full-subtracter. (We did this in Figure 6-8.) After all the bits fall through the gates, the output looks like 15 in binary code (1111). This is the *largest* number a 4-bit output can display, and doesn't look anything like a 1 (we should be getting a negative 1 from this computation). How can we tell if this 15 is really a 15 or a negative 1? In addition to the four (difference) outputs of the subtracter, there's also a borrow output left over at the left-hand end of the circuit. If this is ON, it means that we are still trying to borrow, and there aren't any more places to borrow from. This means that the an-

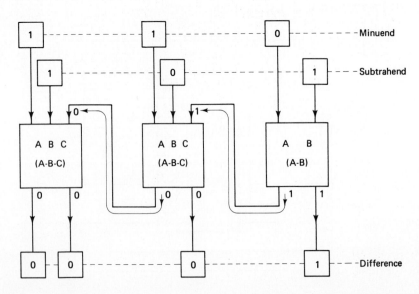

FIGURE 6-6 Three-bit parallel full-subtracter.

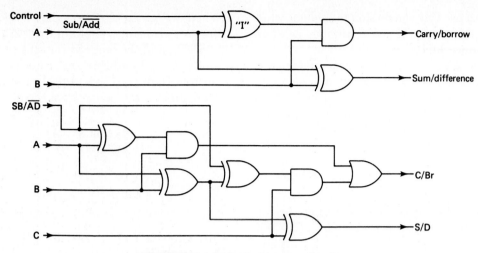

FIGURE 6-7   Adder/subtracter circuits.

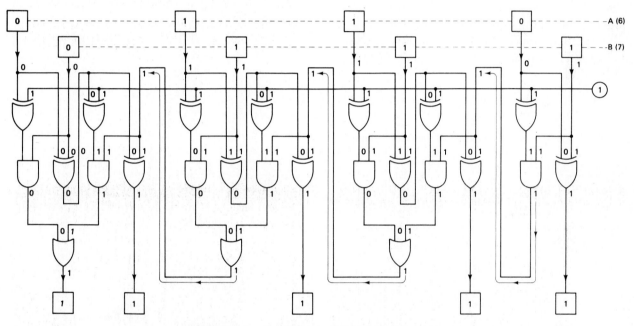

FIGURE 6-8   Four-bit parallel adder/subtracter.

swer is negative. Now, knowing that the answer is negative, we have to recognize that the same binary code is not used for negative numbers as we use for positive numbers. 1111 is a negative 0001, in the **twos'-complement code.** You can find the *two's complement* of any number by carrying out two steps.

1. Invert all bits of the number.
2. Add 1 to the result of step 1.

Let's try this on 0001, so that we can see what negative 1 should look like in twos'-complement form:

Step 1.   0001 becomes 1110. (invert)
Step 2.   1110 becomes 1111. (add 1)

According to this, a negative 0001 should look like a 1111 and if we see a minus sign (if the borrow bit is ON), we know that the (difference) outputs are in twos'-complement code.

To reverse the process (when you have a number in code and you'd like to decode it into ordinary binary to see what it is), you carry out these two steps:

1. Subtract 1 from the code.
2. Invert all bits of the result from step 1.

You can use this conversion to find out what the binary value of a negative number is when the borrow bit is ON.

In case this seems impractical and exotic, and you wonder if it's worth knowing, we'd like to point out that every microprocessor we've seen uses twos'-complement arithmetic to represent negative numbers, so this stuff is really out there in the real world.

EXAMPLE 6-1

What would be the result if an 8-bit parallel full subtracter were used to subtract binary 10000000 minus binary 10000001? (We expect −1, but how will it look?)

Solution: We remember that each of the full-subtracter's stages gives outputs for the problem "A minus B minus (borrow from last digit subtracted)" according to the following truth tables.

1. (If No borrow from previous digit)

	A	B	(from the next digit)	Difference	Borrow
a.	0	0		0	0
b.	0	1		1	1
c.	1	0		1	0
d.	1	1		0	0

2. (If borrow from previous digit)

	A	B	(from the next digit)	Difference	Borrow
a.	0	0		1	1
b.	0	1		0	1
c.	1	0		0	0
d.	1	1		1	1

So we go to work on the rightmost digit of the subtraction problem, "zero, take away one (no borrow from a previous digit)," which is case 1 (no previous digit exists) and line 1b. The result is "one, and borrow a one from the next digit."

$$
\begin{array}{r}
10000000 \quad \text{(A)} \\
-\ \underline{10000001} \quad \text{(B)} \\
1 \quad \text{(difference)}
\end{array}
$$
↑
(borrow next)

When we go to the next digit, we have three things to consider, the minuend (A), the subtrahend (B), and the borrow from the previous digit. This is case 2 and line 2a, "zero, take away zero (borrow from previous digit)," and the result is "one, and borrow a one from the next digit."

$$
\begin{array}{r}
10000000 \quad \text{(A)} \\
-\ \underline{10000001} \quad \text{(B)} \\
11 \quad \text{(difference)}
\end{array}
$$
↑
(borrow next)

You can see that the next five digits are going to be exactly like the last one, "zero, take away zero (borrow from previous digit)," so we repeat case 2a until we get to the leftmost digit, which is a different case.

$$
\begin{array}{r}
10000000 \quad \text{(A)} \\
-\ \underline{10000001} \quad \text{(B)} \\
1111111 \quad \text{(difference)}
\end{array}
$$
↑
(borrow next)

We have case 2d, "one, take away one (borrow from previous digit)," for which the result is "one, and borrow a one from the next digit." But *there is no next digit!* What do we do? We put down the answer in the Difference row, of course, but we keep track of the fact that the answer is really *less than zero*, and note that in this case, it isn't a true binary number at all. It's in *twos'-complement* code.

$$
\begin{array}{r}
10000000 \quad \text{(A)} \\
-\ \underline{10000001} \quad \text{(B)} \\
11111111 \quad \text{(difference)}
\end{array}
$$
↑
(borrow next)

The number is *negative one* (twos' complement), which is correct.

6-2 PARITY DETECTOR/GENERATOR

6-2.1 Parity Codes and Their Purpose

Another use for the XOR gate that applies some of the principles of the adder we already studied is the odd or even parity detector. **Parity** is a concept developed early in the history of computing machines to correct errors in data transmitted from one place to another. Normally, when a code is used to represent an instruction (machine code), there is no way to tell from the code itself whether the symbol received is the same as the symbol transmitted. Suppose that the letter C lost or gained 1 bit in the process of being transferred within a digital network. In most codes, the letter would be transformed into another letter, like B or G, or into some other legitimate symbol which happens to differ from C in only one place. If errors like this happen often, we have a problem.

Do errors like this happen often in digital systems? If one bit is lost or gained every million operations, is this going to seriously affect anything? After all, 99.9999% of the symbols are being transmitted reliably.

You bet it is! In computers (yes, even microcomputers!) data symbols may be transmitted and received more than 1 million times a second. If, on the average, a bit fails once in 1 million operations, the computer goes "crash!" once every second or so. Bad news, right? Of course, if the reliability of the computer is much higher than 99.9999%, it will crash only once every 10 seconds or 100 seconds, or maybe once every 1000 seconds. Can we expect to have computers run for *hours* at a stretch without messing up?

Actually, we can. Reliability in digital solid state is very good. A computer can run for days without a **glitch** (without messing up), and that means errors are occurring only once every hundred *billion* operations, roughly.

But can we expect computers to run forever without glitches? Even if the parts that make up the computer are perfect by themselves, we can't prevent glitches. One reason is *cosmic rays* (a quaint name for ionizing radiation from space). If you've ever taken a good look at the physical packaging integrated circuits come in, you probably noticed that they're always in black plastic packages, metal cans, or ceramic containers. The one thing these have in common is that they don't let light in. If light falls on a silicon integrated circuit, every junction and semiconductor device in the circuit becomes a silicon solar cell in miniature. These devices not only pass current everywhere when there was no current before, but they generate their own electricity! Computer chips work best in the dark!

No problem, right? They're all packaged in light-tight containers, so why worry? Well, there are other things besides light that have the same effect on silicon. Particles with enough energy can penetrate right through metal, ceramic, or plastic packaging. In fact, the "atomic" radiation from space can penetrate through many feet of concrete (and particles called "neutrinos" can zip right through a whole planet!). The silicon junctions inside a computer chip are fine detectors for this kind of radiation. Too bad if you wanted the junction to be part of an AND gate in the arithmetic-logic unit of a microprocessor CPU; it's going to play "radiation detector" while the particle zaps through it.

So if there isn't any way to stop glitches, seldom as they might be, perhaps there's a way to catch one when it occurs.

6-2.2 Odd-Parity and Even-Parity Codes

Odd-parity code is a code used in the computer world in which only the patterns with an odd number of 1s are used. For instance, in a code called ASCII, the letter B is represented by the binary number 1000010. The letter C is represented by the binary number 1000011. The ASCII code for B has **even parity** since it contains two 1s, and 2 is an even number; the ASCII code for C has *odd parity* since it contains three 1s, and 3 is an odd number. By themselves, the ASCII codes for letters are half even and half odd parity. To make all the ASCII codes odd parity, we add an additional bit to the codes in the place marked "P" below.

$$B = P1000010 \qquad C = P1000011$$

When the rest of the code for B has even parity, the P bit is turned on, making the code B = 1 1000010. There are three 1s in this code, so it has odd parity. For the letter C, the 7 bits of the ASCII code already have odd parity, so the P bit is turned off, giving C = 0 1000011. This code still has odd parity since adding the 0 didn't change the bit count at all, and there are still three 1s.

6-2.3 Parity Detector and Parity Generator (Made with XOR Gates)

With a bit of "smarts," you can decide whether the P bit (often called a **check bit**) should be ON or OFF for any seven-digit ASCII code, in order to make an 8-bit odd-parity ASCII code out of it. How does a circuit develop the same "smarts" to know when a 7-bit binary symbol has odd or even parity? Figure 6-9 shows a circuit to do this. It looks a lot simpler than you might expect, considering how complicated an odd or even bit count is. Actually, the circuit is an adder with only a

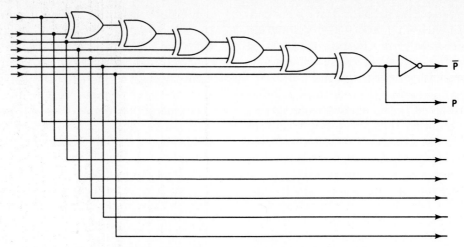

FIGURE 6-9 Parity generator for odd and even parity.

sum output (units' position). The adder converts the code into a bit count (how many 1s there are in the code). In binary, it's much easier to tell if the count from the adder is odd or even than it is in decimal. Odd binary numbers have the bit in their units' position ON, and even numbers don't. The output of the **parity detector** is summing all the bits at its input, and the sum tells the bit count. Now, if all we want of the bit count is to know whether a number is odd or even, we won't want all the carry bits, just the units' position (sum). This is what the parity detector gives us. It's basically a 7-input full-adder (a very full adder?) stripped down at its outputs to show only the 1-bit of the result. The 1-bit will be ON if the count is odd or OFF if the count is even. Now it happens that we want the P bit OFF if the count is odd and ON if the count is even—exactly the opposite of what the XOR circuit does, so an inverted output (a double-rail output) has been added to the diagram to provide the correct logic for our P bit. Since we have two outputs from this

parity generator circuit, we have two kinds of P bits, an *even-parity* $\bar{P}$ *bit* and an *odd-parity P bit*. If the 7 bits of the ASCII input are combined with the $\bar{P}$ output to make an 8-bit **byte**, the byte will have odd parity.

Actually, the circuit in Figure 6-9 is a parity generator. It has been used to detect the parity of a seven-digit binary code, but its output has been added to the 7-bit code as an eighth bit, which makes the code odd parity. This is the circuit used to create the odd-parity code in the first place, before it is transmitted. When it is received, a circuit must be present that can count the parity of an 8-bit byte, and determine whether it is odd or even when it arrives. The circuit in Figure 6-10 does this. It works exactly like the XOR circuit in Figure 6-9, except that it has eight inputs. With these inputs, the circuit determines whether the bit sum of the 8-bit byte is an odd sum or an even sum. If the sum is odd, the output is HIGH, and if the sum is even, the output is LOW. This can be used in the receiver to trigger

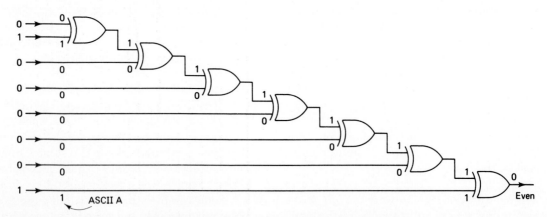

FIGURE 6-10 Parity detector showing even parity.

a circuit that says: "Send that one by me again—I don't think I saw it right" to the transmitter circuit. When the receiver can "talk back" to the transmitter this way, they call it **handshaking** between the two circuits.

How much does this improve the reliability of the transmission? Statistically, if the circuit fails on a bit once every 1 million bytes, the chance that the next bit will fail is still 1 in a million. This means that **parity checking**, which can "catch" an error when one bit fails, but not when two fail, will only get "fooled" every million times the first bit fails, because the chance of a second bit failing in the same byte is 1 in a million. Now suppose that such checking is employed inside the CPU of a computer where bytes fly around at rates near 1 million a second (this is done in the IBM-PC and 'clones') and a one-bit glitch happens every second (the CPUs around are generally a lot more reliable than this). With parity checking, the glitches would "fool" the parity checker only every 1 million glitches, which is every 1 million seconds. Failing once every 1 million seconds instead of every second is a millionfold improvement in reliability (the number of seconds between failures is called the *mean time between failures*, MTBF). You couldn't get anywhere running programs that fail every second, but 1 million seconds is about 12 days, 24 hours a day, without breaks. Few computer programs need that amount of time.

Current-generation microprocessors have much better reliability than 1 in a million, and don't contain parity checking for every internal data transfer, but the parity test circuit is in the CPU, and is commonly used to check data coming into the computer system from outside (peripheral) devices. There are also, despite improved hardware reliability, still the same radiation-induced errors in memory chips, CPUs, and so on. (These are called *soft failures* in the microprocessor community.) In present designs, systems that employ parity error checking are commonly included in microcomputer devices.

EXAMPLE 6-2

How reliable is a system that transmits data with one error in every million bytes if a check bit and parity checking is added to the system? If data are transmitted by this system at 1.544 million bits per second (Mbps), how often will it fail to transmit a byte correctly:
  (a) If no parity checking is used?
  (b) If parity checking is used?

Solution: If a check bit is added and parity checking is done, the chance that a bad byte "gets by" the parity check is the same as the chance that the byte has *two* errors in it. What is the likelihood of that happening?

Since we know that the chance a byte has *one* error in it is 1 in a million, what is the chance that a byte with an error in it (which is already a one-in-a-million chance) has a *second* error in it? The chance of a second error is also 1 in a million (it is absolutely unaffected by the first error). That means that a byte with a *second* error occurs once in every million bytes containing *one* error, or once in every million million bytes.

Now, how often does the system fail to transmit properly at 1.544 Mbps in case a (no parity checking)? Since a byte is 8 bits, without parity, the number of bytes per second is

$$\frac{1,544,000}{8}$$

which is 193,000 bytes per second. If 1 million bytes are transmitted, on the average, before a failure occurs, how long is the time between failures? At a rate of 193,000 bytes per second, the time to transmit a million bytes is

$$\frac{1,000,000}{193,000}$$

which is about 5.18 seconds. The system fails, in round numbers, every 5 seconds. (Suddenly, one error in a million doesn't sound so reliable!)

In case b (with parity checking), we are transmitting 9 bits per byte (8 bits plus parity), so the number of bytes per second is

$$\frac{1,544,000}{9}$$

which is 171,555 bytes per second (somewhat slower). On the average, 1 million million (1,000,000,000,000) bytes are transmitted before a failure occurs. At 171,555 bytes per second, the time to transmit 1,000,000,000,000 bytes is

$$\frac{1,000,000,000,000}{171,555}$$

which is a bit more than 5,829,014 seconds. You don't generally deal in lengths of time that long by measuring them in seconds—but it is more than two months (67.46 days). That's a much better MTBF (mean time between failures) than 5 seconds.

6-3 COMPARATOR CIRCUIT (MADE WITH XOR, XNOR, AND AND GATES)

In Chapter 3 we introduced the exclusive-OR gate and mentioned that it was possible to define what the XOR and XNOR did by saying: "The XOR's output is ON when its inputs are unequal—the XNOR's output is ON when its inputs are equal." This leads to the application for which the XOR gate is best suited: comparing two numbers to see if they are equal or unequal. Figure 6-11 shows a circuit that does this: There are two 4-bit numbers—A0, A1, A2, and A3 and B0, B1, B2, and B3—which are being compared by this comparator. If A0 matches B0, gate G0 will have a LOW output. If A1 matches B1, gate G1 will have a LOW output, and so on up to G3. If the outputs of all four gates are LOW, it means that all four bits match, bit for bit. Attached to the four gate outputs is a gate whose output will be LOW only if all four inputs are LOW, an OR gate. This combination is actually being used as an AND gate for 0s; when inputs G0 *and* G1 *and* G2 *and* G3 are *all* LOW, the output is LOW, and at no other time. This comparator has an active LOW output, which becomes active (LOW) when the two 4-bit inputs match in all places. A second way of developing the same circuit is shown in Figure 6-12.

The transformation begins with the circuit from Figure 6-11. In the second step [Figure 6-12(b)], the DeMorganized version of the OR gate has been put in place of the OR. Since the DeMorganized version of the gate is an AND with all its inputs and outputs inverted, the circuit has not changed logically, but has a different appearance. In the third step [Figure 6-12(c)], we've "slipped" the inverter bubbles down to the other end of each input wire. That doesn't change the logic, either, but now we have a circuit with four XNORs and a NAND. This circuit will be LOW at its output (just like the circuit in Figure 6-11) when the input numbers match. To get a comparator with an active-HIGH output (HIGH when the two 4-bit inputs match), all that's needed is to drop the "bubble" off the NAND at its output. This is done in the fourth step [Figure 6-12(d)], and the circuit that results is an active-HIGH output comparator, made of four XNOR gates and an AND gate. There are several other ways to make this circuit, but they can be made by just DeMorganizing things, and moving inverters up and down the wires (the old "elastic wire trick"). Finally, if you want to compare numbers larger than 4 bits to each other, all that's needed is an OR, NAND or AND gate with more inputs, and a larger number of XOR or XNORs, one for each bit of the word size (if you're comparing 4-bit numbers, the word size of the circuit is four; if you're comparing 8-bit numbers, the word size of the comparator is eight). The circuits in Figures 6-11 and 6-12 are all 4-bit comparators, with either active-LOW or active-HIGH outputs. A symbolic representation of the comparators is shown alongside Figure 6-11 and the final diagram of Figure 6-12. An 8-bit comparator is found in all microprocessors whose word size is eight. It is usually used for processing data arriving at the computer from a remote location.

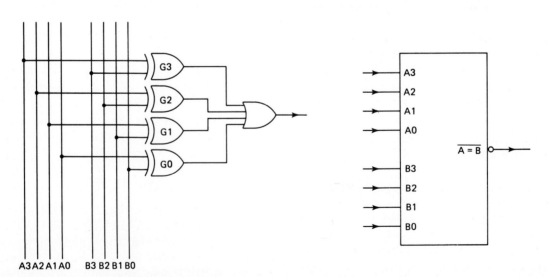

FIGURE 6-11 Four-bit comparator using XOR and OR gates.

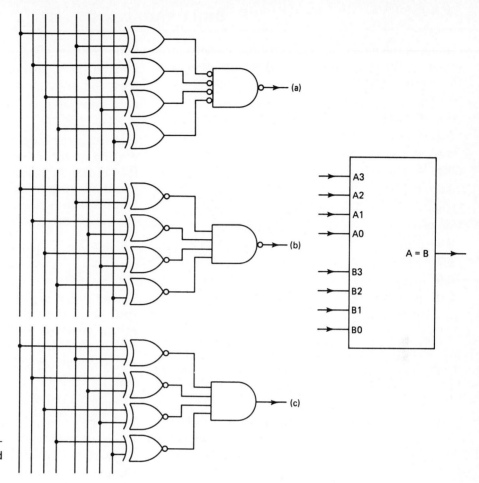

FIGURE 6-12 Four-bit comparator using XNOR and AND gates.

QUESTIONS

6-1. How many half-adders are in a full-adder?

6-2. How many inputs does a full-adder circuit have, and what are they called?

6-3. Determine what the outputs will be given for the inputs shown in the following figure.

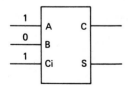

6-4. Do the following binary addition problems.

(a)	101	(b)	110	(c)	011
	+011		+100		+101

6-5. Do the following binary subtraction problems.

(a)	1000	(b)	101	(c)	110
	− 11		− 10		−101

6-6. Pick one addition and one subtraction problem from Questions 6-4 and 6-5 and explain each step, including the borrows and carries.

6-7. Why are full-adders and full-subtracters necessary?

6-8. How many inputs and outputs are there in a full-subtracter, and what are they called?

6-9. Find the two's complement of the following 4-bit binary numbers.
(a) 0101   (b) 1101   (c) 1000

6-10. Convert the following codes to odd parity by indicating a value for the P bit.
(a) P1001101 (b) P1110001 (c) P1101011

6-11. Explain why the transmission of data using only odd parity (or even parity) improves reliability.

6-12. Why would the core of a running nuclear reactor be a bad place for solid-state logic control circuitry? Would parity checking help?

Complete the following one-digit binary addition problems. (Show the answers as though they were the outputs of a binary half-adder or a binary full-adder.)

6-1. $0 + 0 =$

6-2. $0 + 1 =$

6-3. $1 + 1 =$

6-4. $1 + 0 + 1 =$

6-5. $1 + 1 + 1 =$

Complete the following multidigit binary addition problems.

6-6. $101 + 010 =$

6-7. $101 + 111 =$

6-8. $0100 + 0101 =$

6-9. $0111 + 1101 =$

6-10. $10001001 + 00111001 =$

On the following half-adder symbol, find each input or output named, and identify it by writing down its letter. The inputs and outputs for a binary addition example are shown to aid you in identifying the outputs correctly.

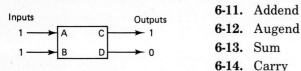

6-11. Addend

6-12. Augend

6-13. Sum

6-14. Carry

Complete the following one-digit binary subtraction problems. (Show the answers as though they were the outputs of a binary half-subtracter or a binary full-subtracter.)

6-15. $1 - 0 =$

6-16. $1 - 1 =$

6-17. $0 - 1 =$

6-18. $0 - 0 - 1 =$

6-19. $0 - 1 - 1 =$

Complete the following multidigit binary subtraction problems.

6-20. $1101 - 0101 =$

6-21. $1101 - 0011 =$

6-22. $1011 - 1101 =$

6-23. $00110110 - 00010011 =$

6-24. $00000001 - 00000010 =$

On the following half-subtracter symbol, find each input or output named and identify it by writing down its letter. The inputs and outputs for a binary subtraction example are shown to aid you in identifying the outputs correctly. The circuit performs the calculation $(A - B)$.

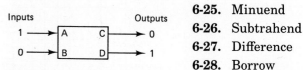

6-25. Minuend

6-26. Subtrahend

6-27. Difference

6-28. Borrow

The following four 8-bit numbers are *negative* results of using a full-subtracter. As a result, they are in *twos'-complement* code. Convert them to true binary and write down (in decimal) what negative number each one is.

6-29. 11111011

6-30. 11111101

6-31. 11111111

6-32. 10000000

Convert the following numbers into odd parity. Change the check bit "C" into a 0 or a 1 according to what's needed for odd parity.

6-33. C01000001

6-34. C01000011

6-35. C00110000

6-36. C00110001

Convert the following numbers into even parity. Change the check bit "C" into a 0 or a 1 according to what's needed for even parity.

6-37. C01000001

6-38. C01000011

6-39. C00110000

6-40. C00110001

7
BINARY-BASED NUMBER CODES

CHAPTER OBJECTIVES

By the time you finish this chapter, you will be able to:

1. Identify 12 different types of binary codes which are used by computers as symbols of other quantities, such as *alphabetic characters*, *integers* (signed and unsigned), and *floating-point numbers*.
2. Convert binary numbers into *octal* numbers, and vice versa.
3. Convert binary numbers into *hexadecimal* numbers, and vice versa.
4. Convert decimal numbers into BCD (*binary-coded decimal*) numbers, and vice versa.
5. Understand how the results of binary addition are "adjusted" to permit addition of BCD numbers.
6. Convert alphabetic or text characters into *Hollerith* and *BCDIC* codes, and vica versa.
7. Convert alphabetic or text characters into EBCDIC code, and vice versa.
8. Convert alphabetic or text characters into ASCII code, and vice versa.
9. Identify the fact that other codes may be used for digital transmission of text characters.
10. Understand how *ones'-complement* and *twos'-complement* codes may be used to represent numbers digitally, and how to add and subtract using them.
11. Convert a true-binary number into *ones'-complement* or *twos'-complement code*, and vice versa.
12. See how a *floating-point number* uses the parts called *mantissa* and *exponent* to represent a number in scientific or engineering notation.
13. Convert 8-bit groups to *octal byte* code, and vice versa.
14. Convert 8-bit groups into *hexadecimal byte* code, and vice versa.

Binary numbers are found in computer systems, stored in the memory device of the computer. There are several kinds of binary number codes that may be stored in the same memory device. The same pattern of 1s and 0s may have a completely different meaning in different places. A code may represent a number, pure and simple, in binary code. In another place, the same pattern of 1s and 0s may represent a letter of the alphabet, a punctuation mark, or even a numeral in the decimal system, rather than a binary value. (It was this sort of code we made into odd parity and even parity in Chapter 6, and this is part of the reason we've decided to follow up with a description of codes in this chapter.) In yet another application, the number stored in the computer's memory may be a binary-code number with a plus or minus sign and decimal point, or an exponent indicating where to shift the decimal point.

Each of these codes uses a combination of binary 1s and 0s in a different way. In some cases, the meaning of a 1 or 0 in a specific place changes according to the code being used. In this chapter we explore a number of these codes and how the 1s and 0s are given different meanings in each one.

7-1 BINARY, OCTAL, BCD, AND HEX NUMBERING

7-1.1 Binary Code

The first application of binary numbers is . . . as numbers! In the interior of a computer memory, a group of binary bits occupies a common location called an **address**. Just how many bits there are at an address depends on the design of the computer, but the simplest way to use these bits is to assign one bit a place value of 1, another bit a place value of 2, another 4, another 8, another 16, and so on, giving each next bit double the value of the last. This is the **simple binary code**, a positional code described in Chapter 6 when addition and subtraction were discussed. For many microprocessors, a size of 8 bits is the natural **word length** of each memory address. A **word** with 8 bits will have binary place values from 1 to 128. A combination of these bits could stand for any binary number from 0 (no bits on) to 255 (all 8 bits on). This method of coding is called **simple unsigned binary**. Only positive, whole numbers (positive integers) are used in this system. If you want to represent a decimal number this way, all you have to do is convert it to binary the way we did in Chapter 6, and if it comes out less than

eight digits, "pad out" the remaining places with 0s. If the number has more than eight places, we have a problem. We either need a way to put together two 8-bit pieces (bytes), or a computer with a larger word length.

7-1.2 Octal Code

We know that the switches inside the computer memory are made from digital logic gates. This means that numbers must be composed of 1s and 0s, since digital gates are either ON or OFF. No matter what number system happens to be convenient for us (decimal, usually) the numbers inside computers (and any other digital logic circuit) are binary, really. Having said that, we still have good reasons for showing the contents of a computer's memory in number systems that are not binary. For example, if you want to print the number 127 on a sheet of paper in the binary system, it's 01111111. That takes up a lot of space, and a lot of ink, on a sheet of paper. In fact, binary numbers take up more space than any other number system, and that costs money. Worse, binary numbers are hard to read, just like decimal numbers with a lot of digits. Try reading eight-digit decimal numbers quickly, and keeping all the hundreds, thousands, and millions straight!

One thing we do with long decimal numbers is add commas to make reading more easy. Reading 12,345,678 is not as difficult as 12345678, since the millions, thousands, and units are separated. If we do the same thing with binary numbers, 127 becomes 01,111,111. Each group of three binary digits has a name (Table 7-1). We call 000 zero and 011 three, for instance. When we read 01111111 (127) as 01,111,111, we call the three groups of digits separated by commas one, seven, seven. The name of each group is the number you would read if the group stood alone as a binary number. Reading 01,111,111 as one, seven, seven—or 177, for short—is actually creating a shorter form of the binary number. This

TABLE 7-1

Names of 3-bit groups

Digit name	3-Bit binary	Digit symbol
Zero	000	0
One	001	1
Two	010	2
Three	011	3
Four	100	4
Five	101	5
Six	110	6
Seven	111	7

short form of binary is convenient for printing, because it takes only three symbols (1-7-7) to represent a number that's eight symbols long in binary. It's also better than the decimal 127 if you want to get the original binary back again. Converting the decimal number 127 to 01111111 is—to put it frankly—a headache, but converting the 177 short form back to 001,111,111 (the extra 0 makes no difference) means just having to know the three-digit binary number for things from zero through seven. If we only need to know the binary for eight different symbols, this means we're using a number system with the base eight, called **octal**.

Just for a little practice, let's use the binary form of the decimal number 56, which we worked out in Chapter 5. It's 00111000. Breaking that up the way we break up long decimal numbers, the number becomes 00,111,000. You can see right away that we don't need the 00 at the front, but we want to keep the number in a form that "fits" an 8-bit computer. This means that we'll have a three-digit octal number every time we convert a byte of binary, whether we need all three digits or not. From 00,111,000 we get zero, seven, zero, by reading each group as a 4-2-1 binary code. Octal 070 takes more space than decimal 56, but is easier to reconstruct into binary if you need to. It also preserves the word size by showing that the 0 in front is a significant part of the number.

To go in the opposite direction, take the octal number 236. It's 010,011,110 according to Table 7-1. Since this number has nine digits, and we're supposing that it is stored in a digital machine with a word size of 8 bits, let's chop off the front 0 of our binary number, leaving 10,011,110, an eight-digit binary number. We've reconstructed what the binary number with eight digits looks like from an octal number with only three digits. We've done it with a lot less work than the conversion from its decimal value would take, and the number (236) doesn't really take any more room to print than its decimal value

(158) takes. (*Note:* Octal numbers usually look a little "larger" than they do in decimal, but have about as many digits most of the time.)

7-1.3 Hexadecimal Code

The octal number is formed from the binary code by grouping bits by threes, then reading each group by itself as a number from 000 (zero) to 111 (seven). Since only eight groups are possible, the octal number is base-eight. Octal numbers will never contain an 8 or a 9, since these digits cannot be written in 3 bits. Otherwise, octal numbers resemble decimal numbers pretty closely.

Now we ask whether there is an even more compact way to write the binary contents of a computer memory (or any digital device). The decimal system is a bit more compact than the octal system. Some three-digit octal numbers can be written with only two decimal digits, some four-digit octal numbers can be written with only three decimal digits, and so on . . . but *decimal* is awkward. To "see" what the binary for a decimal number is, we need a cumbersome conversion process, while octal is conveniently converted to binary by just looking up a 3-bit number for each octal digit. Is there another way to write binary that is as easy to convert as octal, but takes up less space?

Yes. The key is to change the way we break up the binary number using commas. When we took 00111000 and broke it apart into 00,111,000, we put a comma every third place. Suppose that we put a comma every fourth place, breaking the number up into 0011, 1000—two groups of 4 *bits*. Each 4-bit group has a name. In this case, 0011 is called three, and 1000 is called eight, if we read the group as an 8-4-2-1 code. The number 0011,1000 can be written "38" using this form of notation. The names for all the 4-bit groups can be written "38" using this form of notation. The names for all the 4-bit groups are given in Table 7-2.

TABLE 7-2
Names of 4-bit groups

Digit name	4-Bit binary	Digit symbol	Digit name	4-Bit binary	Digit symbol
Zero	0000	0	Eight	1000	8
One	0001	1	Nine	1001	9
Two	0010	2	Ten	1010	A
Three	0011	3	Eleven	1011	B
Four	0100	4	Twelve	1100	C
Five	0101	5	Thirteen	1101	D
Six	0110	6	Fourteen	1110	E
Seven	0111	7	Fifteen	1111	F

Notice that there are 16 possible groups of 4 bits. This makes it impossible to use a decimal symbol for every name, since there are only 10 decimal symbols. We had to use letters of the alphabet for numbers from 10 up to 15. This allows us to still use one symbol for each name, but the "numbers" formed this way will not always look like decimal numbers. Patterns like 3F and D1 are possible, and numbers like BE might not even look like numbers at all. This is one of the little sacrifices we have to make for the sake of compactness. Once we decide to use this system of writing binary numbers, every 8-bit byte becomes a two-digit **hexadecimal** (base sixteen) code that takes only two print characters to write. Since print shops have letters as well as numbers in their type cases, printing lists of binary information using the hexadecimal system is no hardship on the typesetters. Since every eight digits of binary is compressed into two digits of hexadecimal, there is a saving on printing cost over binary, decimal, and octal. The only inconvenience is for the person who must read the code, because numbers like BE are not "numbers" in our normal experience. Once we get over the shock of dealing with "funny" numbers like BE and D7, the two-symbol representation of each 8-bit byte is quite easy to work with. To "see" what binary code is "hidden" inside each *hex* (short for hexadecimal) pattern, the reader just looks up the 4-bit pattern for each symbol—the same process we use for octal, only with a larger lookup table.

We've used the number BE as an example of a hex number. Let's see what a BE is by looking up its symbols. B is the symbol for 1011 and E is the symbol for 1110. Putting these together, we get BE as 1011,1110. This is the decimal number 190. You can see that BE takes up less space than either its binary or decimal form. For this reason, and the fact that hex is more easily converted to binary than decimal is, most listings of numbers in computer memory are printed in hex. Whenever magazine articles or book chapters include listings of the code inside a computer memory, the code—which is really binary—is almost always printed in hex to save space. There are also a number of 8-bit small computers with hex keyboards for number entry. This means that loading code into the computer will take two keystrokes on the keyboard instead of eight switch operations for every byte of code.

One more example. What does the number 127 look like in the computer's memory? What does 127 look like when it's converted to hex?

If you're clever and sneaky, you'll remember that we gave you the binary value of 127 a few paragraphs back, and you won't have to convert it from decimal to binary. 127 is 01111111 in binary. Now, what do we do to make it hex? Right! We break it up every four places with a comma. 0111,1111 is the result. Now, we must look up the names of these two 4-bit groups—0111 is seven (7) and 1111 is fifteen (F). The decimal number 127 comes out as 7F in hex. It doesn't look much like 127, but it takes up less space, and can be converted back to binary by anyone reading a listing a lot easier than decimal. Why convert it back to binary? Because in the computer, the numbers really are binary, and you might need to know just which bits are supposed to be ON or OFF when you're troubleshooting a sick computer—even though the printed listing in your hands is in hex.

EXAMPLE 7-1

Convert the binary number 01110110 into octal and hexadecimal code.

Solution: To convert to octal:

1. Break the number up into groups of three, counting to the left from the "binary point."

    01,110,110.

2. Fill out the leftmost group (if necessary) with zeros, to make three digits, and convert each 3-bit group into its one-digit equivalent (since numbers larger than 7 are impossible, this is really not decimal; it's octal).

    001,110,110.
    $\downarrow$ $\quad$ $\downarrow$ $\quad$ $\downarrow$
    1 $\quad$ 6 $\quad$ 6

The octal value of binary 01110110 is 166.

To convert to hexadecimal:

1. Break the number up into groups of four, counting to the left from the "binary point."

    0111,0110.

2. Fill out the leftmost group (if necessary) with zeros, to make four digits, and convert each 4-bit group into its one-digit decimal equivalent. If the 4-bit binary group is 10 or bigger, replace the group

with letters of the alphabet, according to the following rule

1010	(ten)	= A
1011	(eleven)	= B
1100	(twelve)	= C
1101	(thirteen)	= D
1110	(fourteen)	= E
1111	(fifteen)	= F

$$0111,0110.$$
$$\downarrow \quad \downarrow$$
$$7 \quad 6$$

The hexadecimal value of binary 01110110 is 76.

7-1.4 BCD Numbers

Octal and hexadecimal codes as discussed in this chapter serve only one purpose. They are shorthand for the binary code that is really inside the digital logic circuit. In all cases, the 8-bit code inside the circuit is organized the same way; each position is twice the value of the place to its right, until you run out of places. All numbers are represented as a combination of 1s, 2s, 4s, 8s, 16s, and so on. This underlying *data structure* is the same whether we choose to represent it on paper using octal or hex symbols. The only basic inconvenience of the simple unsigned binary form of data is that simple binary and the decimal system of numbering really don't mix. Conversion between the decimal system we are all comfortable with, and the binary system natural to digital logic circuits, is awkward and clumsy. It's possible, though, to write two digits of hex in every byte or binary—shouldn't it be possible to write two digits of decimal in a byte instead?

Of course, the answer is yes. The two digits of a hex byte could be limited so that the numbers from 10 up are just never used. Some two-digit combinations like 38 (0011,1000) would still be around, but others like BE (1011,1110) would be gone. The last 4 bits (4-bit groups are called **nibbles**) would stand for the ones' place of decimal numbering, and the first 4 bits would be the tens' place. In this code, numbers with 8 bits (a byte) would be made of 1s, 2s, 4s, 8s, 10s, 20s, 40s, and 80s as shown below:

80s	40s	20s	10s	8s	4s	2s	1s
0	0	1	1,	1	0	0	0

In discussing this binary code, we said that the numbers 0011,1000 were 38. That's the same conversion we did in going from binary to hex, but in this case the 38 means thirty-eight in dec-

imal. Notice that the 0011,1000 has one 20, and 10, and an 8. This adds up to decimal 38. The system described above is a different data structure from the simple binary described before, because it is **binary-coded decimal** (BCD). Each 4-bit group (nibble) is a separate decimal place. The number 1986, for example, would be encoded in this sytem as:

$$1 \quad 9 \quad 8 \quad 6$$
$$0001,1001,1000,0110$$

where each nibble is read separately as one decimal digit, and numbers larger than 1001, although possible, are never used. In this BCD code, numbers like 1010, 1011, and 1111 are *illegal* code.

Most microprocessors, and all larger computers, contain circuitry for handling arithmetic in this code, or conversion from simple binary into this code. Such conversion is called a **decimal adjust** operation.

7-2 BCD ADDER/BCD ARITHMETIC

7-2.1 BCD Adder Circuit

Arithmetic circuits that used simple binary numbers were discussed in Chapter 6. At this point, it seems appropriate to see how circuits would be devised that could do the same arithmetic operations with BCD numbers. The new data structure of a BCD byte, and the differences between the binary and decimal systems of numbers, give rise to several problems. Does the designer of the computer need to construct two adder circuits, one for BCD and the other for binary, so that the computer can handle both kinds of numbers? How does the computer recognize when a hex nibble needs to be converted to a BCD nibble, and when no conversion is needed? For instance, hex 8 is also BCD 8—and needs no conversion—but hex C is going to become BCD 12—and something must be done to C (1100) to make it 12 (0001,0010).

To solve the first problem first, microcomputers and minicomputers do not generally add BCD numbers with a BCD adder and binary numbers with a binary adder. The same binary adder does both operations. This means that the computer has a "blind spot" with regard to the BCD-ness or binaryness of the numbers it's adding. It cannot tell which kind of number it has. It's the *programmer* who must tell the computer when to convert the sum to BCD after two BCD numbers have been added. It turns out that the solution to the second problem (the decimal ad-

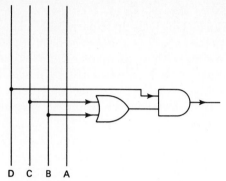

FIGURE 7-1 "Ten-or-bigger" detector.

just circuit) is used to handle the first problem (adding BCD numbers without a new adder).

Figure 7-1 shows a circuit we call a **10-or-bigger detector**. When a nibble made of 4 bits (we're calling these bits D, C, B, and A) is 10 or larger, the output of this circuit goes ON. The logic is fairly simple. If a number contains an 8 (D) and also a 2 (B) it is 10. If we find an 8 combined with either a 2 or a 4 or some combination of the two, the number is at least 10, and probably more. This is expressed in Boolean as

$$T = D \cdot (B + C)$$

Now suppose that we used a computer's binary adder to add the BCD bytes 05 and 07 together. The BCD 05 (0000,0101) looks just like a hex 05 (0000,0101), and the BCD 07 (0000,0111) looks

just like a hex 07 (0000,0111). The binary adder in the microprocessor doesn't know the difference, and adds "05 + 07" and gets a hex twelve 0C as its resulting byte (0000,1100). This is exactly the result a binary adder is supposed to get, but the result is an illegal code for a BCD result. First, the decimal adjust circuit must "recognize" this fact; then it must do something about it.

The circuit in Figure 7-1 recognizes that the code is 10 or bigger when it sees that D and C are ON. That's enough to turn the output ON. Now, we need a circuit that does something when that output is ON to make the hex 0C (0000,1100) into the correct BCD code for a twelve, which is 12 (0001,0010). The logical thing to do seems to be to add something to 0000,1100 that makes it 0001,0010. The difference between these numbers (dealing with them strictly as simple binary code) is six. If we add binary six (0110) to the 0C code, we get 12. It turns out that binary six is a sort of magic "fudge factor" that makes every 10-or-bigger nibble into its correct BCD counterpart.

(*Digression:* If you've never heard of a fudge factor, here's an example of one: You know the reading of the voltmeter is supposed to be 5 V, but you keep measuring 4.8 V. You move your head *way* over to one side, until the needle lines up with the "5," and write down "5 V." That's *cheating*; a *fudge factor* is a number you use to

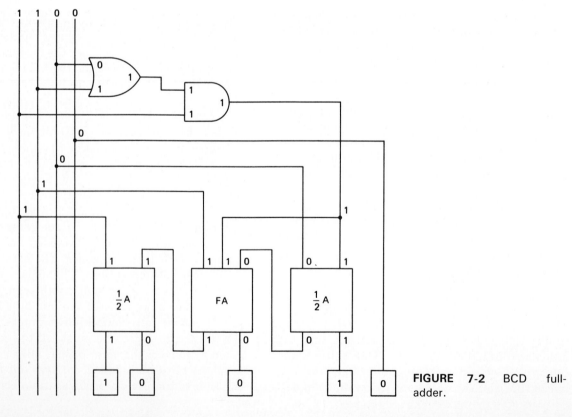

FIGURE 7-2 BCD full-adder.

cheat, so that you make reality agree with what you want to see.)

Now, how are we going to add a 6 to the number we started with, to get it "fudged"? Figure 7-2 shows a circuit that can be used for this purpose. It's the 10-or-bigger circuit we just saw, connected to an adder circuit. If the output of the 10-or-bigger gate is 0, we're adding 0 to the 2s and 4s place (no effect) and the rest of the numbers go through the adder with nothing added. If the output of the 10-or-bigger gate is 1, we're adding 1s to the 2s and 4s place (adding 6) and the rest of the numbers coming into the adder come out with 6 added. Since the fudge factor—six—doesn't have anything in the ones' place, there's no adder there. (Nothing gets added at the units' position when you add 6.) This is shown in Figure 7-2 for the number 0C converted to 12. We begin with the number 1100 (C) at the top of the diagram. The 10-or-bigger detector says "aha!" and its output turns ON. The 110 bits that enter the adder are added with these two 1s (110 + 11) and the sum (1001) drops out the bottom. The result contains 5 bits. Four of them (0010) are in the units' nibble and the carry goes to the tens' place. The number 1,0010 is 12, BCD.

For larger numbers—8 bits, for instance—there is more elaborate circuitry needed, but the circuit for one nibble (Figure 7-2) contains all the elements necessary to make the larger circuits.

EXAMPLE 7-2

Add the BCD numbers 1001,0110 (96) plus 0010,1000 (28). Show where a decimal adjust operation is done to get the correct answer.

Solution: Of course, 96 + 28 = 124. We expect the BCD answer to have the appearance of 124 = 0001,0010,0100. To get the result in a computer whose only hardware adds in binary, however, the results of the binary adder must be "adjusted":

Binary addition:

$$
\begin{array}{r}
1001,0110 \\
+\ \underline{0010,1000} \\
1011,1110 \\
\downarrow \qquad \downarrow \\
\text{(eleven)} \quad \text{(fourteen)}
\end{array}
$$

(Eleventy-fourteen is decidedly *not* a legitimate BCD answer!)

1. Add 0110 (six) to the least-significant nibble if it's 10-or-bigger, and needs "adjustment."

$$
\begin{array}{r}
1001,0110 \\
+\ \underline{0010,1000} \\
1011,1110\ \leftarrow \text{fourteen} \\
+\ \underline{0110}\ \leftarrow \text{adjustment} \\
1\ 0100 \\
\uparrow \\
\text{(BCD carry)}
\end{array}
$$

2. Add any BCD carry to the next-most-significant nibble; if it is now 10-or-bigger, add 0110 (six) to it as before.

$$
\begin{array}{r}
1001,0110 \\
+\ \underline{0010,1000} \\
\text{eleven} \rightarrow \quad 1011,1110 \\
\downarrow\ +\ \underline{0110} \\
\text{BCD carry} \rightarrow \quad +\ \underline{1}\ 0100 \\
\text{twelve} \rightarrow \quad 1100 \quad \downarrow \\
\text{adjustment} \rightarrow \quad +\ \underline{0110} \quad \downarrow \\
1 \qquad 0010\ 0100 \\
\downarrow \\
\text{(BCD carry)}
\end{array}
$$

Filling out the leftmost BCD digit to 4 bits gives

0001,0010,0100   (96 + 28 = 124)

7-3 BCDIC AND HOLLERITH CODES

Numbers stored in the memory device of a computer may represent exactly what they look like—numbers—but they may also have other meanings to the user or designer of digital systems. Alphabetic information is needed for communication more often than pure numbers. As with numbers, there is not just one "right" way to represent an alphabetic symbol. Some of these methods derive from early data storage and communications media that were around before there were computers. One is used by IBM, and practically no one else—if IBM weren't such an important sector of the computer and digital industry, there would be little reason to discuss this *alphameric* (*alphab*etic-nu*meric*) *code*. The BCDIC and EBCDIC (BCD Interchange Code and Expanded BCD Interchange Code) are used in various IBM machines and in IBM-compatible equipment made by other manufacturers. Their names include the BCD abbreviation, and they contain the one-nibble codes for the one-digit decimal numbers we called BCD codes. Unlike BCD nibbles stored in microprocessors, these BCD patterns include codes for letters and punctuation marks that really have nothing to do with the BCD code we talked about before. Both of these

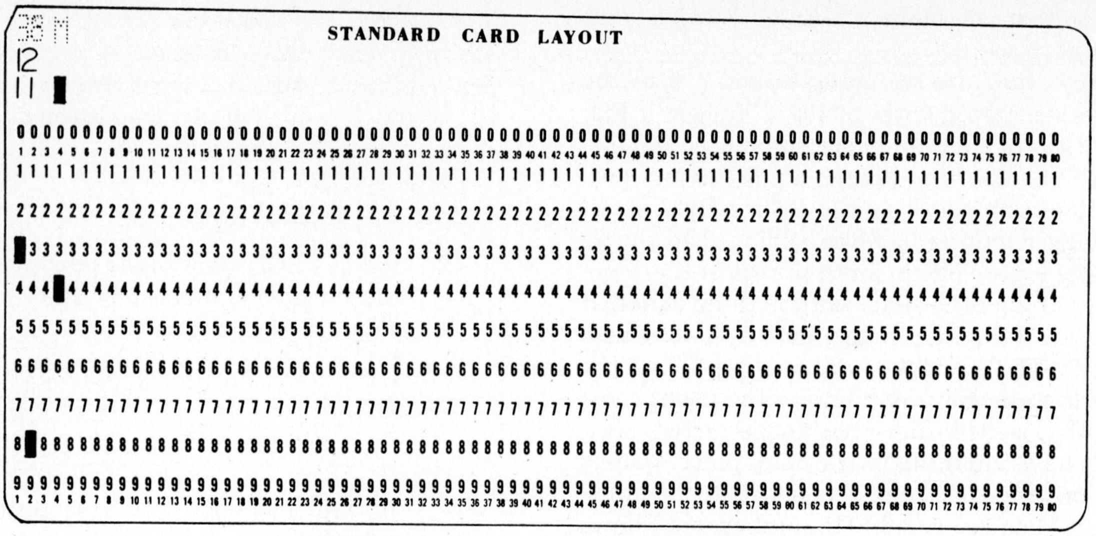

FIGURE 7-3 Hollerith punchcard example.

codes started with the punchcard, and the Hollerith code developed for it in the 1880s by Herman Hollerith. We normally call Hollerith punchcards "computer cards," because they have become associated with computing machines in the present day. Punchcards were around for a long time before computers, though, and were used with machines called *unit-record machines* decades before the first computer was built. They contain a code which works in the following way.

There are 80 columns on a punchcard, in which one or more holes may be punched. If one hole is punched in a column, it is usually a number. There are 12 places where a hole may be punched. Nine of these are called *numeric punches*, and are indicated on most punchcards by having printed numbers in them from 1 to 9. If a hole is punched in one of these places in a column, that's the code for the decimal number indicated. Figure 7-3 shows a "3" punched in its

first (leftmost) column, and an "8" in the next column over. The remaining three rows are called *zone punches*. They contain a "0" and two positions at the top of the card which are usually unmarked, called the "11" and "12" rows. By themselves, these positions represent the number zero and two punctuation marks. They are usually used in combination with one of the numeric punches to make letters of the alphabet. When two holes are punched in a column, one a zone and one a numeric punch, the code is an *alphabetic symbol*. Three punches represent a punctuation mark or special character, usually a zone punch, a numeric punch from 1 to 7, and an "8" punch. Four or more punches were originally illegal in Hollerith code, because they were thought to weaken the card too much for it to pass safely through the card reader. This code follows a simple pattern, shown in the chart in Figure 7-4.

Hollerith punchcode chart/BCDIC chart
numeric punches used in the symbol

	No #	Numeric 1	Numeric 2	Numeric 3	Numeric 4	Numeric 5	Numeric 6	Numeric 7	Numeric 8	Numeric 9	Zone bits	Binary code
Z o n e	No zones ƀ	1	2	3	4	5	6	7	8	9	00	$\overline{B} \cdot \overline{A}$
p u n c h e s	12 zone &	A	B	C	D	E	F	G	H	I	11	$B \cdot A$
	11 zone —	J	K	L	M	N	O	P	Q	R	10	$B \cdot \overline{A}$
u s e d	0 zone 0	/	S	T	U	V	W	X	Y	Z	01	$\overline{B} \cdot A$
	Numeric bits	0001	0010	0011	0100	0101	0110	0111	1000	1001		
	Binary code	$\overline{8} \cdot \overline{4} \cdot \overline{2} \cdot 1$	$\overline{8} \cdot \overline{4} \cdot 2 \cdot \overline{1}$	$\overline{8} \cdot \overline{4} \cdot 2 \cdot 1$	$\overline{8} \cdot 4 \cdot \overline{2} \cdot \overline{1}$	$\overline{8} \cdot 4 \cdot \overline{2} \cdot 1$	$\overline{8} \cdot 4 \cdot 2 \cdot \overline{1}$	$\overline{8} \cdot 4 \cdot 2 \cdot 1$	$8 \cdot \overline{4} \cdot \overline{2} \cdot \overline{1}$	$8 \cdot \overline{4} \cdot \overline{2} \cdot 1$		

In the symbol

FIGURE 7-4 Hollerith/BCDIC code chart.

The numeric punches alone are the codes for numbers and a few punctuation marks (&, -). The numeric + zone punches provide the first, second, and third nine letters of the alphabet. What's that, you say? There aren't that many letters in the alphabet? You're right, and Hollerith stuck the symbol "/" in the middle of his alphabet to round out the difference.

From this, we get the BCDIC and EBCDIC codes used in computers. The **BCD Interchange Code** (BCDIC) is a 6-bit code which uses 4-bit BCD code for the numeric punch (the 8-4-2-1 bits) and two bits (the B and A bits) for the zone punch. (A 6-bit code is sometimes called a *six-level code*.) The values of the numeric and zone bits for each character of the alphabet are found on the same chart as the Hollerith code. One example, to see how this chart is used: To find the Hollerith code for the letter M, search out the zone punch row and numeric punch column that cross where the M is. When you have them, you'll see that the numeric punch in an M is a "4" punch, and the zone punch is an "11" punch. To find the BCDIC code for M, find the zone bits and numeric bits listed at the opposite end of the row and column from the punches. The zone bits for the "11" zone are B = 1, A = 0; and the numeric bits for the "4" punch are 8 = 0, 4 = 1, 2 = 0, and 1 = 0. Put together, the BCDIC code looks like:

$$
\begin{array}{cccccc}
B & A & 8 & 4 & 2 & 1 \\
1 & 0 & 0 & 1 & 0 & 0
\end{array}
$$

for the letter M.

There is an **Expanded BCD Interchange**

Code (EBCDIC) which uses all 8 bits of a byte (an eight-level code), and permits uppercase and lowercase letters as well as numbers and a variety of punctuation symbols and control characters. Its structure is basically similar to the BCDIC, containing zone and numeric bits, but with a greater variety of zone punch combinations in the expanded Hollerith code, four zone bits are used to represent the combinations of two, three (and even *four* using numeric "9" as a zone punch!) zone punches, providing a possible 16 "shifts" for each numeric punch code. (Maybe the Hollerith cards got *stronger* between 1890 and the development of EBCDIC, permitting more punches?)

A chart showing the EBCDIC code and its structure (related to the Hollerith punches on a punchcard) is given in Figure 7-5.

The letter M, represented in EBCDIC code, is:

$$
\begin{array}{cccc|cccc}
\multicolumn{4}{c}{\text{Zone Bits}} & \multicolumn{4}{c}{\text{Numeric Bits}} \\
D & C & \bar{B} & \bar{A} & 8 & 4 & 2 & 1 \\
1 & 1 & 0 & 1 & 0 & 1 & 0 & 0
\end{array}
$$

Note that the B and A bits are just exactly inverted in the EBCDIC code from BCDIC's B and A bits. This is the reason why they've been indicated in inverted Boolean form in this example.

BCDIC and EBCDIC codes are an improvement on the 12-bit punch codes used on cards. The card codes never use all 12 punch positions at once (for reasons of mechanical strength) and are inefficient (with 12 positions, 4096 different codes are possible, but fewer than 100 are actually

Some characters of the expanded B.C.D. interchange code (EBCDIC)

No #	Numeric 1	Numeric 2	Numeric 3	Numeric 4	Numeric 5	Numeric 6	Numeric 7	Numeric 8	Numeric 9	Zone bits	Binary code
No zone ø	1	2	3	4	5	6	7	8	9	111	$D \cdot C \cdot B \cdot A$
12 zone &	A	B	C	D	E	F	G	H	I	1100	$D \cdot C \cdot \bar{B} \cdot \bar{A}$
11 zone −	J	K	L	M	N	O	P	Q	R	1101	$D \cdot C \cdot \bar{B} \cdot A$
0 zones 0	/	S	T	U	V	W	X	Y	Z	1110	$D \cdot C \cdot B \cdot \bar{A}$
12 · 0 zones	a	b	c	d	e	f	h	g	i	1000	$D \cdot \bar{C} \cdot \bar{B} \cdot \bar{A}$
12 · 11 zones	j	k	l	m	n	o	p	q	r	1001	$D \cdot \bar{C} \cdot \bar{B} \cdot A$
11 · 0 zones	?	s	t	u	v	w	x	y	z	1010	$D \cdot \bar{C} \cdot B \cdot \bar{A}$
Numeric bits	0001	0010	0011	0100	0101	0110	0111	1000	1001		
Binary code	$\bar{8} \cdot \bar{4} \cdot \bar{2} \cdot 1$	$\bar{8} \cdot \bar{4} \cdot 2 \cdot \bar{1}$	$\bar{8} \cdot \bar{4} \cdot 2 \cdot 1$	$\bar{8} \cdot 4 \cdot \bar{2} \cdot \bar{1}$	$\bar{8} \cdot 4 \cdot \bar{2} \cdot 1$	$\bar{8} \cdot 4 \cdot 2 \cdot \bar{1}$	$\bar{8} \cdot 4 \cdot 2 \cdot 1$	$8 \cdot \bar{4} \cdot \bar{2} \cdot \bar{1}$	$8 \cdot \bar{4} \cdot \bar{2} \cdot 1$		

FIGURE 7-5 Hollerith/EBCDIC code chart.

used). In BCDIC, the four numeric and two zone bits are used to "compress" all the information available in one, two, or three punches of 12 possible punches. This reduces the number of circuits needed to store bits in the computer by half, if BCDIC is used instead of storing Hollerith directly. The EBCDIC code uses four numeric and four zone bits to compress the information available in one, two, three, or four punches. Again, reduction of the 12 bits of Hollerith (never all used) to 8 bits of EBCDIC is a considerable savings in circuit components (only two-thirds as many are needed for EBCDIC as for direct storage of Hollerith). Of course, this also makes computers that use these codes to store Hollerith information cheaper than they'd be otherwise, a fairly good reason for their use.

7-4 ASCII CODE

The **American Standard Code for Information Interchange** (ASCII) is a binary code for alphameric symbols that was defined by an agreement between most manufacturers of digital equipment in the United States. Practically everyone uses it for all data communication, with the exception of IBM (a rather outstanding exception!). It is not developed from Hollerith punch-card codes as BCDIC and EBCDIC are, but has some structural similarity to those codes. ASCII is a seven-level code (each symbol is represented by 7 bits). As with BCDIC and EBCDIC, the front bits of the ASCII character define different shifts or types of characters. There are four types of ASCII codes defined by the front 2 bits of the code: control characters, numerals/special characters, uppercase alphabetic characters, and lowercase alphabetic characters. Each of these four types contains 32 symbols defined by the back 5 bits of the character. A chart of the ASCII code is shown in Figure 7-6. From this chart we can find the code for G. The front 2 bits are identified at the left end of the row G is in, and the back 5 bits at the top of the column G is in. For G, the front bits are 10 and the back bits are 00111, making the code:

$$G = 10 \; 00111 \qquad \text{uppercase type} \atop \text{seventh letter}$$

whereas a lowercase letter g would have the code

$$g = 11 \; 00111 \qquad \text{lowercase type} \atop \text{seventh letter}$$

and the number 7 would have the code

$$7 \times 01 \; 10111 \qquad \text{numeric type} \atop \text{seventh number}$$

Of course, if a parity bit is added to the ASCII code (usually in front of the rest of the bits), it becomes an eight-level code.

7-5 OTHER ALPHAMERIC CODES

7-5.1 Morse Code

Historically, the first code used commercially for data communications by ON and OFF digital signals, **Morse code** was (and is) used for telegraph messages. Devised by Samuel F. B. Morse, inventor of the first commercially practical telegraph system, it is really a trinary system, with signals sent using short ON, long ON, and OFF levels. Although still used in some places (wireless telegraphy, etc.), Morse code has no significant importance in the digital data communications field.

7-5.2 Baudot or Murray Code

Baudot was an inventor whose pioneering efforts led to the printing telegraph, forerunner of to-

"back" 5 bits	0 0 0 0 0	0 0 0 0 1	0 0 0 1 0	0 0 0 1 1	0 0 1 0 0	0 0 1 0 1	0 0 1 1 0	0 0 1 1 1	0 1 0 0 0	0 1 0 0 1	0 1 0 1 0	0 1 0 1 1	0 1 1 0 0	0 1 1 0 1	0 1 1 1 0	0 1 1 1 1	1 0 0 0 0	1 0 0 0 1	1 0 0 1 0	1 0 0 1 1	1 0 1 0 0	1 0 1 0 1	1 0 1 1 0	1 0 1 1 1	1 1 0 0 0	1 1 0 0 1	1 1 0 1 0	1 1 0 1 1	1 1 1 0 0	1 1 1 0 1	1 1 1 1 0	1 1 1 1 1
"front" 2 bits																																
00	Control codes for typewriter control which don't print anything																															
01	␢	!	"	#	$	%	&	'	(	)	*	+	,	-	.	/	0	1	2	3	4	5	6	7	8	9	:	;	<	=	>	?
10	@	A	B	C	D	E	F	G	H	I	J	K	L	M	N	O	P	Q	R	S	T	U	V	W	X	Y	Z	[	/	]	^	←
11	'	a	b	c	d	e	f	g	h	i	j	k	l	m	n	o	p	q	r	s	t	u	v	w	x	y	z	;	|	:	~	DEL

American Standard Code for Information Interchange (ASCII)

FIGURE 7-6 ASCII code chart.

day's data terminals and Teletype machines. The **Baudot code** still used by older Teletype machines was not devised by Baudot, but by Murray. It's a five-level code which manages to represent the letters and the numbers by using a "figures shift" (numeric) and "letters shift" (alphabetic) symbol before each field of characters. Since numbers and letters are usually bunched in groups, this scheme works fairly well, using shift codes only once in a while. Sixty characters are possible in Murray (Baudot) code, but transmission becomes severely slowed down if data are transmitted containing both numbers and letters on each line, because many shift characters must also be sent.

Modern Teletype machines do not use Baudot code because of its inefficiency and limited number of characters—they use the ASCII code, instead. The only reason for knowing about Baudot code is the off chance that you might bump into one of the ancient "newsroom" Teletypes still chugging along at some remote location—they're noisy, but virtually indestructible. (See Chapter 16 for further details.)

7-6 NUMERIC CODES: TWOS' COMPLEMENT (SIGNED NUMBERS)

So far, we've looked at codes where a number represents a number, and where a number represents a letter. The numbers we looked at were all positive integers—whole numbers without a + or − sign in front of them—but a digital computer must have ways of handling negative numbers as well as positive ones. In Chapter 6 we discussed subtraction, and how a negative number coming from a subtracter would look when it found out there were no more places from which to borrow a 1. We found that the answer (a negative number) came out in a different code (twos'-complement code) than a positive number would.

All microprocessor chips produce twos'-complement code when they arrive at a negative answer. So do minicomputers, with ALU (arithmetic logic unit) chips made of TTL or ECL logic. The really big mainframes have hardware that can produce anything they darn well please—cost is no object—but even the biggest digital systems usually proceed to subtract numbers by one of the following three methods:

1. Add one number to the *ones' complement** of the other number. Take the bit at the carry of the answer, and carry it

around to the units' position, then add it on there. This is called the **end-around-carry method**, and it works. It's also easier to build with hardware than it is to explain in words. The answer is normal binary if it's positive and in ones'-complement code if it's negative.

2. Add one number to the twos' complement† of the other number. The answer is normal binary if it's positive, and in twos'-complement code if it's negative.

3. Use a subtracter or adder/subtracter circuit. The answer is normal binary if it's positive and in twos'-complement code if it's negative.

When numbers are subtracted by methods 1 or 2, the same adder used to add is also used to subtract. This isn't much of a savings, though, since circuits must be added to convert one of the two numbers into its ones'- or twos'-complement form whenever subtraction is desired. To do this, a *controlled inverter* (XOR) circuit must be added for each bit of the number. Method 3 has the controlled inverter circuits already built into the adder, if we use an adder/subtracter. It's really the same as method 2, when all the pieces of the circuit fall into place.

The ones' complement is one way of representing negative numbers, and the twos' complement is another different—but equally good—method. In methods 1 and 2, subtraction proceeds by adding a negative number. That's the same as subtracting a positive number, provided that we accept the fact that twos' complement and ones' complement numbers are negative.

Although we described how to convert a binary number to its twos' complement in Chapter 6, we'll repeat the steps here, together with a description of the ones' complement:

To convert a binary number to its ones' complement:

1. Invert all the bits of the number.

To convert a binary number to its twos' complement:

2. Add 1 to its ones' complement.

And vice versa, to convert a twos' complement to ones' complement:

1. Subtract 1 from the twos' complement.

---

*Ones' complement = invert all the bits.

†Twos' complement = invert all bits and add 1.

To convert a ones' complement to binary:

2. Invert all the bits of the number.

That was a little different from our description in Chapter 6, wasn't it?

Let's suppose that you're the designer of a digital system. Once you've settled on the method you'll use for showing negative numbers—let's use the twos' complement—you still need a way to tell whether a number is a large positive one or a small negative one. Here's what we mean: Suppose that you have the binary code 11111000, and you know it's not a letter in ASCII or EBCDIC code—it's a number. You also know that the digital system you've designed uses twos' complement numbers for negative values. Now, you have a problem: Is 11111000 a positive 248 or a negative 8? If you just add up the place values of the bits, assuming the number to be an unsigned binary integer, the number is 248 just as it appears. If you treat it as the twos' complement of something (you assume it's negative), you subtract 1 from the number—getting 11110111—and invert all the bits—getting 00001000—which tells you this number is an 8, in twos'-complement code, of course. Now, which is it?

There's no way to tell, just going on what you've got in the example. To know for sure, you need more information, which we're about to give you.

People who design computers that use signed binary numbers in the twos'-complement system use the front bit to tell if the number is positive or negative. As long as the front bit is 0, the number's positive. If the front bit is 1, the number is negative, and it's in twos'-complement form. The front bit of the number becomes its *sign bit*, and the rest of the bits give a signed value from 0 to 127 (positive) and −1 to −128 (negative).

Of course this limits the number of positive numbers you can write, because only numbers from 00000000 to 01111111 are positive. Anything from 10000000 to 11111111 is negative and in twos'-complement code, so there are only 128 positive numbers and 128 negative numbers possible in this system. Unsigned numbers existed in 256 combinations, but they were all positive. We had to give something up in order to gain something. To get negative numbers as well as positive ones in one byte, we had to cut down on the number of positive codes possible. We chop them up half and half, giving half the values negative signs and the other half positive signs. The same idea is used in 16-bit machines—for number

from −32,768 to +32,767—and in 32-bit machines—for numbers from −2,147,483,648 to +2,147,483,647.

7-7 MANTISSA AND EXPONENT (FLOATING-POINT NUMBERS)

With the twos'-complement system, we have signed numbers, but they're still all integers. How does a computer handle numbers like $2\frac{1}{3}$ or 22.75 or 6.71×10^{-3}?

7-7.1 Binary Fractions

To write a *binary number* with a *fraction*, it's necessary to know what's on the other side of the *binary point*. Up to now, we've stayed on the left side of the binary point (a decimal point in a binary number) and used whole numbers. The place values on the right-hand side of the point are organized like this:

$$
\begin{array}{ccccccccc}
16 & 8 & 4 & 2 & 1 & & \frac{1}{2} & \frac{1}{4} & \frac{1}{8} \cdots \\
1 & 0 & 1 & 1 & 0 & . & 1 & 1 & 0 \\
& & & & & & \text{point} & &
\end{array}
$$

for the decimal number 22.75. Notice that as you go left each place value gets twice as big, and as you go right, each place value gets half as big. The only new thing we did is go past the "ones" place, finding the "halves," "quarters," and "eighths" places, and so on.

Any number can be converted from decimal to binary, even fractions—but a number like $2\frac{1}{3}$ takes an infinite number of decimal places to write. It also takes an infinite number of binary places. Writing this fraction (in either decimal or binary), we have to decide how many places we'll go before we stop. We could decide to have 8 bits of fraction and 8 bits of integer. Each number would take two bytes of storage space in a computer memory. There would be no need to store the "." point if we knew which byte was the integer part and which was the fraction part—an important advantage.

In real computer applications a one-byte integer can't be any larger than 255 (127, if it's signed). That limits the things we can do with the computer pretty severely. If our number is less than 0.004, decimal, it can't fit in the fraction byte either. Numbers are usually stored in computers using more than two bytes. Four bytes give a signed number whose decimal value has nine decimal places. This is about the same as a good pocket calculator.

Instead of using half the bytes for the integer part of a number and the other half for the fraction part, most schemes store a number in **binary scientific notation**. Rather than trying to explain every possible scheme used (there are many) we'll just look at one scheme: Four bytes contain a number called the **mantissa**. It has the signed value of the first 31 significant binary places of the number (nine decimal places), but doesn't say where to put the binary point. An additional byte contains a signed number called the **exponent**. It says how many places to the left (if it's negative) or right (if it's positive) the binary point should be moved from the front of the number. Let's look at an example:

Mantissa

11001111 00000000 00000000 00000000

Exponent

00000111

First, we look at the front bit of the mantissa. This tells us that the number is negative. Then we convert from twos' complement to binary to see what it is:

Step 1. Subtract 1.

$$
\begin{array}{r}
11001111 \quad 00000000 \quad 00000000 \quad 00000000 \\
-1 \\
\hline
11001110 \quad 11111111 \quad 11111111 \quad 11111111
\end{array}
$$

Step 2. Invert all bits.

00110001 00000000 00000000 00000000

Now, we remember that the binary point is at the front of the number, and that it's negative:

−.00110001 00000000 00000000 00000000

Right now, the number is "one-eighth" plus "one-sixteenth" plus "one-256th" (decimal value = −0.1914), but we remember that the exponent says that the binary point must be moved (unless it's 00000000). The exponent has a positive value (whew!) and is a seven in binary code. That means we should move the binary point seven places to the right.

−0011000.1 00000000 00000000 00000000

This number has an integer part (binary 11000 = decimal 24) and a fraction part (binary .1 = decimal .5). Put together, the number is negative 24.5.

We'd like to point out three things before you worry too much about this. First, there are hundreds of variations on these codes, so don't bother to memorize this one—it's probably not the one you'll end up with, no matter what machine you work on. Second, this kind of number, called a **floating-point variable**, is about as complicated a beast as you'll find hiding inside a computer. Everything else is simpler than this. Third, if you are planning to become a technician who repairs the computer, it's not too likely that you'll have to know how floating-point numbers look inside the computer, until you have to troubleshoot everything in a memory bit by bit. By that time, you'll certainly have been trained enough on the system you're working with to know how its floating-point numbers are stored.

EXAMPLE 7-3

Two numbers are stored in a computer as

(a) 01111111 11111111 11111111

  11111111 00000111

and

(b) 11111111 11111111 11111111

  11111111 00000111

If they are floating-point numbers with a signed, four-byte mantissa and a signed, one-byte binary exponent, which number is larger?

Solution: This problem is easier than it appears. The key concept here is that both the mantissa and exponent are *signed* numbers. Examining the numbers, we see that they are both identical everywhere *except the first digit of the mantissa.* In number a, this digit is a 0, and the number is *positive.* In number b, this digit is a 1, and the number is *negative.* Since negative numbers are less than zero, and positive numbers are greater than zero, number a, the positive one, is the larger of the two numbers.

7-8 OCTAL BYTE CODE

It is common to represent binary numbers in 8-bit groups called *bytes*. This is done even for 16- and 32-bit computers. To write the **octal byte code** for a binary number:

1. Break the number up into groups of 8 bits.

2. Separate each 8-bit group, with commas, as you do with large decimal numbers.

3. "Digest" the groups between commas into octal digits. There should be three octal digits for each 8-bit group.

For example, the decimal number 32,767 will look different in octal byte code than it looks in simple octal code. If the number 32,767 is written in binary and converted directly to simple octal, it looks like this:

$$32{,}767 \text{ (decimal)} = 111111111111111$$

$$111111111111111 = 111{,}111{,}111{,}111{,}111.$$

$$32{,}767 \text{ (decimal)} = 7 \quad 7 \quad 7 \quad 7 \quad 7$$

But the number 77777 (octal) doesn't show how each byte of the number 32,767 (decimal) looks inside the machine. The number 32,767 is too big to fit inside one byte, and we can't see from the 77777 which 7s go into the front byte and which go into the back byte.

Converting the number 32,767 (decimal) into octal bytes proceeds like this:

$$32{,}767 \text{ (decimal)} = 1111111 \quad 11111111$$

$$1111111 \quad 11111111 = 1{,}111{,}111. \quad 11{,}111{,}111.$$

$$32{,}767 \text{ (decimal)} = 1 \ 7 \quad 7 \quad 3 \ 7 \quad 7$$

which shows that in octal byte code, everything isn't 7s, and the front byte contains 177 (octal) while the back byte contains 377 (octal). A two-byte octal number like this may be written 177,377, with a comma to show the byte boundary between each 8-bit group.

7-9 HEXADECIMAL BYTE CODE

For reasons described above, numbers are also represented in hex according to bytes. To write the hexadecimal byte code for a binary number:

1. Break the number into groups of 8 bits.
2. Divide each byte produced by step 1 into groups of 4 bits with commas.
3. "Digest" the numbers between commas into hex digits. There should be two digits for each byte.

We'll do this for the number 32,767 again.

$$32{,}767 \text{ (decimal)} = 111111111111111$$

$$111111111111111 = 1111111 \quad 11111111$$

$$1111111 \quad 11111111 = 111{,}1111. \quad 1111{,}1111.$$

$$32{,}767 \text{ (decimal)} = 7 \quad F \quad F \quad F$$

We find that 32,767 (decimal) is 7F,FF (hex). Unlike the octal byte code, numbers in hex byte code look the same as hex numbers in general. 32,767 is 7FFF in hex even when it's not broken into bytes. Making hex into hex byte code is as simple as putting a comma every two digits. Maybe that's why hex is more popular than octal for writing listings of binary computer programs (the fact that it costs less to print hex is no handicap, either!).

QUESTIONS

7-1. Convert to binary from decimal.
(a) 87 (b) 142 (c) 1987 (d) 63 (e) 212

7-2. Name several major reasons for using hex and octal code to represent binary numbers.

7-3. Convert from binary to octal.
(a) 1001011   (b) 1101100
(c) 1101110   (d) 00110110
(e) 11111111

7-4. Convert from octal to binary.
(a) 346 (b) 067 (c) 303 (d) 257 (e) 170

7-5. Convert the binary numbers in Question 7-4 to hexadecimal.

7-6. Convert from hex to binary.
(a) A1   (b) BE4
(c) 2CD   (d) FEED
(e) ACE

7-7. What decimal number does the BCD code

0001,1001,1000,0111

stand for?

7-8. Add the BCD numbers. (Show all steps.)
(a)   1001 0100   (b)   1000 1000
    +0011 0101   +0101 0010
(c)   0100 1001
    +0111 1000

7-9. Use the Hollerith code chart (Figure 7-4) to read the punch codes on the document below.

7-13. Convert the following numbers from signed twos' complement to decimal.
(a) 11111110   (b) 11110100
(c) 10000000   (d) 01111110

7-10. Why does it make more sense to store the information on punchcards in a computer's memory with BCDIC or EBCDIC rather than directly in Hollerith code?

7-11. Which of the following alphameric binary codes are developed from the structure of Hollerith code?
(a) EBCDIC   (b) ASCII
(c) BCDIC   (d) Baudot

7-12. Does the ASCII code contain zone bits and numeric bits?

7-14. What is the decimal value of the floating-point number below? It has a two-byte signed true binary mantissa and a one-byte signed twos'-complement exponent. The most significant bit of the mantissa is assumed to be a 1 in the units' position, but it's used as a sign bit to indicate the sign of the number.

10000000 00000000   00000011

7-15 Find the octal byte code for the decimal number 5000.

DRILL PROBLEMS

Convert from binary to octal.
7-1. 011000011
7-2. 110110110
7-3. 01110110
7-4. 11010011

Convert from octal to binary.
7-5. 303
7-6. 46
7-7. 377
7-8. 166

Convert from binary to hexadecimal.
7-9. 10101011
7-10. 01001111
7-11. 00111101
7-12. 00010010

Convert from hexadecimal to binary.
7-13. A1
7-14. 3D
7-15. CB
7-16. 16

Convert from decimal to BCD.
7-17. 25
7-18. 76
7-19. 1993
7-20. 2020

Convert from BCD to decimal.
7-21. 0001,1001,1001,0011
7-22. 0010,0101
7-23. 10011001
7-24. 00110010,00010000

Identify the following characters in Hollerith code.

7-25. A 12-5 punch combination

7-26. A 6 punch, all by itself

Convert the following to Hollerith code.

7-27. The letter R

7-28. The letter Z

Convert the following characters into BCDIC and EBCDIC code.

7-29. 9

7-30. Z

7-31. R

7-32. I

Convert the following characters into ASCII code.

7-33. P

7-34. A

7-35. 0

7-36. Ⱦ (a blank space, typed by pressing the spacebar on a typewriter keyboard)

Convert the following signed, 8-bit binary numbers into their decimal values. Negative numbers should be written with a minus sign in front of them, as in $11111000 = -7$.

7-37. 00100101

7-38. 10100101

7-39. 10000111

7-40. 11111000

Convert the following decimal numbers to binary.

7-41. 2.5

7-42. 193.75

7-43. 0.375

7-44. 180.015625

Convert the following binary numbers to decimal (roundoff to the nearest one-hundredth).

7-45. 00011010.01000011

7-46. 01000000.10100011

7-47. 00001100.00111111

7-48. 01100011.11111111

Convert the following from octal byte code to binary and decimal.

7-49. 377,377

7-50. 003,350

8

D/A AND A/D AS PARALLEL (ASYNCHRONOUS) CIRCUITS

CHAPTER OBJECTIVES

By the time you finish this chapter, you will be able to:

1. Define the terms *analog* and *digital*.

2. Describe the processes of *digital-to-analog* (D/A) and *analog-to-digital* (A/D) conversion.

3. Describe how a *code wheel* or *shaft encoder* converts rotational position into digital code.

4. Define what is meant by the term *resolution*.

5. Explain why *Gray code* is better than binary code for *disk* and *shaft encoders*.

6. Identify the circuit for *Gray-to-binary* conversion, and vice versa.

7. Describe or design the two basic types of *resistive ladder D/A converter*.

8. Describe how a *flash converter* performs A/D conversion.

9. Understand the operation of an analog-input, digital-output "logic gate" called an *analog comparator*.

10. Describe how "flash code" output from a flash converter is converted into true binary code.

Even if the rotation is very fast, this "momentary" mistake—called a *glitch*—is thousands of times slower than the response time of a logic gate. Another glitch occurs if the B detector picks up the 2-bit before the A detector loses the 1-bit. In this case a 1 and a 2—a three—appears between the numbers 1 and 2 as the wheel rotates. There are glitches between the numbers 1 and 2, also between 3 and 4, between 5 and 6, and between 7 and 0. Because these glitches occur as the detectors cross over the boundaries between sectors, they are sometimes called *crossover slivers*.

To eliminate the crossover sliver problem, a code that has 1s and 0s, but only changes one bit at each crossing, was devised by Baudot and Gray for early printing telegraphs. The **Gray code** (Baudot's name is associated with a different code) is shown in Figure 8-3. The code wheel and shaft encoder for the Gray code are included with this diagram for comparison with Figure 8-2.

Inspect the table of Gray code in the figure. It's immediately apparent that this isn't standard binary code. Remember the technique for counting in binary that we showed you earlier? The same approach can be used to count in Gray code, for as many bits as you like. Reading the A column of the Gray table, you can see the pattern 0110 repeated. In the B column, each 0 and each 1 of the A column is repeated two times. In the C column, each 0 and each 1 of the 0110 pattern are repeated four times, except that the pattern doesn't have enough lines to get finished. If you can see the similarity between the way we

counted a truth table in binary, and the way we count in Gray code, there should be little trouble in generating Gray code for numbers as large as you want.

EXAMPLE 8-1

What is the resolution of an 8-bit shaft encoder? (Assume that all possible 8-bit patterns are present.)

Solution: The number of possible 8-bit patterns (in either binary or Gray code) is

$$2^8 = 256$$

The resolution may be measured in degrees:

$$\frac{360°}{256} = 1.40625°$$

Or in radians:

$$\frac{2\pi}{256} = 0.0245436926 \text{ radian}$$

Or in parts of a whole circle:

$$1 \text{ part in } 256$$

Or in percent of a whole circle:

$$\frac{100\%}{256} = 0.390625\%$$

8-3 GRAY-TO-BINARY DECODER CIRCUIT (USING XOR)

It's a bit inconvenient to have position-sensor circuits sending information to the inputs of a digital system in Gray code. The reason it's inconvenient is that standard digital circuits don't count, compute, or communicate in Gray code. Since standard digital devices use binary code, it's nice to know that the circuit for converting Gray to standard binary is simple. Figure 8-4 shows a circuit for this conversion, using XOR gates. To convert Gray to binary, the leftmost bit of Gray needs no conversion (compare the C bit of the Gray code wheel with the C bit of the binary wheel). The rest of the bits are converted by XORing each Gray bit with the binary bit to its left. Since the bit at the extreme left is Gray code or binary, it's one of the inputs to an XOR. Each of the other XORs combine a Gray bit with the output of the XOR to its left. Decoders for 3, 4, and 8 bits are shown. It should be easy to see from these diagrams how a Gray decoder could be designed for any number of bits.

C	B	A
0	0	0
0	0	1
0	1	1
0	1	0
1	1	0
1	1	1
1	0	1
1	0	0

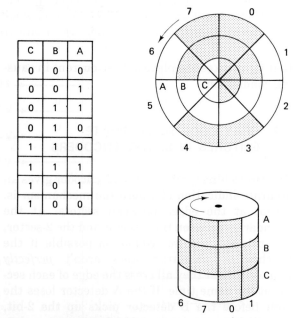

FIGURE 8-3 Gray-code code wheel.

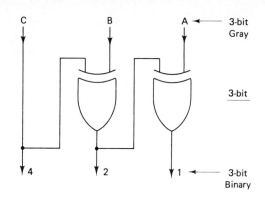

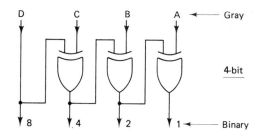

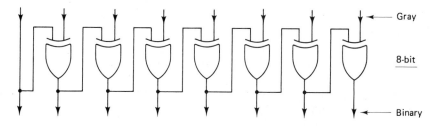

FIGURE 8-4 Gray-to-binary converters.

8-4 LADDER-TYPE D/A CONVERTERS

In the section just completed, we saw a way to convert the position of an object into a binary-coded signal. The rotational position of the wheel or shaft is an analog quantity. The wheel can occupy an infinite number of positions between "unrotated" and "fully rotated." Our readout—the binary number from the detectors and decoder—has only a limited number of values. We've lost information in the conversion from analog to digital. Suppose that we seek to reverse the process, converting binary code from digital into an analog quantity (for instance, voltage) and apply the analog signal to an analog display device (for instance, a voltmeter). If we do things right, we should be able to turn a code wheel to produce a binary number, then put the binary number through a D/A (digital-to-analog) converter that makes it into a voltage, then attach the voltage to a voltmeter, and the needle of the voltmeter should move exactly like the code wheel. As we turn the code wheel to the left, the

meter needle moves left, and as we turn the code wheel to the right, the meter needle moves right.

This isn't exactly what happens. The meter needle moves to the right and left in "jumps." It only stays at rest in certain positions, and spends the rest of its time jumping from one of these positions to the next. It doesn't move smoothly from one position to another. This is normal with D/A conversion systems. The binary codes jump from one number to another as the code wheel is rotated, and the inbetween positions don't show up in the code. The output of any D/A circuit will be "stepwise" instead of smooth. If there are a lot of numbers, very close together, the illusion of smooth, continuous motion can be created.

8-4.1 Summing Amplifier (Op Amp)

Before we can make a D/A converter, we need to know about a few of the "building blocks" that are used to make one. In Figure 8-5, we have a circuit called an **operational amplifier**—an *op amp* for short—which is represented by a trian-

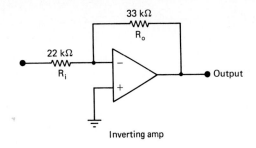

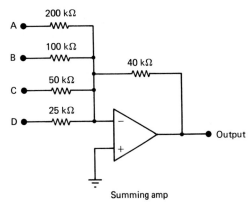

FIGURE 8-5 Op-amp D/A converter.

gle. In the circuit, the "+" and "−" mark the inputs of the op amp, and its output is at the point of the triangle. Like all amplifiers, this one's job is to take in a signal with a small amount of electrical energy and put out a signal with a large amount of electrical energy. We won't bother explaining how this one does it, but we will tell you that the extra energy comes from a *power supply* attached to the $V+$ and $V-$ terminals. You never get something for nothing, even in electronics.

The op amp in this circuit has a lot more amplification than we're using. The two resistors, R_i and R_o, are used to cut down the amplification to a small figure. The figure you get if you divide R_o/R_i (called the *amplification factor*) tells how many times the voltage *in* is multiplied to get the voltage *out*. It follows the rule:

$$\text{voltage out} = \text{voltage in} \times \frac{R_o}{R_i}$$

If you choose the right resistors, you can amplify the voltage in by any number you like, as long as the voltage you want out isn't any larger than the power-supply voltage (not more positive than $V+$ or more negative than $V-$).

What does all this have to do with D/A? At the bottom of Figure 8-5 is a slightly fancier op amp called a **summing amplifier**. It has four R_i resistors instead of just one, so there are four places where a signal may be put into the circuit

and amplified. The circuit at the top of Figure 8-5 has an R_o resistor of 33 kΩ and an R_i resistor of 22 kΩ. To see how much the input voltage will be amplified before it comes out, we take $R_o/R_i = 33/22$ and get 1.5. The voltage out will be 1.5 times as large as the voltage in. It will also be negative if the input is positive, which is indicated by the − sign where the input resistor is attached. If the voltage in is 5 V, the voltage out will be $-(5 \times 1.5)$, which is -7.5 V.

For the summing amplifier, each input is amplified by its own input resistor's R_o/R_i factor. The sum of all the voltages you would get from each resistor alone is added at the output of the amplifier. That's why it's called a summing amplifier. This is useful to us, because the R_i resistors in this circuit have been chosen very carefully. If a 5-V HIGH logic signal comes into the amp at the input called A, it will be amplified by $R_o/R_i = 40/200 = 0.2$ and the output voltage will be $-(5 \times 0.2)$, which is -1 V. The B input will be amplified by $R_o/R_i = 40/100 = 0.4$ and a 5-V input will produce -2 V out. The C input has an amplification factor of -0.8, and produces -4 V out for a 5-V input. Finally, the D input is amplified by -1.6, and produces -8 V out when the input is 5 V.

Put together, the output is -8, -4, -2 and -1 V when the input is 8, 4, 2 or 1 in binary code (provided that a logic 1 is 5 V). If several of the 8-4-2-1 bits at the inputs are HIGH, the output will be the sum of the -8s, -4s, -2s, and -1s that have HIGH inputs. If 1001 is applied to the inputs (a 9 in binary) the output will be -9 V. This summing amplifier is a digital-to-analog (D/A) converter because its inputs have been weighted to produce the right outputs for 8-4-2-1 code.

EXAMPLE 8-2

What is the voltage output of the op-amp circuit D/A converter shown below if the input is: HIGH = 10 V, LOW = 0, and the input code is DCBA = 1001?

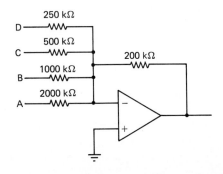

Solution: Signals arriving at D, C, B, and A are amplified by a voltage gain equal to the output resistor divided by the input resistor at D, C, B, or A, according to which signal is being amplified. Voltage gains calculated by this method are

$$Av(D) = \frac{200}{250} = 0.8$$

$$Av(C) = \frac{200}{500} = 0.4$$

$$Av(B) = \frac{200}{1000} = 0.2$$

$$Av(A) = \frac{200}{2000} = 0.1$$

Since the signals are all 10 V, or nothing at all, and the circuit is a summing amplifier, the amplified signals are

$$0.8(10 \text{ V}) = 8 \text{ V}$$

$$0.1(10 \text{ V}) = 1 \text{ V}$$

and the amplified output is

$$8 \text{ V} + 1 \text{ V} = 9 \text{ V}$$

8-4.2 R/8, R/4, R/2, R Ladder Network

The resistor networks used for the D, C, B, and A inputs of the summing amp are a fine D/A converter all by themselves, if you don't mind small voltages. Figure 8-6 shows the same four resistors used as a D/A converter with outputs in millivolts (thousandths of a volt). Without the op amp, this resistor network is called a **ladder**. Each resistor, from the top down, is one-half as big as its upstairs neighbor. If we call the top resistor R, the resistor attached to B has a value of

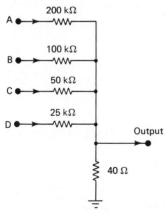

FIGURE 8-6 8-4-2-1 D/A ladder without op amp.

R/2, the resistor at C is R/4, and the resistor at D is R/8. For a D/A with more inputs, we would add more resistors and just keep going with the sequence R/16, R/32, and so on.

Let's look at how this works without an amplifier. The current each resistor will pass to ground is E/R = 5 V/R according to Ohm's law. As long as any HIGH input is 5 V, the current through the four resistors will be

$$\text{A current} = 25 \ \mu\text{A}$$

$$\text{B current} = 50 \ \mu\text{A}$$

$$\text{C current} = 100 \ \mu\text{A}$$

$$\text{D current} = 200 \ \mu\text{A}$$

and if the input is LOW, there's no voltage and no current. The 40-Ω resistor in the ground path is so small by comparison to R, R/2, R/4 and R/8 that it can be ignored. Any voltage drop across the 40-Ω resistor is so small compared to 5 V that the current isn't affected, even when all four currents flow through the 40-Ω resistor. The result is that the total current flowing through the 40-Ω resistor is the sum of the A, B, C, and D currents listed above, for whatever inputs are HIGH.

The voltage drop across the 40-Ω resistor is small, but not unimportant. If a binary 1 input is applied to the ladder, a current of 25 μA flows through the 40-Ω resistor, producing a voltage drop of 1 mV according to Ohm's law (current × ohms = volts). Each input produces 1, 2, 4 or 8 mV, according to which one it is. A group of inputs together, for instance ABCD = 1001, produces a sum voltage, in this case, 1 + 8 = 9 mV.

Either the ladder network or the summing amp can be used to produce a voltage output that matches its code input (a millivoltage in the ladder's case). The only reason for adding the op amp to the circuit is to get larger voltage and more power.

8-4.3 R, 2R Network

Figure 8-7 shows another type of ladder network. Its advantage is that there are only two sizes of resistors needed to make it. Apart from the fact that it does things a different way, it accomplishes the same task as the ladder described in the preceding section, and can be used as the input of an op amp circuit just as the other ladder was.

R, 2R networks in several sizes are shown in Figure 8-7, so that you can see the pattern re-

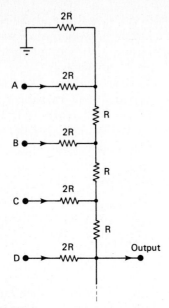

FIGURE 8-7 2-1-2-1 D/A ladder network.

quired to construct a ladder of any size using the R, 2R method.

Suppose that we use one of these ladder networks to convert the binary sector numbers from a rotating code wheel into voltages. If we use the code wheel in Figure 8-3 and the Gray-to-binary decoder in Figure 8-4, we'll get reliable, glitch-free binary code to feed our D/A converter. Then, we attach 1, 2, and 4 to the A, B, and C inputs of the D/A converter in Figure 8-5, and attach the output to a voltmeter. What you get is Figure 8-8. If you turn the knob on which the code wheel is mounted, the voltmeter needle will turn in step with the knob. When you rotate the knob to the right, the voltmeter needle will jump to the right in 1-V steps. If you turn the knob back, the voltmeter needle will jump back in 1-V steps.

If you try to build one of these, you'll find that there are problems. You can't get 50- and 25-kΩ resistors. The nearest things generally available are 51- and 24-kΩ resistors, and these are only within 5% of the value they're marked. What

do you do when the real parts aren't exact? You get uneven steps. The steps aren't all 1 V. If your resistors are off badly enough, the "8" voltage might be less than the "7" voltage, which is the sum of three voltages, the "4" voltage, the "2" voltage, and the "1" voltage, *all* of which have errors if the resistors are off. If, for example, the "8" resistor is 10% too high, and the "4," "2," and "1" resistors are all 10% too low, the "8" voltage will be −7.2 V, and "7" voltage will be −7.7 V. In this case, it's impossible to tell what's a "7" and what's an "8." You'd be more likely to guess wrong than right. How much "slop" can we tolerate in the ladder's resistors?

If there are eight steps, our D/A circuit has steps that go up or down 1 part in 8 (12.5%). If some of the resistors are 6.25% high and others are 6.25% low, we could end up with the "7 or 8?" problem just described. We must have resistors less then 6.25% off if we always want a higher code to produce a higher voltage on the "staircase."

If our resistors are 5% off, some high and some low, there is a good chance that the steps on our voltmeter will keep going up as we turn the knob to the right, and going down as we turn the knob to the left. They may not all be even steps, but the accuracy won't be too bad. If we use 10% resistors, all bets are off. The error each resistor causes in any voltage must be less than half the smallest step. We say that the **tolerance** of the resistors must be less than half the resolution of the D/A converter. If the steps are uneven, we say that the converter has poor **linearity**.

8-5 FLASH CONVERSION A/D LADDER

At the beginning of this chapter, we saw how analog motion and position is converted into digital binary code with code wheels and shaft encoders. Then we saw how digital binary code is converted

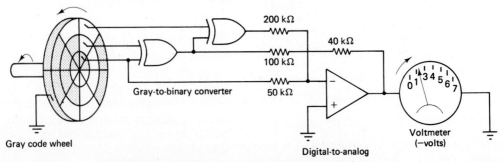

FIGURE 8-8 Analog-to-digital, digital-to-analog system.

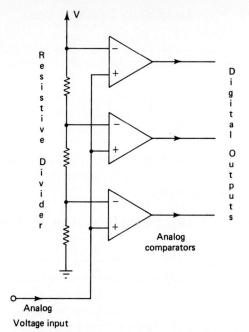

FIGURE 8-9   A/D flash converter.

into analog voltages by D/A ladder networks and summing amplifiers. The D/A converter started with a digital (electrical) signal and converted it into an analog (electrical) signal, but the A/D converter did not start and end with electrical energy. In this section we see how to make an A/D converter which begins with an analog (electrical) signal and ends with a digital (electrical) signal (a voltage). This circuit is called a **flash converter**, made with a new type of gate called an *analog comparator* (Figure 8-9).

8-5.1   Analog Comparator (Op Amp)

The circuit shown in Figure 8-10 is an op amp used as a type of logic gate. The + and − inputs of the op amp are labeled A and B. Op amps are actually *differential amplifiers*—their output is equal to the difference between the voltage at A and B, times an amplification factor:

$$\text{voltage out} = [(\text{voltage A}) - (\text{voltage B})] \times K$$

where the amplification factor, K, is somewhere between 10,000 and 1,000,000 for most op amps. If the voltage at A is 3 V and the voltage at B is 2 V, and K (the *open-loop gain*) is 100,000, we should have an output voltage of 100,000 V. An exciting possibility! The sparks would be dozens of times as large as the whole integrated-circuit package, and the mushroom cloud where the op amp was should be visible for blocks.

That's not what happens! For one thing, the positive voltage from the power supply ($V+$) is only 5 V (a TTL logic 1). You can't get more volts out of the op amp than the supply can give. Instead of 100,000 V, the output is 5 V (maximum) when A is 1 V larger than B. Since the K of the op amp is so large, almost any voltage at A that's bigger than B will produce a HIGH output. The difference between A and B can be as little as 0.00005 V, in our example, and the output is still HIGH.

If this **analog comparator** is a gate whose output is HIGH when A has a higher voltage than B, what conditions will produce a LOW output? Consider what happens if the B input is 3 V and the A input is 2 V. If the same K (100,000) applies to negative outputs, the output in this case "wants" to be −100,000 V. Since the negative output can't be more negative than the negative ($V-$) power-supply voltage, the real output isn't −100,000 V. It's close to zero, and can be considered a LOW (TTL) logic state.

Now we have something that's an analog circuit at its input but has a digital output that's HIGH or LOW. Think of the analog comparator as an "A bigger than B" gate whose output is HIGH when the stated condition is TRUE and LOW otherwise.

8-5.2   Divider Network and Flash Converter

Since we now have a gate that can compare voltages, we need some voltages to compare to. In Figure 8-11, a *voltage divider* splits the 5-V $+V$ voltage into six levels—5, 4, 3, 2, 1, and 0 V—of which we'll use the 3-, 2-, and 1-V levels. The analog input our A/D converter is measuring goes to the A inputs of four analog comparators. If the analog input voltage is near 0 V, the outputs of all three *comparators* are LOW. If the analog input is larger than 1 V (even a little larger!), the "1" comparator switches ON—for 2 V the "2" comparator switches ON, and so forth.

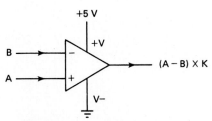

FIGURE 8-10   Analog comparator.

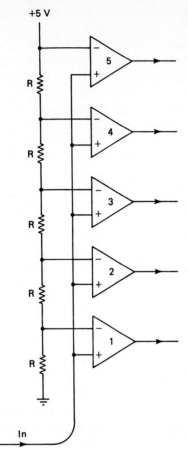

FIGURE 8-11 Five-level flash converter.

Since all the comparator gates work in parallel, and the conversion time to convert the 2-V input to a "11" or the 3-V input to a "111" is as fast as the op amps can switch—in a "flash"—this circuit is called a *flash converter* (Figure 8-12).

8-5.3 Decoder for Flash-Converter Code to Binary

There's one problem with the flash converter's outputs:

$$0 \text{ V} = 000 \qquad 2 \text{ V} = 011$$
$$1 \text{ V} = 001 \qquad 3 \text{ V} = 111$$

This isn't binary code, it's Roman numerals! Fortunately, there's a simple circuit that will perform the conversion:

$$000\text{---}00 \qquad 011\text{---}10$$
$$001\text{---}01 \qquad 111\text{---}11$$

What is it? A full-adder, of course! All that needs to be done to the "Roman numerals" of this flash converter is to add the 1s. If you use a converter with more than three outputs (more than three analog comparators), you'll start using really fancy adders to convert the code, but no matter whether there are three or a hundred analog comparators, all the converters are doing is adding.

Suppose that you want to make a flash converter that can read out numbers from 1 to 100 V, in 1-V steps. You'll need 100 analog comparators and a voltage divider string of series resistors with 100 resistors. Then you'll need a very full adder circuit that can add 100 inputs! There must be an easier way—and there is. We'll be seeing how this can be done much more simply after we go through the next few chapters and find out about counters.

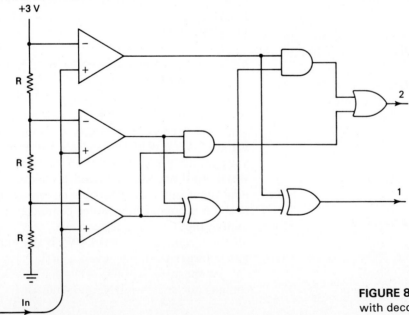

FIGURE 8-12 Three-level flash converter with decoding.

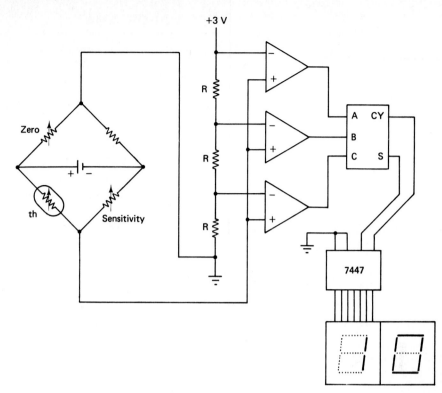

+3 V

R

R

R

A CY
B
C S

7447

FIGURE 8-13 Digital thermometer with flash converter.

8-5.4 A Digital Thermometer Using a Flash Converter

The circuit in Figure 8-13 is a simple example of a practical use of the flash converter. It uses a thermistor to develop a positive voltage that increases as the temperature goes up. A little judicious "fudging" with the variable resistor (sensitivity) should make it possible for a temperature in the range 0 to 10° Celsius to produce a voltage from 0 to 1 V, a temperature from 10 to 20°C to produce a voltage from 1 to 2 V, and so on. The flash converter and full-adder make the voltage into binary code; the 7447 TTL decoder and the seven-segment readout show the voltage as a 0, 1, 2, or 3. We faked a zero after the front digit of the digital thermometer, although the readout only shows temperature to the nearest 10°C. By the way, 30°C is a warm summer day and 0°C is freezing, so this range would be fine for an outdoor thermometer (but make it waterproof!).

QUESTIONS

8-1. Describe the difference between analog and digital.

8-2. Define A/D and D/A.

8-3. Describe the difference between brush and optical detectors used to read shaft or disk encoders.

8-4. What is the resolution of a 5-bit shaft encoder or a five-track code wheel?

8-5. Define "crossover slivers" and describe how Gray code on a shaft or disk encoder helps eliminate this problem.

8-6. Draw the schematic of a 5-bit Gray-to-binary converter.

8-7. Draw the schematic of a D/A like the one in Figure 8-5, but with 5 bits of input.

Make the LSB produce steps half as big as the LSB (bit A) in Figure 8-5.

8-8. Draw the schematic for a D/A like the one in Question 8-7, but use an R, 2R ladder network.

8-9. Explain the relationship between the desired resolution of a D/A converter and the tolerance of resistors it contains.

8-10. Describe the effect of poor linearity on a D/A circuit's output waveform.

8-11. Draw a circuit for an eight-level A/D flash converter.

8-12. If you want to make a digital thermometer that reads from 0 to 100°C in 1°C steps, how many analog comparators must there be in the flash converter?

Make a table of 4-bit binary and Gray code, and answer the following four questions.

8-1. What is the next binary number after 1001?

8-2. What is the next Gray number after 1001?

8-3. Fill in the code wheels.

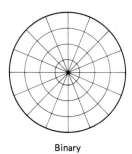 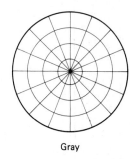

Binary   Gray

8-4. What is the resolution, in degrees, of either of the code wheels shown in Problem 8-3?

A resistive ladder D/A converter is shown below. The next three questions are about the circuit shown.

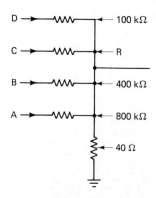

8-5. What is the value of the missing resistor?

8-6. What is the ladder's resolution, in volts?

8-7. What tolerance, in percent, should the resistors be? Remember, resistors only come in specific tolerance values, such as 20%, 10%, 5%, 2%, 1%. Which of these would be appropriate to use for this ladder?

The following six questions are about a resistive-divider type of flash converter (as described in Section 8-5).

8-8. How many resistors are needed for the range 0 to 5 V with a resolution of 1 V?

8-9. How many comparators are needed for the range 0 to 5 V with a resolution of 1 V?

8-10. How many resistors are needed for the range 0 to 5 V with a resolution of 0.1 V?

8-11. How many comparators are needed for the range 0 to 5 V with a resolution of 0.1 V?

8-12. What percent tolerance would you need for the resistors in a flash converter with the range 0 to 5 V and 1-V resolution?

8-13. What percent tolerance would you need for the resistors in a flash converter with the range 0 to 5 V and 0.1-V resolution?

Troubleshooting problems: Suppose that a binary shaft encoder works normally when tested at low speed (turning the shaft by hand) but sends erroneous codes when operating at its full rotating speed. You suspect that one of the bit-input sensors is failing to read bits in its track, so you attach the output of the shaft encoder to a D/A converter and observe the waveform on an oscilloscope while the encoder is revolving. You compare the waveform from a good encoder to the defective one. Since the binary numbers count upward normally, this is what the scope waveform for the good encoder looks like:

The bad encoder produces the waveform below, instead (the good waveform is shown as a dotted line, for comparison). Answer the following five questions about the defective decoder.

8-14. What bit is probably bad? (*Hint:* See which groups of steps on the "staircase" have "dropped" and how many steps there are in each dropped group.)

8-15. How many sectors must the code wheel have (total)?

8-16. How many tracks must the code wheel have (total)?

8-17. What is its resolution, in parts of a complete circle?

8-18. What is its resolution, in degrees?

Troubleshooting problems: Suppose that a binary shaft encoder works normally when tested at low speed (turning the shaft by hand) but sends erroneous codes when operating at its full rotating speed. You suspect that one of the bit-input sensors is failing to read bits in its track, so you attach the output of the shaft encoder to a D/A converter and observe the waveform on an oscilloscope while the encoder is revolving. You compare the waveform from a good encoder to the defective one. Since the binary numbers count upward normally, this is what the scope waveform for the good encoder looks like:

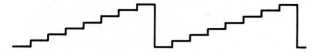

The bad encoder produces the following waveform, instead (the good waveform is shown as a dotted line, for comparison). Answer the following six questions about the defective decoder.

8-19. What bit is probably bad? (*Hint:* See which groups of steps on the staircase have dropped and how many steps there are in each dropped group.)

8-20. How many sectors must the code wheel have (total)?

8-21. How many tracks must the code wheel have (total)?

8-22. What is its resolution, in parts of a complete circle?

8-23. What is its resolution, in degrees?

8-24. What tolerance, in percent, should the resistors in the D/A have in this case?

The National Semiconductor LM3914 is a 10-level flash converter. It is usually attached to an LED bar-graph display with 10 lamps. If the circuit just lights all 10 lamps at an input level of 4 V, and none of the lamps are lit at 0 V:

8-25. What is the flash converter's resolution, in volts (what is the value of each single step)?

8-26. How many separate conditions are possible on the display?

8-27. How many resistors must the 3914's voltage divider contain?

8-28. What percent tolerance must the resistors in the voltage divider have?

8-29. How many comparators must the LM3914 contain?

8-30. Is the output of the 3914 a true binary code?

9
FAMILIES OF LOGIC

In Chapter 1 we saw several kinds of devices that could be used as solid-state switches (devices that could behave like a switch without any moving parts). The bipolar transistor and the MOSFET are used in the two most popular families of logic devices, TTL and CMOS. In this chapter we explore the various ways of making logic devices using the bipolar transistor, the MOSFET, and other types of solid-state switching devices. We promised earlier that we'd show you what's used in the real world to make the gates we've just drawn symbols for. Now is the time to see how these gates are constructed, and how they work.

9.1   DIODE LOGIC *AND* AND *OR* GATES

We begin in order of historical development, with the earliest solid-state switching devices developed. The switching diode has two states. If the *bias* (voltage polarity) applied to the diode makes the *anode* end more positive than the *cathode* end, current will flow through the diode. If the anode is *not* more positive than the cathode, no significant current will flow. The condition that permits current to flow in the diode is called **forward bias**; the condition in which current is blocked is called **reverse bias** (see Figure 9-1). For switching purposes, think of forward bias as a short-circuit condition (we'll replace the diode with a wire in the equivalent diagrams). Think of reverse bias as an open-circuit condition (we'll replace the diode with a gap in the current path in the equivalent diagrams).

Figure 9-2 shows a diode AND circuit. A logic 1 ON condition in a voltage-operated system like this one is a positive voltage (the positive voltage of the power supply). From this fact, we get the name **positive logic** for circuits that work this way. The logic 0 OFF condition in this circuit is a no-voltage condition (the zero volt or ground voltage of the power supply). (Any device whose ON condition is more positive than its OFF condition is called positive logic.)

In Figure 9-2, the two inputs can be attached to HIGH or LOW voltage inputs (+5 V and

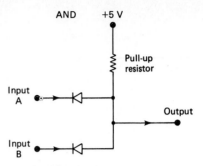

FIGURE 9-2   Diode AND gate.

0 V, respectively). Early logic circuits using diodes were not "5-volt logic," but could be run on this voltage as easily as any other. Since the 5-volt logic is the standard today, we felt that it made sense to use that supply voltage in our diode examples.

There are four different ways that inputs can be applied, and they are shown in Figure 9-3. Each circuit diagram has an equivalent diagram to its right, showing how the diodes would be "switched" by the voltage conditions applied to them. In Fig 9-3(a)–(c), there is always a conduction path that permits current flow from the high potential to the lower potential through the **pull-up resistor**. The voltage at the output of the circuit will be LOW because there's a direct conduction path between output and ground in each case. In Figure 9-3(d), there's *no* path to ground through any of the diodes. There *is* a conduction path from the +5-V source to the output, and for this reason, the output will be HIGH.

Although the path that conducts a logic 1 to the output has much more resistance than the path connecting the output to a logic 0 (when one of the inputs is a 0), the output will still be able to drive a logic 1 to any load attached to the output terminal, unless it has less resistance than the pull-up resistor. Driving a logic-0 to any load attached to the output will be easy, because the path that conducts a LOW to the output has practically no resistance, and can drive practically anything.

Figure 9-4 is the schematic diagram of a diode OR circuit. Like the AND circuit, it is also a positive logic circuit, using +5 V for logic 1 and 0 V (ground) for a logic 0.

Notice that the OR circuit has its diodes facing the inputs in the opposite direction to the diodes in the AND gate. The AND gate's diodes switched into a conducting state when the input was LOW. The OR gate's diodes will switch into a conducting state when the input is HIGH.

Figure 9-5 shows the action of the diode OR gate with the same four combinations of inputs

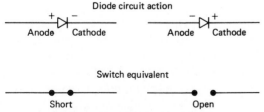

FIGURE 9-1   Diode-circuit action.

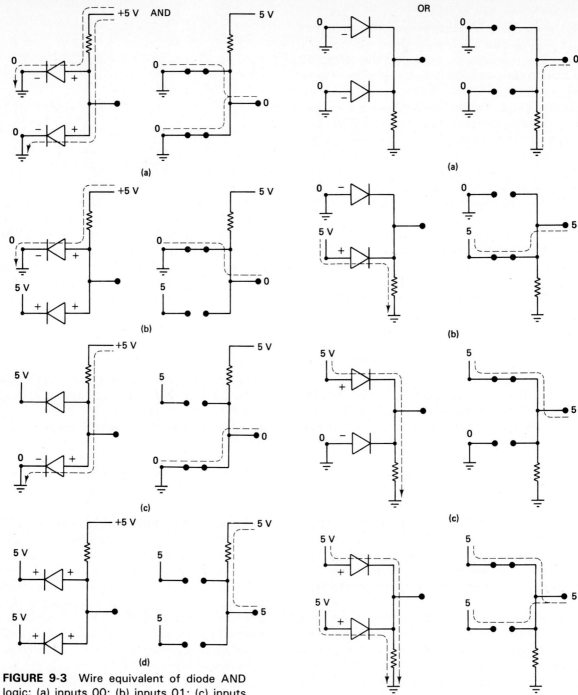

FIGURE 9-3 Wire equivalent of diode AND logic: (a) inputs 00; (b) inputs 01; (c) inputs 10; (d) inputs 11.

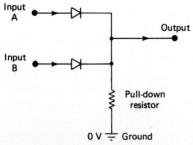

FIGURE 9-4 Diode OR gate.

FIGURE 9-5 Wire equivalent of diode OR logic: (a) inputs 00; (b) inputs 01; (c) inputs 10; (d) inputs 11.

we applied to the AND gate in Figure 9-3. The results are quite different. Since the diodes are conducting between the inputs and output when an input is HIGH, the output is HIGH whenever any input is HIGH. The only time an output is LOW is when both diodes are cut off [the condition in Figure 9-5(a)] and do not conduct. The out-

A	B	AND
0	0	0.7 V
0	1	0.7 V
1	0	0.7 V
1	1	5 V

A	B	OR
0	0	0 V
0	1	4.3 V
1	0	4.3 V
1	1	4.3 V

FIGURE 9-6 Voltage tables for diode AND/OR logic.

put is then LOW because a **pull-down resistor** is conducting a LOW voltage (from the ground) to the output. The LOW output is not traveling through as good a conduction path as the HIGH outputs do, but should still be able to drive logic 0 to a load, unless the load has less resistance than the pull-down resistor.

The truth tables for the diode logic circuits are shown in Figure 9-6, showing voltages instead of logic 1s and logic 0s. Note that there is a small drop across the diodes; they are not perfect conductors. Depending on the material they are made of, diodes may have various voltage drops. In the example, we used silicon diodes, which drop 0.7 V when conducting. This accounts for the fact that the 0 outputs of the AND gate and the 1 outputs of the OR gate have "lost" a bit of the

LOW or HIGH level they would have had with *perfect* conductors in the place of diodes. For now, we'll define 0.7 V and 4.3 V as LOW and HIGH because 0.7 V is a lot closer to 0 than it is to 5 V and 4.3 V is a lot closer to 1 than it is to 0 V.

Another way of looking at the two diode logic circuits is to picture the diodes as they are shown in Figure 9-7. The diodes with input signals going in to the diodes' cathode end will conduct when the input has a more negative polarity than the rest of the circuit. This can only happen when the input is a logic 0. We can say that the diodes whose "arrow" is pointed *toward the input* will conduct 0s only. The diodes whose inputs come in to the anode will conduct when the input has a *more positive* polarity than the rest of the circuit. In this case, we can say that the diodes whose arrow is pointed *away from the input* will conduct 1s only.

Think of the diodes pointed in to the circuit as conductors for 1s, and the diodes pointed out of the circuit as conductors for 0s. The AND gate works because the diodes between the inputs and the output conduct 0s only. This is one of the definitions of the AND gate: "A 0 comes out whenever a 0 goes in. If no 0s go in, the output will be

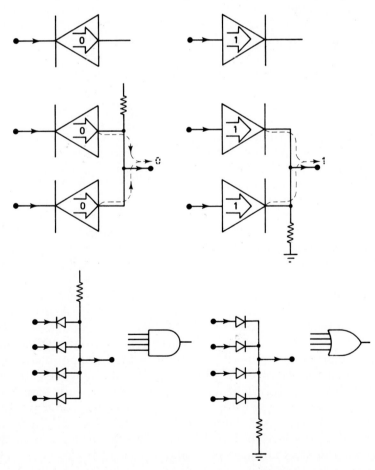

FIGURE 9-7 Diode AND and OR junctions.

a 1." (In this circuit, the pull-up resistor takes care of the last condition.) The OR gate works because the diodes between the inputs and the output conduct 1s only. This is the definition of the OR logic gate: "A 1 comes out whenever a 1 goes in. If no 1s go in, the output will be a 0," (In the OR circuit, the pull-down resistor takes care of the 0 output condition.)

This alternative way of looking at the AND and OR circuits provides insight into how a circuit with more than two inputs could be constructed. Since any AND circuit gives an output 0 when a 0 comes into an input (no matter how many inputs there are), more inputs could be added to the circuit by just connecting more diodes between each input and the output, aimed so that they conduct 0s coming into each input. Since any OR circuit gives an output 1 whenever a 1 comes into any input, just add more diodes into the circuit, directed to conduct 1s. This is shown at the bottom of Figure 9-7.

There is no *inverter* shown in this series of diode logic circuits, because it isn't possible to make a diode inverter. There also isn't any *amplification* possible in diode logic, and this becomes important when there's a small loss in each stage of a logic network. Cascading together lots of ANDs, for example, would result in the 0 level drifting higher and higher as you pass the signals through each consecutive layer of logic. This is bad. If a logic signal 0 has to pass through six diode AND gates, it'll come out the other end at a voltage of 4.2 V. That's a logic 1. The only way to avoid this problem is to have some sort of switching amplifier capable of correcting for the voltage shift produced by each diode. Without the inverting and amplifying functions, diode logic cannot be considered as a complete family of logic devices. It is not possible to make all logic functions using combinations of diode logic gates alone.

A note of no great importance: If you switch the definition of 1 and 0 on Figures 9-2 and 9-4 from positive to negative, and make the " − " lead logic 1 level, you get negative logic. When this is done, Figure 9-2 suddenly turns from a positive logic AND gate into a negative logic OR gate. Figure 9-4 is transformed from a postive logic OR into a negative logic AND. Try checking this out by redrawing the diagrams in Figures 9-3 and 9-5 with the new polarities, checking out the bias on the diodes to complete the equivalent diagrams with their new shorted and open switching states.

You might ask: "Since this is the first type of solid-state logic that was developed, isn't it obsolete now?" If you did ask that question, you'd

be right. Nobody's building computers made of diode logic anymore. The only reasons to bother studying diode logic are: (1) almost all of the other bipolar junction logic families work using the same principles we see here, and (2) there are some shortcuts that you can do using a diode or two to replace a whole integrated circuit. We'll find out about these shortcuts when we move on to other, more advanced families of logic.

9-2 DIODE–TRANSISTOR LOGIC

Diode logic had two drawbacks; there was no inverter in diode logic, and the diode-logic circuit had no amplification capability, so signals would lose some of their strength going through each stage of the circuits we used for AND and OR. There is another problem that arises because no amplification is possible in **diode logic**. Without amplification, the signal at the output will have to get all its power from the signal going in. For the OR gate, this means the 1s coming into the input will have to be the source of all the current that goes out the output. An amplifier can produce an output with more current than its input. This would be useful in a logic circuit, because an output often has to drive more gates "down the line." Generally, there is more than one output to drive, and that means that the input going to the diode logic gate must be capable of driving enough current for two, three, and even more gates in a large logic network. The addition of amplifiers to the diode-logic (DL) circuit makes both amplification and inversion possible.

If you do a quick "flashback" to our introduction to inverters, you'll find that a single transistor can be made into an inverter (either bipolar or MOSFET). The transistor is also an amplifier. It requires much less current at the transistor's input to switch it on than the output develops when the transistor conducts. Putting a *bipolar junction transistor* into the circuit with diode logic makes a complete family of logic devices possible. Not only can we get the AND, OR, and NOT functions necessary to make other possible logic devices, but we can get a "fringe benefit"—amplification. The output of a **diode–transistor logic (DTL) gate** can drive more current than you put in at the input. The ability to drive more than one gate "down the line" is called the **fanout** of the gate. For example, a transistor circuit that can drive 10 times as much current at its output as you have to put in at its input has a fanout of 10. This means that a gate like this can drive 10 more gates just like itself, because its output gives 10 times as much current as its input needs.

Diode-logic gates had a fanout of 1 (maybe a little less than 1 because of losses inside the gate). DTL gates can have fanouts limited only by the choice of transistors and resistors in the schematic. Fanouts of 10 are quite common and fanout values can be much higher than that.

Figure 9-8 is a circuit diagram for a diode–transistor logic NAND gate. It combines a diode-logic AND gate with a transistor inverter. Since we have seen each of these circuits before, individually, the idea of putting them together shouldn't be any surprise. A gate-logic diagram accompanies the DTL schematic to show where each of the parts fits in the symbolic logic diagram for a NAND gate.

If you recall how the diode AND circuit works, there'll be no need to repeat the description here. The only question about how the AND circuit manages to operate the transistor is: If the 0s coming out of the diode circuit are actually not 0 V, but 0.7 V instead, how does the transistor's input manage to stay OFF when its voltage isn't really 0? *Answer:* Use a silicon transistor (the same material as the diodes) and add another silicon diode to join the AND output to the transistor input. The voltage from the diode-logic AND is 0.7 V when LOW. This is the voltage needed to get current to pass through a single silicon junction (a diode). The 0.7 V can pass current through one diode (junction) to the 0 V input, but does not develop enough voltage at "X" to enable current to pass through two diodes. Since current to the

transistor inverter will need to get through two junctions (one in the connecting diode and one in the transistor itself) to reach ground (0 V) level, but only needs to pass through one junction to reach ground (the 0 V input) at the input of the diode AND, *all* the current takes the path with one junction and none enters the transistor. When the diode AND has an authentic logic 1 at its output (5 V), there is plenty of voltage to get through the two junctions and put base current into the inverter. If you did that flashback to the description of the inverter that we suggested earlier, you'd see how the presence of base current and the absence of base current affect the transistor inverter's output.

There was another type of transistor switching circuit that we described in the early going. It was a *normally open* switch, whose output turned ON when its input had base current and turned OFF when base current stopped. This is a *noninverter* circuit (called a *buffer*). It can be added to a diode logic AND or OR gate to increase its fanout without doing an inverter function. This is shown in Figure 9-9, which represents a DTL AND gate with increased fanout. In addition to the improvement in drive capacity, the DTL AND has a HIGH output of nearly 5 V, and a LOW output of 0 V. These values are nearer to the "ideal" values for logic 1 and logic 0 than the AND gate's output using diodes alone.

There is no reason why diode AND gates should be the only gates with transistors attached (to make DTL NANDs and ANDs). A diode OR gate can be used to drive a **bipolar transis-**

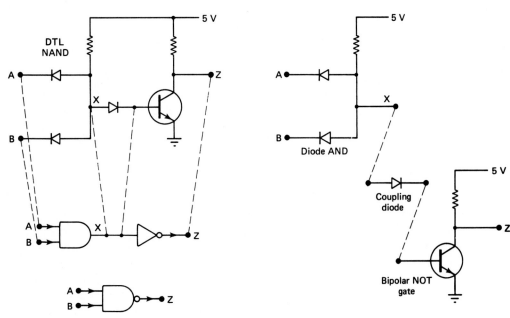

FIGURE 9-8   DTL NAND.

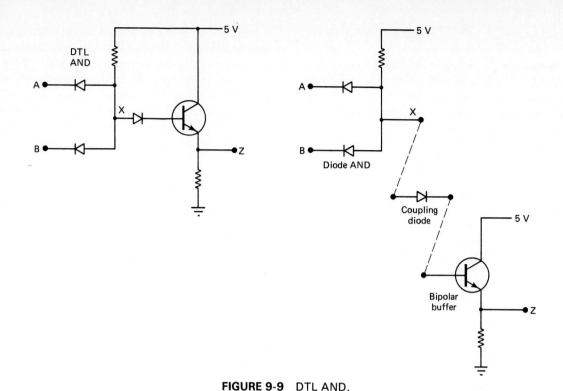

FIGURE 9-9 DTL AND.

tor inverter; this makes a DTL NOR circuit [Figure 9-10(a)]. If the output of a diode OR is used to drive a buffer, the logic circuit is still an OR gate, but it now has more fanout, and its HIGH and LOW outputs are closer to the ideal values.

DTL circuits are still available from a few sources as *integrated circuits* (ICs). Their lifetime as discrete circuits (made of separate diodes, resistors, and transistors wired together on circuit boards) is over, and as ICs, they are pretty much obsolete. The reasons for studying them are:

1. They're still around in places, and you never know when a DTL circuit might turn up for you (the digital technician) to repair.

2. The principles that make DTL work are the same principles on which the TTL

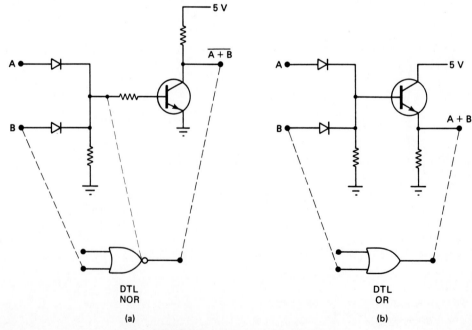

(a)

(b)

FIGURE 9-10 DTL NOR and OR.

(transistor–transistor logic) circuits that replaced them still work.

TTL circuits abound in microcomputers, minicomputers, and all sorts of specialized digital logic circuits.

9-3 TRANSISTOR–TRANSISTOR LOGIC

Transistor–transistor logic (TTL) is the most popular family of bipolar gates. There are more types of gates available in standard TTL and Schottky TTL than in all of the families of bipolar logic previously described. TTL has made the DL, DTL, and RTL families of logic obsolete. The working principles by which TTL operates build on the DL and DTL circuits to which TTL is the successor.

Part of the reason TTL is so successful is the fact that a greater variety of gates is available in *standard TTL* than in all the other families. There is also a second reason why TTL is successful, and this has to do with design characteristics. All standard TTL gates have inputs with identical electrical characteristics, no matter what gate is being considered. The output characteristics of standard TTL gates also match for all gates. Designers love this kind of thing. On the other hand, DTL AND and OR gates, for example, have different input characteristics; the 0 input to an AND (DTL) gate causes current to flow, and the 1 input cuts current off, while in the OR (DTL) gate, just the opposite happens—0 inputs cause no current to flow, but 1 inputs do. DTL AND and NAND gates also have different output characteristics; AND (DTL) gates have a strong pull-up and a weak pull-down, while NAND (DTL) gates have a weak pull-up and a strong pull-down. If two gates of different logic types in the same family of logic have such different behavior, the designer has to be constantly aware of these differences in putting together systems. If all the gates behave the same way, that's one worry the designer does not have.

A third reason why TTL is so successful is a result of the popularity it already enjoys. The price of TTL gates is lower than that of the other families of gate logic. Economists call this the Law of Supply and Demand. The more readily available and mass-produced an item is, the less it costs. As the costs of development and tooling up standard TTL production lines was paid off, the cost of production lowered and the price of the devices came down. TTL devices now are the cheapest, and most readily available, of all bipolar logic families.

9-3.1 Standard TTL

The "typical" gate of standard TTL logic is the NAND gate. All other TTL gates contain the features found in the NAND gate with only a few variations. Figure 9-11 is the schematic diagram for a TTL NAND gate, showing its relationship to earlier families of logic. The inputs of the NAND gate are the base-emitter junctions of transistor Q_1. Inputs A and B are drawn in the schematic diagram as arrows because they are, in fact, diodes. The base-emitter junctions of the Q_1 transistor are junctions of P and N type silicon exactly the same as a couple of silicon diodes connected between the A and B and the 4-kΩ resistor. A transistor like Q_1 has a third junction, the base-collector junction, which also functions as a diode, this one being connected between the 4-kΩ resistor and the base of Q_2. A diagram showing the direction of this junction's polarity is drawn beneath Q_1. By replacing the transistor's junctions with individual diodes, we can see that Q_1's junctions are connected in the same way to the (4-kΩ) pull-up resistor as the junctions in a diode-logic AND gate (the "front end" of a DTL AND or NAND gate) with a coupling diode, as seen in Figures 9-8 and 9-9. In fact, except for the fact that Q_1 is a transistor with amplification capability, in Figure 9-11 there is no difference in the junction-and-resistor circuit of the TTL NAND input and the front end of Figure 9-8 or Figure 9-9.

The input of the TTL circuit is very much like the input of a DTL circuit, but the output is very different from the output of DTL NAND and AND gates. Q_2 looks like a combination of the DTL NAND with its pull-up resistor and the DTL AND with its pull-down resistor. The circuit Q_2 is in has two outputs, one where an inverter would have an output, and the other where a buffer would have an output. In analog jargon, this type of circuit is called a **phase splitter** or a **split-load phase inverter** (although in analog circuits the two resistors, 1 kΩ and 1.6 kΩ, are usually the same size). In digital use, this circuit is intended to put out an inverted signal on the topmost output, and a noninverted signal on the bottom output. The Q_3 and Q_4 transistors, together with the 115-Ω resistor and the diode, are called a **totem pole** output. This rather humorous name comes from the stack of different devices "sitting on each other's shoulders," which

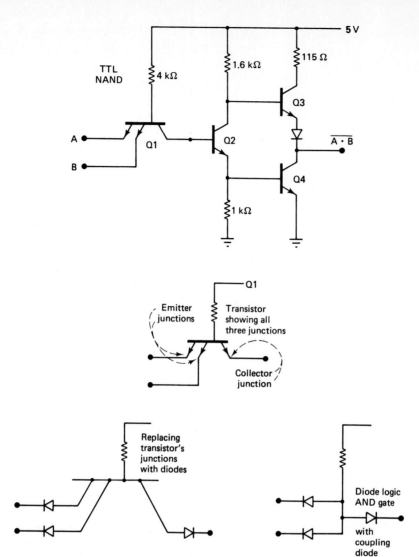

FIGURE 9-11 TTL NAND showing similarity to DTL.

made someone think of the creatures stacked atop one another in a real totem pole.

Figure 9-12 shows how the NAND circuit operates for two cases: one where the output will be LOW and another where the output will be HIGH. Each case shows the current flow or conduction in three current paths. The current "wants" to get from the +5-V supply to the ground (to go from HIGH to LOW) and has three paths to travel.

Figure 9-12(a) shows what happens when both inputs are HIGH. Current wants to get to ground through R_1. The two possible ways current can travel are through the left junctions (the inputs) or the right junction (shown with a dotted arrowhead). Since the inputs are not connected to ground, there is no way for current to travel to ground through them, and no current can flow. There is a current path that reaches ground on the right-hand side. The R_1 current (shown as a dotted line) passes through a junction in Q_1, a sec-

ond junction in Q_2, and a third junction in Q_4. The current coming out of Q_2 actually has two ways it can go: through the R_3 resistor or through the forward-biased Q_4 base–emitter junction. Most of the current flows through the Q_4 path, and the R_3 current has been left off the diagram, although there is a small amount of current flowing there.

Notice that the R_1 current path provides a base current for Q_2. That switches on the Q_2 transistor, and permits a second current to reach ground. This second current path is the current that wants to get to the ground through R_2 (shown as a line of alternating dots and dashes). Both currents together travel to Q_4.

In Q_4, the R_1 and R_2 current paths provide base current to the Q_4 transistor. It is switched on *very* strongly by so much base current; the current is many times the current needed to saturate the transistor (make it a short circuit) and Q_4 provides a very good conduction path between

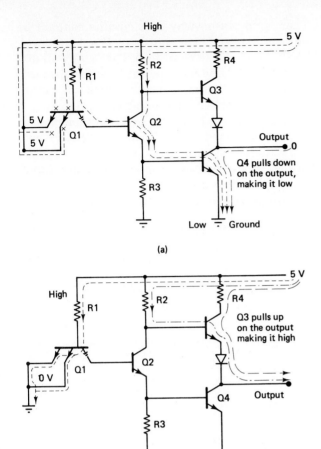

(a)

(b)

FIGURE 9-12 TTL NAND action: (a) inputs HIGH; (b) inputs LOW.

the output terminal and the ground. Q_4 is the pull-down transistor. When it is *saturated* like this, the output is connected to the LOW level voltage by a nearly 0-Ω path. The dashed line indicates a path that current could take through Q_4 if the output were attached to some source of current.

The current that wants to reach ground through the R_4 resistor cannot get through Q_3. Since Q_3 is the pull-up transistor, and it has no base current (as shown in the diagram), it does not conduct.

Notice that the totem pole arrangement provides an **active pull-up** and an **active pull-down**. That means that the circuit that switches a 1 to the output has an active switch (transistor Q_3) and the circuit that switches a 0 to the output also has an active switch (transistor Q_4). A **passive pull-up** is a resistor that conducts a 1 or a 0 to an output as long as there is no better conductor connected to the other logic level. This was used in DTL and RTL circuits as the standard method of pull-up and pull-down. The active pull

up of TTL logic not only connects a 1 or 0 to the output, but disconnects the opposite logic level at the same time.

But is the current path through Q_3, in fact, really disconnected? Why is the (dotted-and-dashed line) current through R_2 going into Q_2 and not into the base of Q_3? There are *two* ways current could reach the ground from R_2. Current through Q_2 must pass through *one* base–emitter junction to reach Q_4, whereas current entering the base of Q_3 would have to pass through *two* junctions to arrive at Q_4 (the base–emitter junction of Q_3 and the diode in series with Q_3's emitter). The path with one junction is "easier." In fact, once current starts to flow through Q_2, the voltage cannot rise to a level high enough to push any current through Q_3 and the diode. That's why only the pull-down transistor is conducting, making the output LOW.

Figure 9-12(b) shows what happens when both inputs are LOW. Current still wants to get to ground through R_1. The two possible ways current can travel are through the left junctions (the inputs) or the right junction (shown with a dotted arrowhead). This time, the input junctions are connected to ground, and current can reach ground more easily through each input junction than through the Q_1 base–collector junction, the Q_2 base–emitter junction, and the Q_4 base–emitter junction. It takes 2.1 V to push current through three junctions, and only 0.7 V to push current through one junction. Current will flow through the left junctions and not through the right junctions, so there will be no base current for Q_2.

Without a base current, the Q_2 transistor is *open*. No current can pass through it. In Figure 9-12(a), we considered that there were two ways for the current through resistor R_2 (alternating dots and dashes) to reach ground. The path through Q_2 has been closed off. The only remaining path for the current from R_2 to reach ground is the one shown in the picture, provided that the output is connected to something that eventually reaches ground.

In the absence of any current through Q_2, there will be no current into the base of Q_4. With Q_4 (the pull-down transistor) switched off, no connection to the LOW logic level exists, while Q_3 (the pull-up transistor) will pass HIGH logic level to the output.

You may have noticed that in these diagrams, unlike the DTL logic circuits, the transistors don't have little circles around them. There is a reason for this; we really didn't just forget the circle. Transistor schematics with the junctions enclosed in a circle indicate a discrete transistor, each bipolar device enclosed in its own package. Without the circles, a bunch of transistors are presumed to be part of an integrated circuit having all the devices in a common package and joined together on the surface of a common piece of silicon. DTL and RTL integrated circuits do exist, but they were first designed as discrete circuits, each transistor, resistor, and diode placed into holes on a printed-circuit board and then soldered to the foils. Integrated circuits were first developed in the early 1960s. At about that time TTL logic was introduced in integrated form.

There's no reason why TTL circuits couldn't be built of discrete logic, but once ICs were developed, there was no point to it. That's why all diagrams of TTL and more recent logic families will be drawn with transistors that lack the circles used in the earlier types of logic.

In Figure 9-13 we see several other types of gates constructed from TTL logic. The NOT gate shows the same configuration as the NAND gate; in fact, it's a NAND gate with only one input. The NOR gate is like the NOT gate at each input; the inputs are parallelled at Q_2, so that the out-

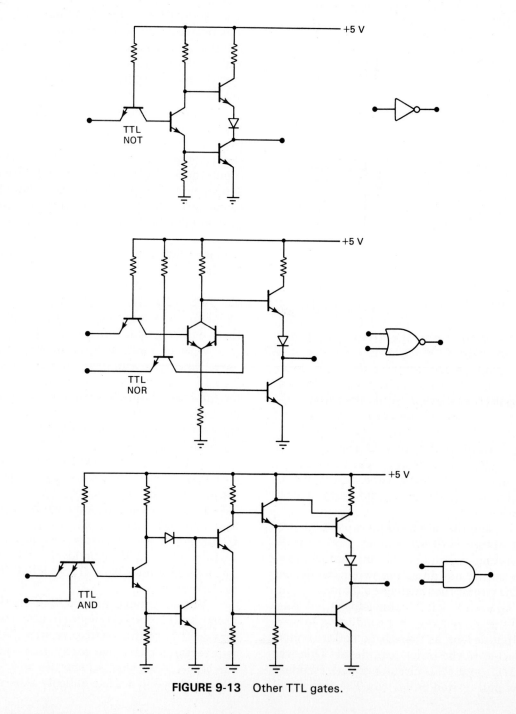

FIGURE 9-13   Other TTL gates.

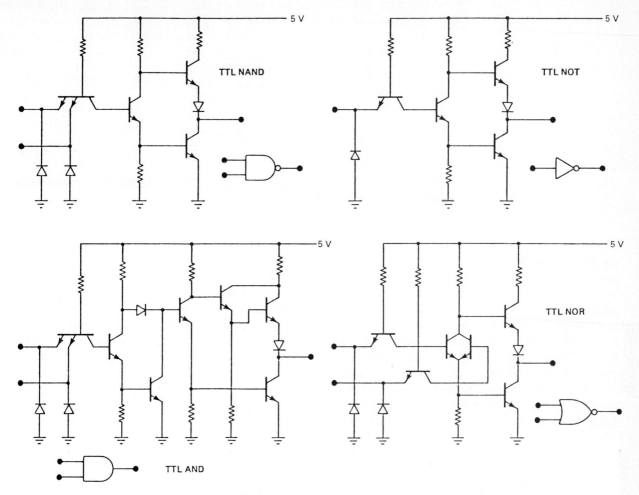

FIGURE 9-14 Various TTL gates.

put can be driven LOW by a HIGH at either input. Finally, the AND gate has extra transistors between Q_2 and the totem pole to provide an extra inversion at the output, making what would otherwise be NAND logic into AND logic.

In Figure 9-14, the same circuits shown in Figures 9-11 and 9-13 are provided with an additional diode at each input. The purpose of each diode is to protect the TTL device against negative voltages at the inputs. This is strictly an "oops . . . did I do that?" problem. Nobody designs negative logic into a TTL input, but negative "spikes" sometimes arise in the signal from unexpected (and undesigned) places. These can cause damage to the Q_1 transistor, so the designers of TTL integrated circuits add these diodes to "chop off" negative pulses if they appear at these inputs.

9-3.2 Open-Collector TTL

Figure 9-15 is the schematic diagram for a NAND gate with open collector (O.C.) outputs. You'll no-

tice that there's something missing from the totem pole output of a standard TTL device: the pull-up transistor and its associated resistor and diode are gone. This means that instead of two outputs, LOW and HIGH, which develop because the output is pulled down and pulled up, there's only one authentic logic-level output. The output of the open-collector TTL gate can pull down, but not up. Where a HIGH output would be the normal result, the O.C. gate's output **floats**. New word: When a voltage is floating, it isn't anywhere—the output isn't pulled up, so it isn't HIGH—the output isn't pulled down, so it isn't LOW. The voltage at the output depends on the type of *load* attached to it. If its load is connected to a HIGH level, the floating output will *float high*. If its load is connected to a LOW level, the floating output will *float low*. If nothing is connected to the output, its voltage state will be uncertain, and may drift to either HIGH or LOW states from time to time as stray electrical noise influences it. Since there is no solid connection to V_{cc} (the +5-V source) or ground when the output

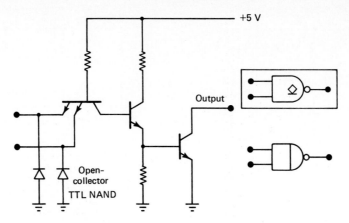

FIGURE 9-15 Open-collector TTL NAND gate.

is floating, the output in this state is also called the **high-impedance state**. "High impedance" is an electrical term for a circuit that doesn't conduct well, so it makes sense to call floating the high-impedance state—but electrical engineers are never satisfied using words like "impedance"; they'd rather just use one symbol, Z. This is why the floating state is also sometimes called the "high-Z" state on technical documentation.

Figure 9-16 shows some other open-collector gates, and the symbolic representation for O.C. logic. The "racing stripe" down the middle of the

gates identifies them as O.C. logic devices,* and the truth tables with X in the output where 1 should be are used to show the floating state.

Why does this type of TTL logic exist? We now have a third state of logic, the floating state, that is used to replace the HIGH state in this family of logic. This third state of logic is useful in circuits where **OR-tie** capability is important. An example is shown in Figure 9-17, where three

*As does the inset symbol in the newer logic gate diagrams shown.

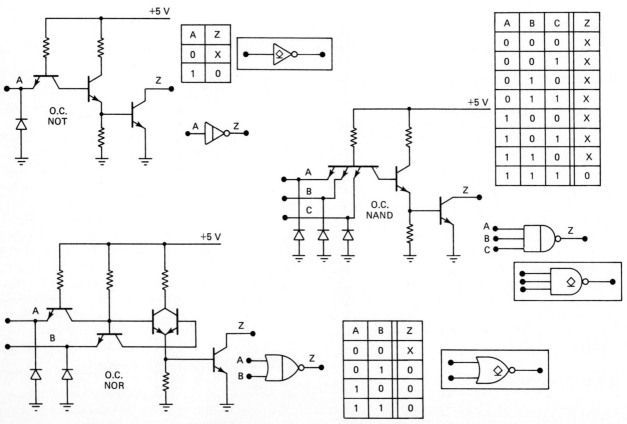

FIGURE 9-16 Other open-collector gates.

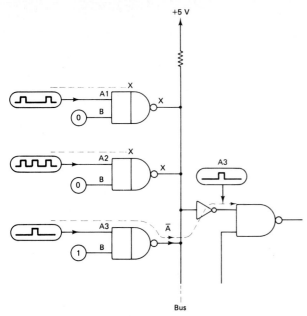

FIGURE 9-17 Open-collector gates on a common bus.

gates are trying to provide signals to the input of a fourth. The three signals can be thought of as being like three people on a telephone party line, all trying to talk to a fourth person. Each must take a turn or the listener won't be able to understand anybody. To take turns, two people must shut up and only one may talk at a time. The digital logic equivalent of this situation requires that one gate of the three may generate 1s and 0s, while the other two gates must "shut up." How does a gate shut up at its output? It won't do for the two "quiet" gates to be LOW, because they'll be "talking." A LOW gate will pull down on the **bus wire** (this is the name of the wire shared by the three *sending* gates and the one *receiving* gate) and two LOW gates "fighting" with the one sending gate will ground out any signal from the gate that is talking. If both quiet gates are HIGH when they shut up, they'll put enough current into the bus wire that the talking gate will burn out its pull-down transistor when it tries to put out a 0. The only way for the two quiet gates to shut up so that the third gate can talk is for the outputs of the quiet gates to float. In this condition, the quiet gates will not fight with the talking gate no matter whether it's sending a HIGH or LOW state. Since the O.C. logic family is the first we've seen that permits a quiet logic state that won't pull down against HIGHs or up against LOWS from the talking gate, O.C. gates will show up often in **bus-oriented** systems, where a large number of sending and receiving devices will share a common wire.

These systems are an important part of digital computers, about which we'll see more later.

The circuit in Figure 9-17 has one additional feature besides the sending gates and the receiving gates attached to the bus wire. This is an **external pull-up resistor** which is attached to the BUS. The open-collector NAND gates are enabled by a HIGH input on their B inputs, and their outputs are disabled in the floating condition. The problem arises for the enabled gate that, even when it is enabled, its output (which is supposed to be sending 1s and 0s) is LOW but never HIGH. When the outputs of the two other gates are floating, and our active gate wants to send a HIGH, there's no pull-up inside the gate to do it. We have to add an external pull-up to provide the HIGH when nobody's sending a LOW. As long as all the gates' outputs are floating, the output will be HIGH because the pull-up resistor pulls up (weakly) on the bus. As soon as any gate's output becomes LOW, the output will become LOW also, because the pull-down transistor can pull down more strongly than the resistor pulls up. Since the LOW can be provided by the first OR the second OR the third sending gate, the term **OR-tied logic** is applied to this type of circuit. Since an OR gate for LOW signals is really an AND gate for HIGH signals, the name **wire-and** is also used to describe this type of connection.

9-3.3 Tri-State TTL

In the open-collector family of logic, we found a third state of logic besides HIGH and LOW. This third state was the floating condition. Open collector logic gained this third state by giving up one condition. In sacrificing the pull-up section of the totem pole output, the HIGH output was also sacrificed. Can a family of logic be designed which has the floating condition without sacrificing the HIGH or LOW output? The answer is called **Tri-State* logic**. The name is an indication that three states are possible in these logic devices. There is no reason why a circuit with a totem pole output can't have a floating condition. What must be done is to switch OFF both the pull-up and the pull-down transistor in the totem pole at the same time. The circuit in Figure 9-18 shows one way in which this can be done. You'll observe that all the normal parts of the TTL NOT gate are present. The addition to the circuit is sketched in with dotted lines. Junctions added to the input and the pull-up transistor's base make it possible to disable these parts of the circuit when the disable

*Tri-State is a trade name of National Semiconductor.

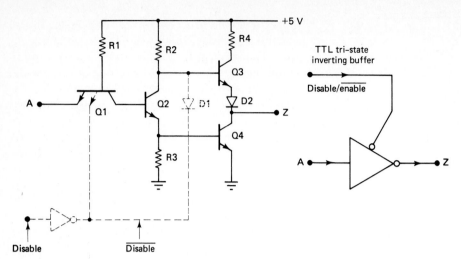

FIGURE 9-18 Tri-State inverting buffer.

input is HIGH. When the Q_1 transistor is disabled, its output to Q_2 can *never* receive current, because the path to the DISABLE wire, which is LOW (ground) is a better conduction path than the path to the base of Q_2. No base current can flow to Q_2 regardless of what the A input is doing. Meanwhile, the diode D_1 attached to the base of Q_3 is able to "sink" current to the ground much more easily than that current could reach the ground through Q_3 and D_2. Since Q_2 is not conducting, there is no base current for Q_4. Now, we know that Q_3 (the pull-up transistor) cannot conduct, because its base current has been taken away by D_1, and Q_4 (the pull-down transistor) cannot conduct because it gets no base current from Q_2. That means that the output is not pulled up and it is not pulled down, either. The output of the Tri-State logic gate must be floating. When the DISABLE input is not HIGH, the junctions do not conduct, and can be ignored in the circuit. If these two junctions are ignored, all that is left in the circuit is the standard TTL NOT gate, which can have either HIGH or LOW outputs, depending on what the input is at A.

The Tri-State logic gates can be identified by the "new" way in which they are enabled and disabled. When enabled, the outputs of the tri-state family may be HIGH or LOW, but when disabled, the outputs always float. It is for this reason that the floating state of logic is sometimes called the **Tri-State condition**, and a disabled gate of this type is said to be **Tri-Stated**. The symbolic diagram for a Tri-State gate is included to the right of the schematic in Figure 9.18. You can spot Tri-State logic by the input coming into the side of the gate instead of the back. This input entering the angled face of the inverting buffer is always the enable/disable input. The input entering the back is the data that will be passed through to the output (in this case, it will be inverted). It is for this reason that the Tri-State inverting buffer in Figure 9.18 has a symbolic logic diagram that looks like an inverter with an extra input coming in at an angle. Tri-State buffers are sometimes called **enable gates**.

9-3.4 Schottky (S), Low-Power Schottky (LS), Advanced Schottky (AS), and Advanced Low-Power Schottky (ALS) TTL Logic Devices

Another branch in the family tree of TTL logic devices is the **Schottky TTL** branch. The most common Schottky TTL devices are often called **low-power Schottky**. They require less power than standard TTL devices to operate, and their inputs do not require as much power to drive as the inputs of standard TTL devices. Schottky TTL has replaced standard TTL in many applications where its increased speed and reduced power consumption outweigh its slightly higher cost. For most standard TTL devices, a Schottky version exists which has the same pinout (can be plugged directly into a socket where a standard TTL device was before), and costs about the same.

As you might have deduced from the statement above, Schottky devices are faster than standard TTL. This is a result of clamping the base–collector junctions of standard bipolar transistors with a Schottky diode (Figure 9-19). The result of doing this is threefold. First, the diode across the base-collector junction prevents the transistor from operating in a fully saturated state. This, we'll see later, is an important source of delays in bipolar switching. Second, the Schottky junctions in the transistor itself switch faster than standard silicon junctions because they have less resistance, and it is a combination

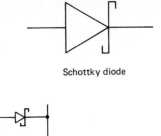

Schottky diode

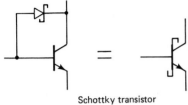

Schottky transistor

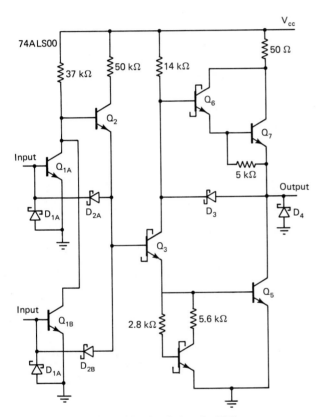

FIGURE 9-19 Schottky TTL.

only feature they have that is superior to standard TTL devices. The amount of power they consume is considerably lower. Since a lower voltage is required to switch each device ON, the power the device consumes is less than a higher-voltage gate. The inputs of Schottky TTL devices normally require about one-fourth of the current that standard TTL devices need to switch. This, combined with the smaller voltage drop across junctions inside the device, produces much lower power consumption in each Schottky gate, hence the name "low-power Schottky (LS) TTL."

Figure 9-19 shows the schematic symbols for a Schottky diode, a Schottky-clamped transistor, and the schematic diagram of an **ALS (advanced low-power Schottky)** NAND gate. The ALS circuit does away with the multiemitter transistor so characteristic of other TTL inputs, but in terms of junctions, the input current passes through PN and NP junctions oriented the same way as in standard TTL. ALS and AS TTL are about twice as fast as LS and S (low-power Schottky and Schottky) TTL circuits. Schottky TTL devices in a schematic can be distinguished by the base of the transistor (which incorporates the Schottky diode's cathode symbol) and in the parts identification number by the insertion of the letters "S," "LS," "AS," or "ALS." The 7400 NAND gate of standard TTL, for instance, becomes a 74S00, a 75LS00, a 74AS00, or a 74ALS00 in these Schottky logic families.

9-3.5 Schmitt Trigger TTL

One family of TTL logic exists which has *analog* as well as digital circuitry inside it. The **Schmitt trigger** is an analog circuit that responds to the voltage put into it by developing a HIGH or LOW state at its output. The Schmitt circuit "triggers" in the following way. There are two important voltage levels at the input of the Schmitt trigger called the **upper trip point** and the **lower trip point**. If the input voltage rises above the upper trip point, the Schmitt trigger will switch into a certain state (let's call it "ON") and will stay there even if the voltage drops below the upper trip point. If the voltage drops below the lower trip point, the Schmitt trigger will switch into its OFF state, and stay there until the input again rises above the upper trip point. The upper trip point is 1.2 V higher than the lower trip point (0.8 V), and this means that an input will have to "swing" up and down several volts to turn the Schmitt trigger ON and then OFF again. Small "wiggles" in the input will not change the state of the Schmitt trigger.

of the resistance and capacitance in semiconductor devices which delays the rise and fall of voltages being switched ON and OFF. If you reduce the resistance, the speed improves. Third, the Schottky junction begins to conduct ("switches on") at a lower voltage than do standard silicon junctions. This means that the voltage will not have to rise for as long a time to switch on a Schottky junction as it would rise to switch on a standard silicon junction. That's just a matter of common sense; if you don't have far to go, it won't take long to get there.

The speed of Schottky TTL devices is not the

It is for this reason that the Schmitt trigger family of logic is used in noisy environments, where wiggles appear on voltage levels that should be flat. Normal TTL logic has little immunity to this noise. If a voltage applied to the input of a standard TTL gate is near the 1/0 dividing line for the gate, the wiggles on the voltage will switch the gate from LOW to HIGH with every wiggle that's big enough to cross the dividing line. A Schmitt trigger TTL gate has a Schmitt trigger built into each input. Since there is no 1/0 dividing line for a Schmitt trigger, the wiggles that change a 1 to a 0 will have to swing all the way from the upper trip point to the lower trip point, a distance of several volts, before the gate will switch. This reluctance to switch is called **hysteresis**, a name which indicates a device has a memory for the last state it was in, and will not change easily to a new state. Hysteresis is found in magnetic switching devices, where it is described with a diagram called a **hysteresis loop**. The hysteresis loop is inserted into the middle of a gate diagram to indicate a gate with hysteresis (a Schmitt trigger gate). Schmitt trigger gates are used to filter noise out of digital signals, and restore the "crisp" 1s and 0s the signals should have (Figure 9-20).

The purpose of using Schmitt trigger logic is to ignore the small wiggles that represent electrical interference, but accept the large wiggles that represent real transitions between 1 and 0. The *hysteresis voltage* or **backlash** (the voltage difference between the upper trip point and the lower trip point) is responsible for the noise immunity of this family. If a voltage swings a greater distance than the backlash, it's a 1/0 or 0/1 transition. If the voltage swing is less than the backlash, it will be ignored by the input of the Schmitt-triggered logic gate.

9-4 MOS LOGIC

The devices described in Sections 9-2 and 9-3 are all based on the bipolar type of switch described in Chapter 1. There is another type of solid-state switching device described in that chapter right after the bipolar transistor switch; this is the

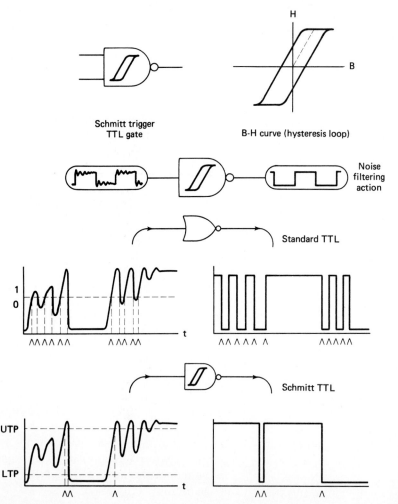

FIGURE 9-20 Schmitt trigger TTL showing noise reduction.

MOS field-effect transistor (MOSFET). Unlike the bipolar transistor, the MOSFET needs no current at its inputs to operate. The voltage sets up an electric field in the semiconductor which permits the device to pass or block current. Switching the N-channel MOSFET in Figure 1-7 is a matter of applying a positive voltage to the gate of the FET (switch ON) or not applying a positive gate voltage to the FET (switch OFF). This is positive logic, as we have defined it earlier in the chapter. Negative logic is also possible with various types of MOSFET logic, and we'll explore this aspect of logic as we go through this section. It's worth remembering, though, that positive logic is more standard, and in most logic design, positive logic is preferred.

9-4.1 NMOS (N-Channel MOSFET Logic)

Figure 1-7 illustrates a MOSFET switch—simply a light bulb and a MOSFET in series. Logically speaking, it is a *buffer*, since the output (the bulb) went ON when the voltage at the input was ON. The typical **NMOS** logic gate is shown in Figure 9-21. It's a NOT gate (inverter) with a resistor attached to the source terminal as a pull-up resistor. This circuit inverts because the FET is connected to ground (the 0 logic level). If it conducts, the FET will conduct a 0 to the output. In the absence of FET conduction, the pull-up resis-

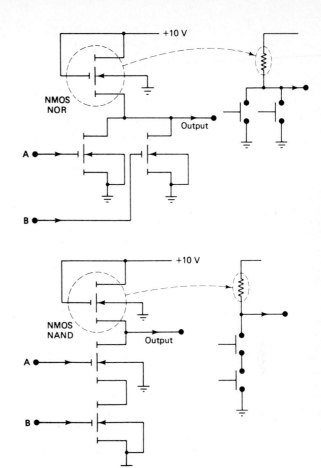

FIGURE 9-22 Other NMOS gates.

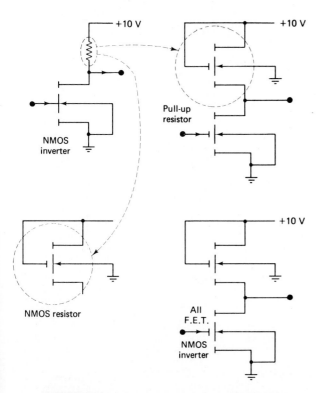

FIGURE 9-21 Components of an NMOS NOT gate.

tor will pull up the output to a HIGH level. Since a HIGH (positive) input makes the FET conduct, and the conducting FET passes a LOW to the output, the circuit is an inverter.

In the DTL, RTL, and TTL circuits, two types of devices were found in the schematic diagrams: semiconductors and resistors. This made integrated-circuit construction more complicated than it needs to be. In the N-MOS circuit, the pull-up resistor can be made, and is made using the same semi-conductor devices as those used for switching. This is shown as a part of Figure 9-21, illustrating how an N-channel MOSFET is used as a pull-up resistor, and how the whole circuit looks when it's made of MOSFETs alone.

Figure 9-22 shows a few NMOS circuits that use the same type of switching for more complex gates than the NOT. To make it clear how they work, a diagram that replaces each FET with a normally open pushbutton switch is drawn next to each gate. The primary design philosophy embodied in these circuits is *active pull-down* and *passive pull-up*. Each gate is made by installing a MOSFET switch to deliver 0s to the output when it is supposed to have 0s.

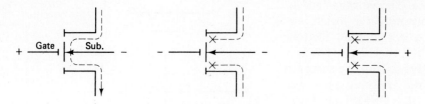

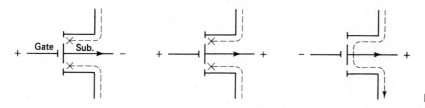

N-channel MOSFETS conduct if their gate is
more positive than their substrate

P-channel MOSFETS conduct if their gate is
more negative than their substrate

FIGURE 9-23 MOSFET operating principles (supplement).

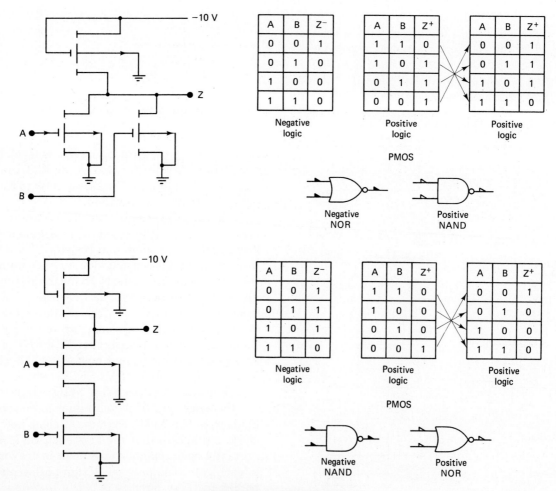

FIGURE 9-24 PMOS gates.

9-4.2 PMOS (P-Channel MOSFET Logic)

P-channel MOSFETs have negative logic or positive logic capability. The simplest circuits that can be made with PMOS devices are negative logic. The voltage used to turn ON a PMOS switch is a negative gate voltage, measured from ground (see Figure 9-23). If the negative power-supply voltage is defined as a HIGH, logic 1 condition and the ground is a logic 0, the circuit has negative logic. Once that definition is accepted, all the gates shown in the diagram (Figure 9-24) are the same as their NMOS counterparts in Figure 9-22. Only the *polarity* of the logic and MOS arrow directions have been changed.

An alternative way to define the action of the gates is to call the ground voltage a logic 1 and the power-supply voltage a logic 0. That makes the circuit positive logic. If *every* 1 and 0 in the truth table for each gate is reversed, we have the truth table for a positive logic interpretation of that gate's action. After shuffling around the lines of the reversed truth table, we can see that we don't get the same gate in positive logic as in negative logic. This is why the circuits in Figure 9-24 are identified twice, as, for example, negative logic AND and positive logic OR. This is one of the features we mentioned back in diode logic, and now we see, again in PMOS logic, that the relationship still holds.

9-4.3 CMOS (Complementary MOS Logic)

A little while back we said that positive logic is preferred to negative logic by most gate designers. This is simply a matter of preference, but it is a very widespread preference. You'll see that the **CMOS** circuits in Figure 9-25 contain both P-channel and N-channel devices, yet they are all defined in terms of positive logic and they all have positive power supplies. This is no accident. The designers of CMOS knew what their customers wanted, and developed positive logic circuits to satisfy their customers' requirements. We won't say that there's no CMOS negative logic around; we just haven't seen any.

Next to TTL circuits, CMOS gate packages are the second most popular family of gate logic. They are not available in quite as many gate types as TTL, and cost a bit more for a comparable package, but they have a significant advantage over TTL devices. The CMOS circuit in Figure 9-25 has an active pull-up and an active pull-down in the schematic (an inverter). The fact that the inputs require no current when HIGH or LOW and that the current path from $+V_{dd}$ to

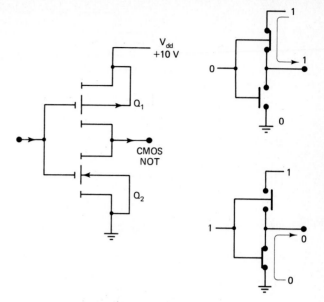

FIGURE 9-25 CMOS NOT gate.

ground is always open at either Q_1 or Q_2 means that the power consumed by each gate is virtually zero. With no power consumed by the gate in either of its stable states (okay, maybe a few nanowatts!), the number of gates that can be put on a piece of silicon without the "chip" overheating is huge. MOS devices on a single square centimeter of silicon can exceed 500,000 in number at current densities, and these densities can be pushed much higher with improvements in miniaturization before the possibilities of further expansion are exhausted.

The circuit in Figure 9-25 works by a combination of pull-up and pull-down. A LOW input turns ON the Q_1 transistor because the gate of Q_1 is more negative than its substrate. At the same time Q_2 is OFF, because its gate is not more positive than its substrate. Since Q_1 is ON, and it is the pull-up transistor, the output is HIGH (a LOW input causes a HIGH output). Now, if we put a HIGH input into Q_1 and Q_2, we find that Q_1 is OFF because its gate is not more negative than its substrate. At the same time Q_2 is ON, because its gate is more positive than its substrate. This means that since Q_2 is now ON and it is the pull-down transistor, the output will be LOW (a HIGH input causes a LOW output). This circuit is an inverter. To make more complex gates, it will be necessary to design the pull-up part of the logic gate to deliver 1s to the output when needed, and to design the pull-down part of the circuit to deliver 0s to the output when they are needed to implement the output in certain lines of the truth table.

Figure 9-26 shows a more complicated

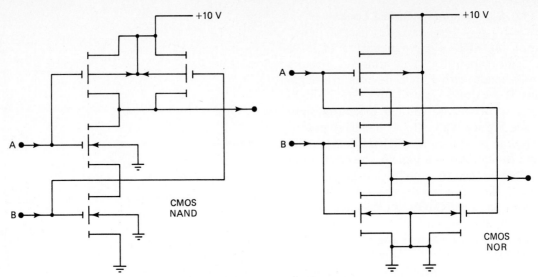

FIGURE 9-26   Other CMOS gates.

CMOS gate, the NAND. Its pull-up section has two *parallel MOSFET* switches driven by 0s—a 0 at either input A OR B will pull up the output. The pull-down section has two series MOSFET switches driven by 1s—a one at both inputs A AND B will pull down on the output. The result: a NAND gate whose output is LOW if A AND B are HIGH, and HIGH when A OR B is LOW.

The design philosophy embodied in the CMOS family of logic devices is active pull-up and active pull-down. To design a CMOS device it will be necessary to provide a switch that conducts a 1 to the output for every HIGH condition, and also separate switches to conduct a 0 to the output for every LOW condition. Two important conditions follow from this; a 1 or 0 must *always* be conducted from somewhere in every case, and there should *never* be a 1 and a 0 conducted to the output at the same time.

A note on the MOS families of logic in general, and CMOS in particular: These devices are extremely "touchy" when subjected to static electricity. They are easily zapped by just being touched. If you touch a MOS device with a few hundred volts of static charge on a fingertip (the kind of static that would make a spark as you touched a doorknob on a dry winter day), the MOS device's inputs have such high resistance that the voltage builds up to a dangerous level before the charge "bleeds off" to the rest of the components in the circuit. These inputs cannot stand 50 V, much less a couple of hundred, and the inputs will be destroyed. Various devices have been added to MOS circuits to bleed off charges at the inputs, with some success, but MOS devices of all kinds

require special handling precautions. (More later.)

TTL devices have a much lower input resistance, and the same static spark that would zap a MOSFET does no harm to the TTL gate because the charge leaks off so quickly that the voltage can't build up to a dangerous level.

9-5   EMITTER-COUPLED LOGIC

We have just finished looking at the two most popular families of logic, TTL and CMOS. They represented the most commonly used family of bipolar logic and the most popular family of MOS logic, respectively. There are other forms of gate logic available which are not so widely used, but have applications for special purposes. The **emitter-coupled logic** (ECL) family of bipolar gates is the fastest bipolar logic available. It costs more than TTL, uses voltages that are not compatible with standard TTL or MOS, and is more power hungry (consumes more watts per gate), but it is several times faster than TTL. In applications where cost is not the only consideration and speed is of the greatest importance, ECL is the natural choice.

The schematic of a typical ECL gate is shown in Figure 9-27. It is an OR/NOR gate with an output for either OR or NOR logic. The symbolic diagram for a gate with two outputs like this is shown next to the schematic. The major important difference between ECL and other families of logic is the fact that the transistors in the ECL circuit are *never saturated*. Transistors are satu-

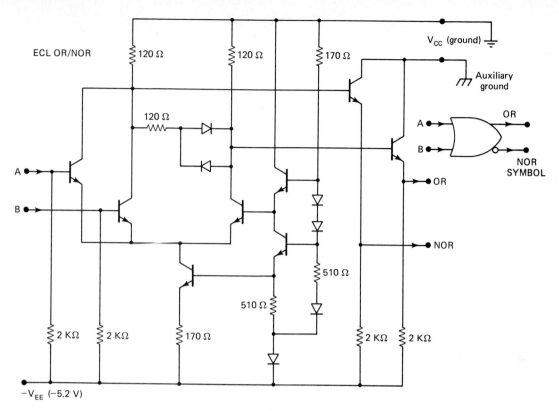

FIGURE 9-27 ECL OR/NOR gate.

rated in RTL, DTL, and TTL logic by being given a larger current at their inputs than the minimum needed to switch the transistor ON. Most of the time, transistors in RTL, DTL, or TTL circuits are either cut off (no base current at all) or they have many times the current needed to switch ON the transistor (they are saturated). In the ECL circuit, the input transistors are coupled together in a differential amplifier with an emitter resistor that makes it impossible to saturate. The transistors are always just over the edge of being switched ON, but are never overdriven. The switching time required to switch the transistors OFF and ON is much less in ECL than TTL because there is a smaller voltage swing between OFF and ON.

One disadvantage that appeared in ECL as more-advanced gates were developed was that the faster they become, the more power they consumed. This limits the number of gates that can be put on a single chip. Because each gate develops so much heat, large-scale integration (LSI) integrated circuits with thousands of ECL devices on a chip would *burn up* if integrated at the same density as MOS devices. It is for this reason that more chips and more silicon must be used to implement any logic function in ECL than in MOS, TTL, or the other families of logic we've

mentioned. This costs more money, so this is why ECL systems are more expensive to build than TTL or MOS systems.

9-6. INTEGRATED INJECTION LOGIC

Integrated injection logic (IIL) is a family of bipolar logic which uses bipolar transistors in a form of open-collector logic. It has nearly the speed of ECL logic and lower power consumption than any other family of bipolar logic (almost as low as MOS). With all these things going for it, it's a puzzle why IIL (also called I-squared-L) logic hasn't caught on better than it has. Developed by Texas Instruments in the 1970s, IIL's only drawbacks are a rather low signal amplitude (more easily swamped by noise?) and possibly a lack of second sources (other than TI) for products of this family. There are a very limited number of gate devices available in this family; almost everything available is a LSI or VLSI device such as a microprocessor or integrated sound-synthesizer chip.

The transistor in Figure 9-28 is an inverter all by itself, if you consider it an open-collector inverter. The two states possible at the output are LOW and floating. By adding the "injector" cir-

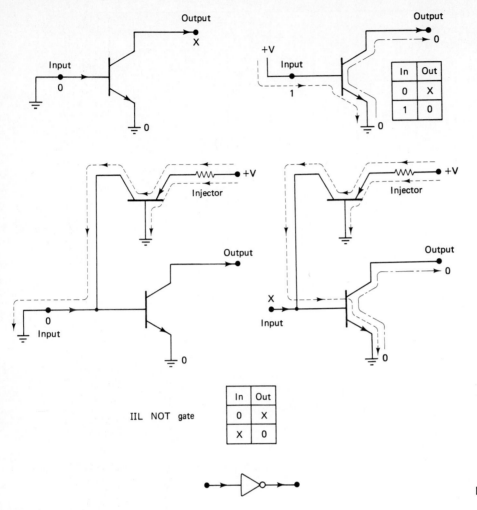

IIL NOT gate

In	Out
0	X
X	0

FIGURE 9-28   IIL NOT gate.

cuit to the base input of the transistor inverter, its input needs only a LOW or float condition to operate. The LOW condition shown provides a current path to ground for the injector current, while the input floating condition forces the injector current to take a path through the base–emitter junction of the transistor, making it conduct. Since the transistor is between ground and the output, when it conducts, the output becomes LOW. A truth table with the input and output conditions for this one-transistor inverter is shown next to the diagram.

In Figure 9-29, the classic circuit of IIL, a NOR gate, is shown, illustrating the other primary technique involved in gate construction with this family of logic: *wire-or* connection. Combining the open-collector inputs and outputs of two IIL inverters by tying the outputs in a wire-or gives a circuit that can pull down its output whenever an input floats. If we just replace the HIGH state of the ordinary NOR with the floating condition everywhere in its truth table, we get the truth table of the IIL NOR. For all practical purposes the floating condition *is* the HIGH

condition in IIL. This tendency of the inputs to "see" a floating state as a HIGH is a direct "clone" of the way TTL operates, as we'll see in the next section. Other gates, built up from the NOR plus inverters, are shown in Figure 9-29 with gate symbols illustrating how the combinations give the logic desired when simplified.

9-7   CHARACTERISTICS OF DIFFERENT LOGIC FAMILIES

In this chapter's discussion of each family of logic, certain details were left out of the survey of these devices. Now we can explore these details as we compare the characteristics of the different families of logic.

From among the gates available in each family, we have selected one example whose "statistics" represent the family. Wherever specific information on voltages, speeds, or other characteristics is compared between families, the sources for these comparisons will be the products listed below.

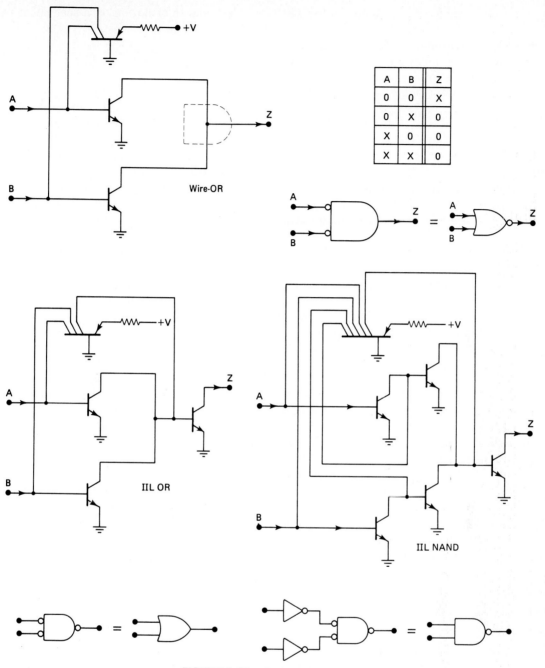

A	B	Z
0	0	X
0	X	0
X	0	0
X	X	0

FIGURE 9-29 Other ILL gates.

Diode logic. The diode-logic gate is not available as an integrated circuit, hence no product type is listed that describes the DL gate. The circuit used to develop the electrical characteristics of the gate is the one shown in Figure 9-2.

Diode–transistor logic. The typical DTL gate is one of the 930 series, a National Semiconductor 946 quad-NAND gate. This is an integrated circuit with four gates in one package, identical to the DTL NAND gate shown in Figure 9-8.

Transistor–transistor logic. The typical standard TTL gate is a Fairchild 7400 quad-NAND gate. (There are zillions of other companies making these, too.) The circuit for a TTL NAND gate of this type is shown in Figure 9-11. There are four of these in the 7400 package.

Open-collector TTL gates are also a part of the 7400 series. While standard TTL gates usually have even numbers, the "exotic" members of the family have odd numbers. A typical open-collector TTL gate is the 7401 quad-NAND. The circuit in Figure 9-15 depicts one of these gates.

Four gates identical to this are found in the 7401 package.

Schottky TTL gates are also a part of the 7400 series, but are identified as Schottky by a special "S" designation: for example, 74S00. A representative Schottky gate is the NAND gate seen in Figure 9-19. The low-power Schottky version is more popular, and is designated "LS" instead of "S." A 74LS00 package contains four LS NANDs in a package that matches the 7400 pin for pin.

As a typical Schmitt trigger gate, we chose the NAND gate shown symbolically in Figure 9-20. The NAND is part of a 74132 quad-NAND Schmitt trigger.

MOSFET logic. NMOS and PMOS gates are used in a variety of multigate and large-scale integrated-circuit applications. There really isn't any such thing as an NMOS or PMOS circuit like a 7400 NAND which you can use *by itself.* These gates are tucked away deep inside complex circuits like microprocessors, where you can't get at them one at a time.

The most common CMOS family is the 4000 series, which has almost as many types of gates as the 7400 TTL family. The RCA 4011 is a quad-NAND identical in function to a 7400 TTL, but different in pinout and electrical characteristics.

A family of CMOS gates exists which is pin compatible with the TTL 7400 series. A typical gate of this series is the 74C00 quad-NAND by Intersil, which has exactly the same connections (pin-for-pin compatible) as a 7400 IC, but contains four of the circuits shown in Figure 9-26. It's basically a 4011 with the pins moved around to look like a 7400. (Either 4011 or 74C00 could be regarded as a typical CMOS, with 4011 a little more typical.)

Emitter-coupled logic. As a typical ECL circuit, we choose the Fairchild ECL 10,000 series, represented by the OR/NOR gate shown in Figure 9-27.

Integrated injection logic. IIL is not fully standarized yet—manufacturers should be consulted about the electrical characteristics of devices made with IIL technology. One device of this type is the TMS 9900 microprocessor from Texas Instruments, which exists in an IIL version.

A "spec sheet" (specification sheet) for one of the devices listed above (a 7400 quad-NAND) is provided (Figure 9-30) to assist you in seeing how information is obtained from one of these documents. We'll see how to interpret various electrical information on a spec sheet in the following sections.

9-7.1 Source Load and Sink Load

The inputs of different logic gates have to be driven in different ways. For example, to fan out a signal to a number of diode OR gates, there would be no problem sending a 0 to as many OR gates as you'd like, since the 0s produce no current, but a 1 (HIGH) level would drive current into each input, and the more inputs attached to the driver, the more current it must provide. This type of fanout situation is called a **source load**. The driving gates must be a source of current (positive charge flow for positive-logic gates).

In the same family of logic, the diode AND gate does not accept any current into its diodes when the inputs are HIGH. There would be no problem sending a 1 to as many diode AND gates as you'd like, but sending 0s would be a very different proposition. Each 0 drains current out of the pull-up resistor attached to the input diode. Since the driver is draining the current out of each input, the more inputs attached to the driver, the more current it will need to be able to sink to ground. This type of fanout situation is called a **sink load**. The driving gate must be able to sink positive charge flow into the ground for all the driven gates' inputs.

Since DTL gates' inputs are diode-logic gates, you'd expect the DTL OR and NOR gates (which have diode-OR front ends) to be source load, and the DTL AND and NAND gates (which have diode-AND front ends) to be *sink load*. If you did expect this, you'd be right.

TTL gates are all based on the diode-AND or DTL-AND type of input; since the DL and DTL ANDs are sink load, so are the TTL types. An illustration of this is shown at the bottom of Figure 9-31. A 1 can be driven to fanout to any number of TTL inputs you want, but a 0 will have to sink current from each input it tries to drive. To make the inputs LOW, TTL fanout must be able to sink a current 10 times the value caused by driving one input. This is shown in the illustration to be 1.1 mA. The drive current of one input is called a **unit load,** and fanouts are measured in multiples of the unit load rather than in terms of current or amperes. Saying that a TTL device has an output with a fanout of 10 is the same as saying that the TTL gate can sink 10 times as much current as is needed to pull down a single input (1.1 mA). Most TTL outputs can sink more current than this, but a fanout of 10 is a safe figure which even the weakest of normal TTL gates'

QUAD 2-Input NAND Gate

Logic and connection diagram

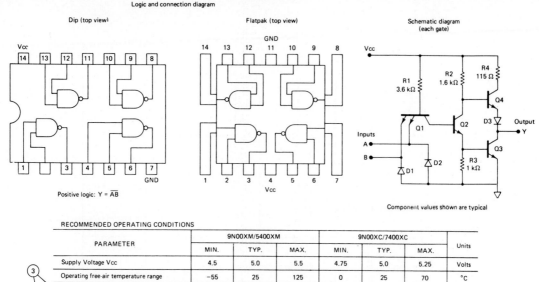

Dip (top view)

Flatpak (top view)

Schematic diagram (each gate)

Positive logic: Y = $\overline{AB}$

Component values shown are typical

RECOMMENDED OPERATING CONDITIONS

PARAMETER	9N00XM/5400XM			9N00XC/7400XC			Units
	MIN.	TYP.	MAX.	MIN.	TYP.	MAX.	
Supply Voltage Vcc	4.5	5.0	5.5	4.75	5.0	5.25	Volts
Operating free-air temperature range	−55	25	125	0	25	70	°C
Normalized Fan-Out from Each Output, N			10			10	U.L.

X = package type; F for Flatpak, D for Ceramic Dip, P for Plastic Dip. See Packaging Information Section for packages available on this product.

ELECTRICAL CHARACTERISTICS OVER OPERATING TEMPERATURE RANGE (Unless Otherwise Noted)

SYMBOL	PARAMETER	LIMITS			UNITS	TEST CONDITIONS (Note 1)	TEST FIGURE
		MIN.	TYP. (Note 2)	MAX.			
V_{IH}	Input HIGH Voltage	2.0			Volts	Guaranteed Input HIGH Voltage	1
V_{IL}	Input LOW Voltage			0.8	Volts	Guaranteed Input LOW Voltage	2
V_{OH}	Output HIGH Voltage	2.4	3.3		Volts	Vcc = MIN., I_{OH} = 0.4 mA, V_{IN} = 0.8 V	2
V_{OL}	Output LOW Voltage		0.22	0.4	Volts	Vcc = MIN., I_{OL} = 16 mA, V_{IN} = 2.0 V	1
I_{IH}	Input HIGH Current			40	μA	Vcc = MAX., V_{IN} = 2.4 V Each Input	4
				1.0	mA	Vcc = MAX., V_{IN} = 5.5 V	
I_{IL}	Input LOW Current			−1.6	mA	Vcc = MAX., V_{IN} = 0.4 V, Each Input	3
I_{OS}	Output Short Circuit Current (Note 3)	−20		−55	mA	9N00/5400 Vcc = MAX.	5
		−18		−55	mA	9N00/7400	
I_{CCH}	Supply Current HIGH		4.0	8.0	mA	Vcc = MAX., V_{IN} = 0 V	6
I_{CCL}	Supply Current LOW		12	22	mA	Vcc = MAX., V_{IN} = 5.0 V	6

SWITCHING CHARACTERISTICS (T_A = 25° C)

SYMBOL	PARAMETER	LIMITS			UNITS	TEST CONDITIONS	TEST FIGURE
		MIN.	TYP.	MAX.			
t_{PLH}	Turn Off Delay Input to Output		11	22	ns	Vcc = 5.0 V C_L = 15 pF R_L = 400 Ω	A
t_{PHL}	Turn On Delay Input to Output		7.0	15	ns		A

NOTES:
(A) For conditions shown as MIN. or MAX., use the appropriate value specified under recommended operating conditions for the applicable device type.
(B) Typical limits are at Vcc = 5.0 V, 25° C.
(C) Note more than one output should be shorted at a time.

Used with permission of Fairchild Semiconductor, a division of Schlumberger Co.
FIGURE 9-30   Data sheet for a typical TTL gate

outputs can pull down. Since the output of the TTL gate's totem pole must "worry" about the amount of current it can pull down, the pull-down circuit is the lesser-resistance better conductor side of the totem pole. We mentioned earlier that the TTL totem pole had a weak pull-up and a strong pull-down; this is why.

In the example spec sheet, there are lines marked (1), containing information about the drive and load characteristics of the logic device.

The lines that tell about the fanout are "I_{OH}" and "I_{OL}." They tell what current this gate's output can drive to a load it is trying to pull up (OH = Output High) and pull down (OL = Output Low).

As an exercise, we want you to obtain spec sheets for as many of the typical devices as you can, and complete Table 9-1 of relative drive and load characteristics. (Don't be surprised, or disappointed, if you can't get a spec sheet for everything.)

TABLE 9-1

Relative drive and load characteristics

	DL ?	DTL 946	Schmitt TTL 74132	TTL 7400	O.C. TTL 7401	LSTTL 74LS00	Tri-State TTL 74126	IIL 9900	ECL 10000	NMOS Z-80	PMOS 5309	CMOS 4011 or 74C00
Unit load fan-in (current)												
Fanout (unit loads)												
I_{OH} (current to another gate if output is HIGH)												
I_{OL} (current from another gate if output is LOW)												
I_{SOURCE} (maximum current that can be used to pull up a load)												
I_{SINK} (maximum current that can be used to pull down a load)												

144

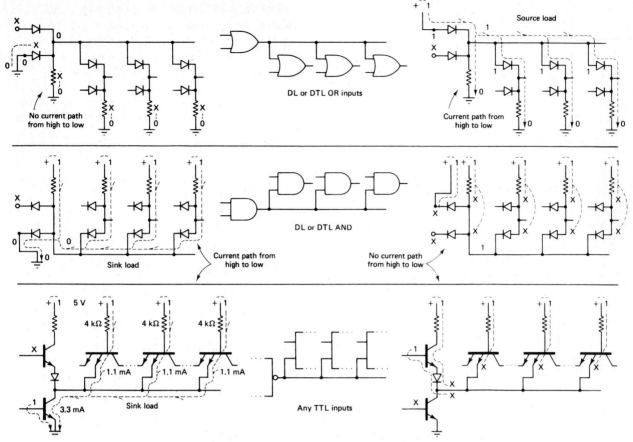

FIGURE 9-31 Source loads and sink loads.

9-7.2 Speed

The comparative speeds of the different families of logic we've discussed are found on the spec sheet on lines marked (2), called propagation delay, rise time, and fall time.

The **propagation delay** is the time it takes an input to produce an effect at the output. It's the time a signal takes to get all the way through the gate, from input to output. Propagation delay time is often different for inputs that cause a "rising" output (called t_{PLH}) and a "falling" output (t_{PHL}).

Rise time and **fall time** are the times it takes an output to "rise" from a LOW to HIGH or "fall" from a HIGH to a LOW. These times depend *only* on the characteristics of the output transistors, not on the size of the total gate.

The fastest of the logic types is bipolar, and the fastest of all bipolar types is ECL. With MOS logic as with bipolar, as the logic becomes faster, it also consumes more power per gate. This is even true of a single gate being operated at various frequencies up to its maximum speed, especially true of CMOS devices. As the switching rate of the CMOS devices is pushed up toward the limit, the low power consumption of these gates becomes less and less of an advantage over bipolar logic. For very fast switching rates, CMOS is consuming power like a TTL gate, and at the top speeds TTL can handle, CMOS can't perform at all.

A new generation of logic devices based on Josephson junction superconductors has been developed which is hundreds of times as fast as TTL and dozens of times as fast as ECL. It also consumes practically no electrical power when running. Its drawback? (There's gotta be a catch, right?) To make it work, the gates (called SQUID, for semiconductor quantum interference devices) must be cooled to within a few degrees of absolute zero. This is hundreds of degrees below zero Fahrenheit or Celsius, and requires one heckuva refrigerator to operate. Not likely to show up in pocket calculators tomorrow, unless you've got good frostbite insurance and don't mind carrying a Frigidaire on your belt.

Seriously, this development promises very advanced computing machinery, on a "cost is no object" basis, for large institutions and the military. Some day, if someone invents a room-temperature superconductor, this breakthrough may affect everyone personally ("ordinary people" such as you and I may have one!). In space, where temperatures near absolute zero can be gotten by just putting something in a shadow for a while, cryogenic circuits like this might find their natural habitat (look for these in future space projects?)

9-7.3 Impedance

The information on the spec sheet marked (3) tells the impedance characteristics of the inputs and outputs of each of the families of logic we've discussed.

Inputs. The impedance of inputs to bipolar and MOSFET logic families are very different. Bipolar inputs have a few thousand ohms of impedance (resistance). The TTL input of a NAND gate, for instance, has a 4-kΩ resistor between the 5-V supply and the Q_1 base. Current flowing when an input (TTL) is driven by a 0 is limited by this 4-kΩ resistance and the impedance of the base–emitter junction, which is small. MOSFET inputs have billions of ohms of impedance. The gate of a MOSFET is separated from all the rest of the semiconductor contacts by a layer of *silicon dioxide* (chemically the same as glass). Even though this layer is only a few millionths of an inch thick, it's a dandy insulator, and is close enough to a complete open circuit that the difference can be ignored.

Since bipolar inputs have low impedance, they generally sink or source fairly large currents. Static MOSFETs require virtually no current, just enough to charge the gate material positive or negative. This means that the gate looks like one plate of a capacitor. At high frequencies (high switching speeds) MOSFET impedance becomes quite low. Current surging into and out of the gates of rapidly charging and discharging FETs can be a source of considerable power consumption in MOS circuits running at high speeds.

Inputs that are sink loads generally float high. That means when nothing is connected to the input terminal; the gate inside acts the same as when its input is connected HIGH. This makes sense when we realize that a DTL NAND or TTL gate of any type has no current through its input terminals when they are HIGH. There would also be no current if the terminals weren't connected to anything. Thus, the circuits with sink load inputs "think" that a disconnected input is a 1 and act the same as if the input were HIGH. It's not a good idea to merely "leave a wire hanging"

when an input should be HIGH; the floating wire will pick up stray electrical interference like an antenna, and will cause the gate to respond to false 1s and 0s. An unused input for a sink load logic gate should be tied to a real 1 or 0.

Inputs that are source loads will float low. Since universal logic with TTL-compatible inputs is designed as sink load logic, there aren't as many gates which float low as those which float high when their inputs are disconnected. DTL OR and NOR gates also have source load inputs, and they act as though their input is LOW when it is disconnected from everything. The same rule applies to letting an input float instead of connecting it LOW. You can do it, but you take your chances, especially in noisy environments near electrical motors, relays, and other devices.

Outputs. Outputs of TTL devices have been described as having weak pull-ups and strong pull-downs. The HIGH (5-V) level at the top of a standard TTL totem pole must pass through two junctions and a 115- to 130-Ω resistor to reach the output. By contrast, the LOW (ground) level at the bottom of the totem pole needs to pass through only one forward-biased junction. In a saturated transistor, this is virtually no resistance. Such outputs can fan out to 10 unit loads.

DL and DTL devices which were built on discrete logic boards could have different fanouts and fan-ins (the size of a *unit load* is a fan-in of 1) by just changing a resistor or two. Thus, their output impedances could be anything that worked. This turns out to be between a few ohms and a few hundred thousand ohms. Generally, DL and DTL outputs were designed to fanout to 10 or more gates of the same type. RTL outputs with a pull-up resistor have strong pull-downs and weak pull-ups. RTL outputs with a pull-down resistor have strong pull-ups and weak pull-downs.

The same is true for the outputs of DTL NAND, NOR, and NOTs which have pull-up resistors. They have strong pull-downs (the switching transistors) and weak pull-ups (the resistors). For the DTL AND, OR, and BUFFER, whose output is an emitter-follower transistor with a pull-down resistor, there is a strong pull-up and weak pull-down.

MOSFET outputs have fairly high impedances: CMOS circuits like those shown in Figure 9-26 clearly have equal pull-up and pull-down capability. In LSI and microprocessor applications the circuits (usually NMOS) can drive one TTL unit load or a large number of MOS unit loads. ECL impedances are quite low compared to TTL.

IIL input impedances, although no higher than standard bipolar devices, are used as sink load with floating and LOW inputs, so the current is limited by the injector, which is designed to provide very small currents to outside drivers. From the outside, the IIL family of logic has many of the characteristics of TTL logic with much higher impedances. The outputs of IIL gates compare most directly with open-collector TTL devices, which have a strong pull-down and no pull-up.

From the spec sheets you obtained to complete the exercise at the end of Section 9-7.1, fill in Table 9-2.

9-7.4 Voltage Levels

The characteristic voltages which represent a 0 or a 1 at the input or the output of each type of gate are shown in the spec sheet on lines marked (4). The items labeled "V_{IH}" and "V_{IL}" describe the input voltages that the gate "recognizes" as a LOW (IL = Input Low) and a HIGH (IH = Input High). (LO = *no more than* V_{IL}; HI = *no less than* V_{IH})

The items labeled "V_{OH}" and "V_{OL}" are the voltages the output puts out when it's trying to be HIGH (OH = Output High) or LOW (OL = Output Low). (HI = *at least* V_{OH}; LO = *at most* V_{OL})

Complete Table 9-3 using information from the spec sheets used for the previous exercises.

9-7.5 Noise Immunity

Noise immunity can be inferred from the difference between HIGH input and ouput levels. If the voltage IH and the voltage OH are 2 V apart, the noise immunity is 2 V. The highest noise—an extra signal superimposed on the "authentic" logic levels—that the inputs could receive without getting "confused" is

$$V_{noise} = V_{IH} - V_{OH}$$

Fill in the rest of Table 9-4 using information obtained from your collection of spec sheets.

EXAMPLE 9-1

Suppose that we have a long cable carrying a signal from a 7400 NAND gate's output to a 7404 NOT gate's input. We are worried because the cable is not shielded, and it passes through an area near motors and other sources of electrical "noise." How much peak noise can be tolerated on the cable connecting the 7400 to the 7404?

Solution: We consult the 7400 and 7404 data sheets in a TTL databook. We find that a 7400 TTL gate output has $V_{OL} = +0.4$ V and $V_{OH} = +2.4$ V (worst-case values), and a 7404 TTL gate input has $V_{IL} = +0.8$ V and $V_{IH} = +2.0$ V (worst-case values).

The *noise margin* for a LOW is

$$V_{IL} - V_{OL} = 0.8 \text{ V} - 0.4 \text{ V} = 0.4 \text{ V}$$

This means that a positive peak (or pulse) added to a LOW output of +0.4 V cannot be more than +0.4 V, or the total will exceed +0.8 V, and the receiving 7404 gate may not see it as a LOW.

The *noise margin* for a HIGH is

$$V_{IH} - V_{OH} = 2.0 \text{ V} - 2.4 \text{ V} = -0.4 \text{ V}$$

This means that a negative peak (or pulse) added to a HIGH output of +2.4 V cannot be more negative than −0.4 V, or the total will be less than +2.0 V, and the receiving 7404 gate may not see it as a HIGH.

For these TTL gates, the peak noise tolerance (for the weakest gates that passed testing) is 0.4 V, peak, in either polarity.

9-7.6 Static Sensitivity

The MOS families of logic are especially sensitive to damage by static electricity. There is little likelihood of damage to bipolar devices due to static. Relative sensitivity to static is partially a function of input impedance. The high impedance of the MOSFET gate insulation layer is responsible for the development of dangerously high voltages when a small static charge flows into the FET inputs. The same charge flowing into the inputs of a bipolar device would pass through its low-impedance input without developing dangerous levels of voltage.

Figure 9-32 shows what happens if a static spark jumps through a $\frac{1}{4}$-cm gap from your fingertip to an input of a TTL [Figure 9-32(a)] and a CMOS device [Figure 9-32(b)]. At the moment the spark jumps, the TTL and CMOS ground connections are connected to 0 V. For a spark to jump that distance requires about 5000 V. There is resistance in the air gap of about 1 gigohm (1 billion ohms). The resistance of the silicon dioxide insulation layer between the CMOS inputs and the substrate is about 9 GΩ. The resistance between the TTL input and the +5-V level is 4 kΩ (4000 Ω). In the CMOS gate, the voltage drop across the air gap is 500 V and the drop across the insulation layer is 4500 V. Since the insula-

TABLE 9-2

Impedance characteristics

	DL ?	DTL 946	Schmitt TTL 74132	TTL 7400	O.C. TTL 7401	LSTTL 74LS00	Tri-State TTL 74126	IIL 9900	ECL 10000	NMOS Z-80	PMOS 5309	CMOS 4011 or 74C00
Output LOW impedance												
Output HIGH impedance												
Input LOW impedance $\left(\frac{V_{SOURCE} - V_{IL}}{I_{IL}}\right)$												
Input HIGH impedance $\left(\frac{V_{IH}}{I_{IH}}\right)$												
Source load or sink load? (or neither?)												
Inputs: float HIGH? or float LOW? (or neither?)												Strong pull-ups and weak pull-downs (source load) or strong pull-down and weak pull-up (sink load)

148

TABLE 9-3

Comparative voltage levels

	DL ?	DTL 946	Schmitt TTL 74132	TTL 7400	O.C. TTL 7401	LSTTL 74LS00	Tri-State TTL 74126	IIL 9900	ECL 10000	NMOS Z-80	PMOS 5309	CMOS 4011 or 74C00
$V+$ (positive power-supply voltage)												
$V-$ (negative power-supply voltage)												
V_{IH} [lowest input 1 voltage level (worst case)]												
V_{IL} [highest input 0 voltage level (worst case)]												
V_{OH} [lowest output 1 voltage level (worst case)]												
V_{OL} [highest output 0 voltage level (worst case)]												

TABLE 9-4

Noise immunity characteristics

	DL ?	DTL 946	Schmitt TTL 74132	TTL 7400	O. C. TTL 7401	LSTTL 74LS00	Tri-State TTL 74126	IIL 9900	ECL 10000	NMOS Z-80	PMOS 5309	Schmitt CMOS 4093 or 74C132
Noise immunity volts (if listed)												
Noise immunity % of V_{source}												
Logic 1 noise immunity ($V_{OH} - V_{IH}$)												
Logic 0 noise immunity ($V_{IL} - V_{OL}$)												
Upper trip point (Schmitt trigger logic only)												
Lower trip point (Schmitt trigger logic only)												

9-7. The current at A is _____ (larger, smaller) than when A = 1.

Fill in each blank assuming that the input A is HIGH.

9-8. Transistor Q_1 is _____. (saturated, cut off)

9-9. Transistor Q_2 is _____. (saturated, cut off)

9-10. Transistor Q_3 is _____. (saturated, cut off)

9-11. Transistor Q_4 is _____. (saturated, cut off)

9-12. Output Z is _____. (pulled up, pulled down)

9-13. Output Z is _____. (HIGH, LOW)

9-14. The current at A is _____ (larger, smaller) than when A = 0.

9-15. The gate above is a _____ (AND, OR, NAND, NOR, NOT, BUFFER)

9-16. TTL devices like this have strong _____ (pull-up, pull-down) and weak _____ (pull-up, pull-down) outputs.

9-17. If the transistors in the TTL device shown here are part of an integrated circuit, are they drawn correctly? _____ (yes, no)

For questions 9-18 through 9-30, refer to the following CMOS gate schematic diagram:

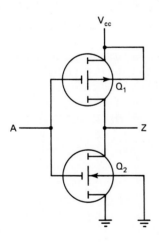

Fill in each blank assuming that the input A is LOW.

9-18. Transistor Q_1 is _____. (saturated, cut off)

9-19. Transistor Q_2 is _____. (saturated, cut off)

9-20. Output Z is _____. (pulled up, pulled down)

9-21. Output Z is _____. (HIGH, LOW)

Fill in each blank assuming that the input A is HIGH.

9-22. Transistor Q_1 is _____. (saturated, cut off)

9-23. Transistor Q_2 is _____. (saturated, cut off)

9-24. Output Z is _____. (pulled up, pulled down)

9-25. Output Z is _____. (HIGH, LOW)

9-26. The input impedance of the CMOS gate above is _____ (more, less) than that of the TTL gate shown earlier.

9-27. What does the C in CMOS stand for? _____

9-28. The gate above is a _____. (AND, OR, NAND, NOR, NOT, buffer)

9-29. Are the outputs of either the TTL or CMOS examples above ever in the Tri-State, or high-impedance, state? _____ (yes, no)

9-30. The fastest of all electronic logic devices is the Josephson junction, but unlike transistors and diodes, it is not a semiconductor. The Josephson junction is a _____.

END-OF CHAPTER LESSON: TESTING WITH SIGNAL INJECTOR AND OSCILLOSCOPE/LOGIC PROBE (TROUBLESHOOTING TTL-COMPATIBLE LOGIC FAMILIES)

1. *Look before you leap.* In many cases, a visual inspection of the circuit board or section of the system you think is not working will reveal what is wrong. Testing for opens and shorts with a continuity tester—often just the ohmeter in your VOM—may make advanced testing with computerized logic analyzers unnecessary, if you know where to look.

2. *Shake 'n bake?* In environments where socketed chips are exposed to vibration, chips may actually be shaken out of their sockets! An empty socket is a dead giveaway in even the most cursory of visual inspections. A chip that is loose in its socket but still held in by a few pins (although not enough to make it work) is a tougher one to spot. In a socketed board, especially one in portable or mobile equipment, *always* start by pushing all the chips down in their sockets until they bottom out. Before doing any fancy troubleshooting, power-up the system to see if that fixed it.

3. *When nothing obvious shows up visually, signal-trace.* When a signal is injected at an input and does not appear at an output where is it *supposed* to, the fault is not always with the chip.

The socket it is placed in, or the connections to the circuit board (if it is soldered in), may be defective. For example, suppose that a multiplexer's channel-select inputs BA are sent the binary code 10 (two) and the output does not match the input at channel 2. There are several possibilities. The output could match the input of another channel—let's say, channel 3—or it could be stuck HIGH or stuck LOW, or be floating.

a. *Output from wrong channel (wrong address).* If an output signal from channel 3 is coming out when channel 2 should be selected, it may be that the IC pin going to select line A is loose or disconnected from the socket contact or circuit board trace. If this pin floats (disconnected) and is "seen" as a HIGH condition—which is normal for TTL—the chip will "see" the binary select code at BA as 11 (three), even though 10 (two) may be the actual code on the traces leading to those pins.

b. *Output stuck HIGH.* An open connection at the channel 2 input pin, for instance, would be interpreted as a HIGH input by the chip logic. This could be correctly delivered to the output, but would result in a flat output stuck HIGH, although a waveform might be measurable on the input 2 trace leading to the integrated circuit.

c. *Output stuck LOW or floating.* Another symptom may have the output stuck LOW. This might happen if the multiplexer unit is *not enabled*. The active-LOW enable signal may have a disconnected pin, which would then float high, and the chip would not be enabled. Its output would *never* be enabled in that case, no matter what happened at the select lines or channel inputs. It would be LOW at all time, in simple multiplexers, or floating in the case of Tri-State multiplexer outputs.

4. *Is it the chip?* If you measure the signals on the traces leading into the socket or position on the board where the chip is soldered-in, do not assume that the chip is at fault. Always measure the signal by placing the point of the probe tip against the "shoulder" of the pin where it bends to go into the IC package itself. A signal that can be measured on the foil or trace leading to the IC socket, but is not found (or not the same) on the IC pin itself, is an indication of an open or incomplete connection. In TTL logic, such as the 74153 mentioned earlier, its normal response is to float HIGH at an unconnected input. The output would then be stuck HIGH. A disconnected output pin may reveal the correct output at the "shoulder" of the pin, although the foil trace leading away from the chip may be flat zero on the oscilloscope. That disconnected trace may look like a 1 to the next chip even though it doesn't to an oscilloscope or logic probe.

A logic probe or oscilloscope will not always show a LOW (but disconnected) pin, as a 1. In fact, even with the probe tip at the shoulder of the chip, the appearance of a disconnected LOW input may look OK. With the power off, you can check the "solidness" of each contact, using an ohmmeter as a **continuity tester**. Between the foil trace and the shoulder of the chip, the resistance should measure zero. If it is *not* zero, the connection is bad. Remove the chip to see if it has been "stuffed" into the socket or circuit board so that this pin is bent. This is a common fault with automatic chip-stuffing equipment, and even more common with *human* chip stuffers! If the pin is not bent, the socket may be faulty; another socket is the cure for that. If the chip is not socketed and the pin can clearly be seen protruding through the foil side of the circuit board, try reheating and resoldering the connection. If there is a cracked pin, it will probably fall off when this is tried. If a loose pin or cold-solder connection is at fault, this will cure the problem. Often, a second visual inspection (now that you have an informed eye and a suspicious mind) will reveal where an open between the foil and the pin is located. A broken trace on a circuit board may be bridged with a short length of wire soldered to the remaining foil on either side of the break.

If continuity testing does not reveal any disconnections between input pins and the signal traces that lead to them, and all the important inputs necessary to get an output are there, a "no output" condition indicates a defective chip. If all signals have been traced OK right up to the shoulder of each input pin, and if the output is faulty right at the shoulder of the output pin, *then*, and *only then*, do you replace the chip!

5. *Socketed boards and tarnished pins.* Many socketed circuit boards have been "repaired" by replacing chips in the sockets, with the result being a (temporary) restoration of circuit function. The original DIP package wasn't faulty but had formed oxide (an insulating layer of tarnish) on its pins. A fresh chip, with clean pins, worked fine in the same socket—until tarnish formed on its pins. Meanwhile, a perfectly good chip with tarnished pins was trashed when all that was needed was a good cleaning.

Often, simply unsocketing the DIP from the board (as in an inspection for bent pins folded under the package) results in scraping enough oxide off the pins to restore a good contact when the chip is resocketed. Occasionally, pressing down with the logic probe tip or oscilloscope tip re-

stores contact momentarily. The circuit then works while you test it, and dies after you stop (it may even work fine until the day after it's been returned, which causes your stock as a repairperson to plummet like you wouldn't believe). Whenever faced with a socketed chip or chips in a circuit board, and an intermittent or "phantom" problem, unsocket, inspect, and reseat chips, using a nonresidue contact cleaner on any tarnished-looking contacts.

Caution: Do *not* use contact cleaner that contains a lubricant! These are often oils, and at 5-V levels are better insulators than the tarnish they remove. In some cases, the lubricant actually conducts electrically; when it gets into places that *shouldn't* conduct, things can get really bad!

6. *Don't assume TTL characteristics.* If the equipment you're troubleshooting contains chips that are not TTL, mixed in with ones that are, don't assume that NMOS or CMOS chips called "TTL compatible" really have exactly the same characteristics as TTL, such as "floating-HIGH" inputs. Some "TTL-compatible" chips have MOS inputs that float LOW, instead of HIGH, when disconnected. The bipolar NE555 timer and the CMOS TLC555, for instance, are pin-for-pin compatible. When I disconnected the active-LOW trigger pin of an NE555, it floated HIGH (not triggered) while the disconnected TLC 555 input floated LOW, and triggered itself continuously! The spec sheet packed with the chip, of course, said nothing about this little discrepancy; they couldn't imagine anyone needing to know it! Clearly, the one wasn't an exact replacement for the other in *all* circuits.

PART II
Synchronous Circuits

10
INTRODUCTION TO FLIP-FLOPS

CHAPTER OBJECTIVES

By the time you finish this chapter, you will be able to:

1. Describe the behavior and characteristics of the following latching elements:
   a. NAND latch (R-S latch, nor gated)
   b. NOR latch
   c. Clocked (or gated) R-S latch
   d. R-S master/slave flip-flop
   e. D-type latch
   f. J-K flip-flop (without master/slave)
   g. J-K master/slave flip-flop
   h. T-type flip-flop
   i. Other edge-triggered flip-flops
   j. Bistable, monostable, and astable multivibrators
2. Identify symbolic diagrams and truth tables for the above-named items.
3. Describe some applications for the above-named items.

In previous chapters we looked at circuits that used logic gates for various applications. One thing that all of these circuits had in common was the fact that a certain input would cause an appropriate output, but as soon as the input was removed, the circuit would forget what it was doing. Circuits like this could all be made of *momentary contact switches*. To understand why this type of switching logic can not do all the things we might want a logical system to do, let's imagine what would happen if we used a momentary contact switch to turn on the lights in a room:

If the light switch in your room is a push-button, you can light up the room when you come home by pressing on the button. When you release the button, the room gets dark again. The only way to get anything done is to (1) have the thing you want to do within arm's reach of the light switch and only do things that need one hand, or (2) only do things with two hands that can be done in the dark.

This is pretty silly. What's really needed is a light switch that you can turn ON that will remember to stay ON when you take your finger off the switch. Switches like this are called **latching switches**.

A good latching switch will also remember to stay OFF after you turn if off and walk away.

We would like to have a solid-state "no-moving-parts" circuit that does what a latching light switch does. The key to this type of circuit is its *memory* capability. How we can make a circuit that will remember anything, we'll see in a minute. Before we look at the logic gates that do this, however, it's helpful to think about one of the ways people remember something when they want to be able to repeat it exactly at a later time. Human beings have a *short-term memory* that can recall a lot of very detailed information for a few minutes, but forgets more and more detail as time goes by. The *long-term memory* of a human being can hold detail a lot longer, with a lot less forgetting, but the original learning takes a lot more work. How does a human being remember something in exact detail for more than a few minutes without going to the effort of memorizing it?

Picture a small boy, sent to the store by his mother, told to buy two loaves of bread and a gallon of milk. He didn't write his shopping list down on a sheet of paper, and there are a lot of distractions on the way to the store. How is he going to remember what to buy all the way to the store? He repeats to himself, "Two loaves of bread and a gallon of milk—Two loaves of bread and a gallon of milk—Two loaves of bread and a gallon of milk," all the way to the store. The things he does

and the distractions along the way won't make him forget, as long as he repeats this to himself all the time until he gets to the store.

So what (you say to yourself) does this have to do with digital logic circuits?

What the little boy is doing, the engineer would call *feedback*. Part of the information the boy wants to remember is *output* (spoken), and carried by the air back to his ears, where it is *input* (heard). As long as the boy repeats what he hears, and hears what he repeats, he won't forget.

10-1 NAND LATCH (R-S FLIP-FLOP)

In a circuit, the same things have to happen. There must be an original input (what the boy's Mom tells him), which produces an output (the little boy repeats what he heard). Then there must be feedback (the little boy hears his own voice, and his short-term memory remembers the information until he repeats it again). Digital logic circuits also have a sort of short-term memory based partially on the *propagation delay time* that it takes for an *output* to change after the *input* changes.

10-1.1 Latch Action

Look at the *NAND latch circuit* in Figure 10-1 (also called a *flip-flop*). There is a set of *primary inputs* (called S and R) used to operate the NAND gates. (Think of the signals that enter these wires as the boy's original instructions from his Mom.)

The outputs (called Q and $\overline{Q}$, for reasons we'll explain later) are fed back into a set of *feedback inputs* on the NAND gates, and used to keep the NANDS in the same logic state as the primary inputs. (We'll run through this in detail in a moment.) Since the outputs last for a little while after the input goes away the feedback keeps repeating the message to the NAND gates after the primary inputs go away.

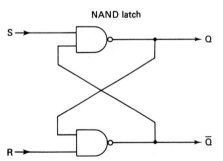

FIGURE 10-1 Cross-coupled NAND latch (7400IC).

To look at how this happens in detail, there is one important feature of the NAND gate you'll need to keep in mind: If any input to a NAND becomes LOW, the output will automatically become HIGH. Another way of saying this is to say that the input to the NAND is *active* LOW. A LOW input will always make a 1 come out.

Keeping this in mind, think of S and R as active LOW inputs. If S, for instance, is a HIGH level, there's no way to tell what's going to come out. That depends on the other input. If S is LOW, the output of the NAND gate, Q, is going to be HIGH, no matter what the other input is doing. As long as we know that there's a LOW at S or R, we know that the respective outputs, Q or $\overline{Q}$, will be HIGH.

The two NAND gates in the **latch circuit** are *cross-coupled* (each output crosses over to the other gate's input). As an example of what the latch circuit does, suppose that inputs for the circuit are S = LOW and R = HIGH. When S = LOW, that makes Q = HIGH. Since both inputs to the bottom NAND gate are HIGH, its output, $\overline{Q}$, is LOW. (Notice that Q is the opposite of $\overline{Q}$.) This state, with the output Q ON is called **set**. The latch is set when its output is ON. Whenever we refer to "the" output of a latch, we're talking about Q unless we say otherwise.

Remove the LOW input from S (and let it become HIGH). The circuit should stay set. Since we want this circuit to behave like a light switch, the LOW at S is like the push on the switch that turns the light ON. Removing the LOW at S is like taking away the finger that's pushing the button. The light should still stay ON. We expect the latch to stay ON at Q. Let's see if it does (Figure 10-2).

The output of $\overline{Q}$ is LOW, and is connected to the feedback input of the top gate. Since the top NAND gate has two inputs, a LOW at S isn't

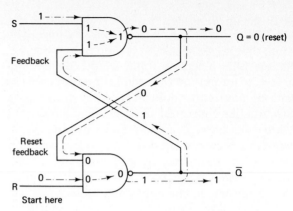

FIGURE 10-3 Cross-coupled NAND latch (RESET).

needed after the other input receives a LOW from $\overline{Q}$. This means that the *NAND latch* will stay set even though the active (LOW) input at S is gone.

Figure 10-3 shows what happens when S = HIGH and R = LOW. When R = LOW, the output $\overline{Q}$ = HIGH. Actually, the circuit is symmetrical; if you flip over the diagrams in Figure 10-2, and keep the letters S, R, Q, and $\overline{Q}$ where they were, you'd have Figure 10-3. Notice that when everything's done, Q is OFF. The state of the latch when its output is OFF is called a **reset** condition.

Is the reset condition "remember"? Since the LOW input at R is the important input, making S = HIGH and R = HIGH removes the active (LOW) input that did the resetting. If the R = LOW input lasts long enough for Q to become LOW, then the NAND that was getting R = LOW doesn't need it anymore. The LOW level on Q, which is fed back through the cross coupling, is enough to keep everything **stable**.

What we've seen in Figures 10-2 and 10-3 are two stable conditions in the latch circuit. A condition is stable if it remains when the input that caused it is no longer active. Certainly, the set and reset are stable, because they're both remembered when the S and R inputs are now low.

At this point, you've probably already guessed that the inputs S and R stand for set and reset, respectively. To keep track of things, just remember:

Set = turn Q ON

Reset = turn Q OFF

10-1.2 Latch Equation and Truth Table

If we think of the NAND latch as a new kind of logic gate, one with a memory, there must be a

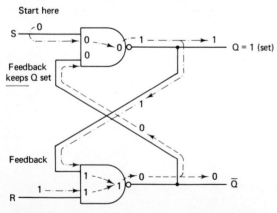

FIGURE 10-2 Cross-coupled NAND latch (SET).

Boolean expression and a truth table for this gate, as we had for all gates in the past. First, to write a Boolean expression that says what a latch does, we need two expressions: one for Q and one for $\overline{Q}$.

The first output, Q, is the output of a NAND gate. We have already seen how to express the behavior of a NAND gate in Boolean algebra. The NAND gate output, Z, for a gate with inputs called A and B was

$$Z = \overline{A \cdot B}$$

In the NAND latch, the inputs for the gate whose output is called Q are S and $\overline{Q}$. This makes the Boolean expression for this gate

$$Q = \overline{S \cdot (\text{not } Q)}$$

According to DeMorgan's theorem, there is a second way to express this output: break the bar and change the sign. This gives us the expression

$$Q = \overline{S} + \overline{(\text{not } Q)}$$

If we examine the expression above, we find $\overline{(\text{not } Q)}$. This is the same as inverting Q twice, which is Q. We can rewrite the expression for the Q output of the latch as

$$Q = \overline{S} + Q$$

Now, let's examine our handiwork. Turning the expression into words, it says: "Q is ON, *if* S is LOW, OR Q *was* ON." Notice our use of the word "was" in the expression. This is probably the first time you have seen a Boolean expression where the *past state* of an input affected the *present* output. It is also probably the first time you have

seen an expression for an output which included the *output itself* as one of the *inputs*. Because this circuit has memory, the past state of its output affects the present state of that output. In the logical expression of the circuit, the past state appears in the expression for the present state.

There is a second NAND, whose output is called $\overline{Q}$. Its inputs are R and Q. The Boolean expression for this gate is

$$\overline{Q} = \overline{R \cdot Q}$$

In its DeMorganized form, this is

$$\overline{Q} = \overline{R} + \overline{Q}$$

This says: "$\overline{Q}$ is ON, *if* R is LOW, OR $\overline{Q}$ *was* ON." Again, we see the use of the word "was." Again, also, the output appears in the expression as an input. This shouldn't be much of a surprise, since we know that the present state of the latch circuit's output depends on its past state. What is true for the Q output should be true for $\overline{Q}$, also.

Figure 10-4 is the truth table for a NAND latch. It shows the effect of inputting 1s and 0s at S and R. There are two lines that have special importance when we compare them to truth tables for gates shown previously. The top line of this truth table has an unusual output shown for Q. The output shows a "?" mark, because the latch is being set and reset at the same time. Latch circuits get confused when you tell them to be both ON and OFF at the same time, and an unstable state results. The truth table's bottom line has an output with a letter in it instead of a 0 or a 1. The output marked with "Q_{t-1}" means that the present Q is the same as the value of Q *was* before the present inputs were applied to the

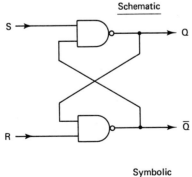

Inputs		Output	Truth table
S	R	Q_t	
0	0	?	Constraint
0	1	1	Set
1	0	0	Reset
1	1	Q_{t-1}	Latch

$$Q = \overline{S} + Q$$
$$\overline{Q} = \overline{R} + \overline{Q}$$
Boolean expression

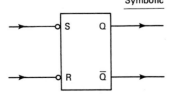

FIGURE 10-4 Latch truth table and symbol.

circuit. This is called a **latch condition**. Since latching is the process of remembering, the line on the truth table where "everything remains the same" is called the "latch condition." On this line, we're neither setting nor resetting the flip-flop, so it remains in whatever state it was before both inputs became 1.

The middle lines of the truth table are labeled "Set" and "Reset." The inputs are active LOW, so the set line is the one on which S is a 0, and the reset line is the one on which R is a 0.

Trying to set and reset the Q output of the flip-flop at the same time would be silly. There's no way for the flip-flop to be ON and OFF at once. That's why the top line of the truth table is labeled "constraint" (which means "avoid doing this"). Nobody would deliberately try to turn a switch ON and OFF at the same time. The latch circuit does have some kind of output when both S and R are active, but it's not stable (doesn't last when the S and R are inactivated).

10-1.3 NAND Latch as a Switch Debouncer

In Figure 10-5 we see an application of the NAND latch to a practical problem. The NAND latch is attached to the two "throws" of a single-pole-double-throw (SPDT) switch. In one position, the switch will conduct a 0 to S. In the other position, the switch will conduct a 0 to R. When the switch is between positions, S and R will be 1, and nothing will happen to the state of the latch; if it was set, it'll remain set; if it was reset, it'll remain reset.

Notice that the constraint condition (where both S and R are 0) never happens. This means that this latch can do only one of three things at any time. It can be setting, resetting, or latching.

The circuit of Figure 10-5 is called a **switch debouncer** because the metallic contacts of an SPDT switch "bounce" as they make contact. In the space of a few milliseconds, the metal contacts may hit, rebound, hit again, rebound again, and hit again hundreds of times before settling down to a steady contact. To a person switching the contacts, this might sound like just one click, but to high-speed digital logic circuits, each bounce lasts long enough to affect the state of the circuitry. A bouncy switch looks, to the digital logic, like a switch being pressed hundreds of times. How can a mechanical switch with metallic contacts be made to operate a circuit just once?

The debouncer has memory. If the switch contacts hit the top of the circuit (S), the Q output

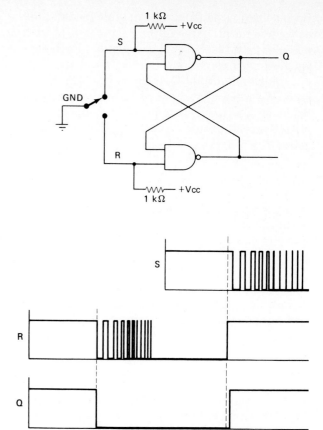

FIGURE 10-5 "Debounced" switch.

will remain HIGH unless the movable contact bounces all the way to the bottom of the circuit. This is unlikely to happen, even with very springy metal in the contacts. Although the switch contacts may bounce in and out of contact with the S terminal, the NAND latch will stay set between bounces. If the switch contacts hit the bottom of the circuit (R), the Q output will remain LOW unless the movable contact bounces all the way to the top terminal. Since the NAND latch will remain reset even though the switch bounces in and out of contact with the R terminal, the LOW output at Q will not bounce, even though the switch does. This is how a single, sharp-edged LOW-to-HIGH transition or HIGH-to-LOW transition is produced at the output of the debouncer even though the contacts bounce hundreds of times.

The resistors in the picture are attached to the inputs to provide a HIGH logic level when the movable contact of the switch (which is LOW) is not touching an input. These pull-up resistors ensure that there is a reliable logic 1 level at any input that's not connected to GND (a logic 0 level).

10-1.4 Constraints and Propagation Delay

The switch debouncer circuit used the NAND latch for a practical purpose. One of its main features was the fact that only the set, reset, and latch inputs were possible. The switch could set the latch (at the top position), reset the latch (at the bottom position) or latch (neither set nor reset, when in between the top and bottom contacts). There was no way for a constraint input to be applied, since the movable contact could never be at top and bottom at the same time. There are other applications for the latch circuit, but none of them use the constraint condition for anything. Does this mean that, in all other circuits, the constraint condition is impossible, as it is in the switch debouncer?

Unfortunately, no. Look at the truth table and associated diagrams (Figure 10-4). The symbolic diagram of the NAND latch has active LOW inputs. S and R have a "bubble" shown at the place where the line joins the box. This means that to set and reset the latch at the same time, all that must happen is for a 0 to appear at S while there's still a 0 at R (or vice versa). Although no one would do this deliberately, it might happen accidentally, in the following way.

The inputs at S and R are usually driven by other logic gates. These gates do not all have the same propagation delay time, especially if the signals at S and R travel through paths with a different number of logic gates in them. If one input (suppose that it's S) is supposed to receive a 0 when the other input (in this case, R) is supposed to receive a 1 (but R has a 0 until the 1 arrives), what will happen if the 0 arrives *before* the 1? Right! Through no fault of the logic, the

latch is going "berserk" because one signal arrives at a different time than the other. If any circuits are attached to the latch's Q and not-Q outputs, the Q circuits will be activated at the same time as the not-Q circuits. This should never happen!

The NAND latch, like all previous gate circuits, has **asynchronous logic**. The name comes from Greek roots, meaning: "a" (not) "syn" (same) "chron" (time). It means that the inputs can arrive at any time they please. There's nothing to make the inputs arrive at the same time, in a circuit with two inputs, so they generally don't. To solve the constraint problem due to unequal arrival times, we need some way to make the inputs wait until they're all stabilized, and then apply those inputs to the latch.

10-2 CLOCKED NAND LATCH (CLOCKED R-S FLIP-FLOP)

10-2.1 Clocked Logic to Fix Propagation Delay Problem

The circuit shown in Figure 10-6 shows two new kinds of latches. Both have active HIGH inputs. We're going to use one of these circuits to develop a **synchronous** latch, which has all inputs applied at the same (syn) time (chron).

Both Figure 10-6(a) and (b) are set and reset when a logic 1 is applied at S and R. To synchronize the inputs, a logic circuit using two AND gates is used to enable and disable the S and R inputs simultaneously. The signal doing the enabling and disabling is called the **clock**. An active-HIGH NAND latch with clock gating added

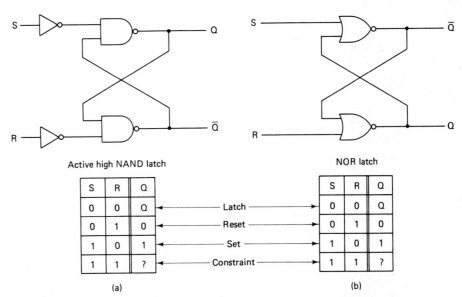

Active high NAND latch

S	R	Q
0	0	Q
0	1	0
1	0	1
1	1	?

NOR latch

S	R	Q
0	0	Q
0	1	0
1	0	1
1	1	?

Latch — Reset — Set — Constraint

FIGURE 10-6 NAND and NOR latches with active-HIGH inputs.

(a) (b)

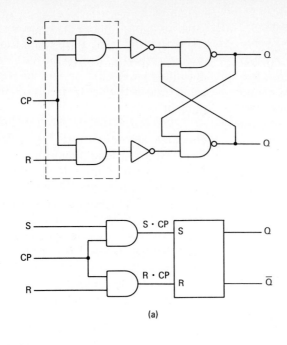

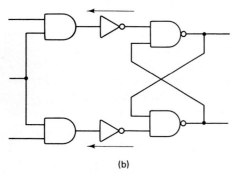

(a)

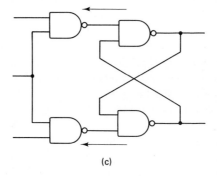

(b)   (c)

FIGURE 10-7   Clocked R-S flip-flop.

is shown in Figure 10-7(a). The same circuit could have been made with the NOR gate latch from Figure 10-6(b), but there's an advantage to the NAND latch that can be seen in the step-by-step development of Figure 10-7(b) and (c). At the end of this progression, you can see that the entire latch can be made using NAND gates only. This is called NAND **implementation**. It is desirable in integrated-circuit fabrication, where a single piece of artwork can be used to mask out thousands of the same (NAND) gates on one wafer. In this type of design, only the interconnects (wires!) have to be changed to make a wafer full of multiplexers, latches, or anything else out of the NANDS. A lot of TTL gate design is done this way.

The *clock gating circuit* is used to make the inputs wait until the 1s and 0s have all had time to arrive (to stabilize). It is normally enabled, or **strobed**, by a signal at a constant frequency (a **clock pulse**), or by a single pulse timed to delay the inputs by a certain interval after the last level

of gates is enabled. The entire circuit with the clock, set, and reset inputs is called a **clocked R-S flip-flop**, and is represented by the symbolic diagram and truth table in Figure 10-8.

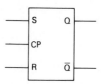

S	R	Output if CP = 1 Q	
0	0	Q_{t-1}	Latch
0	1	0	Reset
1	0	1	Set
1	1	X	Constraint

FIGURE 10-8   Clocked R-S flip-flop symbol and truth table.

EXAMPLE 10-1

What is the Boolean expression for the output of this cross-coupled NOR latch? Write the expression for the Q output.

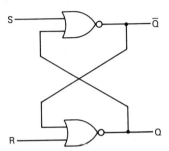

Solution: One important factor to consider in writing the expression for the output is that for any gate's output, the inputs arrived at an earlier time, and the output is developed at a later time. Because the circuit is cross-coupled, the letter that is used to designate an output is also used to designate an input somewhere else. For the letter Q, we'll call the Q that arrives at the input of a gate $Q(t - 1)$, which means "Q, at time $t - 1$." Time $t - 1$ is assumed to be the time the Q signal arrives at the input of a gate but before it has had any chance to affect the output of the gate. When the gates have had a chance to respond, one of them produces a *new* output Q, which we will call $Q(t)$. That means "Q, at time t," and this represents the Q signal coming out of a gate after it has had a chance to respond to all the inputs it has received. With this in mind, we recall that the output expressions for gates in the cross-coupled NAND latch always contained the output letter in the input expression. If the cross-coupled NOR circuit is a latching switch, the same thing will be true. The *new Q* will depend on the *old Q*. We will begin with the Q output:

$$Q(t) = R(t - 1) \text{ NOR } \overline{Q(t - 1)}$$

When we complete the Boolean expression of NOR, we get

$$Q(t) = \overline{R(t - 1) + \overline{Q(t - 1)}}$$

Using DeMorgan's theorem yields

$$Q(t) = \overline{R(t - 1)} \cdot \overline{\overline{Q(t - 1)}}$$

Finally, we simplify by removing double inversions:

$$Q(t) = \overline{R(t - 1)} \cdot Q(t - 1)$$

The new Q *does* depend on the old Q. The expression says: "Q is HIGH if the gate does NOT re-

ceive an active-HIGH R input, and if the old Q was HIGH." This is exactly what we expect from a latch with active-HIGH R (reset) and S (set) inputs.

10-2.2 Action, Truth Table, Symbolic Diagram, etc.

The clocked R-S flip-flop will be set if S = 1 is clocked into the latch. The R-S will reset if R = 1 is clocked into the latch. It's still possible to drive the latch crazy by trying to set and reset it deliberately (by putting 1s into R and S and clocking it), but this is not likely to happen accidentally, at least not if the clock pulse is timed correctly.

A truth table for the clocked R-S flip-flop is the same as the truth table for the NAND latch with active HIGH inputs, except that the clock must be HIGH to set or reset the latch. If the clock is LOW, the AND gates at the inputs are disabled. The outputs of both AND gates are frozen LOW when the ANDs are disabled. Since the rest of the R-S flip-flop has active HIGH inputs, it will neither set nor reset until the clock becomes HIGH, so the latch will remain latched between clock pulses.

10-2.3 Transfer Circuit

Figure 10-9 shows a circuit called a **transfer circuit**. It contains a clocked R-S flip-flop (A) which can be set or reset while the clock is HIGH, and a second R-S flip-flop (B) connected to the clock signal by an inverter. The (B) flip-flop can only be set or reset when the clock is LOW. The two flip-flops are connected together so that any set or reset state clocked into (A) when the clock goes HIGH will be transferred into (B) when the clock goes LOW.

To transfer a *set* from (A) to (B), the outputs of (A) are attached to the inputs of (B) so that Q = 1 and $\overline{Q}$ = 0 of (A) become S = 1 and R = 0 of (B). When these inputs are clocked into the (B) latch (which has active-HIGH inputs), it sets.

To transfer a *reset* from (A) to (B), the outputs of (A) are connected to (B) so that Q = 0 and $\overline{Q}$ = 1 of (A) become S = 0 and R = 1 of (B). When these inputs are clocked into (B), it resets.

10-3 D-TYPE FLIP-FLOP

10-3.1 A Fix for Constraints of Any Kind

At the beginning of this chapter we saw that the R-S latch could be "confused" by an input for set

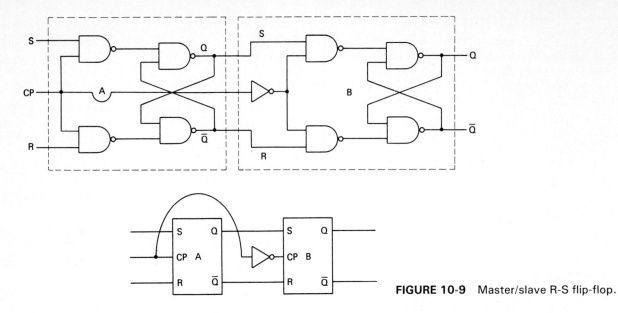

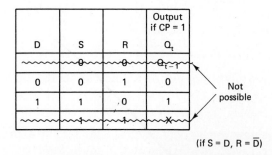

FIGURE 10-9 Master/slave R-S flip-flop.

and reset at the same time. This input causes a constraint condition which is unstable, but the inputs for a constraint are normally not designed into the logic driving the latch. Sometimes, by accident, the logic design of the driving circuits provides a temporary input for the constraint condition. Clocked logic avoids this problem (the propagation delay problem) by forcing the inputs to wait until a clock pulse appears. The delay put into the inputs gives the signals at S and R a chance to arrive, and then be clocked in synchronously.

If you decide you want to, you can still cause a constraint, and an unstable output on the latch, by deliberately applying S and R at the same time. Now, we're going to look at a circuit that can still set and reset, but can't be both because it has only one input. This circuit is called the **D-type flip-flop.**

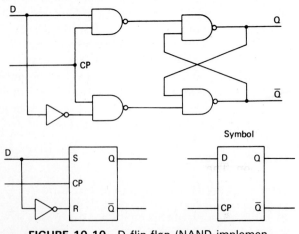

FIGURE 10-10 D flip-flop (NAND implementation).

10-3.2 Action, Truth Table, Symbolic Diagram, etc.

In Figure 10-10, the inputs to a clocked R-S flip-flop have been tied together to a common input point called D. This letter stands for **data,** a name for the 1s and 0s that are stored by the latch in its set and reset conditions. If we think of the latch as having one input (a 0 or a 1) and one output (the Q output) which shows the 0 or 1 stored inside, the D flip-flop has all the signals we need for storing data. If the D input is HIGH, the S is HIGH and the R is LOW (because of the inverter). This combination of S and R will set the latch when it is clocked in. If the D input is LOW, the R is HIGH and the S is LOW, which will reset the latch. To get a 1 at the output (Q) of this flip-flop, clock a 1 in at D, and to get a 0 at the output, clock in a 0 at D.

The truth table for the D flip-flop is actually the truth table for the R-S clocked flip-flop with two lines removed (Figure 10-11). The latch state, with both inputs LOW, and the constraint con-

D	S	R	Output if CP = 1 Q_t
~~0~~	~~0~~	~~0~~	~~Q_{t-1}~~
0	0	1	0
1	1	0	1
~~1~~	~~1~~	~~1~~	~~X~~

(if S = D, R = $\overline{D}$)

FIGURE 10-11 D flip-flop truth table.

dition, with both inputs HIGH, are impossible, because R is always the opposite of S (being attached to S through an inverter). The set and reset conditions are still possible, although only one input is now needed to get them. To latch the D flip-flop, instead of clocking the flip-flop when S and R are both LOW (which is no longer possible), you simply don't clock it at all. When the clock is ON, you'll either set or reset the D flip-flop, but when the clock is off, all data are latched.

D flip-flops are used as delay elements in some logic circuits. Data clocked into a D flip-flop are available at its output as long as new data aren't clocked in, so the D flip-flop can hold a copy of some short-lived data after the wire carrying those data has begun to carry other information. This is an extremely important function in large-scale digital systems and in computers, and the capture and holding of such data is called **latching**, a name that describes the function of the flip-flops.

10-4 J-K FLIP-FLOP WITHOUT MASTER/SLAVE

We decided it was important to design a flip-flop that could be set and reset but not confused by a constraint input. One way to accomplish this was the D flip-flop. The D flip-flop could be set and reset only, but to do this, we limited the flip-flop to only one input, and threw out half the truth table for the latch. This seriously limited the number of things that could be clocked in to the latch. Another solution to the constraint problem is the **J-K flip-flop circuit**, shown in Figure 10-12.

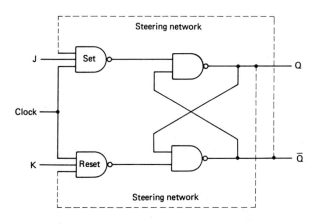

FIGURE 10-12 J-K flip-flop *without* master/slave architecture.

10-4.1 Steering Network and the Toggle State

The J-K flip-flop has the same two set and reset inputs used to operate all the latches up to this point. In this case, there is one additional condition placed on these inputs that makes them different enough to receive new names. The set input is now called J and the reset input is now called K. You can see, from the schematics, that to set this flip-flop, input J must be a 1, *and* the clock must be 1, *and* the output of Q must be 1. To reset this flip-flop, input K must be a 1, *and* the clock must be a 1, *and* the output of Q must be a 1. The additional two wires in the circuit and the additional two inputs to the clock gating part of the circuit are called a **steering network**.

As a result of adding this network to the flip-flop, it's no longer possible to set and reset it at the same time. Only one of the two inputs, J or K, is ever enabled. This means that the constraint input is now impossible. The logic of the J input can be stated as: "You can set me, if I'm not already set at the time the clock is ON." The logic of the K input can be stated as: "You can reset me, if I'm not already reset at the time the clock is ON."

Does this change any of the states that are normally put into a latch? If the J-K is already set, its set input (J) is disabled. So what? If you want to turn the thing ON, it ends up ON after the clock pulse, so who cares how it got that way? (The result, after all, is "you get what you want.") The same thing is true for the reset. If the J-K is already reset, its reset input (K) is disabled, but if you try to reset it, you get what you want after the clock pulse, because there's nothing else happening. Although the reset doesn't get in, the latch doesn't set; if just stays the way it was. *Result:* After clocking in a J or K, the flip-flop ends up set or reset, regardless of the steering network.

Looking at the truth table (Figure 10-13), we see that the first three lines of the truth table (latch, reset, and set) exactly match the truth table for an R-S flip-flop. J takes the place of S and K takes the place of R, but otherwise, nothing is different. The bottom line of the truth table is new. This was the constraint condition for the R-S flip-flop. For the J-K flip-flop it is the **toggle** condition. The output on the bottom line is $\overline{Q}_{t-1}$. This translates as: "After the clock pulse, Q will become what $\overline{Q}$ was before the clock pulse."

Think of the toggle condition like a light switch with a pull-chain. When you pull the chain once, the light goes ON, but when you pull the

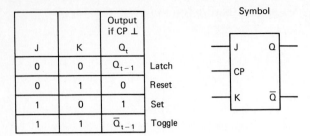

J	K	Output if CP ⊥ Q_t	
0	0	Q_{t-1}	Latch
0	1	0	Reset
1	0	1	Set
1	1	$\overline{Q}_{t-1}$	Toggle

Symbol

FIGURE 10-13 J-K flip-flop truth table.

chain again, the light goes OFF. The light will alternately toggle from OFF to ON, back and forth, although the input (a tug on the chain) isn't any different for the turn-on than it was for the turn-off. In the same way, when the J and K inputs are both active, we are trying to turn the J-K flip-flop ON and OFF at once. Instead of a constraint, we get a toggle, because only one input can be accepted by the latch. The steering network enables only the input that does something new; when the flip-flop is ON, the K input (which resets it) is the only one enabled; it will turn the J-K OFF. When the flip-flop is OFF, the J input (which sets it) is the only one enabled; it will turn the J-K ON. So, for the J-K flip-flop, when both J and K are active, the flip-flop will turn ON if it was OFF, and OFF if it was ON. Remember the light fixture with the pull-chain and think of the toggle condition working like that light fixture, with the clock pulse replacing a pull on the chain.

The symbolic diagram indicates that the J-K flip-flop looks like an R-S with the letters at R and S changed. The clock input indicates that a 1 at the clock will clock in the inputs. We say that the clock is active HIGH, as are the J and K inputs themselves. The outputs are the same as they have been for all other flip-flops.

10-4.2 Racing with Unsuitable Clock Pulses

Figure 10-14 shows how a clock pulse would affect the J-K flip-flop. The timing diagram describes what happens at the same time at CP, $\overline{Q}$, and Q. The toggle takes place when the clock pulse enables the inputs, and the J or K gets in to the latch according to the previous state of Q. Notice what happens [Figure 10-14(b)] when the clock pulse lasts too long. The outputs of the J-K change while the clock is still active, and cause a second toggle. If the clock remains active even longer, the J-K will continue to toggle again and again, since the enable conditions at J and K

change every time outputs get back to the inputs through the steering network. We have a problem here, because the time it takes for the inputs to affect the outputs, and for the outputs to get back to the inputs through the steering network is very short.

If we assume that TTL NAND gates are used, the time it takes a signal to propagate through a TTL NAND is about 7 ns. To get through the two levels of logic at the input gating and the latch might take as little as 14 ns. This could cause the J-K to toggle every 14 ns while the clock is active. The clock pulses have to be shorter than 14 ns or the J-K will toggle repeatedly with a frequency of 71.4 MHz for as long as the clock is ON.

This condition is called **racing**. In racing, the flip-flop is switching back and forth from set to reset as fast as its little heart can beat. It won't keep doing this forever, though, because sooner or later the clock will go OFF. The results aren't very predictable; it's a case of "round and round she goes, and where she stops, nobody knows."

To avoid the racing problem, there are three solutions:

1. Use only very short clock pulses.
2. Slow down the steering network to match the clock.
3. Change the flip-flop so that its outputs don't come out until the clock shuts OFF.

Let's examine these options.

1. Use only very short clock pulses. Unfortunately, this means we need an oscillator that produces pulses whose width is less than 14 ns, no matter how far apart (what frequency) the pulses are. It's not impossible to make a clock oscillator like this, just very tricky, especially if you would like to be able to vary its frequency.

2. Slow down the steering network to match the clock. Presumably, this would mean adding some sort of time-delay element to the steering wires, such as an *RC* charge or discharge circuit. Unfortunately, if you want to change the clock frequency, you'll also have to change the Rs or Cs as the frequency changes. Even if this could be done simply, it turns out that capacitors don't miniaturize very well. For the past several years, the number of components that could be packed into a square centimeter of a silicon has doubled every year. This is due to continuing advances in reducing the size of components. Reducing the size of capacitors, though, reduces their capacitance, or (if you make the di-

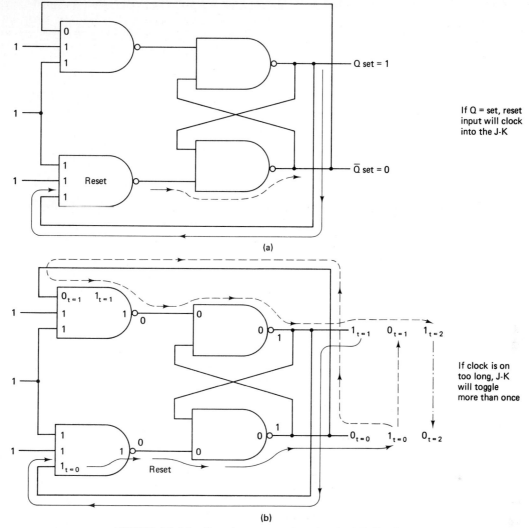

FIGURE 10-14 Steering network action in J-K flip-flop.

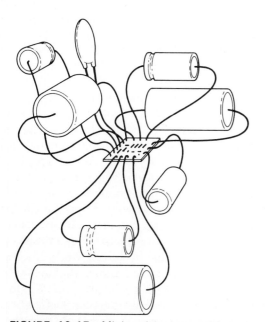

FIGURE 10-15 Miniaturizing capacitive circuits.

electric thinner to compensate) reduces their breakdown voltage rating. To keep a certain capacitance *and* voltage rating, the capacitors must stay the same size. After a while, you'd end up with a circuit that looks like Figure 10-15.

3. Change the flip-flop so that its outputs don't come out until the inputs are no longer active (until the clock pulse is OFF). We made the other solutions sound so hard; we must have been steering you toward this one! It sounds hard too, at first, but turns out better than any of the other solutions.

10-5 J-K MASTER/SLAVE FLIP-FLOP– R-S MASTER/SLAVE

10-5.1 Showing Elimination of Racing and Retention of Toggle State

Figure 10-16 shows a **J-K master/slave flip-flop**. It is actually made of two R-S flip-flops. One (the

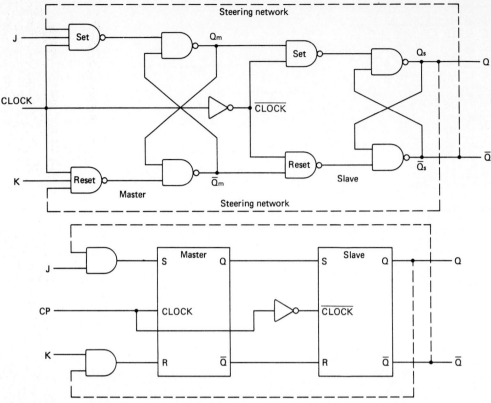

FIGURE 10-16 J-K flip-flop *with* master/slave architecture.

master) is set or reset when the clock is HIGH. The master is connected by two wires to a second flip-flop, the *slave*. You might recognize the hookup as a transfer circuit (seen earlier in the chapter). The slave receives data transferred from the master when the clock is LOW. A steering network connects the output of the slave to the inputs of the master, making set into J and reset into K.

Without the steering network, this circuit would be called an **R-S master/slave flip-flop**. This circuit already fulfills the requirement that the output doesn't come out until the inputs are disabled. When the clock goes from HIGH to LOW, the inputs of the master (which are clocked in by an active HIGH clock) shut off. Then the data in the master transfer to the outputs of the slave (see Figure 10-17).

Adding the J-K steering network to the master/slave R-S flip-flop makes it a master/slave J-K flip-flop. The output of the slave steers the master while the clock is HIGH. If the slave is set, the master can only reset (or latch). If the slave is reset, the master can only set (or latch). While all this is happening to the master, the slave is frozen, because the slave's clock is inactive. This means that the J-K master/slave flip-flop cannot race. Since the master transfers its

state to the slave only after the clock makes a HIGH-to-LOW transition, the slave can only change (this is what caused the racing in the J-K without master/slave) after the inputs of the master have shut OFF. If we want to toggle the J-K master/slave flip-flop, we can apply set (J) and reset (K) inputs at the same time. We will set or reset the flip-flop according to how the inputs are steered, but this time, since the changes in the master cannot influence the slave during the setting or resetting cycle, our clock would have to go through another complete HIGH-LOW cycle to toggle the flip-flop a second time.

This means that the flip-flop will only toggle once for each cycle of the clock, no matter how long the HIGH and LOW parts of the clock cycle last. The flip-flop only sets, resets, toggles (or latches) one-half at a time. The master half of the flip-flop receives its input, and does setting or resetting things during the HIGH half of the clock pulses. The slave half receives data transferred from the master during the LOW half of the clock pulse. The master can't change while it's transferring data to the slave because the master's inputs are shut off while the clock is LOW. On the other hand, the slave has no choice about whether to set or reset when the clock goes LOW; it must set or reset to the same state as the master—it

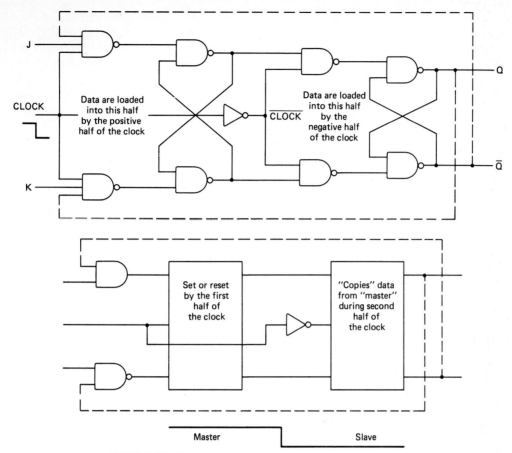

FIGURE 10-17 J-K flip-flop showing clock triggering.

"slavishly" follows whatever the master was doing at the instant the clock went LOW.

10-5.2 Introducing Edge-Triggered Logic

Since the J-K master/slave flip-flop shows a new state at its outputs (which are the Q and $\overline{Q}$ of the slave) only at the instant the clock goes from HIGH to LOW, we say that this J-K is **falling-edge triggered**. The master part of the flip-flop is triggered (enabled) when the clock is at a HIGH level—we call this a **level-triggered** (high) clock. The slave, because of the inverter added to its clock input, is a level-triggered (low) flip-flop. Until we discussed flip-flops with master/slave configuration, all flip-flops we discussed were level-triggered. Now, with the master/slave J-K and R-S types, we see edge-triggered logic. One way to look at this is to imagine that the master/slave R-S is taking a snapshot of the state at its input during the instant the clock goes from HIGH to LOW. The master can be set or reset while the clock is HIGH—in fact, it could set and reset back and forth several times during the HIGH clock—but nothing appears at the slave outputs except the last state held in the slave. At the moment the clock goes LOW, the master's inputs shut off, and the slave's inputs turn on. The state in the master at that instant is transferred to the slave, and appears at the outputs of the slave, which are the ultimate outputs of the whole flip-flop.

Generally, the master outputs of the master/slave J-K flip-flop don't get out to the outside world. Its inputs are J and K, the inputs of the master, and its outputs are the outputs of the slave. The rest of the action going on inside the flip-flop is invisible from outside, so the action of the flip-flop is defined from what can be seen from outside.

For this reason, a master/slave J-K like the one in our diagram (Figure 10-16) is depicted by a symbolic diagram like the one in Figure 10-18.

FIGURE 10-18 J-K flip-flop symbolic diagram.

10-5.3 Action, Truth Table, Symbolic Diagram, etc.

The symbolic diagram shows two new things. First, the clock input has a "bubble" where it enters the rectangle that represents the circuit. This is an active-LOW indication, which we've seen on other logic gates. The fact that we've never shown a flip-flop with an active-LOW clock in earlier parts of this chapter is pure coincidence. They're just as easy to make as active-HIGH clock inputs; we just somehow overlooked designs that used an active-HIGH clock by accident. The second thing that is new in flip-flop diagrams is the small "wedge" inside the rectangle, which represents edge-triggered logic. An input that shows this wedge will do whatever it does just at the instant when a certain logic level begins. In this case, as you can tell, the logic level that triggers the clocking is LOW. We call this input an active-LOW edge-triggered—or more generally, a falling-edge-triggered input. The rest of the inputs and outputs are ones that we've seen before; J and K are still steered set-reset inputs—Q and $\overline{Q}$ are still the TRUE and FALSE outputs for the state of the latch. An invisible added feature of the master/slave logic is the appearance of new states at Q and $\overline{Q}$ as the falling edge occurs on the clock. Although this is a different sort of action, there is no change in the way these outputs are represented in the symbol, or in the appearance of the truth table for this flip-flop.

10-6 T-TYPE FLIP-FLOP

10-6.1 Action of T Flip-Flop, Showing Elimination of All States except Toggle

Earlier, the D-type flip-flop was formed from the more flexible R-S flip-flop, for applications where all the states of the R-S inputs were not needed. A simplified truth table (for a flip-flop with only one input) was produced by dropping certain states out of the truth table. The J-K, too, has a "little brother" which is made out of a J-K by reducing the number of available inputs. The **T-type flip-flop** is the result of restricting the inputs to a J-K to just the one novel state that J-K's share with no other type of flip-flop: the toggle state. By wiring both J and K permanently HIGH, a toggle action happens with every clock pulse. This means that really, the only input to the T is its clock. The only remaining line on the truth table is the flip-flop's toggle state; every time the clock has a falling edge, the T flip-flop's output will change states. A simplified symbol for the T flip-flop, shown in Figure 10-19, has no J and K inputs shown. This can be justified by looking at what would happen to the NAND-gate implementation of a J-K flip-flop if the J and K terminals were removed. The remaining logic at the inputs only needs two conditions at either NAND gate to set or reset the flip-flop. The master part of the flip-flop will set if the clock is a 1 and $\overline{Q}$ is a 1, and the master will be reset if the clock is a 1 and Q is a 1. The state of the master will, of course, transfer itself to the slave as soon as the clock becomes 0. When you think about this, you realize that this is an automatic toggle condition. If the Q output is set, so that it is a 1, then the inputs will force a reset with the next full clock cycle. If the Q output is reset, so that $\overline{Q}$ is a 1, then the inputs will force a set with the next full clock cycle. Q after the clock cycle will always be the opposite of what Q was before the clock cycle. The T flip-flop's action is shown in Figure 10-20. This shows the results of clocking a T flip-flop with several clock cycles in several different ways: A timing diagram shows a waveform at Q which is the result of the clock. Seen side by side, the waveforms make sense—wherever (whenever) the clock has a falling edge, the Q waveform changes from LOW to HIGH or from HIGH to LOW. Another diagram shows the outputs of Q, starting with Q in a LOW condition, alongside a count of the number of clock cycles.

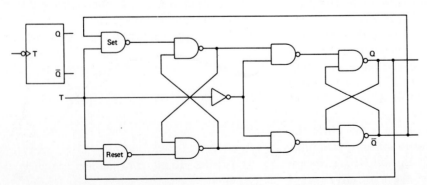

FIGURE 10-19 T flip-flop symbolic diagram and NAND schematic.

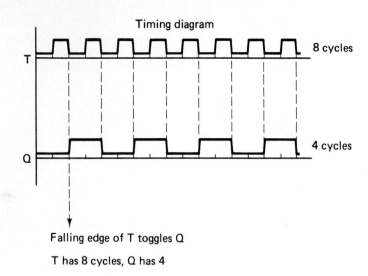

Timing diagram

T

8 cycles

Q

4 cycles

Falling edge of T toggles Q

T has 8 cycles, Q has 4

State table

T cycle number	Q	$\overline{Q}$	
1	0	1	Odd
2	1	0	Even
3	0	1	Odd
4	1	0	Even
5	0	1	Odd
6	1	0	Even
7	0	1	Odd
8	1	0	Even

FIGURE 10-20 ''Divide-by'' and ''odd–even counter'' application of T flip-flop.

10-6.2 Divide-By and Odd-Even Counter Action

By examining the timing diagram, you can see something that wouldn't be so apparent any other way. This is the relationship between the frequency IN and the frequency OUT of the T flip-flop. Looking at the diagram, you can see that the Q frequency is half the CLOCK frequency. Each wave is exactly twice as long at Q as at CLOCK. From this, we get the name **divide-by-2-circuit** for the T flip-flop. Any frequency square wave put into the T input (the clock) will cause a waveform to develop at the output which is the input frequency, divided by 2. The eight cycles IN at T produce the four cycles OUT at Q.

By looking at the **state table** (the output of Q after a certain number of clock cycles) you can see the other aspect of the T flip-flop's action: Every even number of clock cycles, Q is HIGH, and every odd number of clock cycles, $\overline{Q}$ is HIGH. This makes the T flip-flop an **odd–even counter**, and the outputs can be called the *even* output (Q) and the *odd* output ($\overline{Q}$). Q is ON at even numbers in the list, while $\overline{Q}$ is ON for odd numbers.

These aspects of the T flip-flop's behavior will turn out to be useful later, when we look at digital counters.

10-7 EDGE-TRIGGERED FLIP-FLOPS WITHOUT MASTER/SLAVE

10-7.1 ''Differentiating''a Clock Pulse

At the end of Section 10.4 we saw three solutions to the ''racing'' problem in the J-K flip-flop, and we settled on the master/slave configuration as the most reasonable solution. Although the arguments we gave are good justification for the design of J-K master/slave flip-flops, one of the other solutions is also acceptable. This is solution 1: ''Use only very short clock pulses.'' You might wonder how we could use this solution, since we put up some good arguments against it. We said that it would be difficult to design the special clock generator needed to make the flip-flops ''happy,'' so, instead, we should redesign the flip-flop to be ''happy'' with the clock it has. There is a way to do this without using a master/slave design.

This is the trick of ''making the flip-flop happy'' with the clock pulses it has: If the clock pulse is active for too long, *have the flip-flop itself shorten the pulse.* Build a circuit into the flip-flop that takes a long active pulse and turns it into a short pulse that is active for *just the right time* to let the flip-flop set, reset, or toggle *once*, but not long enough to let it toggle twice, for instance.

How is this done? One method is to use a circuit called a **differentiator**, which is a short-term, transient conductor. A capacitor acts this way when you start to charge it. At first (before it is charged), the capacitor acts like a fairly good conductor. After it is charged, the capacitor doesn't conduct at all. This solution to the "racing" problem is to couple the clock pulse into the flip-flop through a capacitor that conducts for *exactly* the right amount of time.

In Figure 10-21, the coupling capacitor makes the useless, racing flip-flop of Figure 10-14 into a working, edge-triggered flip-flop, because the rising edge of the clock charges the capacitor long enough to let the inputs reach the output, and no longer. Selecting the right value for the capacitor is the job of the design engineers involved in chip design, and once it is done correctly, all chips manufactured to the same specifications will work equally well.

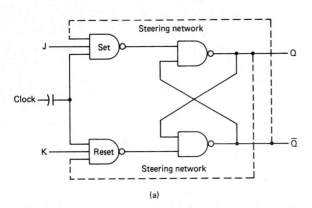

(a)

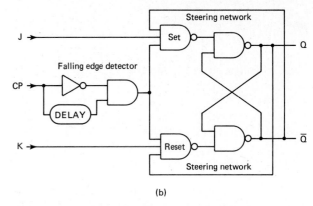

(b)

FIGURE 10-21 J-K flip-flop with a differentiated clock pulse.

EXAMPLE 10-2

Compare a master/slave J-K flip-flop, as defined in Section 10-5.2, with an edge-triggered J-K flip-

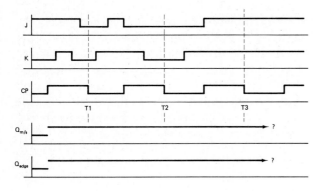

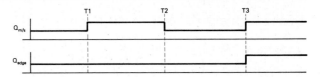

flop, as defined in Section 10-7.1 (with a capacitor-differentiated clock). For the following timing diagram, what difference, if any, would you find in the Q output for each type of flip-flop? Assume that both flip-flops change Q output states at the falling edge.

Solution: *Q (master/slave).* The master/slave transfers the state in the master [established during the time the clock (CP) is HIGH] into the

slave, just after the HIGH-to-LOW transition of the clock. At times T_1 and T_2 on the timing diagram above, the last active input to the master is J = 1, K = 0 (SET) (before T_1) and J = 0, K = 1 (RESET) (before T_2), although neither J nor K is active at the moment of the falling edge. The master, being a latch, holds the last state until the falling edge, so the slave is SET at T_1 and RESET at T_2. At time T_3, Q toggles, since both inputs J and K are active throughout the clock cycle (master toggles, then slave toggles).

Q (edge-triggered). The (differentiated) edge-triggered flip-flop responds to the inputs present at J and K at the time of the HIGH-to-LOW transition of the clock. At times T_1 and T_2, both J and K are inactive. Since there is no master enabled during the time the clock is HIGH, enabling J or K during that time has *no effect*. The only state that Q (edge) "sees" is the state at the instant of the HIGH-to-LOW transition. At time T_3, Q toggles, because the J and K inputs are both active at that moment.

10-7.2   An Edge Detector

Another way to shorten a clock cycle's pulses is to build a circuit that makes a short pulse every time it "sees" an edge. One circuit that can do this is shown in Figure 10-22. It is an XOR gate that has both inputs tied together. Normally, that is a circuit whose output would always be LOW, because the XOR gate's output is LOW when its inputs match. In the schematic, you can see a **delay line** placed on one of the input lines to the XOR. With this delay, the inputs *don't match right away*. When a new state comes in the input line A at time t, the XOR output goes HIGH, because A is different from B (hasn't changed yet). Then at time t + delay, the state arrives at input

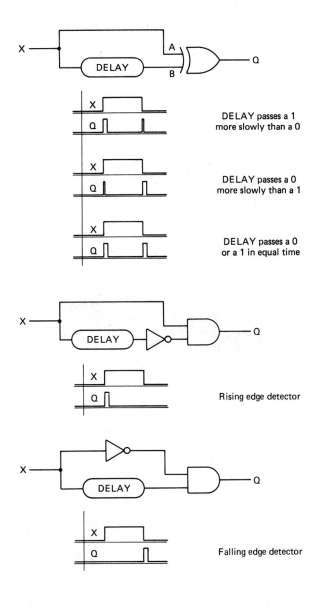

FIGURE 10-22   Edge-detector and timing diagrams.

line B, and the XOR output goes LOW. The output of this **edge-detector circuit** is a HIGH pulse that's HIGH for a time equal to "delay."

"That's an interesting circuit," I hear you say, "but what does it have to do with flip-flops?" One answer is that it can be set up to enable the clock of the J-K flip-flop for "just the right time," as we did with the differentiator before. If the delay line gives a different delay passing a HIGH than passing a LOW (which is not as difficult to do as it might sound), we can have the falling edge operate the flip-flop, while the rising edge produces a pulse too short to do anything. Then we would have a falling-edge-triggered J-K flip-flop. Or, if we wanted to, we could have the rising edge operate the flip-flop, and the falling-edge delay could be ignored, making a rising-edge-triggered J-K flip-flop. Design of a delay line that delays HIGH signals a different amount than it delays LOW signals is outside the scope of this course, but could be discussed in a course on analog electronics.

Finally, let's suppose that an edge detector has been designed which produces equally wide pulses for both a HIGH entering the delay line and a LOW entering the delay line. This circuit really isn't much use for building flip-flops, since a J-K flip-flop triggered by this circuit would toggle *twice* during every cycle of the clock, instead of once. Nobody wants that. But suppose you want a circuit that *doubles the frequency of the clock*, the way a T-type flip-flop divides by 2, only backwards. This is the circuit. You can see it on the timing diagram.

10-8   OTHER MULTIVIBRATORS

10-8.1   Bistable Multivibrators

You are probably wondering why the heading of this section is "Other Multivibrators." "What's a multivibrator, anyway?" you might ask, "And why did you say "other"? Have we already talked about multivibrators, and I missed it, somehow?" No, you didn't miss anything, but nevertheless, we *have* been discussing multivibrators. This entire chapter has been about a type of multivibrator called the **bistable multivibrator**. The common name for a bistable multivibrator is a *flip-flop*. It was a multivibrator all along, and we never said so (forgive us!).

The flip-flop is *bistable*: the Greek prefix *bi* means "two," and it has two stable states. The states are called SET and RESET. These are states that will be held (will be stable) if the inputs to the multivibrator become inactive. We've seen this type of action going all the way back to the cross-coupled NAND latch, and the reason for it is the feedback provided by the two crossed wires in the middle of the circuit. As long as these wires continue to conduct the output of one NAND gate to the input of the other, the circuit will continue to "remember" its state. If SET, it will remember to stay SET; if RESET, it will remember to stay RESET.

10-8.2 Monostable and Astable Multivibrators

Now, we are going to look at two circuits that don't remember very well. To make it simple to visualize how these devices work, we'll develop them out of the cross-coupled latch circuit that showed up first in Figure 10-1. We see the bistable device in Figure 10-23(a). It's a NAND latch. In Figure 10-23(b), one of the cross-coupling wires has been replaced with a coupling capacitor. We already know that this is only a temporary conductor. Its effect on the action of this circuit is that it can't remember to stay SET very long. Suppose that we SET the latch with a LOW at S, and then, after a second LOW comes into the upper NAND gate from the cross-coupling network, we inactivate S (make it HIGH). The *RC* time of the circuit charging the capacitor (C) determines how long the LOW will appear at the upper NAND gate's second input. When the capacitor charges to a voltage that's no longer LOW, the latch will stop being SET. Essentially, the latch forgets to remain SET after an amount of time that depends on the *RC* charge circuit.

Figure 10-23(b) is a **monostable multivibrator**. It has *one* stable state, in this case, RESET. When it is SET, it stays SET for a while, but then it "forgets" and goes back to being RESET. It is called *monostable*; the Greek prefix for "one" is *mono*, and it has one stable state.

Figure 10-23(c) is an **astable multivibrator**. It has capacitor coupling on *both* of its cross-coupling paths. This means that it can't remember to stay SET *or* RESET for very long. It "can't make up its mind"! Instead of staying SET or RESET, it constantly SETs, then forgets how to stay SET, and RESETs, then forgets how to stay RESET and SETs again. This action results in a *square wave* continually being generated at the circuit's output, Q. The circuit really doesn't need inputs S and R, anymore; it will SET and RESET by itself, staying SET as long as C_1 lets it remem-

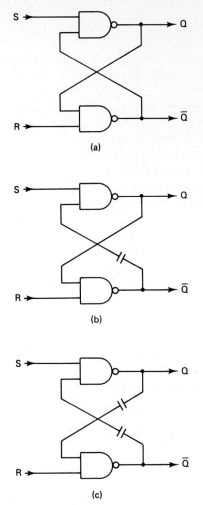

FIGURE 10-23 Multivibrators: (a) bistable; (b) monostable; (c) astable.

ber to stay SET, then going to RESET for as long as C_2 lets it remember to stay RESET.

While the monostable circuit uses S to **trigger** its output (Q) to become active, the astable circuit doesn't need any trigger at all. We could remove the S and R input lines without altering the action of the circuit at all. Because *none* of the states (SET or RESET) is stable, the circuit is called **astable**, from the Greek prefix *a*, meaning "not." Astable means "not stable."

10-8.3 The 555 Timer

The most common way to make monostable and astable devices is not by modifying flip-flops. Generally, an integrated circuit device called a **timer** is used to design these devices. Since the amount of time you want the device to "remember" to stay SET or RESET is an analog quan-

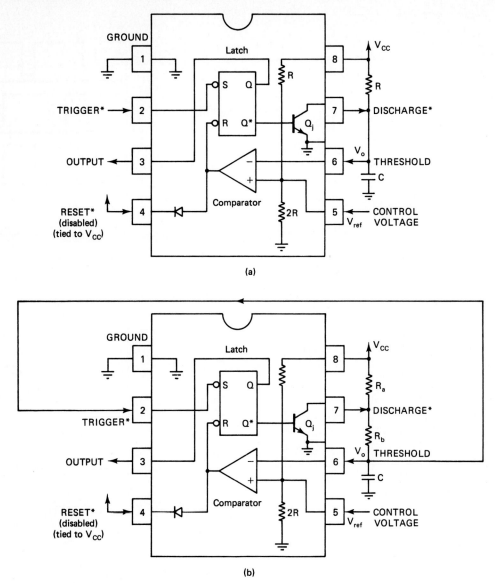

FIGURE 10-24 Multivibrators made with the 555 timer: (a) monostable; (b) astable.

tity, these are generally classed as *analog* circuits, not digital. The most popular integrated-circuit timer is a device called the **555 timer**. Because it is an analog circuit, you will not find it included in digital databooks along with logic gates: (There are specific monostable circuits that fit into the 7400-series, such as the 74123, that *are* included in digital databooks, but the 555 is not.)

Figure 10-24(a) shows a block-diagram model of the 555 timer used as a monostable multivibrator, and Figure 10-24(b) shows how it is used as an astable multivibrator. In Figure 10-24(a), you trigger the monostable (which is also called a **one-shot**) through pin 2. Here's what happens:

1. Pin 2 (trigger) *sets* the latch, turning off transistor Q_1 and allowing capacitor C to be charged through R. Pin 3 (output) is now HIGH.

2. Capacitor voltage V_c rises until it becomes larger than V_{ref} ($\frac{2}{3}V_{cc}$).

3. When V_c is larger than V_{ref}, the output of the comparator switches from HIGH (+) to LOW (−).

4. The comparator's LOW (−) output *resets* the latch. Pin 3 (output) is now LOW.

5. The *reset* latch turns on transistor Q_1, which discharges capacitor C through the short circuit between pins 6 and 7.

(The circuit is now ready to be triggered again.)

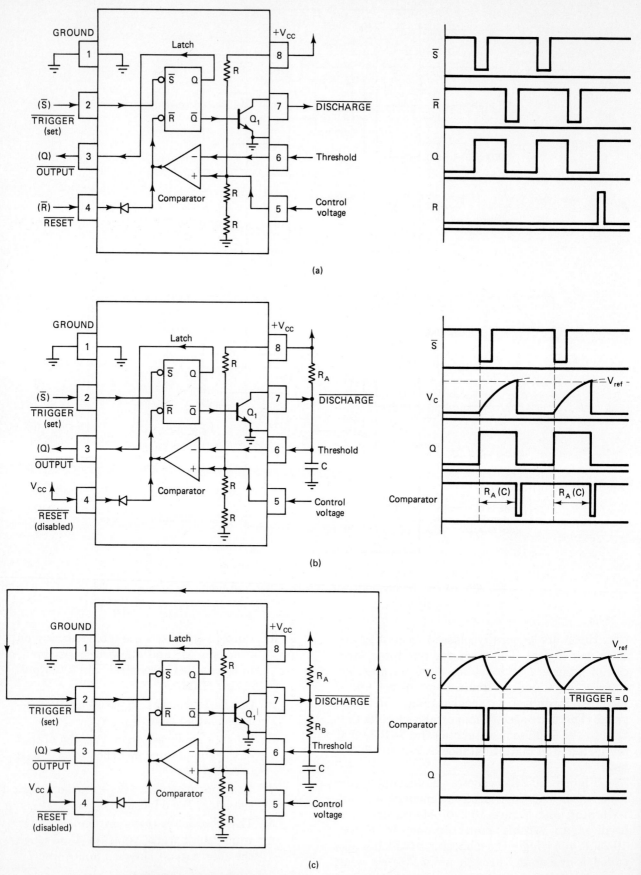

FIGURE 10-25 555 timer circuits and timing diagrams: (a) bistable; (b) monostable; (c) astable.

In Figure 10-24(b), the astable (which is also called a **square-wave oscillator**) starts with $V_c = 0$ at the start. The following sequence of actions then takes place:

1. Pin 2 (trigger) *sets* the latch, turning off transistor Q_1 and allowing capacitor C to be charged through R_A and R_B. Pin 3 (output) is now HIGH.

2. Capacitor voltage V_c rises until it becomes larger than V_{ref} $(\frac{2}{3}V_{cc})$.

3. When V_c is larger than V_{ref}, the output of the comparator switches from HIGH (+) to LOW (−).

4. The comparator's LOW (−) output *resets* the latch. Pin 3 (output) is now LOW.

5. The *reset* latch turns on transistor Q_1, which discharges capacitor C through resistor R_B.

6. When C is discharged, $V_c = 0$ and the LOW connected from V_c to pin 2 triggers the active-LOW trigger input.

(The circuit is now back on line 1, and everything starts over.)

Figure 10-25 shows timing diagrams for the 555 timer used as a bistable (a), monostable (b), and astable (c) multivibrator. Practically nobody ever uses a 555 as a bistable latch, but it can be done, as shown in this figure. Compare the figure and its timing diagrams to the descriptions in the preceding two paragraphs. See if you can identify the points on the waveforms where each of the numbered steps in the descriptions occur.

QUESTIONS

10-1. For the circuit shown here:

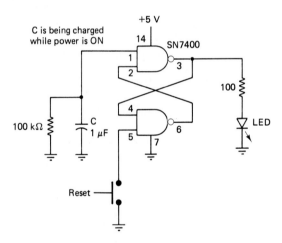

(a) If the reset pushbutton is operated, what state should the LED have afterward, ON (1) or OFF (0)?

(b) If capacitor C discharges, what state will the LED be put into when the power returns, ON (1) or OFF (0)?

10-2. For the symbolic R-S flip-flop in the following figure:

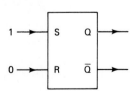

(a) What is Q? (0 or 1?)
(b) What is $\overline{Q}$? (0 or 1?)

10-3. For the schematic shown:

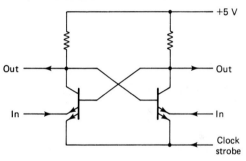

(a) Is there feedback?
(b) Is there cross-coupling?
(c) Could this be a flip-flop?

10-4. For the circuit appearing here:

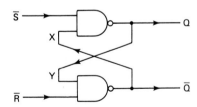

(a) Is the set input active HIGH or active LOW? (0 = LOW; 1 = HIGH)
(b) Is this a synchronous or an asynchronous latch?
(c) Does the circuit have a memory capability for its logic state?
(d) Is there cross-coupling in this circuit?
(e) If S and R are both LOW, is the resulting output stable?
(f) Could this circuit be used in a debounced switch?

10-5. Refer to the figure in Question 10-4 and select from the choices at the right:
- **(a)** Which inputs are the primary inputs?
- **(b)** Which inputs are the feedback inputs?

   (1) S
   (2) R
   (3) X
   (4) Y

10-6. Match the circuit of the figure at the top below with an equivalent NOR/NAND circuit. (1, 2 or 3)

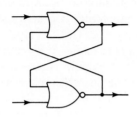

(1)
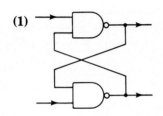

(2)
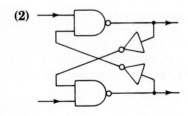

(3)
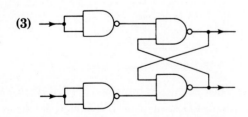

10-7. Which problem does not affect synchronous logic?
- **(a)** Constraints caused by propagation delay time
- **(b)** Racing caused by too-long strobe pulses

10-8. Clocked logic is:
- **(a)** Synchronous
- **(b)** Asynchronous
- **(c)** Not strobed

10-9. How many inputs does a D flip-flop have, not counting the clock input?

10-10. How many inputs does a clocked R-S flip-flop have, not counting the clock input?

(True/False questions)

10-11. R-S flip-flops can only be made using NAND gates.

10-12. A D flip-flop remains in a latch condition when it is not strobed.

10-13. A D flip-flop does not need to be strobed to be set or reset through its D input.

10-14. A J-K flip-flop is an R-S flip-flop with a steering network added; it cannot receive a constraint condition from its inputs once this is added.

10-15. A master/slave J-K flip-flop uses a transfer circuit to make racing impossible.

10-16. This symbolic diagram shows:
- **(a)** A synchronous latch. (T/F?)
- **(b)** An asynchronous latch. (T/F?)

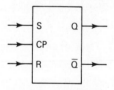

- **(c)** A latch with a clock that is (active LOW or active HIGH)?
- **(d)** S and R inputs that are (active LOW or active HIGH)?

10-17. For the following symbolic diagram:

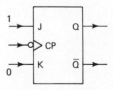

- **(a)** J and K are (active LOW or active HIGH)?
- **(b)** The clock (CP) is (level- or edge-triggered)?

(c) The flip-flop probably is a master/slave type of circuit. (T/F?)

(d) J and K are the flip-flop's (synchronous, asynchronous) inputs.

(e) If the inputs at J and K are clocked in, Q will become (0 or 1)?

(f) Is a constraint condition possible in this flip-flop?

10-18. For the D flip-flop below (use 0 = No; 1 = Yes):

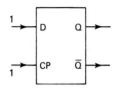

(a) Is it edge-triggered?

(b) What will be the state at Q? (0 or 1?)

(c) What will be the state at $\bar{Q}$? (0 or 1?)

10-19. For the T flip-flop shown:

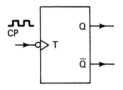

(a) What will be the state at Q after seven strobe pulses? (Assume that Q = 0 was the state before any pulses arrived.) (Answer 0 or 1.)

(b) If CP has a frequency of 30 Hz, what will be the frequency at Q?

10-20. Match the three diagrams (a, b, c) with the types of multivibrator listed (1, 2, 3).

    (1) Astable multivibrator
    (2) Monostable multivibrator
    (3) Bistable multivibrator

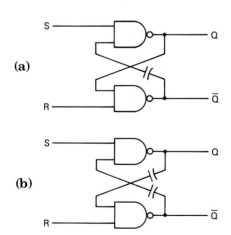

(a)

(b)

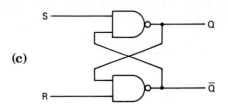

(c)

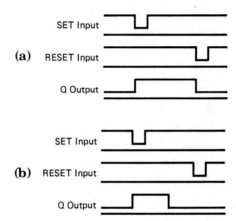

10-21. Identify the timing diagrams shown (a, b) with a type of multivibrator. Choose answers from among the choices given (1, 2, 3).

    (1) Astable multivibrator
    (2) Monostable multivibrator
    (3) Bistable multivibrator

10-22. The 555 timer IC contains a flip-flop and a comparator. What is the 555 most commonly used for?

(a) As a bistable (flip-flop)

(b) As an astable (oscillator)

(c) As a monostable (one-shot)

(d) Both B and C

(e) A, B, and C

10-23. If you are looking for the specifications of an NE555 or TLC555 timer (the TTL and CMOS versions of the same circuit), would you look in a digital databook or an analog databook?

(a) Digital   (b) Analog

10-24. Draw a schematic for a digital circuit that will multiply the frequency of incoming pulses by 8. Assume that the incoming frequency is between 100 and 133 kHz. Use the "edge detector" circuit from this chapter, and select R and C for an *RC* delay gate so that the pulses will be long enough to ensure a 50% duty cycle for a frequency in the middle of this range. (*Hint:* Eight = 2 × 2 × 2.)

Questions 10-1 through 10-20 refer to the following cross-coupled NAND latch:

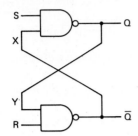

If input S goes LOW and R goes HIGH:

10-1. Output Q goes_____. (high, low, can't tell without X)

10-2. Feedback input Y goes_____. (high, low, can't tell without X)

10-3. Output $\overline{Q}$ goes_____. (high, low, can't tell without Y)

10-4. Feedback input X goes_____. (high, low, can't tell without Y)

10-5. The state of the latch goes_____. (set, reset, latched, cannot be determined since all the NAND inputs aren't given)

If input R goes LOW and S goes HIGH:

10-6. Output $\overline{Q}$ goes_____. (high, low, can't tell without Y)

10-7. Feedback input X goes_____. (high, low, can't tell without Y)

10-8. Output Q goes_____. (high, low, can't tell without X)

10-9. Feedback input Y goes_____. (high, low, can't tell without X)

10-10. The state of the latch goes_____. (set, reset, latched, cannot be determined since all the NAND inputs aren't given)

If input S goes HIGH and R goes HIGH:

10-11. Output Q goes_____. (high, low, can't tell without X)

10-12. Feedback input Y goes_____. (high, low, can't tell without X)

10-13. Output $\overline{Q}$ goes_____. (high, low, can't tell without Y)

10-14. Feedback input X goes_____. (high, low, can't tell without Y)

10-15. The state of the latch goes_____. (set, reset, latched, cannot be determined since all the NAND inputs aren't given)

If input S goes HIGH and R goes HIGH and X is already LOW:

10-16. Output Q goes_____. (high, low)

10-17. Feedback input Y goes_____. (high, low)

10-18. Output $\overline{Q}$ goes_____. (high, low, can't tell without Y)

10-19. Feedback input X goes_____. (high, low, can't tell without Y)

10-20. The state of the latch goes_____. (set, reset, latched, cannot be determined since all the NAND inputs aren't given)

Questions 10-21 through 10-40 refer to the following cross-coupled NOR latch:

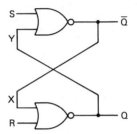

If Input S goes LOW and R goes HIGH:

10-21. Output Q goes_____. (high, low, can't tell without X)

10-22. Feedback input Y goes_____. (high, low, can't tell without X)

10-23. Output $\overline{Q}$ goes_____. (high, low, can't tell without Y)

10-24. Feedback input X goes_____. (high, low, can't tell without Y)

10-25. The state of the latch goes_____. (set, reset, latched, cannot be determined since all the NOR inputs aren't given)

If input R goes LOW and S goes HIGH:

10-26. Output $\overline{Q}$ goes_____. (high, low, can't tell without Y)

10-27. Feedback input X goes_____. (high, low, can't tell without Y)

10-28. Output Q goes_____. (high, low, can't tell without X)

10-29. Feedback input Y goes_____. (high, low, can't tell without X)

10-30. The state of the latch goes_____. (set, reset, latched, cannot be determined since all the NOR inputs aren't given)

If input S goes LOW and R goes LOW:

10-31. Output Q goes_____. (high, low, can't tell without X)

10-32. Feedback input Y goes_____. (high, low, can't tell without X)

10-33. Output $\overline{Q}$ goes_____. (high, low, can't tell without Y)

10-34. Feedback input X goes_____. (high, low, can't tell without Y)

10-35. The state of the latch goes_____. (set, reset, latched, cannot be determined since all the NOR inputs aren't given)

If input S goes LOW and R goes LOW and X is already HIGH:

10-36. Output Q goes_____. (high, low)

10-37. Feedback input Y goes_____. (high, low)

10-38. Output $\overline{Q}$ goes_____. (high, low, can't tell without Y)

10-39. Feedback input X goes_____. (high, low, can't tell without Y)

10-40. The state of the latch goes_____. (set, reset, latched, cannot be determined since all the NOR inputs aren't given)

Questions 10-41 through 10-56 refer to the following clocked RS latch symbol:

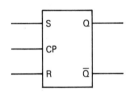

If input S is LOW and R is HIGH and CP goes HIGH:

10-41. Output Q goes_____. (high, low, can't tell without old Q)

10-42. Output $\overline{Q}$ goes_____. (high, low, can't tell without old $\overline{Q}$)

10-43. The state of the latch is_____. (set, reset, latched)

If input R is LOW and S is HIGH and CP goes HIGH:

10-44. Output $\overline{Q}$ goes_____. (high, low, can't tell without old $\overline{Q}$)

10-45. Output Q goes_____. (high, low, can't tell without old Q)

10-46. The state of the latch is_____. (set, reset, latched)

If input S is HIGH and R is HIGH and CP goes HIGH:

10-47. Output Q goes_____. (high, low, can't tell without old Q)

10-48. Output $\overline{Q}$ goes_____. (high, low, can't tell without old $\overline{Q}$)

10-49. The state of the latch is_____. (set, reset, latched)

If CP is LOW:

10-50. Output Q is_____. (high, low, can't tell without old Q)

10-51. Output $\overline{Q}$ is_____. (high, low, can't tell without old $\overline{Q}$)

10-52. The state of the latch is_____. (set, reset, latched)

10-53. The flip-flop's R and S inputs are_____. (asynchronous, synchronous)

10-54. The clock input is active_____. (high, low)

10-55. The R and S inputs are active_____. (high, low)

10-56. The R and S inputs are_____ (edge, level) triggered.

Questions 10-57 through 10-75 refer to the following J-K flip-flop symbol:

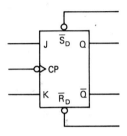

If input K is LOW and J is HIGH and CP goes LOW:

10-57. Output Q goes_____. (high, low, can't tell without old Q)

10-58. Output $\overline{Q}$ goes_____. (high, low, can't tell without old $\overline{Q}$)

10-59. The state of the latch is_____. (set, reset, latch, toggle)

If input J is LOW and K is HIGH and CP goes LOW:

10-60. Output $\overline{Q}$ goes_____. (high, low, can't tell without old $\overline{Q}$)

10-61. Output Q goes_____. (high, low, can't tell without old Q)

10-62. The state of the latch is_____. (set, reset, latch, toggle)

If input K is HIGH and J is HIGH and CP goes LOW:

10-63. Output Q goes_____. (high, low, can't tell without old $\overline{Q}$)

10-64. Output $\overline{Q}$ goes_____. (high, low, can't tell without old Q)

10-65. The state of the latch is_____. (set, reset, latch, toggle)

If CP goes HIGH:

10-66. Output Q is_____. (high, low, can't tell without old Q)

10-67. Output $\overline{Q}$ is_____. (high, low, can't tell without old $\overline{Q}$)

10-68. The state of the latch is _____ . (set, reset, latch, toggle)

10-69. The flip-flop's J and K inputs are _____ . (asynchronous, synchronous)

10-70. The flip-flop's $\overline{R_D}$ and $\overline{S_D}$ inputs are _____ . (asynchronous, synchronous)

10-71. The clock input is active _____ . (high, low)

10-72. The $\overline{R_D}$ and $\overline{S_D}$ inputs are active _____ . (high, low)

10-73. The J and K inputs are active _____ . (high, low)

10-74. The J and K inputs are _____ (edge, level) triggered.

10-75. Can a constraint input be applied through J and K? _____ Through $\overline{R_D}$ and $\overline{S_D}$? _____ .

11

APPLICATIONS OF FLIP-FLOPS (SHIFT REGISTERS/ PARALLEL REGISTERS)

————CHAPTER OBJECTIVES————

By the time you finish this chapter, you will be able to:

1. Describe how D-type flip-flops are used in a *shift register*, and why they must be master/slave or edge-triggered D flip-flops.
2. Describe how a *shift register* is used in a *serial full adder*.
3. Describe how D-type flip-flops or latches are used in *parallel registers*.
4. Describe how *shift registers* are used in *ring counters* and *Johnson counters*.
5. Describe how shift registers are used in *serial-in*, *parallel-out* and *parallel-in*, *serial-out* register/converter circuits.

11-1 SHIFT REGISTERS

11-1.1 D-Type Flip-Flops

Our first application of flip-flops is a circuit called a shift register. We're introducing it here as an application of the D-type flip-flop, although it could be made with other types of flip-flops as well.

A D flip-flop, you'll recall, has just one input called D for setting and resetting instead of two (J-K or R-S). It is set when a 1 is clocked into D, and reset when a 0 is clocked into D. The type of D latch we saw in Chapter 10 was a *level-triggered latch*, shown in Figure 11-1. There is a timing diagram with the picture, showing how we see at Q whatever D is doing while the clock is ON (active-HIGH clock). This is also called a "transparent" latch.

Another type of D latch is possible, shown in Figure 11-2. It is the master/slave version of the D latch (an *edge-triggered D latch*). Like the simple D latch, this one lets us see at Q whatever D is doing when the flip-flop is clocked, but in this case, Q is a "snapshot" of what D was doing at the instant the CP went LOW. You can see this in the timing diagram—compare it with the one in Figure 11-1—where the slave captures the instantaneous state of the master at the falling edge of the clock pulse. The data out doesn't appear until the data input is disabled. We'll find this fact useful when we use D latches to make a shift register.

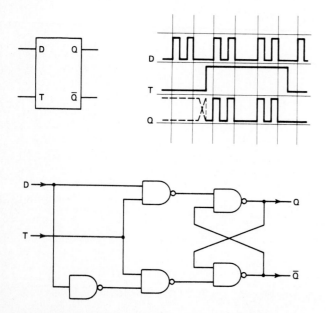

FIGURE 11-1   D flip-flop, NAND implementation, with timing diagram.

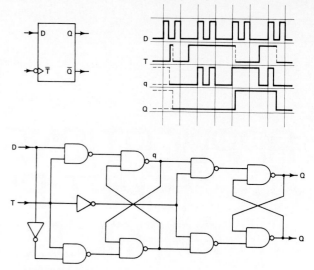

FIGURE 11-2   Edge-triggered D flip-flop, with timing diagram.

11-1.2 Application of the D Flip-Flop

"So what's a shift register?" you ask. A **shift register** is a circuit—made of flip-flops—that holds a bunch of numbers and shifts the numbers over a "distance" of one flip-flop for every clock pulse.

You've probably seen a shift register in action without knowing it. When you look at the display of some pocket calculators, you see the numbers on the display shift over as each new number is keyed in. This is the action of a shift register.

The D flip-flop is used in Figure 11-3 to make an 8-bit shift register. It has eight flip-flops and can hold an eight-digit binary number (one bit to a flip-flop). There is only one input to the shift register from outside. As each number is clocked into the front flip-flop, the other flip-flops transfer numbers in from their neighbors to the left.

We want the shift register to transfer each bit over a distance of one flip-flop for every clock pulse—but will the circuit really do it when we use level-triggered D-type flip-flops? If the clock-pulse is too long, the data bit being clocked into the "front" flip-flop's input comes out its output while the clock is still on. The next flip-flop will receive that bit, and the next, and the next, for as long as the clock is ON. The bit travels a distance of more than one flip-flop—which is not supposed to happen. We call this **racing,** and it's a variation on the same problem we had with the J-K flip-flop (without master/slave). We solved that problem with edge-triggered logic, and that's what we need here, too. To make a shift register, use D flip-flops that are edge-triggered, or you will see the kind of racing shown in Figure 11-4.

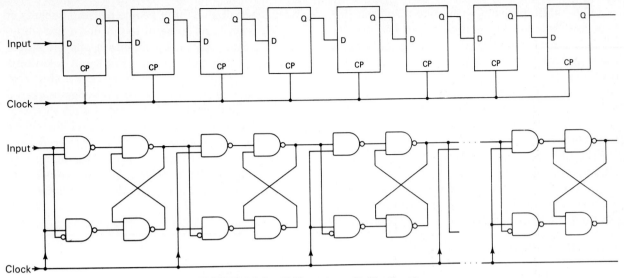

FIGURE 11-3   Shift register (D flip-flops).

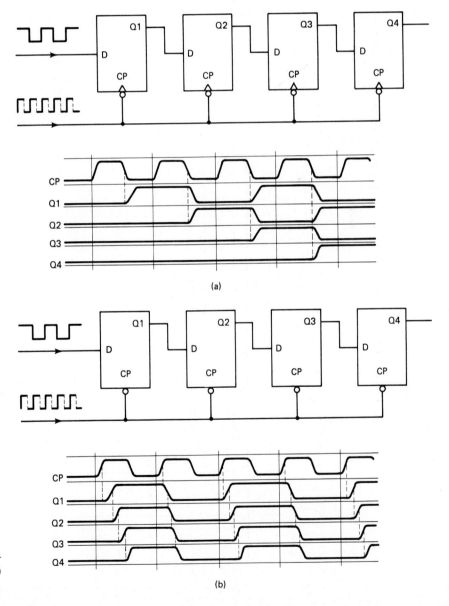

(a)

(b)

FIGURE 11-4 Shift registers: (a) edge-triggered; (b) level-triggered.

The shift register with edge-triggered flip-flops will transfer bits a distance of one flip-flop (and no more) during each clock pulse, because the clock inputs of all eight flip-flops enable while the outputs are still the number from the last time state. The new data bit can't reach the front gate's output until the clock is OFF. That prevents the next gate (a D edge-triggered flip-flop) from receiving the data bit until there's another clock pulse. The register can't race under these circumstances.

This shift register is sometimes also called a "bucket brigade" shift register. It reminded somebody of a chain of people passing buckets of water from hand to hand to put out a fire—only, in this case, the buckets contain one bit apiece and the people are flip-flops. Figure 11-4 contains two *timing diagrams* that show how a shift register made of level-triggered and edge-triggered logic would transfer a 1 through a 4-bit shift register. The timing diagram of Figure 11-4(a) shows how the shift register with edge-triggered logic would work, and Figure 11-4(b) shows racing with level-triggered logic as the input bit transfers to all four flip-flops during the first clock cycle.

EXAMPLE 11-1

Show what happens in an 8-bit shift-right register as the binary number 10000011 is clocked into it serially. Assume that the most-significant bit is clocked-in first, and the least-significant bit last.

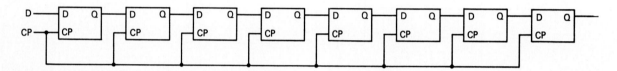

Solution: The eight symbols shown below represent the shift register shown schematically in the diagram above. The number in each box is the Q output of the flip-flop.

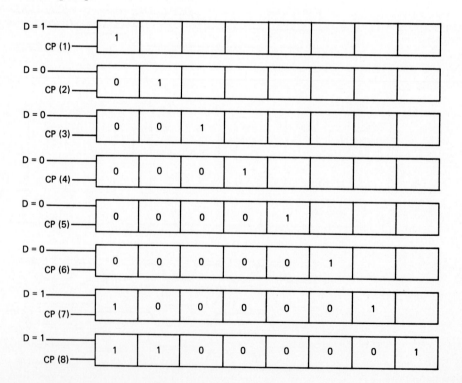

11-2 SERIAL FULL-ADDERS

We introduced adder circuits in an earlier chapter. At that time we found that 3-bit numbers could be added by a full-adder circuit containing three adders, 4-bit numbers could be added by parallel full-adders containing four adders, and so forth. The bigger the numbers got, the more adders we needed to add them. Now, we're going to look at a circuit that can add numbers of any size with only one full-adder circuit, provided that the registers for holding the addend, augend, and sum are shift registers.

In Figure 11-5, the two numbers to be added—the addend (5) and augend (7)—have 3 bits each. They are held in two 3-bit shift registers with the units' bit in the rightmost latch. At the time shown in Figure 11-5(a), the units' bits are going into the adder. Since delay latch C is LOW,

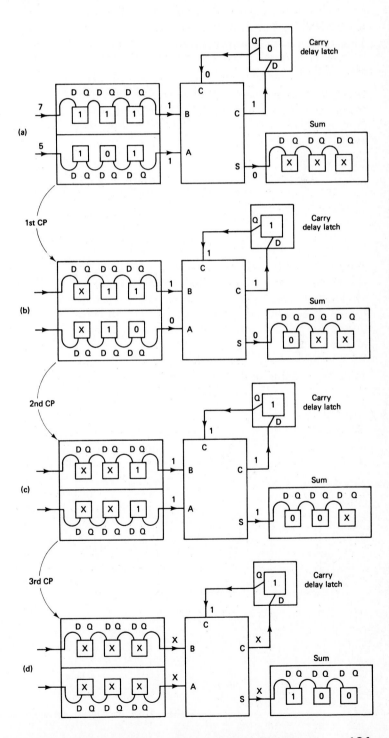

FIGURE 11-5 Serial full-adder action.

the adder adds the sum of 1 plus 1. The output at C and S is 10, the number 2.

After a clock pulse, the C and S outputs have been shifted into the delay latch C and the sum shift register. The addend and augend have also been shifted over, and now the adder has 0 and 1 to add, plus the 1 from the delay latch C. This sum is 2, and the C and S outputs are 1 and 0.

After another clock pulse, the second C and S values have been shifted into the C flip-flop and the sum shift register, and the addend and augend have shifted over once again. The adder has 1 and 1 to add, plus the 1 from the delay latch, C. This sum is 3, so C and S outputs are both 1s.

Finally, after a third clock pulse, the C and S values are shifted into the C flip-flop and the sum shift register. The sum is 100 and the carry (C flip-flop) is 1. The carry is a "fourth" bit of the sum, and if we put them together, the adder has added:

$$\begin{array}{cccc} \text{Addend} & \text{Augend} & \text{C} & \text{Sum} \\ 101 & + \quad 111 & = 1 & 100 \\ (5) & (7) & & (12) \end{array}$$

This circuit is called a **serial full-adder**, since the bits are added one after another (in series). It has a lot fewer adder gates than the parallel adder, but takes a lot more time. To add three-digit numbers, we used three clock pulses, to add four digits would take four clock pulses, and so on.

In the early days of digital electronics, when gates were made of discrete components, even a half-adder was an expensive circuit. Reducing the number of gates in a circuit was more important than having the fastest possible circuit. Because of this, serial logic was the most popular way to design arithmetic circuitry. Today, when circuits can have thousands of gates in a package, the size of one discrete transistor, parallel circuits with high speed have replaced serial logic. In the trade-off between simplicity and speed, speed has won. Therefore, today's arithmetic-logic units in minicomputers and microprocessors do not contain serial logic; they contain parallel full-adders with as many bits as the word size of the computer.

11-3 PARALLEL REGISTERS

11-3.1 8-Bit Latch

Figure 11-6 shows an 8-bit **parallel register**. You can see that each flip-flop has inputs and outputs that are completely separate from one another. The only thing all the latches in the parallel register share is a common clock pulse. When one latch is loaded with a bit, every other latch is being loaded at the same time. To load an 8-bit register, eight inputs are needed; when the register—sometimes called an 8-bit latch—is strobed (clocked), there are eight outputs. A parallel register, like the 8-bit latch, is often used as a "delay" gate to put an 8-bit number on hold for one clock pulse.

Latches are commonly available in 4-bit, 6-bit, and 8-bit sizes.

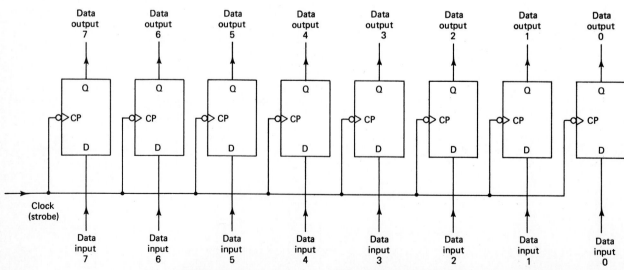

FIGURE 11-6 Parallel-in, parallel-out 8-bit register.

In all diagrams of parallel full-adders in Chapter 6, the registers which held the addend and augend and the registers where the sum and carry went were parallel registers. We just never said what they were.

11-3.2 Parallel versus Serial Signals

Parallel describes the digital "words" whose bits are all available at the same time (in different places).

Serial describes digital "words" whose bits are all available in the same place (at different times).

Parallel data have bits whose "place value" is determined by *where* they are—this is called **space-division multiplex** (SDM).

Serial data have bits whose "place value" is determined by *when* they come along—this is called **time-division multiplex** (TDM).

11-4 SHIFT COUNTERS/ RING COUNTERS

11-4.1 Shift Counter

We said that shift registers could be made of other flip-flops than the D latch. A shift register made with J-K master/slave flip-flops is shown in Figure 11-7(a). The shift counter developed from this shift register uses a steering network like the two connections used to steer a J-K or T flip-flop. Figure 11-7(b) shows the circuit for this counter. Since the shift counter is basically a shift register, we can understand its action in terms of the fact that once a number enters the front end, it will continue along the string of flip-flops with every clock pulse until it reaches the back end. Normally, the shift counter is started at 0000. Each clock pulse will set the front flip-flop as long as the back flip-flop is still a 0. As each 1 is moved

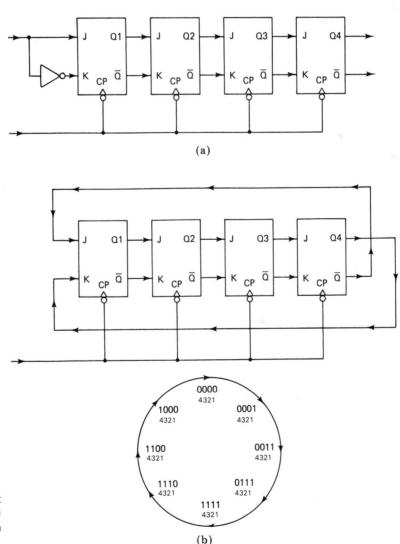

FIGURE 11-7 (a) J-K shift register; (b) shift (Johnson) counter with state transition diagram (STD).

in at the front, it pushes the rest of the stack of 1s toward the back until the last flip-flop has a 1 shifted into it. When a 1 arrives at the back, its outputs (which are the inputs at the front) change from 01 to 10, and the input at the front changes to reset from set. Now the front flip-flop will reset with every clock pulse, and a stack of 0s will push toward the back until the shift register is 0000 and the cycle starts again. This is shown in a state transition diagram alongside the schematic in Figure 11-7(b).

A shift counter is also called a **Johnson counter**. The *Johnson code* produced by the shift counter has twice as many patterns as there are flip-flops in the counter. A shift counter with three flip-flops, for instance, will produce six distinct codes; a shift counter with four flip-flops produces eight distinct codes, a shift counter with five flip-flops produces 10 unique code patterns, and so forth. The number of distinctly different patterns a counter will produce is called its **modulus**. The modulus of a Johnson counter with n flip-flops is $2 \times n$.

11-4.2   Ring Counter

A ring counter is, like a shift counter, made with a shift register as its main part. We could make this circuit with either D or J-K flip-flops, but chose to use J-K types again. In Figure 11-8(a), the schematic of the ring counter may appear identical at first to the shift counter. Closer inspection will reveal the *steering network* that feeds the output of the back flip-flop to the inputs of the front flip-flops are not crossed. This forms a *ring*—shown more clearly in Figure 11-8(b)— that transfers the same bits around, and around, and around.

The ring counter is not normally started at 0000, because the 0s all shift around the ring, but the new 0000 looks just like the old. Normally, the starting number for a ring counter is something like 1000 or 0001 that has a single 1 in it. As the ring is clocked, the 1 circulates around the ring from position to position. The 1 can be used to switch on a sequence of machines, one after another, whose operating time is controlled by the length of the clock pulses. A state transition table is shown alongside the ring counter in Figure 11-8(b).

A ring counter will have as many different codes as there are flip-flops, since the only difference between outputs of the *ring code* is the place where the 1 is. We say that the modulus of the ring counter is n when there are n flip-flops in the counter.

EXAMPLE 11-2

Show what happens in a 4-bit shift-right Johnson counter as it is clocked through its entire modulus. Assume that the starting condition of the counter is with a Q outputs LOW.

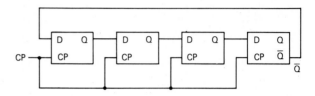

Solution: The eight symbols shown below represent the Johnson counter shown schematically in the diagram above. The number in each box is the Q output of the flip-flop.

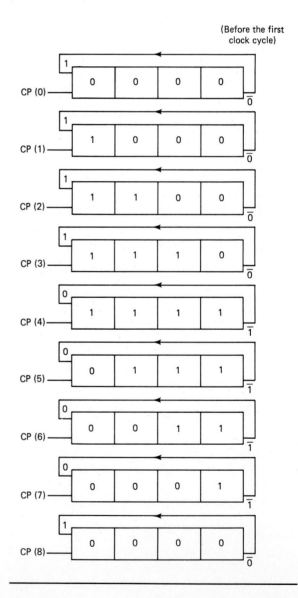

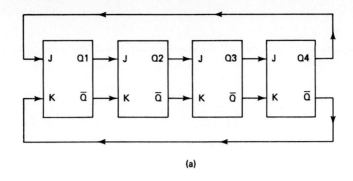

(a)

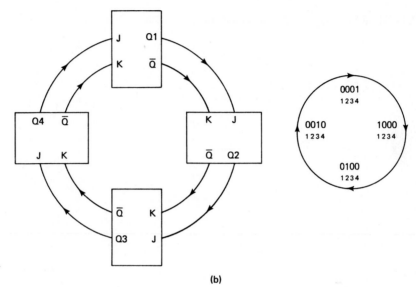

FIGURE 11-8 (a) Ring counter; (b) alternative configuration with STD.

(b)

11-5 PARALLEL-TO-SERIAL CONVERSION

The circuit in Figure 11-9 shows how a number can be loaded into the flip-flops of an 8-bit shift register. This parallel loading of data into every flip-flop of a shift register is called a **broadside load**. The S circle in each flip-flop's input is a switch that changes the connection from the parallel input to the serial input when the control goes from LOW to HIGH. (We've shown two possible switch circuits to fill the circle with in the diagram.)

To convert a parallel data word—with 8 bits in eight different places at the same time—into a serial data word—with 8 bits one after another on the same wire at different times—the **parallel-to-serial converter** is operated in the following way:

1. When the control line is LOW, the parallel inputs get to D, and the eight flip-flops are a parallel register—an 8-bit-

latch—a single clock pulse will load the eight inputs into the eight flip-flops.

2. Next, the control line is made HIGH, and the parallel inputs are "disconnected," while the serial input—connecting each stage to its neighbor—is "connected." That makes the eight flip-flops into a shift register.

3. Now each clock pulse shifts the bits in the register over, bringing a new bit to the output. After eight clock pulses, an 8-bit serial number will have appeared on the output wire.

Parallel data are commonly available at input and output ports of computers, where a bus of wires is available to carry 8-bit or 16-bit data around. Serial data are important where it's necessary to get all the data bits on one wire—a single telephone circuit, for instance—and the speed the bits travel at isn't as important as the cost of additional "matched" circuits. To enable one

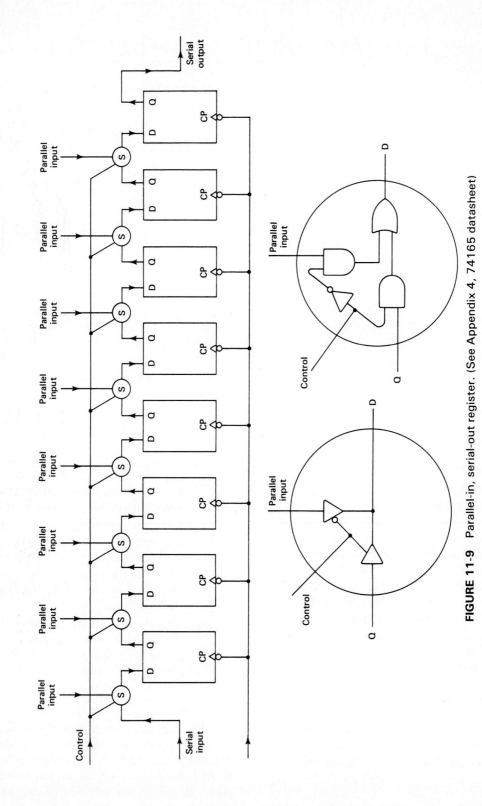

FIGURE 11-9 Parallel-in, serial-out register. (See Appendix 4, 74165 datasheet)

computer to "talk" to another at long distance using telephone lines, the transmitting computer must convert its parallel binary output into serial signals to go down the telephone wire. The parallel-to-serial converter circuit of Figure 11-9 is sometimes called a *transmitter* for this reason. See Appendix 4 for a data sheet of the 74165, a device of this type.

11-6 SERIAL-TO-PARALLEL CONVERSION

The telephone link between computers mentioned in the preceding section is going to need a receiver at the other end of the telephone circuit to convert the serial data on the phone line back into parallel at the computer. A circuit to do this is shown in Figure 11-10. You can immediately see that this circuit is practically identical to the transmitter circuit of Figure 11-9. The only difference in this circuit is that the S switching circuits are at the outputs of each flip-flop, and determine that after eight clock pulses shift in all the bits from the serial input, the outputs of every D flip-flop are enabled to the block marked "parallel device."

The shift register converts data to parallel from serial just by being a shift register. The outputs are all there anyway—although connected to the next flip-flop's D input—so there's no reason why additional wires shouldn't be switched into contact with these 8 bits when the "word" has been completely clocked in after eight clock pulses.

It wouldn't be hard to make the transmitter at the sending end of the telephone link double as a receiver—just put the input bits into the front flip-flop's D input and add buffers from the eight outputs to the parallel device. The compound device that results would be called a **universal asynchronous receiver/transmitter** (UART) and is one of the subjects we discuss in detail in Chapter 22. See Appendix 4 for a data sheet of the 74164, a device of this type.

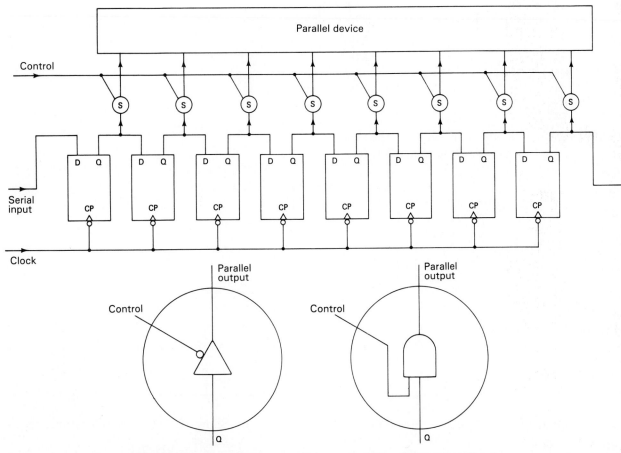

FIGURE 11-10 Serial-in, parallel-out register. (See Appendix 4, 74164 datasheet)

11.1. Name two types of flip-flops that can be used to make shift registers.

11-2. Are the flip-flops used in shift registers edge-triggered or level-triggered types?

11-3. For the following circuit,

complete the timing diagram for outputs A, B, C, and D of the shift register. The inputs and clock are shown on the figure.

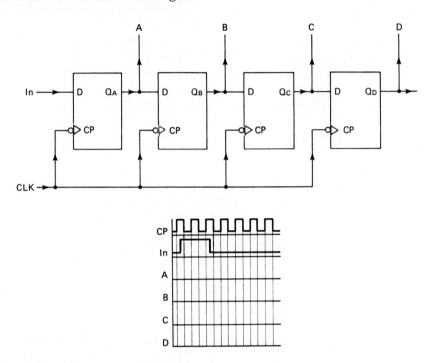

11-4. Redraw the serial full-adder in the sketch according to the instructions that follow.

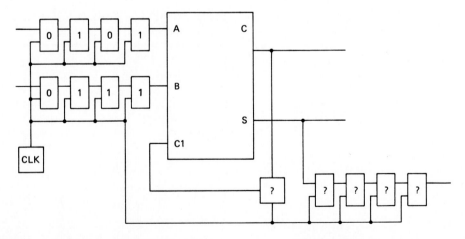

(a) Show the outputs after one clock pulse.

(b) Show the outputs after two clock pulses.

(c) Show the outputs after three clock pulses.

(d) Show the outputs after four clock pulses.

11-5. How many clock cycles are needed to add two 8-bit numbers with a serial full-adder?

11-6. Describe the advantages and disadvantages of parallel and serial adders.

11-7. Draw a schematic for a serial full-adder/subtracter. (*Hint:* See Chapter 6.)

11-8. Does a parallel register handle data that are TDM or SDM?

11-9. Redraw the shift (Johnson) counter in the following figure, to have 2 flipflops.
   (a) Show the outputs after one clock pulse.
   (b) Show the outputs after two clock pulses.
   (c) Show the outputs after three clock pulses.
   (d) Show the outputs after four clock pulses.
   (e) Show the outputs after five clock pulses.

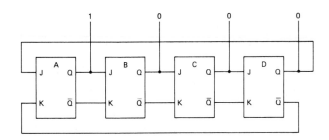

11-10. Work out a state-transition diagram for the circuit of Question 11-9.

11-11. Redraw the ring counter in the following figure, as indicated:
   (a) Show the outputs after one clock pulse.
   (b) Show the outputs after two clock pulses.
   (c) Show the outputs after three clock pulses.
   (d) Show the outputs after four clock pulses.
   (e) Show the outputs after five clock pulses.

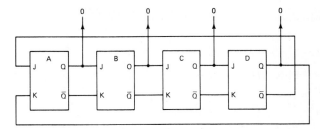

11-12. Work out a state-transition diagram for the circuit in Question 11-11.

11-13. What is the modulus of a four-flip-flop Johnson counter?

11-14. What is the modulus of a four-flip-flop ring counter?

11-15. A shift register with a parallel load input ("broadside load") can convert data from SDM to TDM. (True or False?)

11-16. A parallel-to-serial converter is a shift register with gated outputs. (True or False?)

11-17. A serial-to-parallel converter is a shift register with gated outputs. (True or False?)

Questions 11-1 through 11-10 below refer to the following symbolic diagram:

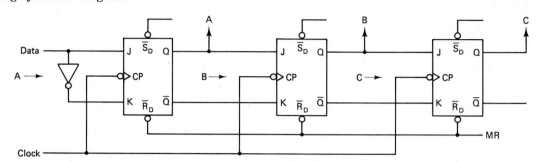

11-1. The circuit is a shift-_____ (right, left) register.

11-2. Data are transferred a distance of _____ (one, two, three) flip-flops with every cycle of the clock.

11-3. To work properly, a shift register should be made with _____ (level, edge, either level or edge)-triggered inputs.

11-4. Data are transferred at the _____ (falling, rising) edge of the clock cycle.

11-5. The MR input _____ (clears, sets) all the flip-flops.

11-6. Data arriving at the flip-flop on the left will apper at output C with a delay of _____ (one, two, three) clock cycles.

11-7. This type of shift register is a _____ (LIFO, FIFO) register.

11-8. Binary numbers arrive at the Data input in _____ (serial, parallel) form.

11-9. Once the register is loaded, the data are available at outputs A, B, and C in _____ (serial, parallel) form.

11-10. A shift register like this could also be made with _____ (D-type, R-S-type, either D- or R-S-type) flip-flops.

Questions 11-11 through 11-20 below refer to the following symbolic diagram:

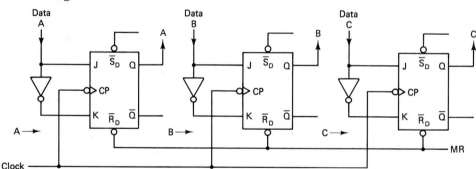

11-11. The circuit is a _____ (shift, parallel) register.

11-12. Data are transferred into _____ (one, two, three) flip-flops with every cycle of the clock.

11-13. To work properly, this type of register should be made with _____ (level, edge, either level or edge)-triggered inputs.

11-14. Data are transferred at the _____ (falling, rising) edge of the clock cycle.

11-15. The MR input _____ (clears, sets) all the flip-flops.

11-16. Data arriving at Data input C will appear at output C with a delay of _____ (one, two, three) clock cycles.

11-17. This register is also called a 3-bit _____ (series, broadside, latch).

11-18. Binary numbers arrive at the Data inputs in _____ (serial, parallel) form.

11-19. Once the register is loaded, the data are available at outputs A, B, and C in _____ (serial, parallel) form.

11-20. A register like this could also be made with _____ (D-type, R-S-type, either D- or R-S-type) flip-flops.

Questions 11-21 through 11-30 refer to the following symbolic diagram:

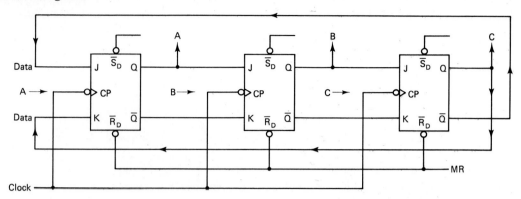

11-21. The circuit shown is a _____ (ring, Johnson) counter.

11-22. If MR is held LOW, the counter _____ (will, will not) be able to count.

11-23. If this counter is reset, the next clock cycle will transfer a _____ (1, 0) into flip-flop A.

11-24. After three clock cycles, A, B, and C are all _____. (1s, 0s)

11-25. If started at zero, this counter will repeat its sequence of outputs every _____ (one, three, six, eight) clock cycles.

11-26. If output C is HIGH, the input to A will be _____. (HIGH, LOW)

11-27. Interchanging the inputs to A would make this a _____ (ring, Johnson) counter.

If the counter is started at zero, which of the timing diagrams shown below is the correct one for:

11-28. Flip-flop A? _____

11-29. Flip-flop B? _____

11-30. Flip-flop C? _____

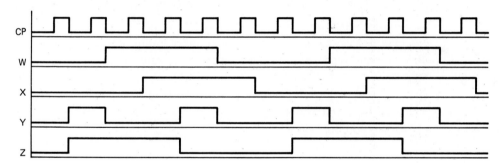

12

MORE APPLICATIONS OF FLIP-FLOPS (COUNTER CIRCUITS)

CHAPTER OBJECTIVES

By the time you finish this chapter, you will be able to:

1. Identify the schematic diagrams of the following counter circuits:
   a. Divider chain
   b. Ripple-type up-counter (binary)
   c. Ripple-type down-counter (binary)
   d. Forced-reset counter (moduli not 2^n)
   e. Synchronous binary counter
   f. Synchronous counter (moduli not 2^n)
2. Describe the modulus and state-transition diagram of any of the above-named types of counters, given the schematic or description of the item.
3. Design the above-named counters for various moduli.
4. Describe the commonly used sources of clock pulses for the above-named counters.

12-1 ODD-EVEN (DIVIDE-BY-2, MOD-2) COUNTER (RECAP: T-TYPE FLIP-FLOP)

In Chapter 11 we saw so many circuits that made use of the D flip-flop that perhaps the title of that chapter should have been "Applications of D Flip-Flops." In this chapter we see so many circuits that use the T flip-flop that we might want to call it "Applications of the T Flip-Flop." Many of these circuits can be made with other flip-flops—and some can't be made with T flip-flops—so we've chosen to title this chapter "Counters." We already saw two circuits called counters in Chapter 11, the ring counter and the shift counter. These are really glorified shift registers rather than counters—they don't count in a familiar binary code. In this chapter we see how to make counters that count numbers in standard binary codes.

What is a digital counter, how does it work, and what uses does it have once we have it working? These are questions we answer in this chapter. We'll begin with a recap of the T flip-flop's action (first described in Chapter 10).

Figure 12-1 shows a T flip-flop being clocked with pulses at its T output. This is the CP input—the clock—for a J-K flip-flop, but the J and K are not used. Instead, the flip-flop has only one possible response to a clock pulse. It must toggle at its output whenever the clock has a falling edge. The timing diagram of the T flip-flop shown in Figure 12-1 is a graph of voltage (logic state) versus time. It shows how the input and output signals are related. Since the output changes with every falling edge of the input, the output has half as many cycles as the input, in the same number of seconds. We call cycles per second (Hertz) the **frequency** of a signal, and the T flip-flop's ability to divide its input frequency in half is the most important thing it does.

In Figure 12-2, a string of T flip-flops has been attached in "cascade." The output of one has been attached to the input of the next, forming a chain of four flip-flops called a **divider string**. The name comes from the effect of the flip-flops on the input frequency.

If one T flip-flop is a divide-by-2 circuit, four of them form a divide-by-16 circuit. Here is the reasoning involved: If the frequency of the incoming signal at flip-flop A is 64 Hz (64 pulses per second), the frequency at output A is 32 Hz, because A is a divide-by-2 device. The input to the B flip-flop is 32 Hz, because the B flip-flop's input is the A flip-flop's output. B divides this by 2

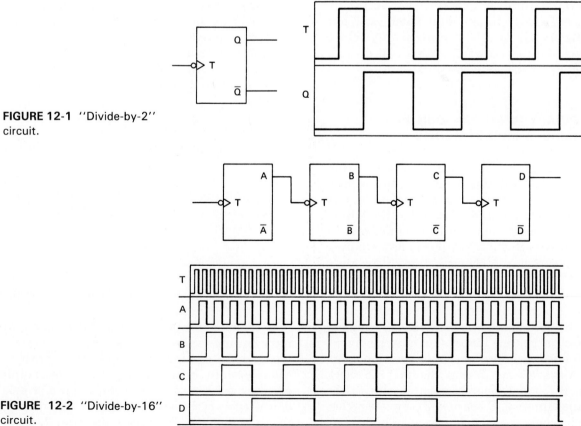

FIGURE 12-1 "Divide-by-2" circuit.

FIGURE 12-2 "Divide-by-16" circuit.

again, and the B output has a frequency of 16 Hz. Similarly, flip-flop C divides by 2 again, and has 8 Hz output. Flip-flop D divides by 2 once more, and has a 4 Hz output. The original 64 Hz put into A causes a 4 Hz output from D, so the overall effect of the divider string has been to divide by 16. The relationship between the number of flip-flops, n, and the divide-by of the string is:

$$\text{divide-by} = 2^n$$

For example, digital watches are often controlled by crystal oscillators that oscillate at 32,768 Hz. This is a large frequency compared to the 1-Hz (one pulse per second) "ticks" that the clock actually "counts" to "keep time." A divider string is included in the clock circuit to reduce the frequency from 32,768 Hz to a manageable 1 Hz. The divide-by-32,768 required "happens" to be exactly

$$32,768 = 2^{15}$$

so that a divider string of 15 T flip-flops in cascade will do the job required.

Why not just use a crystal with a natural frequency of 1 Hz? It turns out that a crystal oscillates in a time period based on its size. A 1-Hz crystal is 32,768 times as big as one with a 32,768-Hz natural frequency. The 32,768-Hz crystal fits easily into a digital watch case, but the 1-Hz crystal doesn't fit conveniently into a city block. Buy a wristwatch with a 1-Hz crystal, and there goes the neighborhood!

12-2 BINARY RIPPLE UP-COUNTERS (MOD-4, MOD-8, MOD-16)

Let's take another look at the divide-by-4 string in Figure 12-2. A timing diagram is included in the figure which shows the state of D, C, B, and A during 48 clock pulses. Notice that the falling edge of each waveform is marked to stand out. At the moment each CP (clock pulse) has a falling edge, the output of A changes states (it toggles). When A has a falling edge, B toggles. When B has a falling edge, C toggles; and so forth.

As you check that each output toggles at its neighbor's falling edge, you find that the number of pulses in the time interval follows the divide-by-2 rule for each flip-flop. There are eight complete A waveforms during the first 16 clock pulses, B has only four complete cycles in the same time interval, C has only 2 complete cycles, and D has only one complete cycle of both LOW

and HIGH states. In each place where there are fewer pulses, there is a lower frequency.

Finally, notice that if you read the 1s and 0s from top to bottom (in DCBA order) in each timing state (each cycle of CP), the four-digit numbers DCBA are binary codes for the numbers from 0 to 15. This is the reason we call this circuit a counter. It counts the number of pulses (clock cycles) it has received through its input in binary code. A state transition diagram (STD) included with the timing diagram shows a decimal values for the numbers the counter produces. Each number is called a state. It represents one of the patterns that comes out at DCBA. On the timing diagram, the number 0 follows the number 15, and the whole sequence of numbers repeats every 16 clock pulses. The STD shows this as a "clock face" with an "hour hand" that goes around once every 16 timing states. Since the 0 follows the 15 on the clock face, the sequence of 16 decimal numbers repeats over and over, just as the real binary numbers from the counter do.

A counter is called a counter because it counts the number of clock-pulses that come in its input, but no counter keeps counting pulses forever. Sooner or later, any counter runs out of new states and has to start using old states over again to count with. The total number of states a counter uses to count is called its *modulus*. This is a name we saw in Chapter 11 associated with shift and ring counters. In this chapter on binary counters, the name still means: "The circuit produces just this many output codes, and no more." For the four-stage counter in Figure 12-2, there are 16 *states*, and we say that this is a **modulus-16 counter**.

Since we already know that Figure 12-2 is a divide-by-16 circuit (when used as a *divider string*), we might wonder if there is any relationship between the divide-by (16) and the modulus (16) of the counter. Of course there is—they're the same thing! Since we already know how to figure the divide-by of a counter with n flip-flops, we know that

$$\text{modulus} = 2^n$$

is the expression for the modulus of a counter that's also a divider string.

We also call this modulus-16 counter a **binary counter** since it counts in binary code, and an **up-counter** since the numbers it produces count upward from 0 to 15. The name **ripple counter** refers to the fact that there is a "ripple-down" effect of the action in one flip-flop affecting the next flip-flop. The clock pulse after 7, for instance, causes A to toggle from 1 to 0, and that's a falling edge, which causes B to toggle from 1 to 0, and that's a falling edge for C which toggles from 1 to 0, which finally toggles flip-flop D from a 0 to a 1. The whole thing proceeds like a stack of falling dominoes, and this gives it the name *ripple effect*, and the counter gets the name *ripple counter* because of it.

If we cascaded five flip-flops in Figure 12-2, we'd have Figure 12-3, which is a mod-32 counter ("mod-32" is short for "modulus-32"). Since this counter can count from 00000 (0) to 11111 (31), the counter counts all the possible 5-bit numbers there are. We said a long time ago that if you count, you can't miss any possible codes. The binary counter counts in the way we meant, and no counter can produce more states with the same number of flip-flops. There are, however, other ways to count that will give just as many numbers, but in a different sequence.

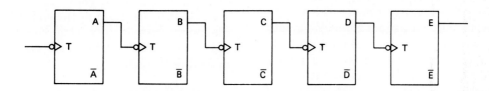

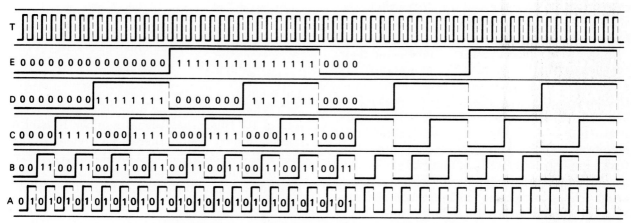

FIGURE 12-3 "Divide-by-32" counter with binary code.

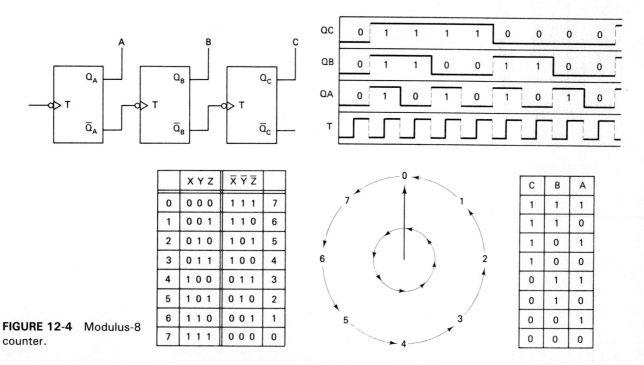

FIGURE 12-4 Modulus-8 counter.

	X Y Z	X̄ Ȳ Z̄	
0	0 0 0	1 1 1	7
1	0 0 1	1 1 0	6
2	0 1 0	1 0 1	5
3	0 1 1	1 0 0	4
4	1 0 0	0 1 1	3
5	1 0 1	0 1 0	2
6	1 1 0	0 0 1	1
7	1 1 1	0 0 0	0

C	B	A
1	1	1
1	1	0
1	0	1
1	0	0
0	1	1
0	1	0
0	0	1
0	0	0

12-3 BINARY RIPPLE DOWN-COUNTERS

Figure 12-4 (page 205) is a counter with three flip-flops connected in cascade. According to what we already know, the counter should have a modulus of 8 if it has a natural binary count.

There is a difference between this counter and the other binary counters we've seen previously. Each T input is attached to the previous flip-flop's $\bar{Q}$ output. Does the counter still count up in natural binary code? Look at the timing diagram. We see the toggle state at the rising edge of each Q. This happens because the not-Q—which is the real clock pulse for B and C—is falling when Q is rising. Reading the outputs A, B, and C, we find a backward sequence of binary numbers. This counter still counts in natural binary, but it counts down. This is the reason it's called a **down-counter** and the clock face in the state transition diagram has its hour hand going backward.

We've included a table of numbers and their complements in Figure 12-4. The **complement** of a number is what we get if we invert every bit in the number (it's also called a *one's complement*). You can see that the table of up-counted numbers from 0 to 7 becomes a table of down-counted numbers from 7 to 0 if we just invert the bits. If we took a binary up-counter and inverted all the outputs, the numbers we would get would be a down-count. The arrangement in Figure 12-4 does the same thing; it just does it a different way.

The table of binary numbers from a counter (shown for the downcounter in Figure 12-4) is called a *state transition table*. It's like the STD, except in binary, and shows what each output (A, B, and C) is doing. It contains the same information as the timing diagram, but does it with numbers instead of waveforms.

12-4 COUNTERS WITH MODULI OTHER THAN 2^n (FEEDBACK AND FORCED-RESET)

12-4.1 Natural Count and Forced-Reset

The counters we've discussed all reset at a certain point. They count up to a certain number, then the next count is zero. We could imagine a counter that counts forever, but is reset by someone pushing a reset button every time the count reaches 16—and this would describe a mod-16 counter. The natural count of a divider string doesn't really get reset at the end of its sequence, the number zero is the normal "next number" for the counter. With divider strings, only certain moduli are possible:

1 flip-flop	modulus $= 2^1 = 2$
2 flip-flops	modulus $= 2^2 = 4$
3 flip-flops	modulus $= 2^3 = 8$
4 flip-flops	modulus $= 2^4 = 16$
5 flip-flops	modulus $= 2^5 = 32$

and so on.

What about making a counter with a modulus of 10? Since the decimal number 10 isn't a natural binary count, there's no divider string that will normally count 10 numbers and return to zero. What we want is a counter called a mod-10 or decade counter. It should count in the sequence:

$$0, 1, 2, 3, 4, 5, 6, 7, 8, 9, 0, 1, 2, \ldots$$

repeating every 10 clock pulses. To do this, we'll use one of the available binary counters, but force it to reset to zero (make all flip-flops LOW) at an earlier time than its natural count would reach zero.

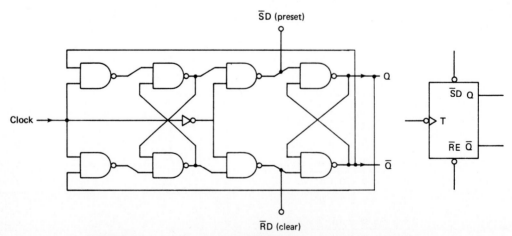

FIGURE 12-5 T flip-flop with PRESET and CLEAR (schematic).

12-4.2 Preset and Clear

To understand how this is accomplished, you need to know a little more about the T flip-flop than we've told you. In Chapter 11 when we discussed the shift counter, we said to start the counter at 0000, but to start the ring counter at 0001 or 1000. You probably wondered how to do that. The answer is provided by a pair of "extra" inputs found on T and J-K flip-flops. These are shown in Figure 12-5.

Figure 12-5 shows a gate diagram of a T flip-flop with two asynchronous inputs called $\bar{S}_D$ and $\bar{R}_D$. They are also called **preset** and **clear**. The preset turns ON the Q output of the flip-flop. The clear turns the Q output OFF. Neither of these actions requires a clock pulse—so we call these inputs asynchronous. The bar over the letters S and R indicates that these inputs are active LOW, and in the symbolic diagram, you can see a "meatball" at each input where the wire enters the symbol. These wires go directly to the inputs of a NAND latch, which are active LOW normally—the "bubble" on each line in the symbolic diagram does not indicate any sort of component between the wire and the flip-flop's "insides," as the NAND-gate diagram makes clear.

12-4.3 Power-On Reset

The **power-on reset (POR) circuit** in Figure 12-6 uses preset and clear inputs to arrange the starting condition of a flip-flop when its power is turned on. Capacitor C in the diagram, like

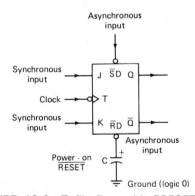

FIGURE 12-6 T flip-flop with PRESET and CLEAR (symbol).

everything else, has no voltage when the power is turned off. When the power is turned on, C is slower in developing a voltage than other parts of the circuit, and for a short time, C is LOW while the rest of the circuit has power. C soon becomes HIGH (its end is marked with a +), but its LOW voltage during the startup lasts long

enough to reset the flip-flop. The POR makes the flip-flop start with a beginning value of 0 whenever power is turned on. The capacitor could easily be attached to preset if you want a 1 at the time power is turned on.

12-4.4 A Forced-Reset Decade Counter

Now that we know how to preset or clear any flip-flop in a counter, it shouldn't be hard to see how that decade counter we discussed before is made. Figure 12-7 begins life as a mod-16 binary ripple up-counter [Figure 12-7(a)]. Since we want to reset all four flip-flops at the same time, we tie all four clear inputs to a common wire (reset) in Figure 12-7(b). At the count of 10, we want to zero the counter instead of letting it count all the way to 15. A decoder for 10 is made with a two-input AND gate in Figure 12-7(c). The decoder will have an active (HIGH) output for numbers 10, 11, 14, and 15, actually, but in an up-counter, 10 will be the first number that comes along. The other numbers won't matter because they'll never be counted—the decoder will clear the flip-flops to 0000 as soon as 10 appears. Since the clear inputs are active *LOW*, the decoder's output is made active LOW to drive it [Figure 12-7(d)]. The **forced-reset decade counter** which we have at the end counts like this:

0, 1, 2, 3, 4, 5, 6, 7, 8, 9, (momentary *ten*),
    0, 1, 2, . . .

Except for the fact that a 10 is present long enough to clear all four flip-flops (about 10 billionths of a second for TTL), the sequence of numbers from the counter is just what a decade counter is supposed to have. If the numbers are displayed on a visual readout device, the "momentary ten" will flash on the display for such a short time that nobody will ever notice it. Although the momentary output is too short for a human being to see, it lasts long enough for other logic gates to recognize. This problem is not unique to the forced-reset types of counter, though, as we'll see in the next section.

12-5 GLITCH STATES IN RIPPLE COUNTERS

We defined the divider string as a ripple counter a while ago. We said that the name comes from the "ripple-down" effect—the effect caused by the CP ripples down from A to B to C to D. It may take a while for all the toggles to happen, so it may take a while for everything to stabilize while a number changes. As an example, we mentioned

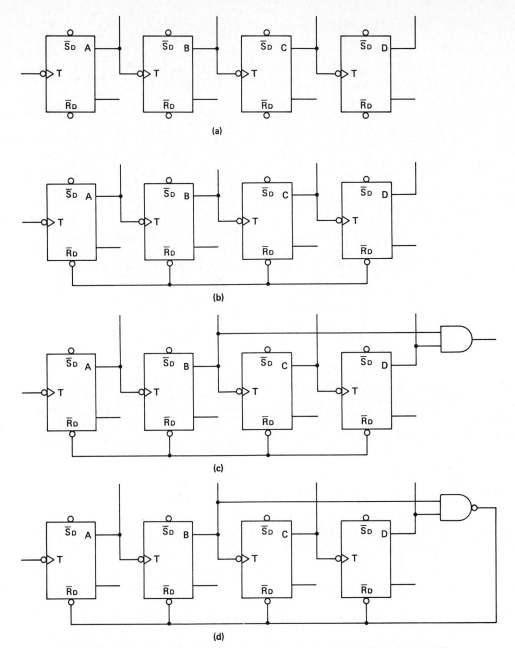

FIGURE 12-7 Development of a forced-reset modulus-10 ripple counter.

earlier what happens when the number 7 changes to 8 in a 4-bit ripple counter:

	D	C	B	A	
7	0	1	1	1	(seven)

CP falling edge

6 →	0	1	1	0	A toggles ⎫
4 →	0	1	0	0	B toggles ⎬ glitches
0 →	0	0	0	0	C toggles ⎭
8	1	0	0	0	D toggles (eight)

When we look at what happens when a ripple counter goes from 7 (0111) to 8 (1000), we see numbers between 7 and 8 that aren't supposed to be there. Certainly, the 6, 4, and 0 don't belong between 7 and 8. When this happened between sectors on a code wheel, we called these false readouts cross-over slivers or glitches. The ripple counters have glitches, too.

We could try making a counter that counts in Gray code, but that isn't easy—flip-flops are naturally binary, not Gray. You might ask: "Doesn't all that stuff happen at once?" The answer is "no"—and it can't happen at once as long as A is the clock for B, B is the clock for C, and

EXAMPLE 21-1

Sketch the schematic, timing diagram, and state-transition diagram for a three-stage ripple-type up-counter. Use T-type flip-flops in your schematic diagram.

Solution:

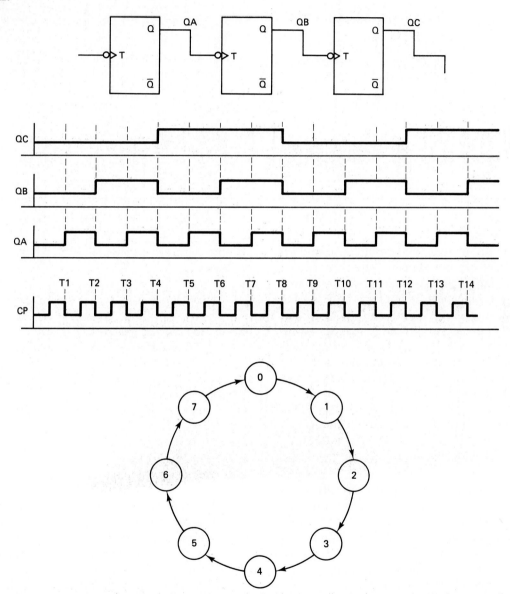

EXAMPLE 12-2

Sketch the schematic of a forced-reset modulus-5 counter made from a modulus-8 ripple-type up-counter.

Solution: The counter has to reset after the count of four so that it has a modulus of 5 (0, 1, 2, 3, 4, 0, 1, ...). To force the counter to zero when it would normally go to five, we build a decoder for the number 101 (five) out of a NAND gate. Technically, a decoder for 101 should have three inputs, and the decoder's output should be active when the inputs are CBA = 101. Since we want to drive the $\overline{R}_D$ lines with the decoder's output,

and they are active LOW, the expression for the decoder is

$$\overline{Reset} = C \cdot \overline{B} \cdot A$$

but we can use a NAND gate whose expression is

$$\overline{Reset} = C \cdot A$$

because the counter is an up-counter, and the first number it will count with C and A high is 101 (five). There is only one other number besides 101 that will produce an output from this NAND gate; that is 111 (seven). Since the counter is reset at the count of five, it will never get as far as seven, so we can use the simpler two-input gate as a decoder.

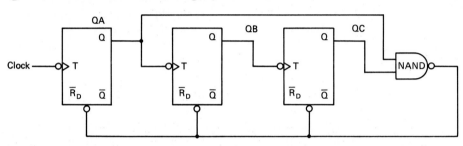

so on. We're up against the law of cause and effect—the effect always happens after the cause. In the case of TTL flip-flops each gate delay is a few dozen nanoseconds. When B is toggled because A has toggled, its output can't toggle until a few dozen nanoseconds *after* A toggles—similarly for all the other flip-flops in the string. Each number in the 7, 6-4-0, 8 sequence lasts long enough to activate several TTL gates down the line from the counter's outputs.

This could mean trouble, since the 0 between 7 and 8 is a phony zero, and some decoder somewhere could be falsely triggered by a 0 between every 7 and 8. For many circuits, the cross-over slivers between numbers aren't important and don't matter—but not for all circuits. In a digital watch or clock, for example, the numbers counted by each counter are meant to be viewed by the human eye. Any glitch states that change faster than the human eye can follow will not be noticed. A ripple counter or divider string is perfectly OK in digital wristwatch applications.

If the same counter is used to trigger a TNT blast after counting from 1 to 7, then back to 0, you wouldn't be very happy if a glitch triggered the explosive between the count of 1 and 2, instead of counting all the way around to 0 again—especially if you are the one setting the timer!

Ripple counters *do* have a glitch-0 between 1 and 2. Boom!

12-6. SYNCHRONOUS COUNTERS TO ELIMINATE GLITCHES

There is another way to count which doesn't depend on something happening at A before something can happen at B. This is called synchronous counting. A synchronous counter has all of its flip-flops clocked at the same time. T flip-flops do not have enough flexibility to do this task, so our synchronous counters will be made mostly of J-K flip-flops.

We might recall that shift and ring counters are clocked everywhere by the same clock pulse. Their flip-flops must all set or reset at the same time, so they are synchronous counters. Since they're not binary and don't have as many states as a binary counter with the same number of flip-flops, the shift and ring counters aren't as important.

To make a binary counter, each flip-flop must toggle every second time its neighbor to the left toggles. This can be done several ways. Two are shown in Figure 12-8.

The first synchronous binary counter in Figure 12-8(a) is made with T flip-flops. To count, some flip-flops must receive a clock pulse through an AND gate. The flip-flops that toggle only on every second, fourth, or eighth clock pulse are synchronized with the other flip-flops because their clock pulses come from the same clock as the A flip-flop. The reason that B, C, and D don't toggle every time the CP line pulses is that the AND gates disable these clock pulses from reaching B, C, and D part of the time. Look at the diagram; you'll see that each AND gate is a decoder for certain output conditions as well as an enable/disable for the CP signal. For example, the C flip-flop will only be clocked if A AND B are both HIGH. This means the first time C can toggle is following a 3 (A and B HIGH).

The second synchronous binary counter in Figure 12-8(b) is made with J-K flip-flops. It uses AND gates somewhat the same way as the circuit with T flip-flops, but instead of enabling and disabling the clock, the AND gates change the input conditions at J and K. If the AND gate's output is HIGH, the J-K will toggle, but when the AND gate's output is LOW, the J-K will latch (no change when clocked). This results in the same action as Figure 12-8(a)—the B, C, and D flip-flops will toggle during the second, fourth, and eighth clock pulses, respectively.

In each case, the result, and the timing diagram, is the same. Where a ripple counter has a "domino effect" of falling edges as 7 becomes 8, the synchronous counter has all four flip-flops toggle at the same instant. This happens because there is a falling-edge clock at the same instant in all four flip-flops, which arrives at a time when all four flip-flops will toggle when they receive a clock pulse. Will a synchronous binary counter go from 7 (0111) to 8 (1000) without glitches?

In Figure 12-8(a), the flip-flops are all T types, so they'll toggle anytime they get a falling edge from the clock. When A, B, and C are HIGH (as they are during 0111), all three AND gates are enabled. The first flip-flop always toggles anyway, and the other three flip-flops toggle this time because their AND gates have the "right" inputs. Since the clock pulse "gets through" to all four T inputs at this moment, 0111 will become 1000 with the next clock pulse. Since all four flip-flops receive the (enabled) clock pulses at the same time, they all change simultaneously, and there are no glitches.

In Figure 12-8(b), all the flip-flops are J-K types in the toggle condition. When A, B, and C are HIGH, all three AND gates have HIGH outputs. The "front" flip-flop in the counter is a T type, and it will always toggle, but flip-flops B, C, and D will toggle because their AND gates have

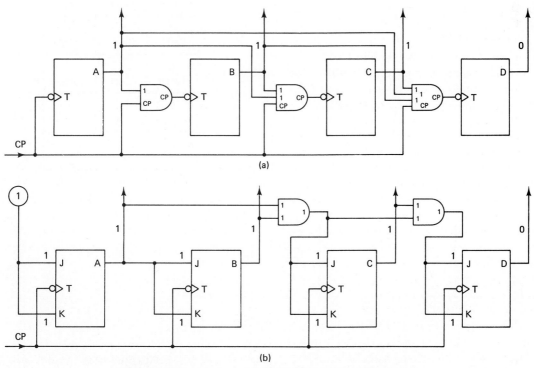

FIGURE 12-8 Binary synchronous modulus-16 counters.

the right inputs. The next clock pulse causes all four flip-flops to toggle, and 0111 becomes 1000 at all four flip-flops simultaneously, with no glitches.

12-7 EXCITATION TABLES AND THEIR USE WITH STATE-TRANSITION DIAGRAMS AND TIMING DIAGRAMS

12-7.1 Predicting What a Synchronous Counter Will Do Next

Synchronous counters are not as simple as ripple counters. For the most part, you can tell what a ripple counter is supposed to do from the number of its T flip-flops or the kind of reset decoder it has. Following a synchronous counter through its sequence of counts to figure out its modulus, or even what number comes next, can get pretty confusing.

A set of steps for figuring out the counter in Figure 12-9 is listed below.

The state table shows the actual results if the counter in Figure 12-9 is analyzed by this method. Step 0 means "pick a number to start counting with"—in this case, 000. Once you decide what number your counter will begin with, step 1 is already done—the outputs are A = 0, B = 0, and C = 0. Step 2, where you figure what's at the inputs, is a case of "following" the 1 s and 0s from each output through the wiring diagram, to see how they steer the inputs down the line. (For example, the AND gate has an output of 0 at the start because its inputs are both 0. This makes the J and K of flip-flop C both LOW.) Once the 1 or 0 at each input has been figured out, it's time to perform step 3—decide if these inputs cause a set, reset, latch, or toggle for their flip-flop. List the result of this determination under the present outputs of the flip-flop, as shown in the state table. It'll help you determine what to put in the next line of the table. Step 4 is where you advance to the next state—the next clock pulse—by going back to step 1 and moving down to the next line on the state table. This is done by looking at the state in each column and the

State table

	C	B	A
	0 Latch	0 Latch	0 Toggle
1CP	0 Latch	0 Toggle	1 Toggle
2CP	0 Latch	1 Latch	0 Toggle
3CP	0 Toggle	1 Toggle	1 Toggle
4CP	1	0	0
	Latch	Latch	Toggle

To Predict:

0. Get started somewhere.
1. Figure outputs.
2. Figure what's at inputs.
3. Decide if inputs will cause a set, reset, latch, or toggle on the next CP.
4. Go back to step 1 after CP.

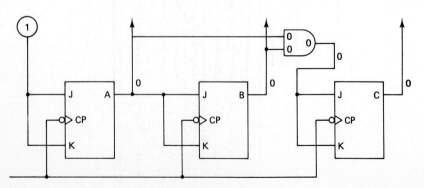

FIGURE 12-9 Mod-8 synchronous binary counter.

set, reset, latch, or toggle that's going to happen to it when there's a clock pulse. Since there is a clock pulse now, you ask, "What happens to this 1 (0) after it toggles (sets, resets, latches)?" You put the answer for each column down on this line. Then we move ahead to step 2, and so on. (Can you tell what numbers should appear at 5CP in the table? Did you guess 101?)

12-7.2 Designing a Synchronous Counter (Optional)

We've decided to made one execption to our basic plan of telling you "how it works" rather than "how to make one." This section can be skipped without losing any understanding of following sections.

Ask yourself the following questions, to begin the design of a synchronous counter:

1. What is the count sequence you want?
2. What state transitions get you from each count to the next?
3. How might we connect the inputs (wiring) to get the set, reset, latch, or toggle we need?

To answer these questions, we need to know how to develop an **excitation table.** The excitation table is a chart something like a truth table in reverse, that answers the question, "What inputs would give me this output?" for every possible state transition.

The J-K flip-flop is the most flexible type for making synchronous counters—every other type of flip-flop can be made from a J-K—so we'll assume J-Ks will be used in all our counter stages. To begin designing a counter out of J-Ks, we need the excitation table for a single J-K flip-flop:

J-K flip-flop's excitation table

Line	Outputs		Inputs	
	Before CP	After CP		
	Q_{t-1}	Q_t	J_{t-1}	K_{t-1}
0	0	0	0	X
1	0	1	1	X
2	1	0	X	1
3	1	1	X	0

On the 0 line of the table we're asking, "If a 0 at Q changes to a 0 after one clock pulse (no change), what inputs at J and K would cause this

sort of action?" There are two answers: a reset (which would turn OFF a flip-flop that's already OFF—no change) or a latch (which means Q doesn't change when there's a clock pulse). Either of these state transitions would result in the same output after a clock pulse. In the excitation table, you can see subscripts under the Qs, the J, and the K. The t − 1 means "before the clock pulse" (the last state), and the t means "after the clock pulse" (the next state). Conditions at J and K that cause the changes at Q when it's clocked must have been at J and K before there was a clock pulse (at time t − 1). The Xs in the table indicate a condition called a "don't care"—where either a 1 or 0 would work equally well.

For line 0, the reset or latch input conditions would both work equally well. To cause a reset, the inputs would have to be J = 0, K = 1. To cause a latch, the inputs would have to be J = 0, K = 0. In either case, J is 0, but K could be either a 1 or a 0. We write J = 0, K = X (don't care) to cover both possibilities.

For line 1 on the table, Q changes from a 0 to a 1. The two state transitions that could cause this are the set or toggle. Inputs for set are J = 1, K = 0, and inputs for toggle are J = 1, K = 1. In either of these cases, J is 1, but K can be either 0 or 1 with the same results. This is indicated at the right-hand side of line 1 by the J = 1, K = X input conditions.

The conditions that give us the results in lines 0, 1, 2, and 3 are:

$$
\begin{array}{l}
\text{line 0: stays off} = \left. \begin{array}{ll} \text{(a)} & \text{Latch} = 0 \quad 0 \\ \text{(b)} & \text{Reset} = 0 \quad 1 \end{array} \right\} \quad \begin{array}{cc} J & K \\ 0 & X \end{array} \\[2em]
\text{line 1: turns on} = \left. \begin{array}{ll} \text{(a)} & \text{Toggle} = 1 \quad 1 \\ \text{(b)} & \text{Set} = 1 \quad 0 \end{array} \right\} \quad 1 \quad X \\[2em]
\text{line 2: turns off} = \left. \begin{array}{ll} \text{(a)} & \text{Toggle} = 1 \quad 1 \\ \text{(b)} & \text{Reset} = 0 \quad 1 \end{array} \right\} \quad X \quad 1 \\[2em]
\text{line 3: stays on} = \left. \begin{array}{ll} \text{(a)} & \text{Latch} = 0 \quad 0 \\ \text{(b)} & \text{Set} = 1 \quad 0 \end{array} \right\} \quad X \quad 0
\end{array}
$$

12-7.3 Design of a Mod-4 Binary Synchronous Counter

To use the excitation table of the J-K flip-flop for designing a counter, let's design a counter that counts in the sequence

$$0, 1, 2, 3, 0, 1, 2, 3, \ldots$$

which is called a **mod-4 binary synchronous counter**.

We'll start by developing the counter's state transition diagram and state table:

STD:

State table

B	A	CP
0	0	0
0	1	1
1	0	2
1	1	3
0	0	4

Here is how the excitation table for the mod-4 synchronous counter looks:

	Outputs				Inputs at time $t - 1$			
	Last state		Next state		B		A	
	B_{t-1}	A_{t-1}	B_t	A_t	J_B	K_B	J_A	K_A
1	0	0	0	1	0	X	1	X
2	0	1	1	0	1	X	X	1
3	1	0	1	1	X	0	1	X
4	1	1	0	0	X	1	X	1

The excitation table was formed from the state table in the following way:

1. Copy lines 0, 1, 2, and 3 of the state table into lines 1, 2, 3, and 4 of the excitation table under the heading "last state."
2. Copy lines 1, 2, 3, and 4 of the state table into lines 1, 2, 3, and 4 of the excitation table under the heading "next state."

On the topmost line:

3. Look at B at times $t - 1$ and t to find how B has changed (0 to 0, 0 to 1, 1 to 0, or 1 to 1) from the table on page 213.
4. On the J-K excitation table, look up the inputs that would cause B to change the way it did. Write these J and K values (0X, 1X, X1, or X0) into columns labeled J_B and K_B on the same line.
5. Look at A at times $t - 1$ and t to find how A has changed.

6. On the J-K excitation table, look up the inputs that would cause A to change the way it did. Write these J and K values into the columns labeled J_A and K_A on the same line.
7. Go to the next line and repeat steps 3 through 6 until all lines are filled in.

Now you've got an excitation table for the mod-4 counter—so now what? Actually, the whole design of the circuit is there, it just doesn't look like it!

The trick is to read down each column of the table, asking yourself for each input: "What's around at time $t - 1$ that has the same 1s and 0s we want at the J or K we're looking at?" For instance, reading down the J_B column, we see 01XX. Since anything will work for the XX part, all we have to do is spot something that's around at time $t - 1$ which starts with 01 (reading down of course!). The A output at time $t - 1$ (reading down) is 0101. This is a match for what we need. A is LOW and HIGH at exactly the times we need J of flip-flop B to be LOW and HIGH. We can say that

$$J_{B_{t-1}} = A_{t-1}$$

This is called a state equation, and we'll see what it's good for in a minute. First, we'll see what the state equations are for all the other inputs.

We look at K_B and see that it reads XX01 while A at the same time ($t - 1$) is 0101. Since it really doesn't matter what the XX are, all we need is something that ends with 01, and A is a match—so

$$K_{B_{t-1}} = A_{t-1}$$

Now we look at J_A and K_A. These are 1X1X and X1X1. If we ignore the don't cares (X), all that's left is 1s. We could go looking for something that has a value of 1 at these times, but since all we ever need is a 1, why not say that

$$J_{A_{t-1}} = 1 \qquad K_{A_{t-1}} = 1$$

Now that we have the state equations for all four J-K inputs, we can finally say what the circuit for our mod-4 counter is going to look like. How do we know that?

Well, it turns out that the state equations aren't just mathematical abstractions like all the algebra equations you've ever seen. These "equa-

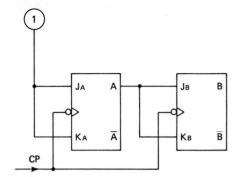

FIGURE 12-10 Mod-4 synchronous binary counter.

Mod-5 excitation table

	Last state			Next state			Inputs at time $t - 1$					
	C_{t-1}	B_{t-1}	A_{t-1}	C_t	B_t	A_t	J_C	K_C	J_B	K_B	J_A	K_A
1	0	0	0	0	0	1	0	X	0	X	1	X
2	0	0	1	0	1	0	0	X	1	X	X	1
3	0	1	0	0	1	1	0	X	X	0	1	X
4	0	1	1	1	0	0	1	X	X	1	X	1
5	1	0	0	0	0	0	X	1	0	X	0	X

tions" actually tell you where to connect the wires! For instance,

$$J_A = 1 \quad K_A = 1 \quad J_B = A \quad K_B = A$$

tell you that both J and K of flip-flop A are connected to a logic 1, and J and K of flip-flop B are connected to the output of A. After we draw two flip-flops and connect them to make this mod-4 counter, it looks like Figure 12-10, which you can compare to Figure 12-9. See the resemblance?

12-8 THE MOD-5, MOD-10, AND MOD-6 SYNCHRONOUS COUNTER AS APPLICATIONS OF THE EXCITATION TABLE

In the preceding section we saw how a counter circuit made of J-K flip-flops could be designed directly from the count sequence of the counter. We used a mod-4 counter that could be made without any fancy feedback or resetting, even using synchronous logic. Now, we'll see how the same approach can be used to design any counter you want (provided that you know what sequence of numbers it should produce).

12-8.1 A Mod-5 Counter

The mod-5 counter we want counts the following sequence of numbers:

$$0, 1, 2, 3, 4, 0, 1, 2, 3, 4, 0, \ldots$$

for as long as it is clocked. The excitation table for this counter will have outputs and inputs for three flip-flops, A, B, and C:

From the Js and Ks we filled in from the J-K table, we can see, for flip-flop A, that K could be 1 at all times, but J is not so simple. In the very last (bottom) row, J is a 0. We must find something at time $t - 1$ that's 1X1X0, reading from top to bottom. We should remember at this point that there's a "not" output for every A, B, or C. What would not-C look like? If C is 00001, then not-C is 11110. It's not-C that matches J of flip-flop A. 11110 matches 1X1X0, since we don't care what replaces the Xs.

Similarly, for flip-flop B, the K input XX01X and matches output A (01010), while J is 01XX0 and also matches A everywhere that's important.

Completing all the state equations, we get

$$J_A = \overline{C} \quad K_A = 1 \quad J_B = A \quad K_B = A$$
$$J_C = A \cdot B \quad K_C = 1$$

Notice the state equations for C. It's simple to handle the K input. Since all you need is a 1 for XXXX1, the K input is a 1 at times (tied HIGH), but when we look for a match for J there's nothing—not A, B, C, or their "not" outputs—that matches 0001X. We need a combination of A, B, and C that will work, and it turns out that the pattern 0001X suggests the designation number of an AND gate, with two inputs, 0001. If the A and B are ANDed together, we get 00010, which is a fine match for 0001X.

This leaves only the actual wiring. Turning the state equations into wiring, we get the circuit in Figure 12-11. There are wires to the HIGH level wherever a J or K is equal to a 1, and a wire to the appropriate A, not-C, and so forth, wherever the state equations suggest that such a connection be made.

12-8.2 A Mod-10 Counter

A mod-10 counter can be made from the mod-5 counter just completed by "daisy-chaining" a mod-2 counter onto it. One method puts the mod-

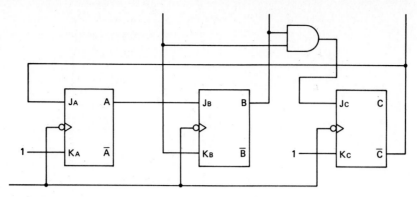

FIGURE 12-11 Mod-5 counter.

2 counter in front of the front end of the mod-5 counter [see Figure 12-12(a)]. This makes it necessary to put 10 pulses into the combined counters before the pattern repeats, because the mod-2 counter is a *divide-by-2* frequency divider, and there must be two pulses into it to get the mod-5 part to change. Since the mod-5 changes every second clock pulse, and the mod-2 changes every clock pulse, there are 10 different *4-bit* codes. This is shown on the chart below. The output of the *mod-5* counter and the *mod-2* counter are put together, and, although you can see the codes from each counter (each of the *mod-5's* codes last two clock pulses), when you put them together, the four-digit codes make the numbers from 0000 to 1001 appear in binary order. The counter not only has 10 states, it counts in a decimal sequence—0, 1, 2, 3, 4, 5, 6, 7, 8, 9, 0, ... —so it's a decade counter.

Mod-5			Mod-2
C	B	A	X
0	0	0	0
0	0	0	1
0	0	1	0
0	0	1	1
0	1	0	0
0	1	0	1
0	1	1	0
0	1	1	1
1	0	0	0
1	0	0	1
0	0	0	0

This decade counter will divide by 10 if the output is taken from the C output of the mod-5. Its output will be HIGH during the counts of "8" and "9," and LOW the rest of the time. The code it counts in is called BCD (for binary-coded decimal) code.

We said that this was one method, and the other is obvious. Put the mod-2 counter at the back end of the mod-5 counter. The resulting combination counter is still mod-10, but isn't a decade counter. Take a look:

Mod-2	Mod-5		
X	A	B	C
0	0	0	0
0	0	0	1
0	0	1	0
0	0	1	1
0	1	0	0
1	0	0	0
1	0	0	1
1	0	1	0
1	0	1	1
1	1	0	0
0	0	0	0

Every count advances the mod-5 counter, but the mod-2 counter only advances when the mod-5 counter goes to ZERO.

This counter—shown in Figure 12-12(b)—is also usable as a *divide-by-10*. Its output is HIGH for five clock pulses and LOW for five clock pulses.

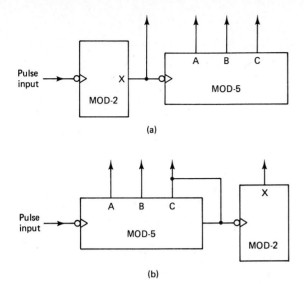

(a)

(b)

FIGURE 12-12 The 7490 counter (a) Decade configuration; (b) biquinary configuration. (See Appendix 4, 7490 datasheet)

This type of count is called a *symmetrical divide-by-10*, and the code it counts in is called *biquinary code*, from the Latin roots BI = 2, QUIN = 5.

This type of counter comes in a TTL package called a 7490. It contains a mod-2 and mod-5 counter, and can be wired to count in either BCD or biquinary code. (See Appendix 4, 7490 data sheet, for more details.)

12-8.3 A Mod-6 Counter

A building block useful in digital-clock circuits is a counter that counts mod-6, like this:

$$0, 1, 2, 3, 4, 5, 0, 1, 2, 3, 4, 5, 0, \ldots$$

It can be designed using the same excitation table approach we used in two previous examples. The table for our mod-6 is:

Mod-6 excitation table

	Last state			Next state			Inputs at time $t-1$					
	C_{t-1}	B_{t-1}	A_{t-1}	C_t	B_t	A_t	J_C	K_C	J_B	K_B	J_A	K_A
1	0	0	0	0	0	1	0	X	0	X	1	X
2	0	0	1	0	1	0	0	X	1	X	X	1
3	0	1	0	0	1	1	0	X	X	0	1	X
4	0	1	1	1	0	0	1	X	X	1	X	1
5	1	0	0	1	0	1	X	0	0	X	1	X
6	1	0	1	0	0	0	X	1	0	X	X	1

and the schematic diagram is shown in Figure 12-13. Can you work out the state equations with this information?

There is also a TTL device called a 7492 which contains a mod-6 and mod-2 counter in one package. They can be put together to count duodecimal (MOD-12) or bihexary code.

12-9 PROGRAMMABLE UP/DOWN-COUNTERS

Figure 12-14 shows a TTL counter chip called the 74193. It is a general-purpose counter in the same way that the J-K is a general-purpose flip-flop—everything can be made from it. The programmable up/down-counter is a 4-bit binary counter with a natural modulus of 16. If clock pulses are put into its UP clock input, the counter will up-

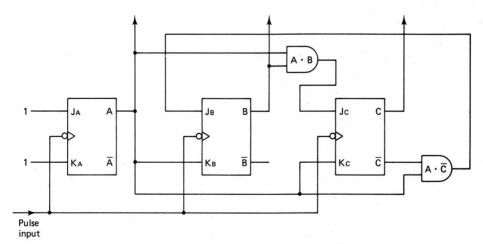

FIGURE 12-13 Synchronous counter (MOD-6).

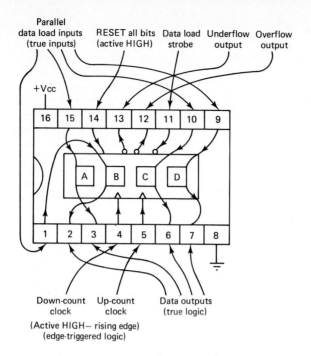

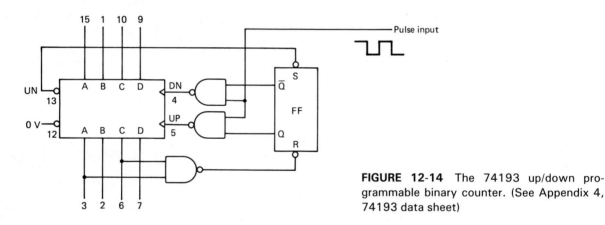

FIGURE 12-14 The 74193 up/down programmable binary counter. (See Appendix 4, 74193 data sheet)

count from 0000 to 1111 in binary code. If clock pulses are put into its DOWN clock input, the counter will down-count from 1111 to 0000 in binary. The load input permits you to clock whatever number you'd like to broadside load into the counter through the DCBA inputs, and it will count from there. The carry and borrow outputs produce a pulse output as the counter overflows (when counting up past 1111) or underflows (when counting down past 0000). By suitable combination of decoding at the counter's DCBA outputs and encoding at its DCBA inputs (which are the bits loaded in during a load), this counter can be made to count numbers in any sequence and in any direction you want. Excitation tables and synchronous design are not necessary if the counter you want is implemented by this approach. At the bottom of Figure 12-14, as an example, we've shown how a counter can be made

from the 74193 that will count in this order:

$$0, 1, 2, 3, 4, 5, 4, 3, 2, 1, 0, 1, 2, 3, 4, 5, 4,$$
$$3, 2, \ldots$$

and so on. This mod-6 counter could be combined with a D/A converter to produce a triangle wave. (See Appendix A, 74193 data sheet for more detail.)

EXAMPLE 12-3

Design a counter circuit that may be loaded with a number up to 999 and will then count down to zero from that number.

Solution: There have to be three counters, one for hundreds, one for tens, and one for units. The hundreds, tens, and units counters must be modulus-10 down-counters. Since all three counters must be loaded with a number, it seems reasonable to use a programmable up-down counter chip in each case. The 74192 programmable up-down decade counter is a good device for this purpose. The 74192 design looks like this:

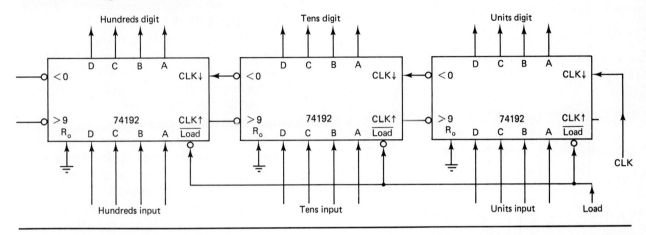

12-10. TIME-BASE GENERATORS FOR CLOCK PULSES

It would be possible to send one-per second clock pulses to a digital clock (for it to count) from a pendulum with a debounced switch attached. Most digital clocks don't do it that way, however. The crystal oscillator mentioned earlier is a much more accurate and reliable mechanism than a pendulum. Figure 12-15 shows a crystal oscillator that would be suitable for use in a digital-clock circuit, and several other pulse sources, in-

cluding an *RC* oscillator and a **one-shot multivibrator** (also called a **monostable multivibrator**) used to "debounce" and "waveshape" the 60-Hz frequency the Edison Company gives us. With a suitable divider string (divided-by-60), this frequency provides a very steady, reliable clock time base, as long as the power is not interrupted.

Of the three types shown, the crystal oscillator is the most precise, the *RC* oscillator the least. The ac line frequency source is fine, but not portable unless your wristwatch has a long extension cord.

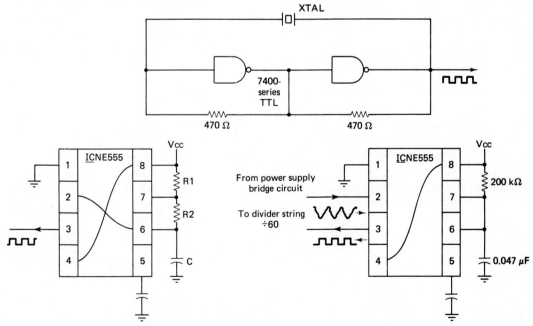

FIGURE 12-15 Oscillators and other time bases.

12-1. Name two types of flip-flops that can be used to make counters.

12-2. Are the flip-flops used in counters edge-triggered types?

12-3. For the circuit shown here, complete the timing diagram for outputs A, B, and C of the divider string. The clock pulses are shown on the diagram.

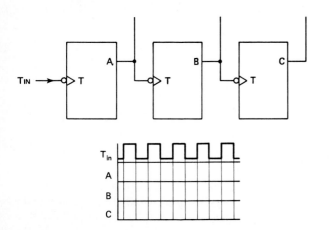

12-4. How many flip-flops are needed for a divide-by-16384 divider string?

12-5. Draw a state transition diagram for the figure shown in Question 12-3.

12-6. What is the modulus of the counter in the figure shown in Question 12-3?

12-7. Redraw the figure shown in Question 12-3 to make it a down-counter.

12-8. Make a state transition table for a binary ripple-down counter made with two T-type flip-flops.

12-9. Define "forced reset" and "power-on reset."

12-10. What do the preset and clear inputs of a flip-flop do? Are they synchronous or asynchronous inputs?

12-11. Draw a schematic for a forced-reset counter that counts modulus 9 (0, 1, 2, 3, 4, 5, 6, 7, 8, 0, . . .).

12-12. The glitch states between the count of 7 and 8 in a binary ripple counter are 6, 4, 0. (True or False?)

12-13. What do terms "synchronous" and "asynchronous" counter mean?

12-14. How does a synchronous counter design eliminate glitch states?

12-15. Make a state table showing last states and next states for a modulus-4 Johnson counter. The table should start with all flip-flops in the reset condition and proceed for five clock cycles.

(Optional section of Chapter 12)

12-16. Make up an excitation table for the Johnson counter described in Question 12-15.

12-17. What does an "X" in the input section of an excitation table mean?

12-18. Find a set of state equations for the Johnson counter described in Question 12-15.

12-19. Draw the schematic of this counter, as developed from the state equations.

12-20. Develop (using the same methods as Questions 12-16 to 12-19) the circuit for a counter that counts (0, 2, 4, 6, 8, 1, 3, 5, 7, 9).

(End of problems for optional section)

12-21. Draw a block diagram showing how a mod-6 and mod-2 counter would be connected for a symmetrical, divide-by-12 circuit.

12-22. There are two ways the "blocks" in a 7490 can be connected to each other. Make state transition diagrams for the output of these two hookups (decade and biquinary).

12-23. Show connections to a 74193 counter to make it count modulus 12 in the following sequence: (1, 2, 3, 4, 5, 6, 7, 8, 9, 10, 11, 12, 1, . . .).

12-24. What common type of time base is most accurate for digital timepieces?

12-25. Why combine a high-frequency crystal oscillator and a divider string instead of using a low-frequency crystal?

Questions 12-1 through 12-10 below refer to the following symbolic diagram:

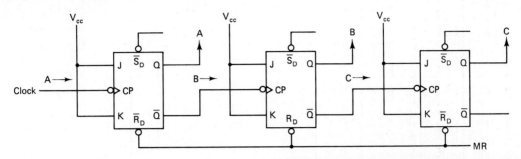

12-1. The counter shown is a _____ (binary, Gray, Johnson) counter.

12-2. The counter shown is a(n) _____ (asynchronous, synchronous) counter.

12-3. The counter shown is a(n) _____ (up, down, up-down) counter.

12-4. The modulus of the counter is _____.

If the frequency of the clock is 4 MHz, what is the frequency at:

12-5. Output A? _____

12-6. Output B? _____

12-7. Output C? _____

12-8. Draw a state transition diagram for the counter.

12-9. Input MR should be _____ (high, low) to let the counter count.

12-10. This counter _____ (does, does not) produce *glitches* as it counts through its natural sequence.

Questions 12-11 through 12-19 below refer to the following symbolic diagram:

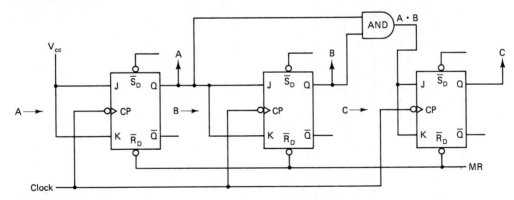

12-11. The counter shown is a _____ (binary, Gray, Johnson) counter.

12-12. The counter shown is a(n) _____ (asynchronous, synchronous) counter.

12-13. The counter shown is a(n) _____ (up, down, up-down) counter.

12-14. The modulus of the counter is _____.

If the frequency of the clock is 8 MHz, what is the frequency at:

12-15. Output A? _____

12-16. Output B? _____

12-17. Output C? _____

12-18. Draw a state transition diagram for the counter.

12-19. This counter _____ (does, does not) produce *glitches* as it counts through its natural sequence.

Questions 12-20 through 12-25 below refer to the following divider string:

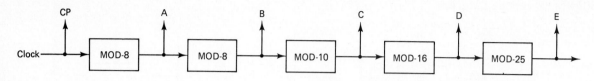

CP   A   B   C   D   E

Clock → MOD-8 → MOD-8 → MOD-10 → MOD-16 → MOD-25 →

If the input frequency at CP is 15.36 MHz, what is:

12-20. The frequency at A? _____

12-21. The frequency at B? _____

12-22. The frequency at C? _____

12-23. The frequency at D? _____

12-24. The frequency at E? _____

12-25. What is the total divide-by of this divider? _____

Questions 12-26 through 12-30 below refer to the following symbolic diagram:

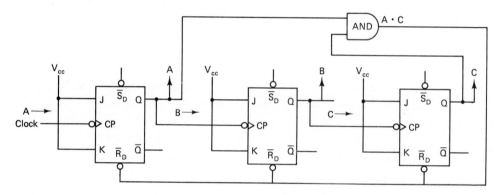

12-26. The counter shown is a(n) _____ (asynchronous, synchronous) counter.

12-27. The counter shown is a(n) _____ (up, down, up-down) counter.

12-28. The modulus of the counter is _____.

12-29. If the clock frequency into A is 15 MHz, what is the frequency at output C? _____

12-30. Draw a state transition diagram for this counter.

13
COUNTER-TYPE A/D CONVERSION

CHAPTER OBJECTIVES

By the time you finish this chapter, you will be able to:

1. Describe how counters are applied to analog-to-digital conversion in the following types of circuits:
   a. Single-slope A/D converter
   b. Tracking A/D converter
   c. Successive-approximation A/D
   d. Ramp-generator-style single-slope A/D converter
   e. VCO-type A/D converter
2. Describe and define the following terms related to A/D conversion.
   a. Resolution
   b. Conversion time
   c. Linearity
3. See how digital waveforms made from analog inputs are used in:
   a. Switching amplifiers
   b. Switching power supplies (switching regulators)

13-1   REVIEW OF FLASH CONVERSION

In Chapter 8 we said: "Suppose that you want to make a flash converter that can read out numbers from 1 to 100 V, in 1-V steps. You'll need 100 analog comparators and a voltage divider string of series resistors with 100 resistors. Then you'll need a very full adder circuit that can add 100 inputs! There must be an easier way—and there is. We'll be seeing how this can be done much more simply after we go through the next few chapters and find out about counters." Well, now we've discussed digital counters, and we can fulfill that promise. Instead of 100 comparators, we need only one if we use the principle of feedback.

The A/D converter is essentially a digital voltmeter. When you put a certain voltage at its input, you get a binary code from its output that's proportional to the voltage. The flash converter did this the "hard way." A separate analog comparator existed for every voltage input you could resolve. To resolve voltages to 1 part in 100, you

needed 100 comparators—but 1 part in 100 is only two decimal places. Suppose that you're designing a digital voltmeter, and you'd like it to read out voltages with four decimal places of accuracy. Four decimal places doesn't sound like much, but to read voltages with four-place accuracy, there must be 10,000 analog comparators in a flash converter! Worse, the code coming out of the comparators must be summed, and even after summing, the code isn't BCD (so it isn't suitable for a decimal readout). The circuit we need for a flash converter becomes uncomfortably huge, even when only a few decimal places' output is desired.

13-2   SINGLE-SLOPE COUNTER-TYPE A/D CONVERSION

Figure 13-1 is a block diagram of a system made out of building blocks we've already discussed. The D/A converter is capable of developing an output whose voltage matches the code put in to

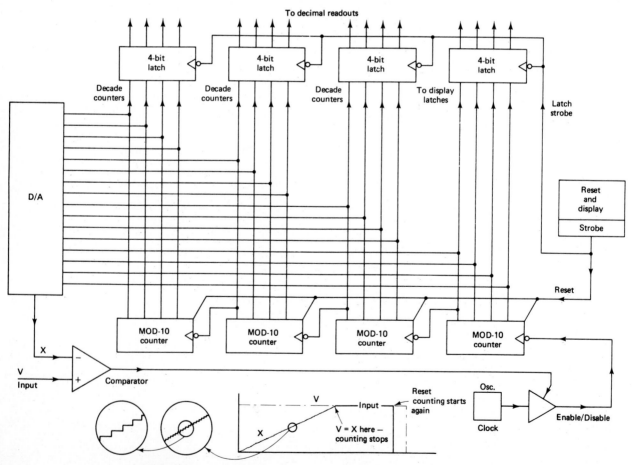

FIGURE 13-1   Four-digit counter-type A/D converter (single slope).

four-place accuracy. This requires precision resistors and is expensive, but not as expensive as the resistors in a flash-conversion ladder would be (10,000 of them!). The counter in the figure is a four-digit decade counter string. If we "daisy-chain" the divide-by-10 output of one decade counter into the clock input of the next, until there are four decade counters wired together, we get a counter whose four parts produce decimal numbers from 0000 to 9999, counting decimals all the way. Attached to the D/A, this counter permits generation of voltages in 10,000 steps from 0000 to 9999 mV. The comparator compares the INPUT with the voltage from the counter-plus-D/A combination. When the counter counts upward from 0000, it sooner or later matches the voltage of the input (provided that the input's less than 10 V). When this happens, the comparator switches from HIGH to LOW. This disables the clock pulses to the counter, and the count "freezes" at that time.

For example, suppose that the voltage input to this circuit is 3472.8 mV. The counter will count up to 3473. At that point, it will stop counting because the D/A voltage—3473 mV—just became larger than the input, and that changes the state of the comparator, which turns off the clock pulses. If we just send the output of the counter to a display at this moment, the readout—3473 mV—will be an accurate reading (to four decimal places) of the input voltage.

Of course, if the input voltage falls, we won't see the change unless we zero the counter and count up all over again to the new voltage. With this type of circuit, which uses an up-counter, we have to up-count just past the input voltage to produce a new reading. This must be done periodically to ensure that the output display "tracks" with changes in the INPUT. How fast can the counter count up again from 0000? That depends on the clock speed and the speed of the logic used in the counter and D/A sections. The time it takes the counter to complete an up-count in the worst case would be the time it takes to count from 0000 to 9999. This is called the **conversion time** of the circuit. As an A/D converter, this circuit will be slower than the flash converter, because it has to count and produce a "staircase-wave" voltage all the way from 0000 to the input voltage level every time a new reading is needed. For a count from 0000 to 9999, with a clock rate based on the slowest part of the circuit, which is probably the analog comparator, the single-slope converter is 10,000 times slower than the flash converter.

Although it is slower, the single-slope converter has a lot fewer parts than the flash converter. Instead of a 10,000-resistor divider string, 10,000 analog comparators, a 10,000-input adder circuit, and a binary-to-BCD converter for four-place decimal numbers, the converter needs only the parts shown in Figure 13-1.

Why is this circuit called a **single-slope** type? The reason seems to be the *ramp wave* that its D/A part produces as it counts. This is a wave that slopes up (a sawtooth curve); since the up-counter can only count in one direction, the wave can have only one slope (upward). You can see this waveform and its relation to the input at the bottom of Figure 13-1.

13-3 TRACKING A/D (COUNTER-TYPE A/D CONVERSION)

The single-slope converter is slow because it must count all the way up from 0000 to the "trip point" before a new reading can be displayed. This follows from the fact that it can count in only one direction. A voltage that increases can be "followed" by the up-counter, but a voltage that decreases can't. To "follow" changes in both directions (without zeroing and counting all over again) we need a counter that counts both up and down. We will use an up/down-counter, programmed to count up when the comparator indicates the D/A has less than the input voltage, and down when the D/A output is more than the input voltage. Now we have a circuit that responds immediately to changes in both directions, and can keep an accurate record of the moment-to-moment voltage as long as the input doesn't go up or down faster than the counter can count.

In Figure 13-2, an up/down-counter replaces the up-counter in Figure 13-1. The output of the analog comparator is used to direct whether the counter is clocked up or down instead of merely disabling the clock. This permits **dual-slope** ramp waves to be generated at the D/A part of the circuit, which can overtake the input waveform no matter whether it rises or falls. As long as the clock makes the counter count fast enough, the up-count will periodically overtake a rising input voltage, and the displays will show rising numbers at each of these times. A down-count will periodically overtake a falling input voltage, and the displays will be updated with decreasing numbers as this happens. This is shown with waveforms at the bottom of Figure 13-2. You can see from the two directions of ramp-wave motion why this is also called a dual-slope A/D converter.

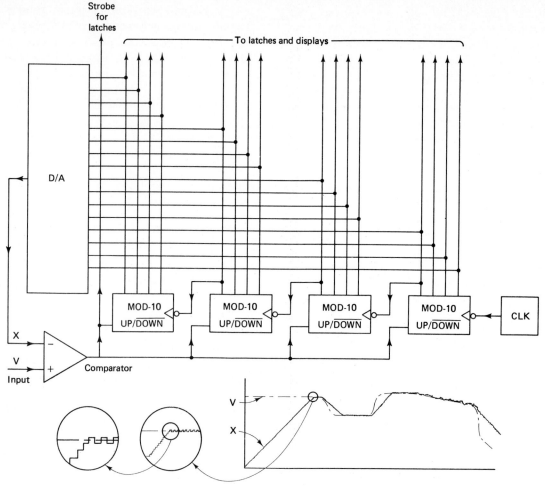

FIGURE 13-2 Four digit counter-type A/D converter (tracking).

13-4 RESOLUTION, CONVERSION TIME, ETC.

You can see that the tracking A/D converter is able to "keep up" and up-date its readings faster than the single-slope converter. We already mentioned that the speed of the clock determines the speed of the converter. This is true because the converter cannot provide new readings unless its counter is clocked. This, in turn, depends on the speed of the analog comparator. If we clock the counter and produce a "staircase wave" that rises faster than the "slew rate" of the op amp used for a comparator, the results won't be accurate. The conversion time needed for four-digit decimals was 10,000 clock pulses in the single-slope converter. The dual-slope converter needs time to count up until the D/A voltage equals the input voltage, but from then on, it's only a few clock pulses either way before the D/A catches up with a changing input. The conversion time in this case would be only a few clock pulses.

The resolution of a counter is usually given in binary digits instead of decimal places, since binary counters are simpler to make than decimal ones. A counter with 8 bits of resolution is one that gives an eight-digit binary number for each voltage input. Since there are 256 different 8-bit codes, the A/D converter which works at 8-bit resolution has an accuracy of 1 part in 256. Since the figure 1/256 is also expressed as 0.4%, we may also say that the A/D has a resolution of 0.4%.

EXAMPLE 13-1

What is the resolution of a 12-bit successive-approximation A/D converter if it will accept inputs between 0 and 5 V?

Solution: It doesn't really matter if it's a successive-approximation A/D converter or any other kind of A/D converter; if it's a 12-bit A/D converter, it can "count" 2^{12} (4096) numbers. The smallest and largest numbers the 12-bit DAC can produce are

$$000000000000 = \quad 0 = 0 \text{ V}$$
$$111111111111 = 4095 = 5 \text{ V}$$

Since 5 V = 4095 in code, the resolution (the smallest amount of change possible) is

$$\frac{5}{4095} = 0.001221 \text{ V}$$

13-5 APPLICATION TO INPUT OF DATA INTO DIGITAL SYSTEMS

The primary use of an A/D converter is as an input device. A digital system that responds to changes in the outside world needs a sensor to detect them with. The A/D circuit is an interface between various transducers which convert physical measurements to voltages, and the digital world inside the system. The conversion from a voltage to a digitally acceptable code is an essential part of this process. Digital devices that receive measurements of some outside world quantity in the form of digital numbers, and then store or process these readings from time to time, are called **data loggers**. Automatic testing procedures in quality assurance and real-time correction and adjustment of product-line equipment are but a few of the myriad of possible uses for data logging systems.

13-6 SUCCESSIVE-APPROXIMATION D/A CONVERSION

In this chapter we have discussed ways in which the voltage at an input can be "digitized" using conventional counters. In the single-slope method, an up-counter counted until it reached a number proportional to the voltage being digitized at the analog input. The counter was then reset for the next sample, and would have to count-up all over again. In the dual-slope method, an up/down counter was used. Once the counter had reached a number proportional to the voltage being digitized, each sample taken thereafter took only the time required to count upward or downward to the new voltage from the last sample taken. If samples were taken "close together" on a varying-voltage waveform, this reduced the amount of time to digitize each sample, as compared to the single-slope method, but the *first* sample was still very time consuming to take.

By using an unconventional counter, we can greatly reduce the time needed to take the first sample. The counter involved in this method of digitizing is a **successive-approximation register**, which we will—from now on—refer to as an **SAR**.

13-6.1 The Successive-Approximation Register

The SAR is a very special type of counter. It follows a very specific set of rules and is controlled by a comparator that matches the analog input voltage to D/A-converted voltage based on what's in the counter. In this regard, control of the counter is just like the dual-slope counter mentioned earlier. However, we said that this is a very special counter. Here's how it works:

0. Start the counter at zero.
1. Check to see if the number in the counter represents a voltage larger than V_{in}. (This can be done by a DAC, as described previously.)
2. If the number in the counter is not larger than V_{in}, turn on the next-lower bit of the counter. (The first time this step is done, this will be the most-significant bit.) If there are no lower bits, go to step 4.
3. If the number in the counter is larger than V_{in}, turn off the current bit and go back to step 1. If there are no lower bits, go to step 4.
4. You're finished. Send the number in the SAR counter to the outputs of this device.

To see how this works (Figure 13-3), suppose that an analog input voltage of 13 V comes into this system, and the system has a 4-bit counter capable of producing voltages from 0 to 15 V in 1-V steps:

A. The starting value of the counter—step 0—is zero. The DAC digitizes this to 0 V. Because 0 V is less than 13 V—step 1—the counter turns on its most-significant bit (binary value = 8)—step 2—and goes back to step 1. V(SAR) = 8 V.
B. Since 8 V is still less than 13 V—step 1, again—the counter turns on its next-lower bit (binary value = 4)—step 2—and goes back to step 1. V(SAR) = 12 V.

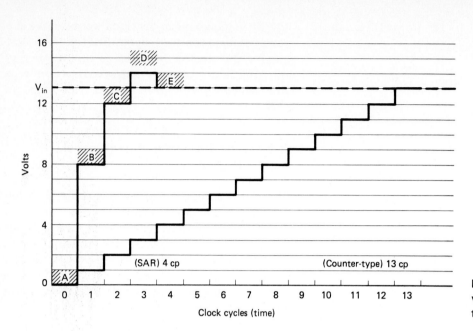

FIGURE 13-3 A/D Converter Waveforms (S.A.R. type)

C. When 12 V is compared to 13 V, it is still less than the analog input—step 1—so the counter turns on the next-lower bit (binary value = 2)—step 2—and goes back to step 1. V(SAR) = 14 V.

D. When 14 V is compared to 13 V, it is larger than the analog input—step 1—and the current bit (binary value = 2) is turned off—step 3—and the counter returns to step 1. V(SAR) = 12 V.

E. When 12 V is compared to 13 V, it is less than the analog input—step 1—so the counter turns on its next-lower bit (binary value = 1)—step 2; there are no lower bits, so go to step 4. V(SAR) = 13 V.

That seems quite complicated, but the comparison actually takes place constantly, and the counter can count each bit on or off with a single clock cycle, so only four clock cycles, or four "counts," actually take place between the start and end of this sequence. In any of the other cases, it would take 13 clock cycles to digitize this value. We have more-than-tripled the speed of the digitizing process.

13-7 RAMP GENERATOR SINGLE-SLOPE A/D CONVERSION

Another way to digitize an incoming analog voltage (Figure 13-4) involves a comparator but doesn't need a D/A converter. Instead, a precision ramp-wave generator, called a **Miller integrator**, produces a voltage that goes up at a constant

rate with time. At the same time, a counter is running, unconnected to the ramp-generator circuit, but started at the same time the ramp wave started. Let's suppose in this case that the ramp generator's voltage goes up 1 V per clock pulse of the counter's clock, and the counter can count from 0 to 15. The comparator stops the ramp generator when its voltage passes the V_{in} voltage level. If the V_{in} voltage is 5 V, it will take five clock cycles for the ramp to match V_{in}. The counter will count to 0101 (= 5) and that number is the digitized value output from the ADC. If the V_{in} voltage is 10 V, the ramp will take twice as long to reach V_{in}, and the counter will have 10 clock cycles to count-in, so its count will be 1010 (= ten). The accuracy of this method is limited only by how accurately the ramp wave rises. If it rises 1 V per clock cycle, its resolution will be 1 V. It can digitize voltages to within 1 V. If the ramp rises 1 mV per clock cycle, its resolution will be 1 mV. As long as the ramp voltage is within 1 mV of the counter's count, even at the largest count, the accuracy of the A/D converter will be 1 mV. If the range of voltages the ramp wave produces is between 0 and 15 V, in the 1-V example, the resolution of the A/D converter is 1 part in 15. In the 1-mV example, the resolution is 1 part in 15,000. Since that also requires the resistors to be accurate to 1 part in 15,000, the 1-mV ramp generator is a high-precision circuit (not to mention expensive!)

The example just described is a single-slope A/D converter. To make a dual-slope (tracking) A/D converter, all you need is a two-way ramp generator that can reverse direction instead of merely stopping.

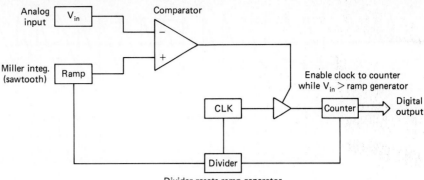

FIGURE 13-4 A/D Converter Block Diagram (single-slope type)

Divider resets ramp generator and counter after a suitable number of clock pulses

13-8 VCO (ANALOG-TO-FREQUENCY) METHOD OF A/D CONVERSION

Another method of analog-to-digital conversion using a counter depends on a linear **voltage-controlled oscillator** (VCO). The VCO produces a frequency output that's proportional to its voltage input. If the number of pulses per second is a direct multiple of the volts (say, 100 pulses per second per volt), all you need to measure a voltage is to count the pulses for a second. It sounds like you could make a 0-to-100 resolution meter with the VCO described, but the linearity is a limiting factor. If the VCO isn't accurate to better than 1%, the resolution of the converter won't be 1% even though its counter can count 100 pulses in a second.

If we can get an accurate VCO whose precision is good for the numbers it counts, we could make a voltmeter like the one in Figure 13-5. In the example, we assume the VCO produces 10 Hz for every volt at its input. All that the VCO has to do to measure a voltage of 3.5 V is produce pulses at 35 Hz. The counter is clocked by the pulses of the VCO 35 times during the second be-

tween reset pulses to the counter. Since two 7490 counters are "daisy-chained" to count these pulses, one counter contains a 3 and the other contains a 5 at the end of 1 second. Then, a reset pulse clears the counter after saving its count in the latch circuit. The latch is decoded and displayed on a readout, and every second the count in the latch is updated at the end of another 1-second count cycle.

Since the two counters (with their latches and displays) can count up to 99, the VCO should have an accuracy of 1% or better for every voltage from 0 to 9.9 V. This is two decimals of accuracy—possible with many VCO circuits—but if you want four decimal places, that makes it necessary for the VCO to be accurate to 0.01%, a degree of precision that's much harder (and much more expensive) to accomplish with a VCO.

The VCO method of conversion is simpler, but inherently more error-prone, than other methods of analog-to-digital conversion.

13-9 PULSE-WIDTH MODULATION

As a final application for all this, we're going to leave this chapter with an example of something done digitally using a circuit like the A/D single-slope converter. What this circuit does is not used digitally, however, but for an analog purpose—to amplify an audio signal with a minimum amount of power wasted in the amplifier.

What you see in Figure 13-6 is called a switching amplifier. Its analog comparator constantly checks whether the A voltage—the audio signal—is higher or lower than the S voltage—a sawtooth wave running at 100 kHz (a frequency much higher than audible sound). If the A wave has a voltage higher than the S wave, the comparator's output is HIGH, and if the A wave is lower than the S wave, the comparator's output

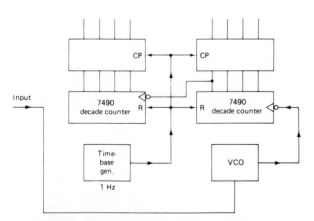

FIGURE 13-5 Using a VCO (voltage-controlled oscillator) for A/D conversion.

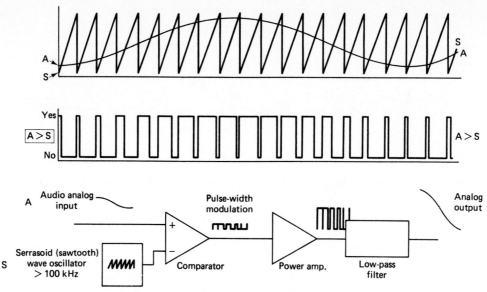

FIGURE 13-6 Switching amplifier.

is LOW. The output wave from the comparator is a **pulse-width-modulated** copy of the audio wave. In the places where the audio wave is higher, the HIGH (positive) part of the pulse wave is present most of the time. In the places where the audio wave is lower, the LOW (zero) part of the pulse wave is present most of the time. Two advantages of amplifying this pulse wave over the original audio wave are:

1. Just about any kind of transistor device can be used—it doesn't have to be linear and have a good proportional response—all it has to do is produce the same output for LOW and HIGH every time it's switched by the pulse wave.

2. The transistor in the amplifier normally runs in a saturated (HIGH) state where it has almost no voltage, or a cutoff (LOW) state where it has almost no current. It's constantly switched to one or the other of these two states, where it develops almost no heat. Since without voltage no heat is generated, and without current no heat is generated, the transistor in the amplifier doesn't waste electrical energy making heat when we want sound.

In Figure 13-6, the comparator comparing the S and A waves is shown, together with a graph of the S and A waves showing where A is bigger or smaller than S. The output of the comparator is shown in a graph below that, and the

amplified output from a cheap transistor amplifier is shown below that. The voltage sent out to the speaker has the 100 kHz filtered out of its signal; the filter sort of "averages" the LOW and HIGH levels, making a waveform at the speaker that's a replica (although with much more power) of the A input wave with which we began.

This isn't, strictly speaking, a digital circuit. Its input and output are analog waves for someone to listen to—but strictly speaking, the signals inside the amplifier are digital (they have only two levels: HIGH and LOW). The only difference between these signals' digital code and the codes we've seen before is that in this case, the number represented by the code is based on how long it stays HIGH or LOW—in fact, the analog level of the wave is determined by the percentage of time the wave is HIGH or LOW. If the pulse wave is stored as 10 bits between S waves, we could see this as a form of digital code. Five 1s and five 0s during a cycle would be a "middle" voltage, nine 1s and one 0 would be a "high" voltage, and one 1 and nine 0s during the interval would be a "low" voltage. Ten possible levels of voltage would be encoded this way—and the pulse code would look very similar to the code out of a flash converter.

More efficient code storage would be possible if the sawtooth S-wave generator were an upcounter controlled D/A converter. The binary code for the voltage at the crossing point would be available at the counter every time the S wave crosses the A wave. In this case, the system would be not only an amplifier but also a true A/D converter.

There is also a type of digital-pulse-code power-supply circuit called a **switching regulator** that uses a comparator to regulate the dc voltage of a circuit by the same method. Power supplies that work this way are called **switching power supplies**, or just *switchers*. They are also noted for their low level of "wasted" power.

QUESTIONS

13-1. How many analog comparators would be needed for an A/D flash converter accurate to four decimal places?

13-2. How many analog comparators are needed for a single-slope counter-type A/D converter accurate to four decimal places?

13-3. Which is faster, the flash converter or single-slope counter-type A/D converter?

13-4. Figure 13-1 uses four 7490 (mod-10) counters. Its resolution is 1 part in 10,000. If 7493 (mod-16) counters were used instead, what would the resolution of the A/D converter be? Assume that all circuit components have sufficient precision to provide accurate numbers with either 7490s or 7493s.

13-5. The tracking A/D converter in Figure 13-2 has two advantages over the single-slope type for "digitizing" complex waveforms. What are these advantages?

13-6. Digital measuring instruments (for example, digital voltmeters) are A/D converters. (True or False?)

13-7. The ADC-80 is a 12-bit A/D converter chip that uses the successive-approximation method. How many steps (clock cycles) will this chip need for the conversion of a voltage into a 12-bit number?

13-8. Measurements made by a measuring instrument, then transmitted to a remote receiver, are called **telemetry**. Describe how telemetry is useful to NASA during a Space Shuttle launch. How do you suppose analog-to-digital conversion fits into this picture?

13-9. (Research question) Suppose that numbers are transmitted from the Space Shuttle in analog form instead of digital. If frequency modulation is used to "encode" the data numbers, the Doppler shift of the shuttle approaching a ground station at 28,200 km per hour will shift the frequencies received. By how much? Will this significantly affect data that are transmitted with six-place accuracy?

13-10. Describe pulse-width modulation. How is it used in the switching amplifier in Figure 13-6?

13-11. What are the major advantages of switching amplifiers and switching power supplies?

13-12. If the ADC-80 12-bit A/D converter mentioned in Question 13-7 has an input voltage which ranges from 0 to 10 V, what is the *resolution* of its output (what fraction of a volt)?

Questions 13-1 through 13-10 below refer to the following timing diagram and graph:

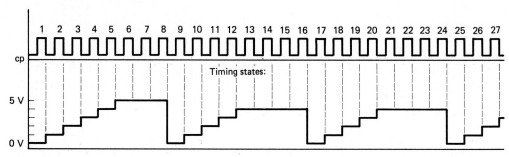

DAC output:

Suppose that these 'scope waveforms were produced by an A/D converter similar to the one shown in Figure 13-1. Based on what you can see during these 27 clock cycles:

13-1. How many samples (conversions) were done during the first 25 clock cycles?

_____

13-2. What was the decimal value of the first example? ___V___

13-3. What was the decimal value of the second sample? ___V___

13-4. The graph indicates that the counter attached to the D/A converter was reset three times during interval shown. The resets were at the falling edge of the clock in each case. When did these resets occur (at the ends of which timing states)?

_____ _____ _____

13-5. This A/D converter has a total range of 0 to _____ volts.

13-6. The resolution of this D/A converter is _____ volts.

13-7. This A/D converter's binary output is _____ (1, 2, 3, 4, . . .) bits in size.

If the clock frequency is 1 MH:

13-8. The "worst-case" conversion time for this A/D converter is _____ μs.

13-9. The sampling rate of this A/D converter would be _____ samples per second.

13-10. At the rate in the preceding question, could this A/D converter "follow" a 4-V peak-to-peak, 500-kHz input? _____ (yes, no)

Questions 13-11 through 13-20 below refer to the following timing diagram and graph:

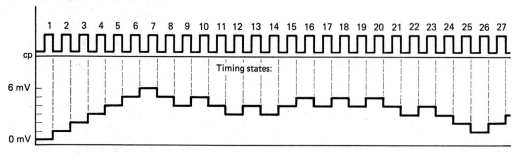

ADC output:

Suppose that these 'scope waveforms were produced by a tracking A/D converter similar to the one shown in Figure 13-2. Based on what you can see during these 27 clock cycles:

13-11. The highest voltage input during the interval was _____ mV.

13-12. The lowest voltage input during the interval was _____ mV.

13-13. The voltage input during timing state 23 was _____ mV.

13-14. This A/D converter contains an _____ (up, down, up-down) counter.

If the A/D converter has an 8-bit binary output:

13-15. The total possible number of "steps" in the diagram is _____ .

13-16. This A/D converter has a total range of 0 to ____ mV.

13-17. The resolution of this D/A converter is _____ mV.

If the duration of a clock cycle is 1 μs:

13-18. The "worst-case" conversion time for this A/D converter is _____ μs.

13-19. Could this A/D converter "follow" a 4-mV peak-to-peak, 100-kHz input? _____ (yes, no)

13-20. Could this A/D converter "follow" a 10-mV peak-to-peak, 100-kHz input _____ (yes, no)

Questions 13-21 through 13-30 below refer to the following timing diagram and graph:

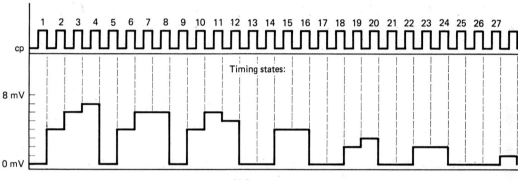

Suppose that these 'scope waveforms were produced by a successive-approximation A/D converter similar to the one described in Section 13-6.1. Based on what you can see during these 27 clock cycles:

13-21. How many samples (conversions) were done during the interval? _____

13-22. The highest-voltage input during the interval was _____ mV.

13-23. The lowest-voltage input during the interval was _____ mV.

13-24. A(n) _____ (1, 2, 4, 8, 16)-bit SAR was used in this A/D converter.

13-25. The number of clock cycles per conversion is _____ . (1, 2, 3, 4, . . .)

13-26. This A/D converter has a total range of 0 to ____ mV.

13-27. The resolution of this D/A converter is _____ mV.

If the duration of a clock cycle is 1 μs:

13-28. The "worst-case" conversion time for this A/D converter is _____ μs.

13-29. Could this A/D converter "follow" a 4-mV peak-to-peak, 100-kHz input? _____ (yes, no)

13-30. Could this A/D converter "follow" a 10-mV peak-to-peak, 100-kHz input? _____ (yes, no)

14
MEMORY DEVICES

A circuit has memory when it can keep the same output state even though the input that caused that output is gone.

The single R-S flip-flop, and later the 4-bit latch, were circuits we discussed, whose memory capacity was their most important feature.

In this chapter we discover what circuits are used to store and recover information in computers, calculators, and other digital systems. The main feature of the digital memory storage devices we'll see in this chapter is their "no moving parts" approach to recording and playback. The most familiar example of a memory device is the tape recorder. It has the capacity to record something and play back that information later—but needs to do it with moving parts. There are several basic ways to make a recording without motors or mechanical motion; we'll see what they are in this chapter.

14-1 CORE MEMORY (TOROIDS)

The earliest type of solid-state memory device—**core memory**—is an example of READ/WRITE (RAM) memory. It uses the same material (ferric oxide—called "ferrite") that is used to record on magnetic recording tape. If you keep in mind how recording and playback are done on tape, memory will be very easy to understand.

14-1.1 Theory of Toroid Action

Recording is called **writing** when it's done in memory. New information replaces old information just the way it does on the tape, when you make a new recording over an old one. Another term for writing is **read-in.** We say that core and tape have *destructive read-in* because they're designed to erase old information before new information is recorded.

Core memory uses remanent magnetism or **permanent magnetism,** which will hold data without any external electromagnet to support it.

It's the permanent magnetism that stores the recorded information in a tape or core that's memorized something.

Why is core memory called core? Normally, when you make an electromagnet, you wrap a wire around what you want to magnetize. It's called a core for the same reason that the middle of an apple is called a core—it's in the middle. Wrapping a wire around the ferrite is awkward and not easily automated, and early in the development of magnetic memory, the core became the outside of the magnet, and the wire unwound to become a straight wire at the center. The magnetic field around a straight wire surrounds it in little rings. To capture that magnetic field and use it for magnetizing ferrite cores, the cores should be little rings surrounding the wire, too. In geometry, a ring shape is called a **toroid,** so the small rings of ferrite we use will be referred to as toroids from now on. Now, instead of having a north pole and a south pole like bar magnets, toroid magnets have a clockwise or counterclockwise magnetic field when the current in the electromagnet is reversed. Of course, a student of digital like yourself should immediately guess that the clockwise and counterclockwise magnetization states are going to be used as 1 and 0 for storing logic numbers!

14-1.2 Coincident-Current Method

How do you control a thousand cores without using a thousand wires?

You could go ahead and use a thousand wires to turn the cores HIGH or LOW. To reverse the currents for a 1 or 0, though, you'd need 2000 (!) transistors attached to the wires—since a transistor can only pass current in one direction.

That's not very good. Let's look at another possibility. Figure 14-1 shows a toroid threaded by two wires called X and Y, and Figure 14-2 shows an array of toroids called a **core plane** threaded with "row" wires (X) and "column" wires (Y). Notice that as long as each toroid is like an AND gate needing a half-current through both wires, a current in an X wire and a Y wire will turn on just one toroid, even though the wires thread 7 toroids altogether. This is because just one of the toroids on the plane has enough current through both its X and Y wires. Using this scheme, 1024 toroids (32 × 32) are controlled using just 32 X wires and 32 Y wires—a total of 64 wires. Even with two transistors to a wire (for 0-current and 1-current), the circuit needs only

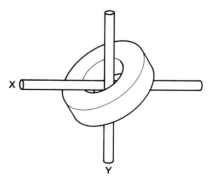

FIGURE 14-1 Toroid in a two-dimensional (2D) core memory.

128 transistors for control instead of the 2000 mentioned before. The proportions get even better when you compare the two control schemes for larger core planes.

This approach is called the **coincident-current** technique. Where two currents coincide (meet) the toroid can be set or reset. A similar approach is used in semiconductor RAM, as well.

14-1.3 X-and-Y Decoder Matrices (2D Memory)

With a thousand separate wires all the toroids could be set or reset at once. With the coincident-current method, only one toroid can be set or reset at a time. This is no big loss, since microcomputers and calculators generally handle only one number at a time.

How is one toroid on the core plane selected? The location of the toroid on a two-dimensional surface is described by two numbers called X (the row) and Y (the column). There is also another way to identify any one of the toroids on the core plane. Each toroid is given a number called its **address.** In Figure 14-2 a small core plane with 16 toroids is shown. The toroids are numbered from 0 to 15. These numbers are the addresses of each toroid and are shown in binary as well as decimal. The X (row) number is shown next to each X wire, and the Y (column) number is shown next to each Y wire. Notice that all the toroids on row 01 have addresses that begin with 01. Notice also that all the toroids on column 01 have addresses that end in 01. In every address, the two bits that start the binary address are the same as the row number and the two bits that end the address are the column number.

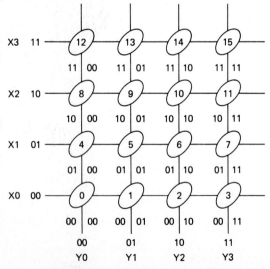

FIGURE 14-2 A 16-bit core plane.

To access a toroid at a certain address, its address must be converted into a row and column number (X and Y). This is easy to do—the row is decoded from the front bits of the address and the column is decoded from the back bits. A circuit that does this is shown in Figure 14-3. The decoders that are used to drive the row wires are called X-decoders and the decoders for the column wires are the Y-decoders.

In the figure, the number 1100 (12) is placed on the address inputs. The 11 turns on the (3) decoder on row 3, and the 00 turns on the (0) decoder on column 0. These decoders turn on the X and Y wires that go through the toroid at address 12. Any other code placed on the address lines will "find" its toroid through the X-and-Y decoders. This same method of selection is used for semiconductor memory, as well.

14-1.4 Sense Wires and Destructive Readout

How do we get the information back from the toroid when it's written there?

When we read a tape (play it back) we do it by moving the tape past an electric conductor called the playback head. The motion of the magnetic field across the conductor in the playback head induces an electric field that causes current to flow—and from there, the electrical circuits take over and "do the job". In the core memory, we have a problem—we have to move the magnetic field without moving the toroid itself.

We're going to have to find a way to move the magnetic field without moving the toroid it is in. This can be done if the field itself can be moved. The only way to do this is to destroy the field (collapse it). As the field collapses, the loss of its magnetism produces an electrical signal, the "legacy" of the toroid's 1 or 0. A collapsing 1 produces a reversed-polarity, larger pulse than a collapsing 0, so it's easy to tell what was in the toroid. A conductor (which serves the same purpose as the playback head in a tape machine) picks up the voltage produced by the collapse of the field. This conductor "senses" the state of the toroid (set or reset) and is called a **sense wire** for that reason.

The readout of information stored in a toroid is called **destructive readout** because it is necessary to destroy what's there to read it. This means the 1 or 0 will have to be put back after it is read. For this reason, amplification is one of the operations performed on the signal received from the sense wire. If amplified, the signal can have enough energy to put back the 1 or 0 and

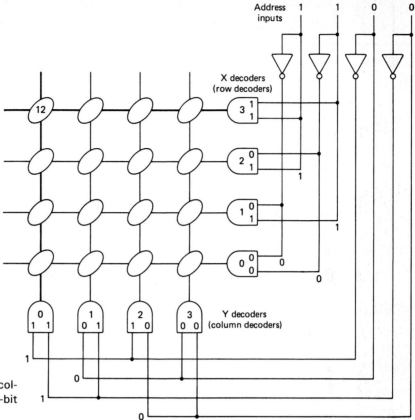

FIGURE 14-3 Row and column decoding for a 16-bit core plane.

have energy left over to be used for digital processing.

    Readout—also called a *read* operation or a playback—is done for the core memory in the following steps;

1. Collapse the bit by sending it a reset current pulse through its X and Y wires.
2. Detect the induced voltage in the sense wire. If there is only a small voltage, the core was already LOW, if there's a large voltage, the core was set (1).
3. Feed back the 1, if there was one, to the set circuits for the location being addressed.

Of course, if there was not a 1, there's not much of a pulse to feed back, or much reason to feed it back—the 0 that's already there has been zeroed by the reset current—but it was a 0 anyway, so that was no change.

    Step 3 is necessary because the toroid-reading process destroys what you've read. This is not normal for any memory device—if a tape recorder worked this way, you would need to record every tape over after playing it once. The "desired" way all memory devices *should* work is:

Destructive read-in = Recording erases old information.

Nondestructive readout = Playback does not affect the information recorded earlier.

14-1.5   3D Memory and Inhibit Wires

Figure 14-4 shows, among other things, a "final" diagram of a single core threaded by four wires: the X, Y, sense, and inhibit wires. We know that the X and Y wires are used to access a location on the core plane for writing or reading what's there, and the sense wire is used to read information. The *inhibit* wire in the diagram is used to write information, but it's only used if there's more than one core plane.

    A **3D core memory** has three dimensions (3D), the X and Y—the number of rows and columns—and the Z dimension, which is how many layers deep the memory is. In a three-dimensional memory, there are several core planes. In Figure 14-4, a memory with *eight core planes* is shown. When you use an address to "access" a location, you access eight toroids at once. The eight toroids share a common X and Y wire, and you can see from the diagram that they really can *not* be accessed separately. Every location is 8 bits

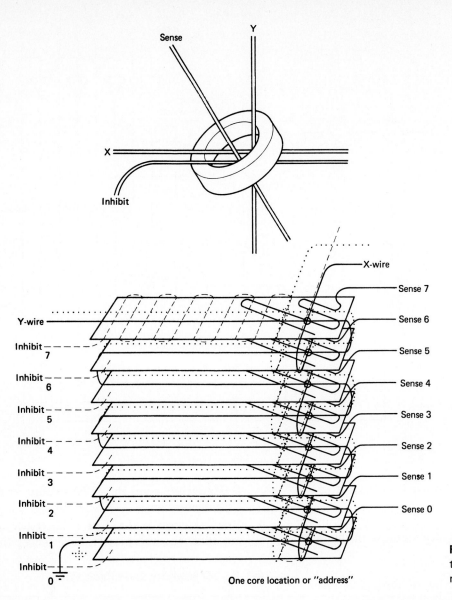

FIGURE 14-4 Toroid in a three-dimensional (3D) core memory.

One core location or "address"

deep in this 3D memory. All eight toroids are part of one address, so if you try to write a 0 or a 1, you'll record the same magnetic field on eight toroids at once. If you want to write something besides 00000000 or 11111111 at this address, you need something more than we've had in previous circuits. This "something" is the **inhibit wire.** Notice that it runs alongside the X wire in our diagram, and there's one inhibit wire for every core plane, passing through every toroid on the plane. On semiconductor RAM chips (used like core planes) this is the function of CHIP EN-ABLE or CHIP INHIBIT leads.

The sequence for writing data into any core location would be something like this:

1. Reset (0) all bits in the address with 0 current pulses in X and Y.

2. Put the word you want to write into the inhibit wires. Energize the wires for the planes where you want a 0. This prevents writing on those planes.

3. Set the bits in the address with 1 current in X and Y (except on inhibited core planes).

Toroids (cores) are not the only kind of flip-flops that can be used to build memories. Semiconductor flip-flops are much cheaper and more compact to use than toroids. Regardless of what technology is used, the same 3D arrangement of rows, columns, and planes is used in computer main memory. We'll find the same concepts whether magnetic cores or semiconductor flip-flops are used in the unit cell (a 1-bit storage flip-flop) of the memory.

EXAMPLE 14-1

How many (total) row and column wires would be needed to control a two-dimensional array of 256K (262,144) latches?

Solution: If the latches are arranged in a two-dimensional array, the number of rows and columns would be equal if the number of memory cells is a *perfect square*. First, we seek to find out if the number 262,144 is a perfect square, by calculating its square root:

$$\sqrt{262,144} = 512$$

That means that the memory cells can be arranged as an array of 512 rows by 512 columns. There would be 512 row wires and 512 column wires, a total of 1024 control wires.

14-2 SEMICONDUCTOR MEMORY (RAM)

There is only one difference between the 2D and 3D core memory and the types of *semiconductor memory* we're going to see in this chapter. Instead of magnetic toroids, the semiconductor memory types use flip-flop circuits to store 1s and 0s electrically.

14-2.1 RAM (Random Access Memory)

Core memory has gone out of use because of its bulkiness and expense compared to semiconductor RAM memory. In the data processing industry, programmers still refer to "core" when talking about any sort of main memory at all in a computer. What they are usually referring to when they talk about "core" is actually semiconductor RAM memory. **RAM** stands for **random access memory**, which refers to the fact that information at any location in the memory device can be found in equal amounts of time. We can illustrate this by comparing two ways of recording music.

A recording on a magnetic tape cassette is an example of **sequential access memory;** items recorded on it are stored one after another. On tape, if you want to "jump" from recording 1 to recording 5, then back to recording 3, it will be necessary to pass through all the intervening recordings to find (access) the recording you want. The farther apart the recordings are on the tape, the longer it will take to "access" the records.

Tape is designed to be played back in the sequence in which it was recorded, and jumping around on the tape in any other order (at random) becomes very time consuming. An equivalent semiconductor logic circuit with sequential access is a **shift register**.

By comparison, a phonograph record is more like random access memory. If you want to listen to recording 5 after recording 1, you do *not* fast-forward through the intervening recordings by spinning the record faster, as you would on tape. You pick up the phonograph needle at the end of recording 1 and set it down at the beginning of recording 5. It would take about the same time to "jump" from any recording to any other recording on the phonograph record. This is the meaning of random access: You can access the records in any order (at random) as fast as accessing them in sequence. The phonograph record is not a very good example of a RAM, however, because it can be played back, but it is impossible to record another piece of music over the one already on the record, as you can do with tape. It is a **read-only memory** device.

In common usage, RAM has come to have a second meaning: **read-and-write memory**. A serial memory, like a shift register, can also be read from, and written to, and a read-only memory has random access, but like the phonograph record, cannot be written over, only played back. RAM memory, then, is a name given to memory that has random access and can be both played back (read) and recorded on (written).

14-2.2 RAM Layout (Writing)

One way to picture a RAM is to think of parallel registers, or latches, like the 8-bit latch described in Section 11-3.1. Because all 8 bits in the latch are clocked at the same time, they transfer an 8-bit parallel **word** into the latch at one time. (A word this size is usually called a **byte**.) Now picture 16 of these 8-bit latches, spread out in an array on a tabletop, in an arrangement that looks like Figure 14-5. If we want to write an 8-bit byte into any one of these latches, we can attach the clock input of each latch to a different output line of a 16-line decoder, like the 74154 (described in Section 5-1 and shown in Figure 5-4). This means that if we put a number between 0000 and 1111 (binary) on the inputs of the decoder, and enable the G_0 and G_1 lines, one of the decoder inputs will become active, and will clock 8 bits of data into one of the latches. Since only one of the decoder's outputs is active at any time, the other latches,

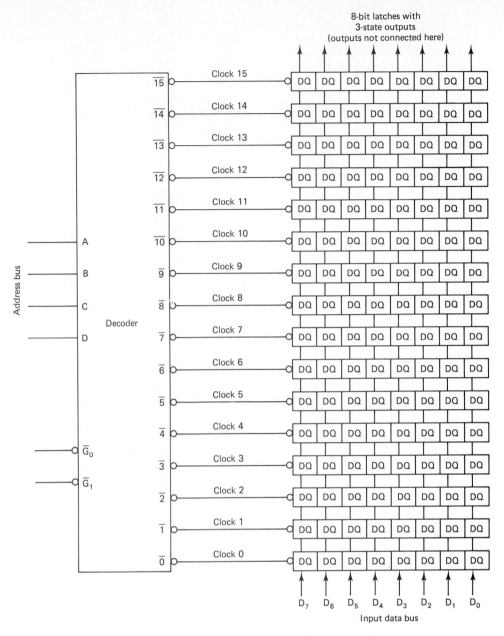

8-bit latches with
3-state outputs
(outputs not connected here)

FIGURE 14-5 A 16 × 8 RAM array with common inputs.

since they are not clocked, will ignore the data waiting at their inputs.

You might wonder why we attached the inputs of all 16 latches to a common group of eight wires. Since the same 8-bit binary byte is attached to all 16 latches at once, how can we control individually what is stored in each latch? The answer lies in the fact that all 16 latches have the same input byte, but *only one latch is clocked*; the others aren't affected. This makes it possible to write 16 different bytes of information into the memory, and hold those 16 bytes, but it must be done one at a time. In each data-record-

ing operation, the **address** of the latch is placed into the decoder, the **data** for that latch are placed on the **input data bus** lines, and after all the 1s and 0s have stabilized on the bus lines, the gate enable (G) inputs of the decoder are made active, and the latch is clocked (strobed) to **write** the data into the latch. A memory like this, with a capacity of sixteen 8-bit words, is referred-to as a **16 × 8 memory**. In referring to the size of a semiconductor memory, the first number is always the number of latches in the memory (sometimes called **memory cells**), and the second number is always the size of each latch (how many

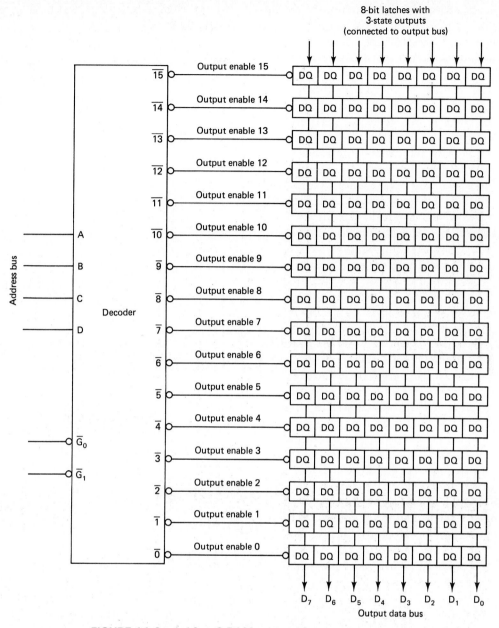

8-bit latches with
3-state outputs
(connected to output bus)

Output enable 15

Output enable 14

Output enable 13

Output enable 12

Output enable 11

Output enable 10

Output enable 9

Output enable 8

Output enable 7

Output enable 6

Output enable 5

Output enable 4

Output enable 3

Output enable 2

Output enable 1

Output enable 0

Address bus

Decoder

D_7 D_6 D_5 D_4 D_3 D_2 D_1 D_0
Output data bus

FIGURE 14-6   A 16 × 8 RAM array with common outputs.

flip-flops or bits are clocked at once). In Chapter 15 we discuss several ways to combine smaller memory device "chips" to make larger memories.

14-2.3 RAM Layout (Reading)

The circuit in Figure 14-5 showed how to get 16 bytes of binary numbers stored into the memory; now we want to get them out. If we tie all the latches' outputs together the way we did with the inputs, we will have a problem. Instead of not being able to read anything in the array (as was the case in Figure 14-5), we will have *too much* coming out of the array at once. The 16 outputs will "fight" with each other to drive the same,

common, eight lines HIGH and LOW at once. Being memory devices, the latches will have outputs active even when the inputs aren't being clocked. In Figure 14-6 we see the solution: each latch has had an 8-bit **Tri-State buffer** attached to its outputs. In this diagram we assume that 16 bytes of data have already been written into the memory. Addressing the buffers with a decoder, we select which data byte will be placed onto the **output data bus**. In this case, since only one decoder output is active, only one Tri-State buffer is enabled, and only one byte of data is actually placed onto the output bus at a time, even though all the outputs are attached to the eight lines in common. In each data-playback operation, the

address is placed on the decoder's inputs, the decoder output's Tri-State buffer is enabled by the decoder when we make the gate (G) inputs active to **read** one output byte, and the **data** appear on the output bus.

Now, of course, in a real semiconductor memory, we don't have two circuit arrays, one for writing and another for reading. We don't actually transplant the 16 latches from one array to the other, so that what has been read can be written. We don't have two separate decoders, one to write and another to read. Generally, there is only *one* decoder—its active output is delivered to the clock when we want to write (strobe data into the latches), and delivered to the tri-state enable when we want to read. The input data bus and output data bus are not always separate, as we will see in later examples of specific memory chips. Finally, as we saw in core memory, when *really* large arrays of memory cells are involved—a few thousand or so—it is more efficient to have two decoders, one for rows and one for columns, to make a two-dimensional memory grid. This type of two-dimensional array is especially well suited to integrated circuits, where everything is constructed on a flat silicon surface. In the type of coincident-current arrangement used with core planes, the clock strobe or output enable would require that the two signals be ANDed together—one for the row and one for the column—on the semiconductor latch. It seems that each bit or flip-flop in the latch should be incredibly complicated, needing row, column, and write enables before it can be clocked, and tri-state enables for the outputs, but that is not so. The unit cells, which take the place of a toroid for a semiconductor memory, are actually quite simple, as we shall see.

14-2.4 TTL Unit Cell

In Figure 14-7, the **unit cell** of a TTL memory is shown. Within the circuit, two transistors form a flip-flop. The cross-coupling between the output of each transistor and the input of its neighbor is similar to the connections in a NAND latch, and the feedback it produces works the same way.

This circuit has two types of input. Data can be input through the S and R inputs as active-HIGH set and reset inputs. The remaining inputs, X and Y, must both be HIGH before a set or a reset can be done. This allows selection of addresses. A large number of these unit cells are arrayed on the surface of a silicon chip, with the X inputs attached to the common wire for all the cells on a given row, and the Y inputs attached to a common wire for all the cells in a given col-

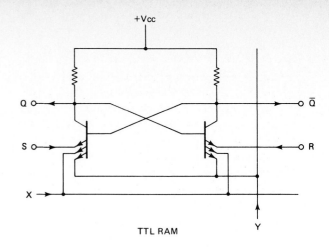

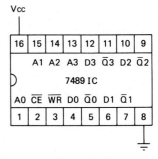

FIGURE 14-7 Bipolar transistor unit cell and TTL RAM.

umn. As you've probably already guessed, TTL X and Y decoder matrices are included in the memory device to select the right X and Y for any cell being addressed.

There are four important things this circuit does:

1. A 1 at S sets the flip-flop, so that its Q output is HIGH.
2. A 1 at R resets the flip-flop, so that its Q output is LOW.
3. The output stays set (reset) when the 1 at S (R) is removed.
4. The flip-flop stays set (reset) when X or Y is not active, but cannot be changed by R (S).

TTL memory like this is called RAM (random access memory). "Random access" means that the cells can be "accessed" for read or write operations in any order at all. Their addresses can be called in numerical order or at random. Core memory works this way, too, but it isn't called RAM. TTL RAM has faster speed, lower cost, and lower power consumption than core. It also takes up less space—so why would anyone use core?

Flip-flops are *volatile*, which means that when their power is cut off, they lose the voltages that keep information "stored" when they are on. The word "volatile" is a chemical term that means "evaporates easily." This is what happens, for instance, when you turn off a pocket calculator with a number on its display, and the display reads 00000000 when you turn it back on. When its power is turned off, the RAM in the calculator forgets.

Core is novolatile. Since its unit cells are permanent magnets, they do not require electricity to "hold" a set or reset state. As long as they were magnetized when power was present, toroids stay magnetized even when the power fails.

Nonvolatile memory can be provided through tape or disk (magnetic) "backup" of programs or data. Unless the computer "saves" every single move it makes on tape while its program is running—an unacceptably slow and cumbersome addition to most programs—there will always be some important information lost when a blackout or power failure occurs. A lot of time is wasted reloading everything back into the computer when power comes back, too. With core memory, the computer "wakes up" with everything just where it was—ready to go the instant power returns. Core memory—considered obsolete for several years—enjoyed a sudden "resurrection" after the second New York power blackout in 1977.

14-2.5 TTL RAM Example: 7489 RAM IC

In Figure 14-7 we see a typical TTL RAM device, the 7489. It contains 16 locations with 4 bits at each address. We can think of it as a 3D memory with four core planes of TTL flip-flops—although they're all laid out on the same surface. The unit cells' S and R have been made into a D-type input, so only one wire is needed to set or reset each bit.

There are 16 connections to the 7489 package. These are:

1. Four D data input terminals (like inhibit wires for core planes): D0, D1, D2, and D3
2. Four not-Q data outputs (like sense wires): $\overline{Q}0$, $\overline{Q}1$, $\overline{Q}2$, and $\overline{Q}3$
3. Four address lines which select the X and Y: A0, A1, A2, and A3
4. A write enable/disable which clocks data in to the flip-flop when they're waiting at D: $\overline{WR}$
5. A chip enable which allows the whole chip to be inhibited like a "super inhibit wire": $\overline{CE}$
6. Power connections for standard TTL (+5 V and ground): $+V_{cc}$ and GND

To write a word of 4 bits into this memory device (if the device is enabled and has power):

1. Apply logic 1s and 0s to the A (address) inputs for the address where you want the word recorded.
2. Apply logic 1s and 0s to the D inputs for the data, which is what you want to record.
3. Make WR LOW. This is like pressing the record switch on a tape recorder.

To read a word of 4 bits from this memory device (if the device is enabled and has power):

1. Make WR HIGH. This is like playback.
2. Apply logic 1s and 0s to the A (address) inputs for the address where you want to examine a word.
3. Observe the data output at the not-Q outputs. This is what you have found in the memory cell being examined.

A memory device that holds only 16 words isn't very impressive, but it's about as large as TTL devices like this can get without overheating. To store larger amounts of information in a similar-sized package, we will have to use MOS devices, described in the following section.

14-2.6 NMOS Unit Cell

The unit cell shown in Figure 14-8 permits the same operations as the TTL unit cell—write, read, select address, and inhibit—but does not consume nearly as much power.

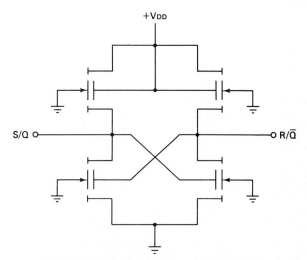

FIGURE 14-8 NMOS memory unit cell.

In the circuit, two N-channel MOSFETs form a flip-flop, and are cross-coupled in the same way as the NAND latch or TTL unit cell. Inputs and outputs are accessed through MOSFET switches that must be given a positive gate voltage (a logic 1) to pass a signal. We didn't show the X and Y selection circuitry and write-enable circuitry. They are used the same way here as in core and TTL RAM, for enabling and access.

MOS RAM consumes so much less power that thousands, tens of thousands, even hundreds of thousands of flip-flops can fit on a chip of 7489 size—and run with less heat. This low power consumption also leads to lower battery drain in battery-operated systems. A battery-operated pocket calculator would only last a few minutes with TTL RAM, but can run for hours or days with NMOS RAM, and for months or years with CMOS.

14-2.7 NMOS Static RAM Example: 2102 IC

The 2102 RAM shown in Figure 14-9 is a typical NMOS memory device. It contains 1024 flip-flops in a package the size of a 7489. Where the 7489 was organized like a 3D memory four layers deep, the 2102 is more like a single core plane with 32 rows and 32 columns. Each address has only one bit, so there is only one data input and one data output on a 2102.

There are 16 connections to the 2102 package. These are:

1. One D data input terminal: D
2. One Q data output: Q
3. Ten address lines which select one of 1024 (2^{10}) locations for access: A0, A1, A2, A3, A4, A5, A6, A7, A8, and A9
4. A read/write control signal which is like a playback/record control button which is "record" in one position, and "playback" in the other: RD/$\overline{\text{WR}}$ or R/$\overline{\text{W}}$

5. A chip enable which allows the whole chip to be inhibited like a "super inhibit wire": $\overline{\text{CE}}$
6. Power connections (+5 V and ground): +V_{cc} and *GND*

Operating the 2102 RAM is exactly like operating the TTL RAM, except that there's only one bit of data per address. NMOS logic is also considerably slower than TTL.

14-2.8 CMOS Unit Cell

In Figure 14-10 we see almost the same circuit as in Figure 14-8, except that the pull-up element on each side of the flip-flop is an active switching element instead of a passive resistor. CMOS is slower and bulkier than NMOS, but uses so much less power that it's sometimes called "nonvolatile" RAM. Although CMOS flip-flops "forget" when their power is shut off, just like TTL and NMOS, the amount of current required to "hold up" the data in a CMOS flip-flop is so small that a charged capacitor can keep data in a CMOS calculator's RAM while the batteries are being changed—even if it takes several minutes.

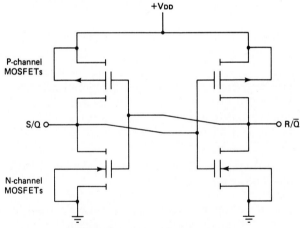

FIGURE 14-10 CMOS unit cell.

14-2.9 MOS Dynamic RAM

When the 2102 RAM was described, it was called a static RAM. We didn't explain what that meant; everything we've talked about so far, volatile or not, is static. There was nothing else to compare it to. Now we come to something called dynamic RAM. The circuit in Figure 14-11 is a dynamic RAM unit cell, and if you look for a flip-flop in the circuit, you won't find one.

The dynamic cell is basically a capacitor with a charge refresh circuit added. At first, a

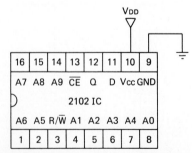

FIGURE 14-9 Typical NMOS RAM.

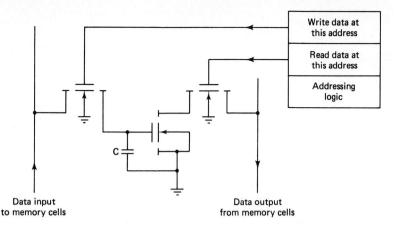

FIGURE 14-11 Dynamic MOS unit cell.

Write data at this address

Read data at this address

Addressing logic

Data input to memory cells

Data output from memory cells

charge is placed on the capacitor (C) part of the cell to represent a 1. As with all real capacitors, the charge begins to "leak off" immediately. If it's not replaced quickly enough, the voltage on the capacitor will fall below logic 1 level, and no circuit using the output of this cell would know it was once HIGH. The refresh circuit must put the charge back before too much of it is lost. This is done when the cell is enabled, and a MOSFET "reads" the charged capacitor as a 1. Since the MOSFET doing the reading is also an amplifier, its output is strong enough to recharge the capacitor and also send a 1 to other places in the circuit. This is a refresh cycle, and it must be repeated often enough to sustain the charge in the capacitance of C. Usually, this means the system using this type of memory must address each cell in the memory every few milliseconds. Special addressing modes that allow all cells on a column or a row to be refreshed by one read operation make this easier. Even with row-address or column-address selection, refresh is a big hassle, and requires either a lot of the computer's time, or a lot of extra circuitry, to handle. The places where dynamic RAM has an advantage over static RAM are large-size memories. Dynamic memory chips can contain many times the number of cells as static, because the individual cells are simpler. In large memories, the number (and cost) of RAM chips you save by going dynamic can be more than the number of extra chips needed for the refresh circuit. In that case, there's an advantage. In smaller memories, the dynamic-RAM-plus-refresh combination may have more chips and cost more than the static RAM it replaces—in that case, there's no advantage. Static RAM will be used instead of dynamic RAM.

TTL logic could be used to read the voltage on C instead of MOS logic. Since TTL inputs pass thousands of times more current than MOS inputs, the input reading the C voltage would discharge it thousands of times faster than a MOS-

FET. Since this would mean thousands of refresh cycles for every one of a MOS RAM, a task that even fast TTL logic isn't capable of, nobody makes TTL dynamic RAM.

14-3 READ-ONLY MEMORY (ROM) AND PLAs

If RAMS are the "magnetic-type" devices of memories, **ROMS** are the "phonograph records." Like phonograph records, ROMS (read-only memories) are not for information you want to erase and record over, but for information that you want to have there whenever you need it—which is the same every time.

Like phonograph records, ROMS are non-volatile. The information recorded on them is permanent (well, semipermanent in some cases), and not easy to get rid of accidentally or deliberately.

14-3.1 Diode-Matrix (Masked) ROM

The unit cell of this ROM (Figure 14-12) is a diode. A 0 is a diode and a 1 is a nondiode. The first-time cost for making a ROM is something in excess of $1000. The ROMS themselves cost only a few cents each, but setting up the first one is the main expense.

One way to make the circuit in Figure 14-12 would be to just build the diodes in where they're needed. There would be a unique circuit for every ROM "recording." That is a simple idea, but it's expensive in practice, since a different design is required for every recording. The original layout of diodes is called a **mask**, and the process of making a ROM from this plan is called **masking**. Another idea is to develop one standardized ROM in a "blank" (all 0s), and put 1s in afterward where they're needed.

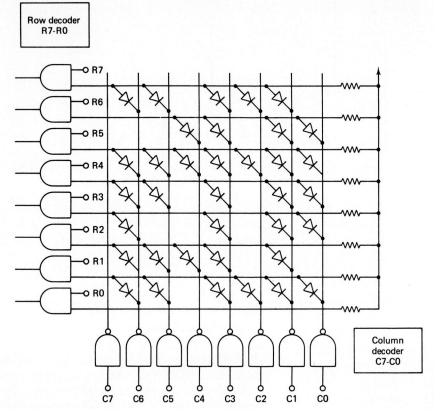

FIGURE 14-12 Diode ROM unit cell and ROM memory array.

Masking this ROM is done by running enough current through the X and Y wires to vaporize the diode at their intersection. Each "zap" is a place where a 0 has been converted to a 1. Since the 1s are unlikely to "heal up" again, the recording is permanent.

To mask the diodes, a current large enough to burn away a silicon junction must be used. This current can't be put in through the X and Y decoders, since it would burn away the junctions in the decoder gates. Without decoders, the ROM wouldn't be usable, so the current must be put into the X and Y wires directly from points inside the decoders. This must be done at the factory, before the ROM package is sealed, and requires some pretty fancy equipment. Once the equipment is set up to mask the ROM—"blow away" the diodes in just the right pattern every time—the recording can be mass-produced.

14-3.2 Fuse-link PROM

Another way of storing 1s and 0s in a permanent form uses two different materials on the ROM. Figure 14-13 shows a unit cell in this type of ROM. The fuse link is made of a material called Nichrome, an alloy of nickel and chromium. Nichrome has a higher resistance and a lower melt-

ing point than the silicon in the rest of the circuit. A current that is safe for the transistor can blow the fuse link. Since the currents used to program the PROM—blow the fuses—are safe for the TTL logic gates, the 1s can be added to a PROM full of 0s from outside. PROMS like this can be "field programmed"—programmed with simple equipment after they've been sealed and sent from the factory. They are sold blank and are programmed by the end user who buys them.

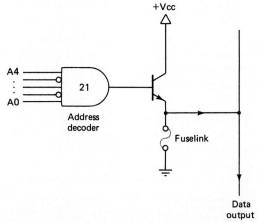

FIGURE 14-13 Fuse-link PROM unit cell (bit 21 in a 32-bit ROM).

If you are the end user, you may find writing a PROM to be a frustrating experience the first time. Your "rig" for "burning PROMS" will be simple electrically, but if you're making a fairly large PROM—say, 1024 bits—you'll find that PROMS have no forgiveness. If you make a mistake, it's graven in stone, and you can't take it back. Making the fuse links "grow back together" is hopeless. You feed the wastebasket, and try again. If you make mistakes, you'll go through a lot of 1K PROMS before you finish one successfully.

14-3.3 UV-Erasable PROM

Here is a forgiving PROM. It is also called an **EPROM** or an **EAROM** (**electrically programmable ROM** or **erasable avalanche ROM**). In Figure 14-14 you see a P-channel MOSFET with its gate floating between two layers of silicon dioxide (chemically, glass is mostly silicon dioxide). The gate would normally be connected

by a wire to "someplace else" where signals come from. In this circuit, however, charge is put on the gate by an electrical breakdown in the silicon dioxide. An overvoltage of two or three times the normal operating voltage is applied to the source and drain of the FET. The shortest path from source to drain is through the gate embedded in the silicon dioxide, but the silicon dioxide must break down first. Fortunately, this is a reversible breakdown. The charge placed on the gate puts its voltage halfway between the source and the drain, electrically. It is more than enough to bias the MOSFET into permanent conduction from the effect of its electrical field.

Information—1s and 0s—is stored in the **UV-PROM**, but it can be erased. The charge on the gate will slowly leak off through the silicon dioxide, but natural discharge will take a long time—perhaps 100 years—before half the voltage is lost this way. A much faster discharge can be done when you want to deliberately erase the information—like when you've made an error. A UV-

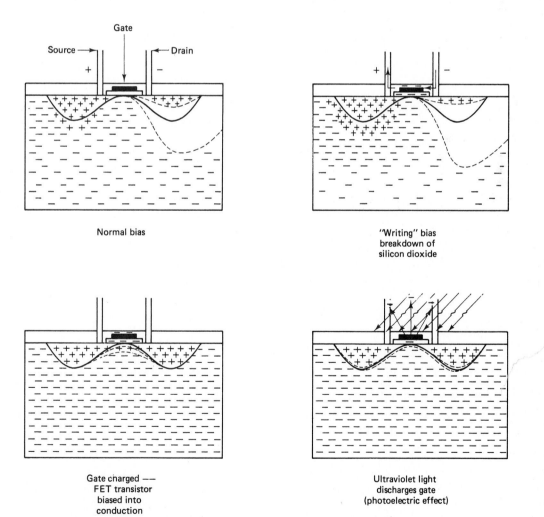

Normal bias

"Writing" bias
breakdown of
silicon dioxide

Gate charged ——
FET transistor
biased into
conduction

Ultraviolet light
discharges gate
(photoelectric effect)

FIGURE 14-14   UV PROM unit cell operating principles.

PROM gets its name from the fact that a quartz "window" on top of the package lets ultraviolet (UV) light reach the chip. When this light strikes the gate, it is able to discharge it. The extra electrons that made the gate negative are driven off through the silicon dioxide by high-energy photons of ultraviolet light. About 25 minutes' exposure to a strong sunlamp will erase all bits stored on a UV-PROM.

It's not possible to erase just the mistakes on a UV-PROM, but at least when the whole thing is erased, you can start writing it over from the beginning—instead of throwing it out.

A type of device called **electrically alterable PROM** is correctable at each address; it works similarly to the UV-PROM, but does not have to be totally erased.

14-3.4 ROMs and Programmable Array Logic Devices

The ROM and PROM types of memory provide nonvolatile memory for computers. This is useful, for instance, for the designer of a pocket calculator who wants the calculator to "remember" that the value of "pi" is 3.1415926535, even though the power has been turned off (or the batteries ran out). To have the calculator remember this fact, the number must be stored so that it will be there whenever the power is turned on. Important mathematical constants like "pi" would be stored in a section of the pocket calculator's memory made of ROM. Generally, the numbers stored in ROM are immune to power failure, but they cannot be easily changed in the course of normal operation as numbers in RAM are.

Another use for ROM memory is as a replacement for a circuit made of random-logic devices. Let's say that we want to have a circuit with four inputs and one output, whose output follows a certain truth table. Instead of using an array of AND, OR, and NOT gates, or wiring the truth table into the inputs of a multiplexer, a ROM could be used in the following way. The truth table of our four-input circuit will have 16 lines, each indicating the state of the output for a particular input pattern of 1s and 0s. We obtain a ROM with 16 internal addresses matching the 16 input lines of the truth table. Since the circuit

has one output, we select a 16 × 1 ROM. In each address, we *burn-in* the 1 or 0 output on the corresponding line of the truth table. When the ROM is programmed, we use its four **address lines** as the four inputs of the circuit; the **data output line** is used for the output of the logic circuit. Since the ROM delivers 1s and 0s when the random-logic circuit would have, it replaces it exactly. This application of ROM memory is used extensively in computers, where complex encoder and decoder arrays can be replaced by a single ROM.

ROM is not always the most efficient way to store a truth table. A **programmable logic array (PAL or PLA)** can provide the same behavior as a complex random-logic circuit, with fewer components than a ROM. We compare ROM and programmable logic arrays in the next few paragraphs.

In Figure 14-15 we see a symbolic diagram for a PLA with two inputs. Like the fuse-link PROM, it uses fusible links to connect or disconnect certain signal paths. The layout of the circuit provides AND, OR, and NOT gates to implement the "minterm sum" form of a Boolean expression. If some of the fuses are blown, the remaining connections would define the inputs needed to obtain a 1 at the output.

In Figure 14-16 a form of symbolic "shorthand" is shown which represents the multiple-input AND with fusible-link connections in a more compact way. The single line into the AND gate on the right represents all three inputs, and each "x" on the line indicates an intact fuse link.

Figure 14-17 is the "shorthand" version of Figure 14-15. The triangular symbols attached to A and NOT-A and to B and NOT-B represent a

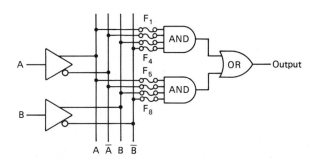

FIGURE 14-15 An unprogrammed PAL has all its fuses intact.

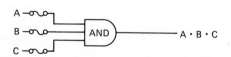

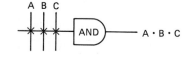

FIGURE 14-16 Shorthand notation for a PAL with intact fuses.

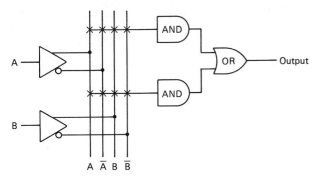

FIGURE 14-17 Shorthand version of Figure 14-17.

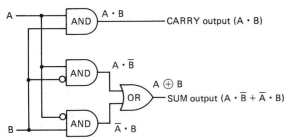

FIGURE 14-18 Conventional random-logic circuit for a half-adder as developed from a truth table.

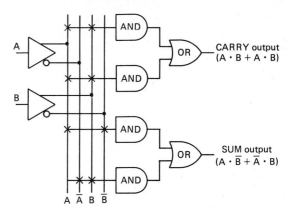

FIGURE 14-19 PAL circuit for the half-adder.

buffer and an inverter to drive a double-rail input. You can see that there are two four-input AND gates with all four inputs connected to the double rails through fusible links. The outputs of the AND gates are OR'ed together through a two-input OR gate. This compact notation will come in handy in later figures where much more complex diagrams are represented.

In Figure 14-18 we show a conventional random-logic circuit for a half-adder as it would look if developed directly from the truth table of the SUM and CARRY outputs, and implemented with AND, OR, and NOT gates. The XOR function of the SUM output requires two inverter (NOT) gates, two two-input AND gates, and an OR gate. The AND function of the CARRY output requires an additional AND gate. If TTL hex inverters and quad, two-input gate ICs were used here, three packages (a 7404, 7408, and 7432) would be needed to build this circuit. Between 25 and 75% of the gates in each package would be "spare gates"—going to waste.

The same half-adder, using a two-input PAL with two outputs, is shown in Figure 14-19. It is the same as two of the circuits in Figure 14-17 and would easily fit in a single IC. You can see that some of the fusible links have been blown out. The remaining connections give the Boolean

expressions for $A \cdot B$ and $A \oplus B$. In the upper part of the PAL implementation, $A \cdot B + A \cdot B$ is the same as $A \cdot B$, and in the lower part, one AND gate's connections to A and NOT-B are intact, while the other AND gate's connections to NOT-A and B are intact.

To get a clearer picture of the structural relationship between ROMs and PAL/PLA devices, let's look at a familiar digital system using the same compact notation we have adopted to look at the PLA. In Figure 14-20, a 4-line-to-16-line decoder, similar to the 74154 of Figure 5-4, is shown in a "shorthand" version. Each AND gate has four inputs connected. Out of a total of eight possible inputs, half are intact and half the fusible links are blown. The AND gate with output 0, for instance, has inputs NOT-A, NOT-B, NOT-C, and NOT-D connected, and the AND gate with output 15 has A, B, C, and D connected.

Figure 14-21 shows how a 16×4 PROM would look in this type of symbolic diagram. Basically, each 4-bit cell of the PROM is a group of four fusible links attached to a different decoder output. Sixteen-input OR gates are used to deliver outputs Q_3, Q_2, Q_1, and Q_0 from any one of the 16 addresses, depending on which AND gate is activated. (Since there is never any need to change the address of a specific cell in a PROM, the inputs of the decoders—the AND gate inputs—are permanent connections, not fusible links.)

Figure 14-22 shows a PLA or **FPLA (field-programmable logic array)** set up to handle four inputs and four outputs. Note its resemblance to a PROM. The major difference is that all the AND inputs (which were used for address selection in the PROM) are fusible links and can be programmed in any way that you please. This provides a greater amount of flexibility.

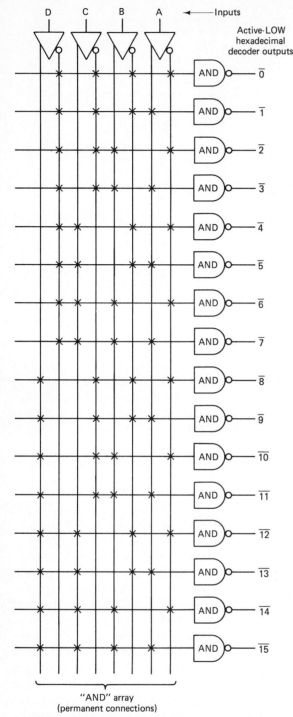

D C B A ← Inputs

Active-LOW
hexadecimal
decoder outputs

AND — $\overline{0}$
AND — $\overline{1}$
AND — $\overline{2}$
AND — $\overline{3}$
AND — $\overline{4}$
AND — $\overline{5}$
AND — $\overline{6}$
AND — $\overline{7}$
AND — $\overline{8}$
AND — $\overline{9}$
AND — $\overline{10}$
AND — $\overline{11}$
AND — $\overline{12}$
AND — $\overline{13}$
AND — $\overline{14}$
AND — $\overline{15}$

"AND" array
(permanent connections)

FIGURE 14-20 Shorthand version of 4-line-to-16-line decoder (compare with Figure 5-4).

The PAL (programmable array logic) device shown in Figure 14-23 is set up to handle four inputs and four outputs. It has permanently connected output connections from the AND gates to the ORs. It is less complicated (although also less flexible) than the PLA or FPLA described previously. Four different conditions (states) can be specified at the inputs, to send a 1 to the output.

In systems where only a few minterms are needed for each output, this PAL may be a more cost-effective solution than the PLA/FPLA.

From these diagrams, we can see that the ROM/PROM is one of a variety of similar structures related to the PLA/FPLA logic array, and that it shares structural features with decoders. From our previous discussion we know that one common use for ROM/PROM and PLA/FPLA/PAL devices is to make decoders and encoders for various specialized codes.

14-4 "EXOTIC" TYPES

This is the area where we look at the "blue sky" developments that are not major components in mini- or microcomputer systems yet, but may be someday.

14-4.1 Magnetic Bubble Memory

This is a nonvolatile, alterable memory that is more like magnetic recording tape than a RAM. It stores 1s and 0s as magnetized areas in a thin layer of a ferromagnetic crystal such as YIG (yttrium-iron garnet). The magnetized areas are movable, and maintain their shape while moving around in the layer of crystal. Figure 14-24 shows a "magnetic picture" of a piece of YIG like those used in bubble memories. In a field-free condition, where no outside magnetism is working on the crystal, its top surface is "zebra-striped" with equal areas of north and south polarity. As an external field is applied in the north direction (to make the top surface of the crystal north), the south areas shrink and the north areas enlarge (these areas are similar to the domains in a toroid). Finally, at some critical level of "bias" field strength, the south stripes break up into small circles of south polarization called "bubbles." The bubbles can be "steered" around on the surface of the crystal by small local fields that attract or repel them.

"Chevrons" are small metal conductors placed on the top surface of the YIG slice. They develop magnetic fields that are used to steer the bubbles around because a magnetic coil around the YIG wafer produces a rotating magnetic field that induces magnetism in each chevron. The magnetism at the edge of each chevron makes the bubbles stream around the chevron in a certain direction. A suitably organized pattern of chevrons can make the bubbles stream around the crystal surface in any path you want, at a rate controlled by how fast the rotating field rotates. The path is usually a zigzag called a "serpentine shift register." The data stored on the bubble

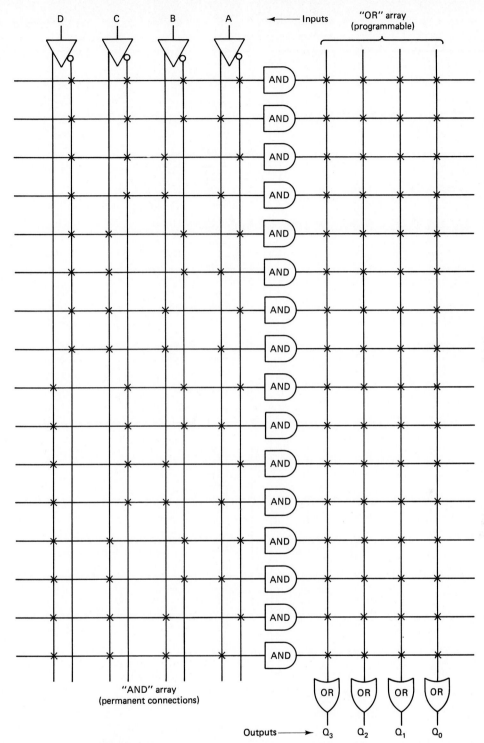

FIGURE 14-21 Shorthand diagram of a PROM (16 × 4).

memory chip's surface are a long line of bits which are written, circulated, and read in serial fashion.

Bubble memory is not a random access device. It works like a shift register, so it is not a replacement for other types of RAM chips. The major probable use of bubble memory is replacement of disk or tape magnetic storage devices which have motors and moving mechanisms to wear out. Since bubbles move around on the YIG wafer without friction, there is nothing to wear out on a bubble memory. Bubble devices became commercially available in early 1977. They have been used to replace fixed-drive disks and paper-tape readers in industrial CNC (computerized numerical control) and military computers.

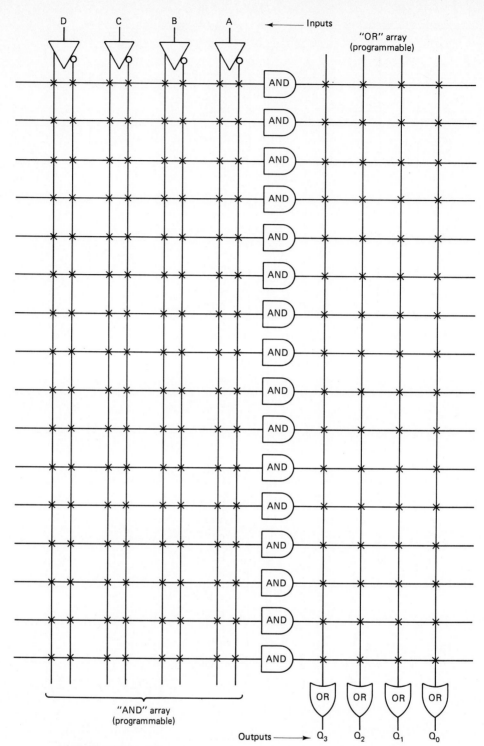

FIGURE 14-22 FPLA (field-programmable logic array) (4 in, 4 out = 16 products).

14-4.2 CCD Memory

CCDs (charge-coupled devices) use packets of electric charge traveling around on the surface of a MOS-type device in the same way the bubble memory uses bubbles. Since charge packets can already be "steered" directly by electrical forces, the control of bits in CCD is simpler than the control of magnetic bubbles, and the technology for producing this type of device is better developed than the magnetic technology of bubble memories. Like bubble memory devices, the CCD mem-

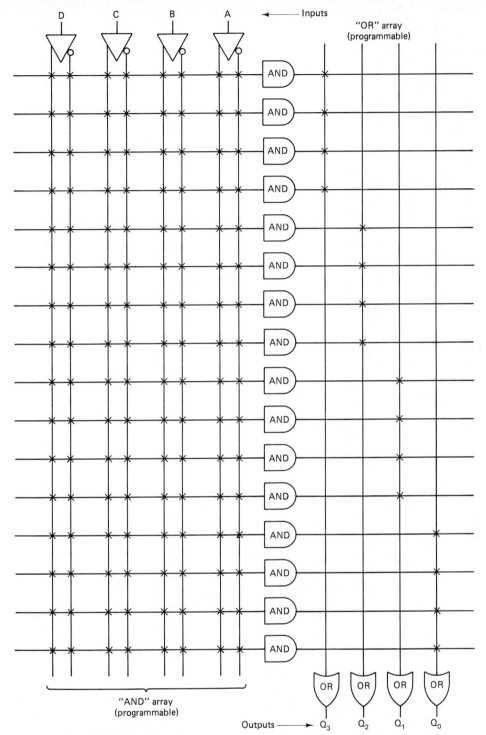

FIGURE 14-23 PAL logic array (4 in, 4 out = 16 products).

ory is a shift register, storing data bits serially rather than by random access. Unlike bubbles, the charge packets of CCD are not permanent, and gradually decay like the charges in a dynamic memory. The loops of data in a CCD must be circulated continuously and refreshed, and are volatile if the power fails. CCDs could be used for very large volume short-term memory—the CCD unit cell takes up even less space than a dynamic MOS RAM—but rather than memory devices, the largest CCD devices right now are being used for a type of tubeless TV camera. Instead of having bits loaded and returned, the CCD camera starts with charged unit cells, discharges them photo-

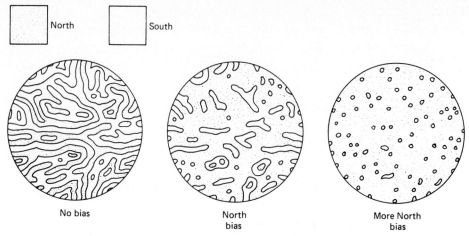

FIGURE 14-24 Bubble memory field pattern.

electrically by focusing a light image on the array, then clocks the analog packets out, producing a TV-video signal.

14-4.3 Josephson-Junction Memory

The *Josephson effect*—discovered by physicist Brian Josephson in 1960—is a "surprise" spin-off of superconductor technology. **Superconductors** are materials that lose all electrical resistance when cooled down to temperatures near *absolute zero* ($-273°C$ or $-459°F$). At these low temperatures, electrons "gang up" in pairs and travel through materials (lead and tin are a couple that work) without resistance. They also acquire some other strange characteristics, including the ability to penetrate through thin layers of insulation without losing voltage (the Josephson effect). Only a fraction of the electrons can "tunnel" through insulation, a fraction that depends on the energy—voltage—of each electron and the thickness of the insulating layer. Those electrons that do get through the insulation get through in pairs and are still superconducting on the other side of the insulation. They proceed without resistance (provided that they're in a superconducting material again).

Below certain voltages, no electrons get through the insulation. Above the threshold voltage, electrons, though only a fraction get through, pass through without resistance. This means that a current, once "pushed" through the insulation, can keep going without voltage. The source of the starting pulse doesn't have to stay on. The current will keep going without it. To turn off the Josephson current, a back-voltage pulse is used to shut off the current. If this sounds like a description of a flip-flop, it is. The unit cell of a *Josephson-junction memory* is a junction with a layer of insulation sandwiched between two layers of superconductor. Called a **SQUID** (superconducting quantum interference device), it not only has the potential to be a smaller flip-flop than anything else—CCDs included—but is a lot faster than anything else. The time it takes the Josephson current to switch ON and OFF is a quantum jump in the picosecond range—10,000 times faster then TTL and 1000 times faster than ECL logic.

Josephson logic's only drawback is the temperature at which it operates. As we mentioned in Chapter 9, the cooling unit for a Josephson-junction computer might be bigger than the computer—although a unit operated behind a sunshade in deep space would probably work just fine.

In late 1986 and early 1987, great strides were being made in the development of high-temperature superconductors. The use of liquid nitrogen as a coolant, at $-196°C$, instead of the far more expensive liquid helium, at $-269°C$, became possible, although it had seemed an unattainable goal for decades. Now, superconduction and the Josephson effect at *room temperature* seem to be within reach.

14-4.4 Holographic (Optical) Crystal Storage

Certain crystal materials are **piezoelectric.** When bent or twisted, the crystal's molecular structure gives rise to a voltage. Usually, this effect is reversible, and a voltage from an electric field will twist or distort the molecular structure of the crystal. Piezooptic crystals distort, and their molecular structure "twists", in the presence of the electric field in a strong beam of light. Some of these crystals also have hysteresis—the twist stays there when the light beam goes out—and can be used to store 1 and 0 states like a 3D memory. The advantage of this type of storage is that, in a transparent crystal like lithium–niobium titanate (linobate), the whole interior of the crystal can be reached by a focused laser beam and set or reset by the laser beam's polarization.

Theoretically, a crystal of linobate could be used to store bits in areas no bigger than the wavelength of light from the laser. A red laser beam with a wavelength of 700 nanometers (there are lasers with smaller wavelengths) could make "dots" of polarized crystal every 2000 nm, let's say, without the dots running together. In a 2-cm^2 crystal—a cube less than an inch on a side—there would be one trillion of these points!

Simple retrieval of the points could be done by "reading" the crystal through a polarizer. Twisted points would block polarized light, whereas untwisted points would not.

The theory is simple, the storage medium (a crystal) is not intricate or complex to produce and is definitely nonvolatile—it wouldn't accidentally be zapped with high-intensity laser beams—and the speed of the storage retrieval mechanism would be the speed of light. It turns out that the engineering problems—beam alignment, focusing, control of crystal position, and uniformity of crystal materials—are formidable. These memories are very experimental at this time and are not expected to be available "off the shelf" for a long time. Barring an unforeseen breakthrough in engineering technology, these are likely to be available only in very large, very expensive installations. Still—who knows about tomorrow?

14-5 DYNAMIC RAM REVISITED

Since dynamic RAM is available in much larger sizes than static RAM, and since "bigger is better" (at least in relation to memory), most micro- and minicomputer systems today are designed using dynamic RAM memory. In Section 14-2.9 we examined the unit cell of **DRAM** and identified some of the terms and concepts that relate to it. There are, however, some basic details in which DRAM differs from static RAM and the examples of TTL and NMOS RAM that were described earlier in the chapter. These are important in understanding the workings of dynamic RAM memory boards and are best described by looking at a specific memory device and then drawing conclusions from the example that can be applied to any size of dynamic RAM.

Let's use the '4256 dynamic RAM as an ex-

ample of DRAM. It is called a 256K × 1 memory, which means that it is organized as 262,144 words of 1 bit each (262,144 memory locations). Even though it is a large chip, there is no way it can be used to make a complete computer memory all by itself. Large memory chips like the '4256 are usually used in groups of 8 or 16 for 8-bit or 16-bit computers. Eight of these chips would probably be used together for a 256K × 8 segment of memory in a PC. Even 16-bit computers often address the memory 8 bits at a time (in *bytes*). A ninth bit may be added to each memory location for **parity,** so that a **memory parity** test may be done on RAM contents to determine if they are still valid.

Like many other VLSI circuits, the '4256 is designed to operate off the same +5-V supply as that used for TTL devices. Although it is definitely a MOS circuit, it is designed to be TTL-compatible. Many computer circuit boards are hybrids containing MOS and TTL devices, and operation off a single power supply simplifies cost and eliminates the need for level-shifting circuitry.

The pin diagram of the '4256 shows (Figure 14-25) that it is only a 16-pin chip. This may seem puzzling in view of the fact that 262,144 is 2^{18}, and you would expect to need 18 address lines to identify any location in the memory. In fact, careful examination of the pinout reveals only nine address lines. That is exactly half the number needed. Is that significant?

In addition to the $+V$ and ground terminals (V_{DD} and V_{SS}, respectively), and the **D** (data-in) and **Q** (data-out) lines, there are three control signals on the remaining pins of the '4256. The **W** (write) signal clocks data into a memory location when it is taken LOW, and is found on any typical read/write memory. The **CAS** and **RAS** signals are new, and are usually present on dynamic RAM devices, although none of the static RAM devices we looked at had them. What they do is use the nine address lines *twice*.

CAS stands for **column address strobe.** **RAS** stands for **row address strobe.** The words

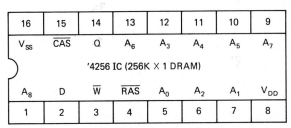

FIGURE 14-25 256K dynamic RAM pin diagram. (See Appendix 4, 4256 datasheet).

"row" and "column" refer to the same kind of rows and columns used in a 2D (two-dimensional) core memory board. Since 262,144 (the size of this memory) is exactly equal to 512×512, the easiest way to arrange the memory is to have 512 rows and 512 columns of unit cells. This requires a row decoder for 512 row-driver lines, and a column decoder for 512 column-driver lines, to address any location in the memory. Each decoder needs nine inputs to identify a number between 0 and 511 (512 numbers). The same nine address inputs, A_0 to A_8, are used to deliver nine-digit numbers to both decoders. First the row address is placed on the bus, and latched into a 9-bit latch which leads to the row decoder. This latch is clocked by a falling-edge trigger at RAS. Then the column address is placed on the bus, and latched into a 9-bit latch which leads to the column decoder. This latch is clocked (strobed) by a falling-edge trigger at CAS. When both row and column addresses have been strobed, the individual row and column are activated, and one of the 262,144 locations is accessed (read or written, depending on W).

DRAM memories commonly need to be **refreshed** every 2 milliseconds or so. The '4256 can go 4 ms before beginning to develop "digital amnesia," but like all DRAM, needs to have its capacitors recharged eventually. Rather than rewriting (refreshing) the charge in each unit cell individually, refresh takes place in an entire row at a time. During a normal read or write operation, the bits on the row being addressed are all refreshed, but the other 511 (!) rows still have to be refreshed within 4 ms. Some microcomputer processors, like the Z-80, contain on-board refresh circuitry that can generate a refresh signal and a row address between other operations of the microcomputer chip. On *really large* integrated circuits like this one, the DRAM chip may even contain its own on-board refresh circuit (called a *hidden* or *transparent* refresh cycle). This may either be done by using a specifically designated REFRESH pin (not present on this example), or by using a combination of CAS and RAS levels not encountered during a normal read or write operation. In any specific memory system, consult the data sheets and timing diagrams for your specific chip to see which method(s) of refresh it uses. See Appendix 4, 4256 data sheet, for more details.

EXAMPLE 14-2

How many address lines does a DRAM chip containing 1M $\times$ 1 of memory require? (1M = 1,048,576.)

Solution: One way to determine the number of address lines is to determine if the number of cells in the memory is an exact power of 2 (it *is*, for most memory ICs), and exactly what power of 2 it is. This can be determined by using logarithms, as follows:

$$\text{power} = \frac{\log(\text{number of cells})}{\log(2)}$$

For an SRAM, the number given by this formula is the number of bits needed to write the largest address in binary; thus it is the number of address lines used to select any memory location. For a DRAM, however, the address is split into a **row address** and a **column address.** The same lines are used to input row address and column address alternately, so DRAMS are only made in sizes for which these numbers are equal. The number given by the power formula above is divided in half, and that number is the number of address lines on the chip for a DRAM. In this case it is

$$\text{lines} = \frac{1}{2} \times \frac{\log(1,048,576)}{\log(2)}$$

which is

$$\text{lines} = \frac{1}{2} \times \frac{6.0206}{0.30103} = \frac{1}{2} \times 20 = 10$$

The DRAM requires 10 address lines.

QUESTIONS

14-1. Define what "writing" and "reading" mean when used to describe operations done in memory circuits.

14-2. How many X and Y wires are needed to select any toroid in a 2D core memory that contains 2304 toroids (48 rows and 48 columns)?

14-3. The X and Y decoders for the memory in Question 14-2 need 12 address lines (six for the X and six for the Y). What is the maximum memory array that could be addressed by these 12 lines?

14-4. How do sense and inhibit wires make reading and writing possible in a 3D core memory?

14-5. How many core planes are needed for a 1K × 8 core memory?

14-6. What is meant by the term "destructive readout" when used to describe the memory-read operation with core memory?

14-7. What is the difference between volatile and nonvolatile memory? Which is TTL RAM, volatile or nonvolatile?

14-8. TTL RAM is faster, cheaper, and more compact than core memory. It also uses less power per "write" and "read." Why would core be used, if at all?

14-9. Does TTL RAM have destructive read-out, like core?

14-10. What is the main advantage of MOS RAM over TTL RAM?

14-11. Draw a diagram showing how 2102 RAMs may be hooked up to work like a 3D memory with eight core planes (1024 × 8).

14-12. The terms "active pull-up," "passive pull-up," "active pull-down," and "passive pull-down" describe how different families of logic work. Examine the TTL, NMOS, and CMOS unit cells in this chapter, and describe the kind of pull-up and pull-down each circuit has.

14-13. The dynamic RAM unit cell does not contain a flip-flop. What does the dynamic cell use instead of a flip-flop?

14-14. Describe why dynamic RAM needs memory refresh.

14-15. What is the unit cell of a masked ROM?

14-16. What is meant by "field-programmable"?

14-17. Suppose that you are programming a PROM for the first time and you are using a manual PROM burner. Would you prefer to learn on a fuse-link PROM or on a UV-erasable EPROM?

14-18. Describe what avalanche breakdown and the photoelectric effect have to do with UV-PROMS.

14-19. What are the advantages of bubble memory and CCD memory over other types (RAM, disk, tape, etc.)?

14-20. The fastest type of memory (or logic of any kind) is Josephson junction technology. What is its main problem?

14-21. What is the highest-density memory storage described in this chapter? What problems does it have?

DRILL PROBLEMS

Questions 14-1 through 14-10 refer to the following core memory block:

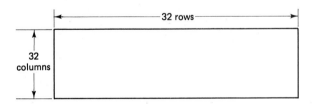

14-1. What is the total number of memory locations in the block? _____

14-2. If power failed, would data stored in this plane be lost? _____

14-3. In order for magnetic core memory to work without moving parts, the data must be read, then rewritten, because reading data is done by _____ (nondestructive, destructive) readout.

14-4. In order for data to be stored in the form of magnetic energy, the material used for toroids must have a _____ (large, small) amount of magnetic *hysteresis*.

14-5. The coincident-current technique used for core toroid selection is similar to _____ (AND, OR, NOT) logic.

If this is a 2D core memory plane:

14-6. How many bits are there in each memory location? _____

14-7. How many read wires (sense wires) are needed to read all the locations in this core plane? _____

If this is a "top view" of a 3D core memory with eight planes:

14-8. How many bits are there in each memory location? _____

14-9. How many read wires (sense wires) are needed to read all the locations in this memory block? _____

14-10. How many inhibit wires are needed to write to all the locations in this memory block? _____

Questions 14-11 through 14-20 refer to the semiconductor RAM depicted in the following pin diagram:

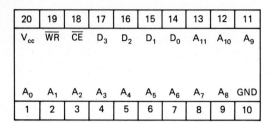

20	19	18	17	16	15	14	13	12	11
V_{cc}	$\overline{WR}$	$\overline{CE}$	D_3	D_2	D_1	D_0	A_{11}	A_{10}	A_9
A_0	A_1	A_2	A_3	A_4	A_5	A_6	A_7	A_8	GND
1	2	3	4	5	6	7	8	9	10

14-11. What is the total number of memory locations in the RAM? _____

14-12. If power failed, would data stored in this RAM be lost? _____

14-13. Data transfer in this RAM is done by _____ (common, separate) input and output data lines. (common I/O, separate I/O)

14-14. To read data, pin 19 should be _____. (low, high)

14-15. To write data, pin 19 should be _____. (low, high)

14-16. To enable this block of memory, make pin 18 _____. (low, high)

14-17. Is this a static RAM or a dynamic RAM? _____

14-18. How many bits are there in each memory location? _____

14-19. To make a 16K × 8 RAM with these chips, _____ (1, 2, 4, etc.) chips would have to be enabled together in each block.

14-20. To make a 16K × 8 RAM with these chips, _____ (1, 2, 4, etc.) blocks of these chips would be enabled by the block decoder.

Questions 14-21 through 14-30 refer to the semiconductor RAM depicted in the following pin diagram:

20	19	18	17	16	15	14	13	12	11
V_{cc}	$\overline{WR}$	$\overline{CE}$	D_3	D_2	D_1	D_0	$\overline{RAS}$	$\overline{CAS}$	A_9
A_0	A_1	A_2	A_3	A_4	A_5	A_6	A_7	A_8	GND
1	2	3	4	5	6	7	8	9	10

14-21. What is the total number of memory locations in the RAM? _____

14-22. If power failed, would data stored in this RAM be lost? _____

14-23. Data transfer in this RAM is done by _____ (common, separate) input and output data lines. (common I/O, separate I/O)

14-24. To read data, pin 19 should be _____. (low, high)

14-25. To write data, pin 19 should be _____. (low, high)

14-26. To enable this block of memory, make pin 18 _____. (low, high)

14-27. Is this a static RAM or a dynamic RAM? _____

14-28. How many bits are there in each memory location? _____

14-29. To make a 512K × 8 RAM with these chips, _____ (1, 2, 4, etc.) chips would have to be enabled together in each block.

14-30. To make a 512K × 8 RAM with these chips, _____ (1, 2, 4, etc.) blocks of these chips would be enabled by the block decoder.

15

BUS-ORIENTED SYSTEMS

CHAPTER OBJECTIVES

By the time you finish this chapter, you will be able to:

1. Describe the functions of the *address bus*, the *data bus*, and the *control bus* in a bus-oriented system.

2. Explain how a *multiplexed display* works, and how buses are used to synchronize action in such a system.

3. Describe the basic block-diagram architecture of a bus-oriented computer, and explain how the actions of the *CPU*, *memory*, and *I/O* sections are coordinated by the buses.

4. Describe what a *program* is.

5. Describe how *address decoding* allows expansion of memory.

6. Describe how a *device decoder* generates a *device select pulse*.

7. Distinguish between *memory-mapped I/O* and *true I/O*.

8. Identify what block of addresses are enabled by a decoder in a memory-expansion system.

9. Identify the characteristics of *common I/O* and *separate I/O*.

10. Describe how *open-collector* and *wire-OR* buffers permit the use of common buses by many devices.

11. Describe how *Tri-State* logic permits the use of common busing.

12. Describe how the use of *multiplexers*, rather than buffers, can be used to permit access to a common bus by multiple devices.

13. Describe some basic considerations in the design and handling of bus-oriented systems like microcomputers.

14. Describe how *input ports* and *output ports* are used.

15. Describe how *parallel port* and PIA integrated circuits are used.

16. Describe the difference between a *parallel port* and a *serial port*.

259

Up to now we have looked at digital systems that have their gates connected pretty much "wherever they need to be." Random connection of gates is all right for small systems, but in large digital networks, where parts of the system must be coordinated precisely, buses are usually used to connect all of the parts of the system together. We'll begin our discussion of **bus-oriented systems**—systems where all the parts are connected together by buses—with a simple system that does a limited task. Later, we'll advance to a discussion of microprocessors and microcomputers, which are also bus-oriented systems.

15-1 MULTIPLEXED-READOUT DISPLAY

The first bus-oriented system we're going to examine is a four-digit counter that displays the four numbers it counts on a readout like those on pocket calculators. In Figure 15-1, there is a schematic representation of a four-digit display (Hewlett-Packard HP 5082-7404). We see four seven-segment digits in one package.

Normally, to control four digits with seven LEDs each, one end of each LED is grounded (the cathode) and a conductor is attached to the anode

end of every LED. Since each segment of the display is an LED, there are 28 (4 × 7) anodes to attach conductors to. If we add a decimal point to each digit, there are four more LEDs and four more anodes to attach conductors to. With the ground conductor this makes 33 conductors for control.

Yet in the 5082, there are only 12 leads (conductors) for control on the package. How is this possible?

To reduce the number of control leads, the 5082 package is multiplexed. Only one digit at a time is used, although there are four altogether. The seven cathodes of each digit are attached to a lead called the digit driver. Since there are four digits, there are four digit drivers. The seven anodes of each digit are attached to seven wires called segment drivers, which are common to all four digits. With an extra segment driver for the decimal point, there are 12 control wires in all. For example, if the number 2 digit driver is made negative, and the a, b, and c segments are positive, a pattern that looks like a 7 will show up at the second position on the display. As we said, only one digit at a time can be displayed, but this 12-wire system can do just as well as the 33-wire arrangement mentioned earlier. The key is to use

Segment drivers

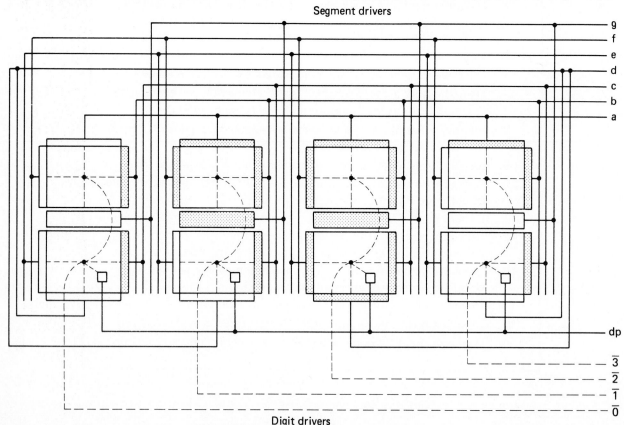

Digit drivers
FIGURE 15-1 Multiplexed four-digit readout.

the persistence of human vision the way motion pictures and television do.

Instead of displaying numbers in all four digits at once, the four digits are used one at a time. By flashing one digit in each of the four places very quickly, over and over, the appearance of four digits at once is created.

15-1.1 Introducing the Concept of the Data Bus, Address Bus, and Control Bus

In the circuit that uses the 5082, the numbers for the display are found in a four-digit counter made of four 7490 decade counters (Figure 15-2). In this diagram we imagine that 1987 pulses have gone into the input called "clock." The BCD value of each digit is present at the outputs of a counter. The leftmost 7490 has the number 0001 (1) and the rightmost has the number 0111 (7). These four digits of BCD are our data words which we want

to display on the readout. We have a display that can only display one digit at a time, although the counter has all four digits at once. We have to do three things to make this work. We must:

1. Select which one of the data words will go to the decoder from moment to moment.

2. Control the display so that the word will always show up at the same place every time.

3. Go to the next word, and the next place on the display, until all four words have been displayed, then repeat the cycle. The "repeats" must happen often enough in a second to give the appearance of continuous illumination.

To make objective 1 happen, the outputs of the counters are attached to four multiplexers. These select which of four inputs will come out the output of the MUX. For instance, there is a

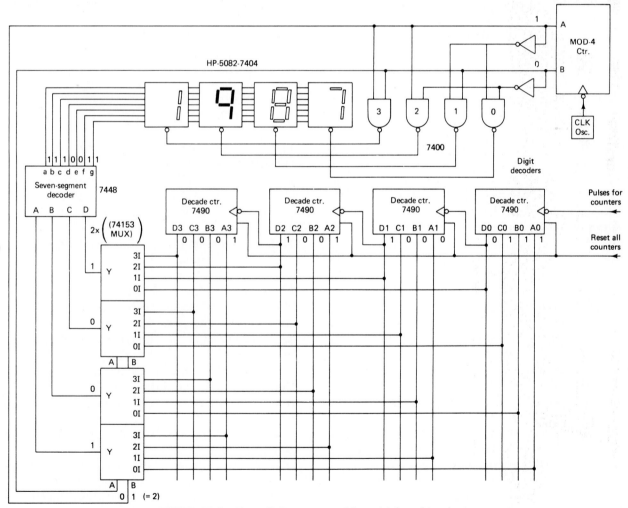

FIGURE 15-2 Four-digit counter with multiplexed readouts.

D output—the 8-bit of each number—at each of four places. We've called these D(3), D(2), D(1), and D(0) to identify which decimal place (10^3, 10^2, 10^1, 10^0) they are coming from. When the MUX whose inputs are D(3), D(2), D(1), and D(0) is given an address of 10 (2) to select input 2, the D that's going to the 7448 decoder is D(2). There are three other MUX circuits, selecting a C, a B, and an A. All of these get the same address information as MUX D, and all of them are selecting their 2 input. At the moment depicted in Figure 15-2, the 4 bits of BCD digit 2 are going to the 7448. The other counters' BCD digits are not "getting through" the MUX circuits.

How does the circuit fulfill objective 2?

At the same time that the address bus selects MUX input 2, address 2 is decoded by the line decoder, which puts a negative (LOW) on digit driver 2. The rest of the digit drivers are HIGH, so it's impossible for segments of any other digit to light. Any code placed on the segment drivers will have to show up at position 2.

It happens that at this instant, the data code going to the 7448—which drives the segment drivers—is 1001. It comes from 7490 number 2. The 7448 decodes the 1001 as a 1111011, which is a HIGH on all segment lines except e. Because the "2" code (address) that selects 7490 2 is also being used to enable digit position, 2, the number from counter 2 can only show up at position 2 on the display. Similarly, counter 3, counter 1, and counter 0 can only be displayed at digit positions 3, 1, and 0, respectively.

The fact that the line decoder and multiplexers use the same address at the same time makes it certain that everything will be synchronized. If we can make the address bus carry numbers that change from 11 to 10 to 01 to 00 (3 to 2 to 1 to 0), the four counters' BCD digits will be "scanned" to the appropriate places in the display, and look like the four-digit 1987 shown at the bottom of the figure.

Doing this—objective 3 on our list—is accomplished by a strobe counter that generates addresses on the address bus. It's a down-counter that counts 3, 2, 1, 0, 3, . . . and keeps repeating. If it repeats fast enough, the digits 1, 9, 8, 7 will be flashed one after another so quickly that the number 1987 will appear to be steady and un-flickering, with all four digits appearing to be present at once. Although each digit is only lit one-fourth of the time, it will appear to be lit steadily if the strobe rate is faster than the eye can follow. A clock rate that strobes the digits so that each one flashes 30 times per second or faster

will do this. The strobe clock only has to "tick" at 120 per second or faster to accomplish this. (The parts in this circuit can actually handle frequencies up to tens of millions per second.)

15-1.2 Synchronization between Parts of the System

The example of a four-digit multiplexed display illustrates a big point about bus-oriented architecture. Systems like this display driver can keep actions in different parts of the system synchronized if the same bus that controls one part of the system—as the address bus controls the multiplexers—also controls other parts of the system—like the line decoders. With the address bus selecting both the place where the digit is coming from and where the digit is going to—the source and destination address—it's impossible for the digits to get off track and end up in the wrong place.

As long as the same bus is connected to all the different parts of the system that need to be synchronized, the information those parts of the circuit get from the bus can keep them in step with one another.

15-2 BLOCK DIAGRAM OF A BUS-ORIENTED COMPUTER

Figure 15-3 is a block diagram of a computer showing three main blocks and three buses connecting them. Each bus coordinates a different sort of activity, and each block handles a special part of the functions of the computer.

15-2.1 Address, Data, and Control Buses: Their Function

Micro- and minicomputers have three basic blocks attached to each other by connection to three buses. The **data bus** carries data—which are numbers that say what some part of the computer has to say to another part of the computer. The **address bus** carries addresses—numbers that say where the data is going to or coming from. The **control bus** is used by the CPU block to activate or deactivate other parts of the system. Control signals are not a binary code like addresses or data. Each one is an independent ON/OFF signal that controls one thing, such as the direction of data travel or the type of **peripheral**—an output or input device—that the CPU is communicating with.

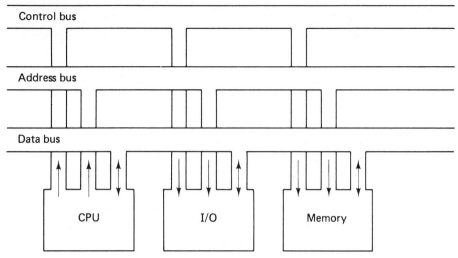

FIGURE 15-3   Block diagram of a bus-oriented computer.

15-2.2   CPU, Memory, and I/O Portions of the System

The three main blocks—the CPU, I/O, and memory blocks—are the central processing unit (CPU), the input and output devices (I/O), and the memory. We know that a memory is a section of the computer whose job is to store and retrieve binary numbers, but we haven't discussed yet why this is necessary, or what the other blocks do.

Let's begin our discussion of the computer by comparing it with something more familiar. You have all had experience with a pocket calculator. Imagine the calculator operation of adding "5 + 2 = 7." If you are operating the calculator, you begin by pressing the keys on the calculator keypad in the sequence "5 + 2 =." The keyboard is an input device. It is a source of two kinds of information used by the calculator. Creating inputs for logic gates, each key of the calculator makes a binary code that represents either a number like 5 or 2—a data word—or an operation like + or =—an **operation code (opcode)**.

How does the calculator keep the number 5 on its display after you take your finger off the key—and for that matter, how does the 5 stay there while you're keying in the +? (The 5 goes away when you hit 2.)

The answer is that the calculator contains a memory section. It needs to store decimal numbers and whatever other **keystrokes** the keyboard user puts in. The + operation, for instance, can't be activated until both the 5 and the 2 have been keyed in—yet the + is keyed before the 2. A memory is needed here to hold the + operation until both the 5 and 2—the addend and the augend—are in the memory. Then the 5 and 2 are added by an **adder circuit,** and the result (7) replaces the last number on the display. This is an output operation.

There you have it. Although this is a calculator, not a full-bore computer, it has all the basic parts: there's an input device, a memory, an output device, and—the part that actually does the adding—a central processor (CPU). With these same parts, we could make a computer with the addition of only one more part.

The calculator's sequence of actions in 5 + 2 = 7 is mostly manual. You have to key in the addend, the augend, (the two operands), and the + sign (the opcode), and then the = sign (which is really an opcode for doing an output of what's been added so far). Nothing's automatic. The calculator won't go and get the numbers on its own and know what operation to do with them; it must be "stepped through" manually.

Computers can do all this automatically; the operations' sequence can be *programmed* in advance, and as long as the computer has been told in what order to push the buttons, it can complete the calculation by itself. A decoder uses the list of instructions—a **program**—to "push the buttons." The outputs of the decoder activate different circuits in the arithmetic and memory sections of the calculator—even the input and output devices. A programmable calculator is one that's able to store up a list of keystrokes—including data and opcodes—and carry out its list of instructions at the touch of a button. There is little real difference between a programmable calculator and a computer.

We'll describe the function of each part of the computer block diagram by comparison with the calculator doing "5 + 2 =."

1. **CPU**   (does the actual computing of 5 + 2)
2. **I/O**   (input gets 5 + 2 = and output shows 7)
3. **Memory**   (holds the data 5, 2 and opcodes +, =)

15-3 ADDRESS DECODING FOR MEMORY AND I/O BLOCKS

The address bus is used by the memory section of the computer to find where things are stored. We already saw a little of this process when we talked about 3D memory. We can take it for granted that the inner workings of any memory device will include X and Y decoders, and we know that these use bits from the address bus to decode the row and column where the information we want is located. The total memory of a computer is usually much larger than any single memory device.

The first thing we'll look at is how the address helps us select which devices will be enabled out of the total memory, and how we put together these devices to make a memory bigger than any one device.

The second place where addresses are used by the computer is the I/O (input/output) section of the computer. In a computer, unlike a pocket calculator, there may be hundreds of input devices instead of just one (the keyboard) and there may be hundreds of output devices instead of just

one (the LED or LCD displays). With hundreds to choose from, the computer needs a way to pick the device being operated. The same address bus used to select memory locations is also used to select which I/O device we want. We'll see how this is done in the following sections.

15-3.1 Expansion of 1K × 1 RAMS (2102) into a 1K × 8 Block

The circuit in Figure 15-4 shows eight memory devices (the 2102 RAM we introduced in an earlier chapter) put together in a 1K memory for an 8-bit computer. The phrase "1K × 8" refers to the fact that 1K (1024) places in the memory can be selected by the address, and each place contains an 8-bit data word (binary number). Each 2102 is a 1K × 1 device, meaning that the 1024 places in the memory each contain a 1-bit data word.

Each memory chip has 10 address lines. These conductors have a number from 0000000000 (0) to 1111111111 (1023) put on them to select one of the 1024 locations. The memory chips are put together like core planes in a 3D core memory. All eight chips use the same address at the same time. Each chip's data input and output represent a different bit of the same word. For instance, if the 8 bits of a word stored in the memory are organized like this:

128	64	32	16	8	4	2	1
D7	D6	D5	D4	D3	D2	D1	D0

then each of the eight 2102 devices represents one of these bits. If we were storing a 01000000 in

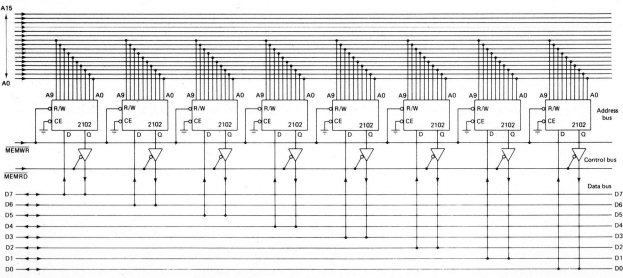

FIGURE 15-4   A 1K × 8 RAM using 2102s.

memory address 30, a 1 would be written into the 30-location of device D6, and 0s would be written in the 30-locations of all the other devices at the same time. All eight are on the address bus in parallel, and their addresses are all the same. We can get back the same 8 bits we stored at any time by going back to the same address on the address bus. With everything synchronized, all eight chips work together every time a memory write or read operation is done. It should be easy to figure out how a 1K × 16 memory could be made with 1K × 1 chips.

15-3.2 Decoding DSPs (Device-Select Pulses) for Memory or I/O Devices

When a computer uses the same address lines to call on memory location 30, output device 30, or input device 30, there is a problem deciding who the 30 is for. It's clear that there are only two ways to tell whether the code on the address bus is for a memory, output, or input. Either never use the same number for all three kinds of device (you have to give up some memory addresses for this) or add some extra signals to the system that tell whether the address is one for a memory, input, or output device.

If you go with the first approach, you've really thrown away some memory addresses in exchange for some input and output devices. Since most computers have a lot more memory addresses than I/O devices, this is usually no big loss. When it's done this way, the input or output devices are said to be **memory mapped.** There couldn't be a memory 30 and an I/O 30 together in this kind of scheme.

The other method (true I/O) uses all addresses for memory read or write operations when memory read or memory write signals are active, and uses the same addresses for input or output when I/O read or I/O write signals are active. These signals are found on a bus of lines called the **control bus.** In this case, there could be more than one "30" in the system.

Figure 15-5(b) shows a device decoder for input device 30 in a memory-mapped system. It generates a strobe pulse called a **device-select pulse** (a DSP) when 30 is present in binary code on the address bus at the same time as memory read is active on the control bus. Notice that what's being done is a memory read operation at address 30. The input device "fakes" the computer into thinking it's read data out of a memory address when the information really comes out of a keyboard, an A/D converter, a punchtype reader—take your pick.

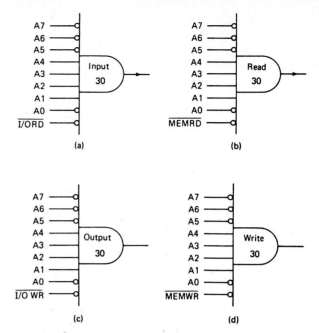

FIGURE 15-5 Device decoders for input: (a) true I/O, (b) memory-mapped I/O; and output: (c) true I/O, (d) memory-mapped I/O.

Figure 15-5(d) shows a device decoder for output device 30 in a memory-mapped system. The only difference in this device is the control signal. This time, the computer "thinks" it's writing into memory, when it's really sending data to an LED display, a printer, a tape recorder—take your pick.

Figure 15-5(a) shows a device decoder for input device 30 in a true I/O system. This system has control signals for I/O as well as memory devices, so the decoder looks for a 30 on the address bus and an I/O read (input) signal. A memory operation at 30 would never have this control signal, so the input device doesn't replace memory 30.

Figure 15-5(c) shows a device decoder that generates a DSP for output device 30. The address is the same in every case, but the control signal is memory write (output) for the output device.

There is one other case where a decoder would have to be attached to address lines to provide a DSP. There's always someone who wants a computer to have a memory bigger than the biggest available device. For example, if the biggest available RAM chip is 16K × 8, someone wants a memory 64K × 8. This means the computer that uses the 64K × 8 memory will have more address bus lines (16) than the address inputs on the 16K × 8 device (14). The wanted memory is four times as large as the real device,

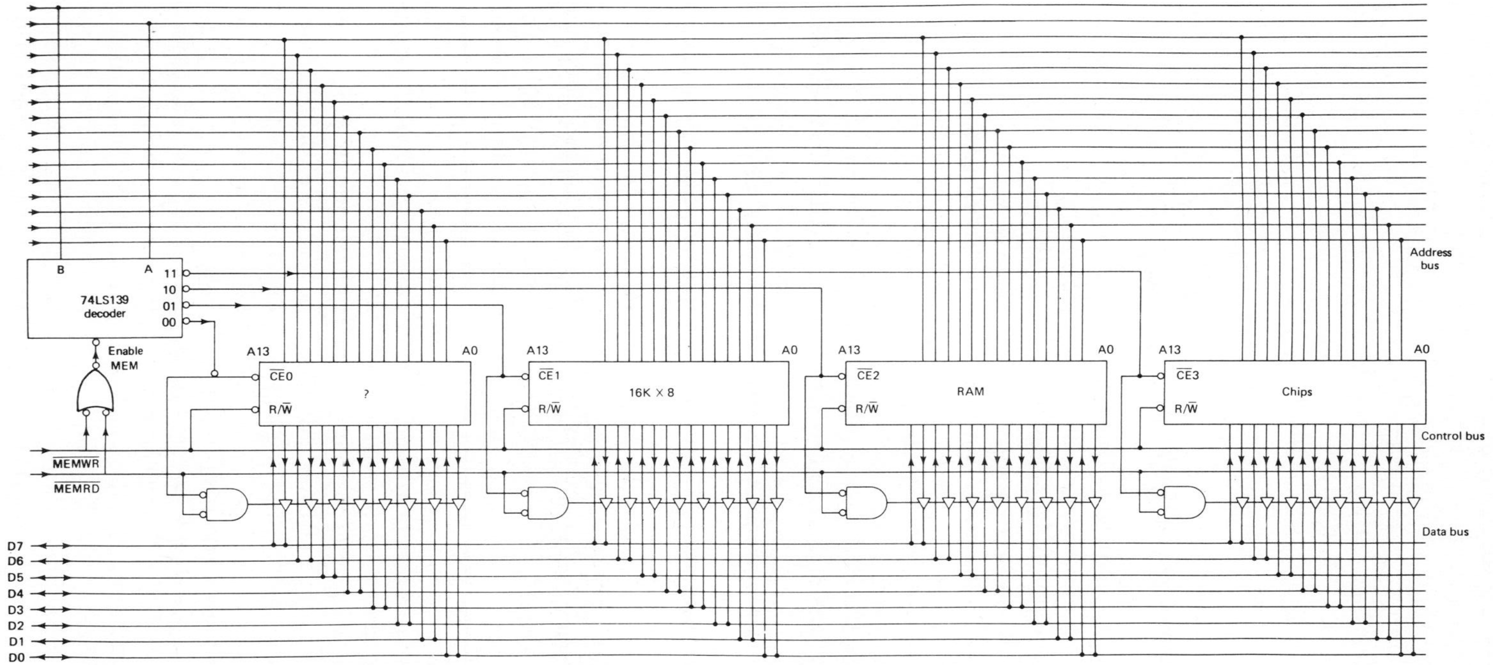

FIGURE 15-6 Expansion of 16K $\times$ 8 RAMs into a 64K $\times$ 8 memory.

so we need four of these devices. When 14 of the available address lines are attached to the 14 address inputs of each device, the two address lines "left over" have four states: 00, 01, 10, and 11. Left to themselves, the four 16K chips attached to the same 14 address lines would all write the same data in four places, and would never contain more than 16K of data (four copies of each byte). This is where a decoder comes to the rescue. By decoding the left over address lines as 0, 1, 2, and 3, the 16K of addresses that start "00XXXXXXXXXXXXXX" will enable only device 0 in Figure 15-6. The first 16K of data will be written there, and no place else, because it's the only device enabled. The addresses that start 01XXXXXXXXXXXXXX will affect only device 1. The device decoders for devices 2 and 3 will assure that only addresses 10XXXXXXXXXXXXXX and 11XXXXXX-XXXXXX will be written in those devices, even though the four devices themselves have only the capability of "seeing" the XXXXXXXXXXXXXX part of the address. Since the decoder and the four device-select pulses it generates, allow 14-bit de-

vices to respond to 16-bit addresses, this is called **memory expansion**.

15-3.3 Expansion of the 2102 Circuit to 16K Using a 74154 Decoder

In Section 15-3.1 we spoke of expansion of a 1K × 1 memory chip into a 1K × 8 memory array. This type of expansion was an expansion of the word size on the data bus. A 2102 chip handles words of 1 bit apiece, but we wanted a 1K memory with words of 8 bits apiece. This required eight 2102s.

What we want to do now is to expand the address bus. Although there are 1024 addressable places in a 2102, we'd like to make a memory out of 2102s with 16K places in it.

In Figure 15-7, a 74154 decoder, shown before in Chapter 5, is used to take the bits of a 16-line address bus and use them to select memory locations in an array of 2102s which only have 10 address lines. Think of the 16-line address bus as selecting memory locations with addresses made from the following binary code:

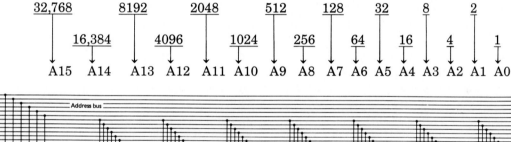

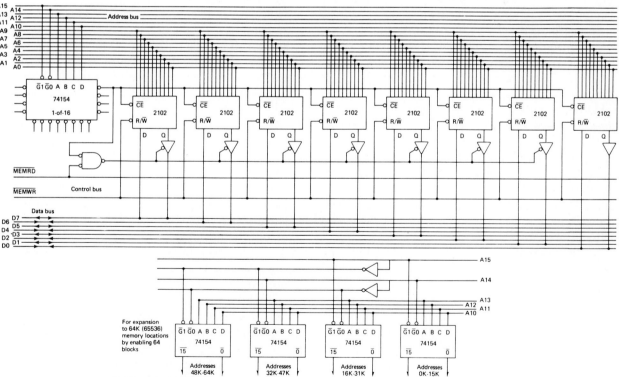

FIGURE 15-7 Expansion of 1K × 1 RAMs into a 16K and 64K × 8 memory. (See Appendix 4 for more details on the 74154.)

Using this code, numbers on the address bus can go from binary

$$0000000000000000 \ (0)$$
$$\text{to} \ \ 1111111111111111 \ (65{,}535).$$

With these numbers, 65,536 different locations could be selected, but this is 64 times the 1K (1024) places there are in a 2102 device. If we build our memory out of 2102s we'll have six address lines left over. What do we do with them?

Although memory devices with a lot more than 1K locations are available, there will always be someone who wants a memory bigger than the biggest available device. A chip enable input is provided on memory chips with memory expansion in mind. On the 2102, this enable input is active low.

There are 10 lines in the address bus that can go directly to all the 2102s. The remaining six lines go to a decoder, where two are used to enable the decoder (when both are LOW) and the rest are decoded by DCBA into numbers from 0000 (0) to (1111) 15. Each of these inputs is used to enable a different block of 2102s.

In this case, we've compromised. With 6 bits, the remaining address lines could be decoded into 64 different numbers. We've chosen to use a decoder that only "sees" the first 16 of these numbers, so we can only handle the first 16K of memory addresses with this one decoder. That's OK; computer designers don't always use the full memory capacity of a computer anyway.

Since only one output of the decoder is active (LOW) at a time, only one block of 2102s is enabled. There are a total of sixteen 1K × 8 blocks that can be enabled. The rest of the disabled chips will still have an output of some kind, and 15 disabled 2102s (whose outputs are all HIGH or LOW) are "fighting" with the output of the one enabled 2102 for control of each data out line. In Chapter 5 we said that *Tri-State* or open-collector logic was necessary in conflicts of this sort. That's why the outputs of each 2102 are attached to a Tri-State buffer which is disabled when the 2102 is disabled.

In this system, memory block 0 is treated as a single device—all its chips are enabled and disabled at the same time—as is true with block 1, 2, and all those up to 16. The decoder's DSP is the chip enable for each 2102 in the block. Memory addresses in the first K (0–1023) of the memory will be found in block 0, the second K (addresses 1024–2047) will be found in block 1, and so forth.

You'll notice that the decoder we used also has enable inputs like the 2102 chips. On the memory chips, the enable was for memory expansion. The enables on the 74154 are also for expansion, and a circuit can be made with four 74154s that can enable a total of 64 blocks. In this example, we only had one 74154, and enabled it when the A(15) and A(14) bits were LOW, but by adding a few inverters, another three 74154s could be activated when 01, 10, and 11 appear on these bits. (A74154 data sheet is available in Appendix 4.)

15-3.4 Memory Devices with Common I/O versus Separate I/O (2114 versus 2102) for Bidirectional Busing

In Figure 15-8, you see a 16K memory made of 2114 memory devices which have a common I/O. In the circuits made with 2102s, there was a separate input data bus and output data bus, because there were separate output and input terminals on each 2102 chip.

On the 2114, the same terminal is used for the input and output of each bit. There are 4 bits in each address, so there are four data pins on the integrated circuit. These terminals are multiplexed. They alternately act as inputs and outputs according to the state of the WR control pin. This scheme is called common I/O because input and output are done at a common point (the name has nothing to do with the I/O of an input/output section of a computer). There is just one data bus—called a **common data bus** or **bidirectional data bus**—in this circuit. Since data goes into or out of the memory devices on the same wires, they are bidirectional, like a reversible express lane on a highway, which is southbound in the morning and northbound in the evening. Data don't travel both ways at the same time on a wire, but they can travel either way, alternately. Signals on the control bus are used to coordinate parts of the system to place data onto, or pick data up off, the data bus.

All of the buffers and separate-bus switching needed for the separate I/O 2102 devices are built in the 2114. Having a single data bus working all the time instead of a double data bus with only one half running at any one time, common I/O systems are more efficient than separate I/O data buses. Early microcomputers used the separate I/O scheme, but this has been largely replaced with common I/O in later designs.

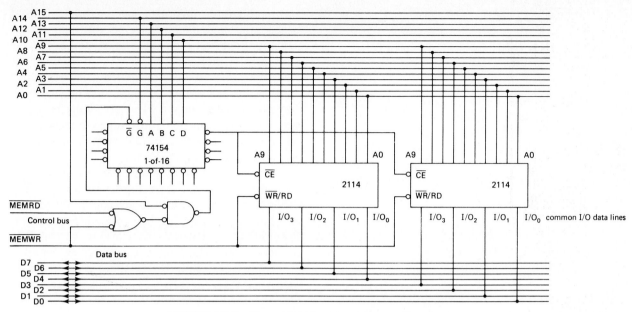

FIGURE 15-8 Use of RAM with common I/O data lines.

15-4 OPEN-COLLECTOR AND WIRE-OR BUSING FOR COMMON BUSES

Attaching all the parts of a digital system to a common bus to synchronize them would be impossible without the ability to float outputs. In any bus-oriented architecture, many circuits share each conductor of the bus. If more than one circuit is sending at the same time, there will be conflict on the bus conductor. When one output wants to pull up and another wants to pull down, they can't both succeed. It's important for only one circuit to be sending and one circuit to be receiving at a time. When the sending circuit's output is driving the bus conductor HIGH or LOW, the rest of the outputs attached to the conductor must be floating.

In Chapter 9 we introduced the concept of open-collector (O.C.) logic to solve this problem. Although the outputs of open-collector logic can't be HIGH, LOW, and floating all at the same time, adding a pull-up resistor to the bus conductor gives results that are just as good.

When one O.C. gate is enabled, its output is LOW and floating at times when other families of logic are LOW and HIGH. Suppose that five O.C. gates are attached to a bus line, three with their outputs driving the line, and two with their inputs driven by the line. When one driving gate is enabled, and the other two driving gates are disabled, the enabled gate pulls down on the bus line and makes it LOW because its pull-down output is stronger than the pull-up resistor we used.

When the enabled gate is not pulling down, nothing else is either, and the pull-up resistor has nothing fighting with it—so the bus line is pulled up and becomes HIGH.

Of course, having O.C. outputs driving signals into the bus won't do any good unless there's only one enabled output. It doesn't matter how many O.C. inputs are enabled, they don't pull up or pull down on the bus that's driving them like outputs do. Also, the amount of pull-up or pull-down from an input attached to the bus (the fan-in) doesn't change when its gate is disabled or enabled. That only affects the outputs. In common busing, the one important thing to remember is that if the gates driving the bus get into a "fight," there's trouble.

Preventing this kind of trouble is the job of the control signals that enable and disable the bus drivers. The logic designed into the system must be mutually exclusive. This means that you can enable one driver OR the second but not both (in a system with two drivers). Driving the bus with multiple open-collector outputs is called **wire-OR**. If more than one is enabled, the outputs are ANDed together, which is not good. The use of wire-OR logic with O.C. devices, it must be remembered, works only as well as the enabling and disabling circuits that control it.

A footnote in the "sneaky trick" department: If you add a diode to the output of a standard logic gate (shown in Figure 15-9), with the cathode connected to the gate's output and the signals sent from the anode outward, the gate

FIGURE 15-9 Adding a diode to make O.C. TTL from standard TTL.

acts just like an open-collector gate, with outputs LOW and floating. If you are doing a circuit "fix" or "kluge" that needs just one gate with a floatable output, and you have some spare gates of nonfloatable types, this is a nice trick to know (germanium diodes work better than silicon diodes for this).

15-5 TRI-STATE LOGIC IN BUS-ORIENTED SYSTEMS

If TTL Tri-State logic is used where O.C. logic was used in the preceding section, there will be no need for pull-up resistors. The Tri-State gate has an authentic HIGH, LOW, and floating condition at its outputs, so the enabled gate can either pull up or pull down on the bus while the other gates float. Of course, if the other gates don't float, their outputs have the same characteristics as standard TTL—two gates "fighting" will resolve their conflict by having the LOW output "win" and the HIGH output "lose," and if three or more gates are fighting, the fight may end with one or several gates being burned out as their pull-down transistors are "blown" by excessive pull-up current. As with O.C. logic, the gates' outputs must activate in a mutually exclusive way. The busing

arrangement is only as good as the logic that keeps only one driver driving and the others floating. Several types of standard devices in the TTL family, including counters and latches, are also available in versions with "Tri-Statable" outputs, for use in bussed systems. The only easy way to make a standard TTL gate into a Tri-State version, when one isn't already available, is to add a Tri-State buffer to its output. There aren't any shortcuts like the one for O.C.

In Figure 15-10 we have redrawn the four-digit display circuit from Figure 15-2 using Tri-State buffers. The counters are connected to buffers that can be disabled. Only one buffer at a time is enabled, and the data that arrives at the 7448 decoder is the data from the enabled device.

Many devices in the MOS logic category are also available in Tri-State varieties. In fact, since O.C. or Tri-State capability is essential to bus utilization, most large-scale integrated (LSI) circuits like MOS microprocessors have outputs that are Tri-Statable (floatable). Bus architecture is also found inside microprocessors and other LSI devices, and of course, enable/disable "float" logic is used in every buffer, latch, and register that attaches to these buses inside the chip.

As mentioned in Chapter 9, IIL logic is open-collector by its nature, and is ideally suited to LSI development using bus-oriented architecture.

Multiplexed Busing. Another completely different approach to bus utilization is to multiplex together the drivers for the bus. We did this when we multiplexed the counters for our four-

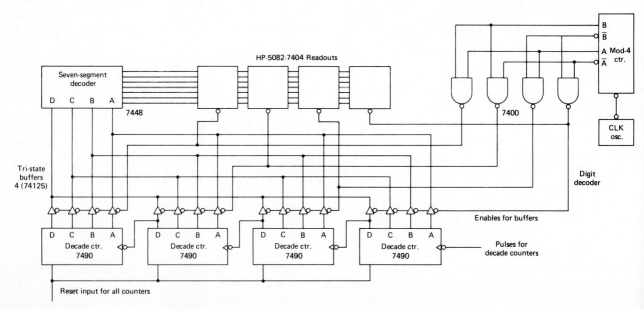

FIGURE 15-10 Four-digit readout multiplexed with buffers.

digit display in the beginning of this chapter. Instead of having each output drive the same bus, each driver drives an input of a multiplexer. In this kind of scheme, a data bus with four driver devices would have a 1-of-4 multiplexer attached to each data bus line. All the multiplexers would share common address (select) lines, so that all the 2-drivers would be connected to the bus at one time, all the 3-drivers at another, and so on.

Some popular microcomputers, like the Radio Shack TRS-80 series use Tri-State buffers for driver selection; others, like Apple computers, have multiplexers in the same places. There doesn't seem to be any advantage of one over the other; the two methods are equivalent and work equally well.

EXAMPLE 15-1

Design a 256K × 8 memory using 64K × 8 SRAM chips with the following logic diagram:

Solution: Four 64K × 8 RAMS are needed for the 256K × 8 memory.

A 256K memory requires 18 address lines. The 64K chips have 16 address lines, A_0 to A_{15}. The least significant 16 bits of the address bus are attached to the corresponding lines on the RAM chips, so that any chip's address lines are in parallel with the corresponding lines of the other three chips. The remaining two lines of the address bus go to a two-line-to-four-line decoder, which is used to enable each of the four chips, according to the state of address lines A_{16} and A_{17}. The four decoder outputs are separately attached to each of the four RAMs.

Addresses between 0 and (64K − 1) will enable memory block 0.

Addresses between 64K and (128K − 1) will enable block 1.

Addresses between 128K and (196K − 1) will enable block 2.

Addresses between 196K and (256K − 1) will enable block 3.

The eight lines of the data bus are attached so that any chip's data lines are in parallel with the corresponding data lines of the other three chips. Only one chip at a time is enabled, so this poses no problem. The write (WR) lines are attached to the control bus of the computer (not shown in this diagram).

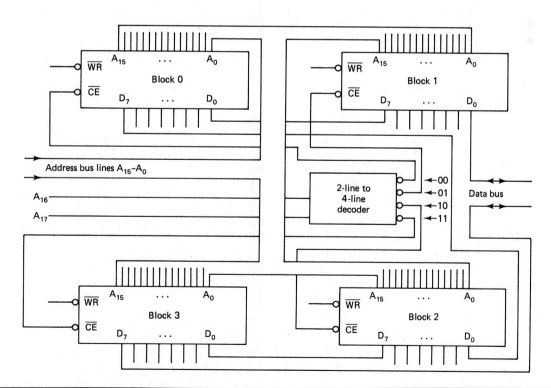

15-6 PRACTICAL CONSIDERATIONS IN BUS-ORIENTED DESIGN

Digital switching circuits sometimes have problems that show up when the circuits are clocked at full speed, but are not evident in static tests with a meter or logic probe. These problems are often caused by noise, extra pulses and spikes that appear on the 1 and 0 logic levels in the circuit. Sometimes, the noise makes itself evident when some gates on a board generate enough noise glitches to make other gates on the board switch incorrectly. In other cases, the noise radiates, and interferes with other circuits (RF interference).

15-6.1 Decoupling Capacitors

TTL devices switching ON and OFF in a very short period of time draw surges of current from the power supply. Even a very well regulated power supply cannot keep this switching from affecting the voltage near the switching chip. To smooth out the "spikes" in the voltage produced by this switching, every second or third TTL package should have a small ceramic disk capacitor (around 0.01 μF) attached to its 5-V and ground connections. This capacitor, called a **de-coupling capacitor,** helps to soak up the noise pulses in the power net near the switching chip.

If a digital circuit board is glitching intermittently with no visible cause, but the board doesn't have much decoupling, try "tack-soldering" a few decouplers to the power leads of some TTL chips. If certain signal lines are causing false switching due to noise, another thing that sometimes helps is adding a pull-up resistor between the line and Vcc. This doesn't reduce the noise, but it makes the 1-level higher. Increasing the separation between the 1-level and the 0-level may "pull the noise up" out of the "gray area" that causes false switching in gates down the line. MOS circuits do not need as much decoupling as TTL, because they draw far less current. MOS LSI devices on circuit boards often use TTL buffers and gates, and run from the same 5-V supply, so these boards should still be decoupled in proportion to the number of TTL devices they contain.

15-6.2 Gridded V_{cc} and Ground

Figure 15-11 shows a circuit board design that uses gridded V_{cc} and ground. This arrangement helps to reduce RF radiation and it cuts down

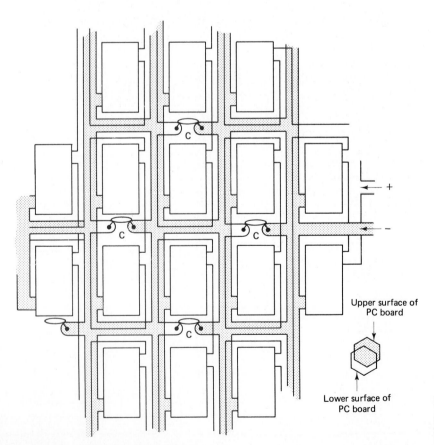

Upper surface of PC board

Lower surface of PC board

FIGURE 15-11 Gridded ground and V_{cc} on a memory board.

noise somewhat, too. You can see decoupling capacitors at some points in the circuit.

If a board without decoupling and gridded power is interfering with some circuit—like a TV—replacing the foil pattern on the board, or rewiring all power conductors, is out of the question. Some things that might help are:

1. *Decoupling.* This is mostly a way to suppress noise but it does reduce RF radiation also.

2. *Faraday shielding.* Enclose the board in some kind of box or cage made of an electrical conductor and connected to ground. This soaks up radiation before it can "get out of the box."

3. *Filtering.* Add a filter to the antenna or power lines of the device being interfered with that blocks the frequency of the interference from the digital system. If it's a TV, the filter also blocks out all the television channels, too bad.

15-6.3 MOS/LSI Handling Precautions for Static

As we mentioned in Chapter 5, MOS devices are especially sensitive to static damage when handled. A whole board full of MOS/LSI devices is just an accident waiting to happen for the inexperienced handler. Merely taking a printed-circuit board full of MOS devices out of its socket can "zap" the board! To safely handle MOS boards:

1. Work surfaces, such as tabletops, should be covered with a grounded, conductive surface. Special plastic or rubber materials are available for this, or a sheet of foil/sheet metal can be placed on the table top. Mats on the floor and static-reducing sprays are also sometimes used in the work area to reduce the buildup of hazardous charges.

2. The ground trace on the board, and a grounding strap around the handler's wrist, should be connected to the same ground as used in step 1 before the board is taken out of its socket.

3. Soldering iron tips can develop an induced voltage that is dangerous to MOS inputs. Ground the tip of the soldering iron before doing any soldering on the circuit board.

4. For shipping, MOS circuit boards should be wrapped in conductive plastic or wrapped in foil. This keeps all charges across the board equal. Static cannot damage MOS components unless voltages are different from one place on the board to another.

5. Unwrapping or unpackaging of MOS/LSI—wrapped as described in step 4—should be done only after you give the charges on your body and the board/conductive wrapper a chance to equalize. Holding the board and wrapper in your hand for 10 to 15 seconds before you unwrap the board should be enough.

In many designs, TTL buffers, drivers, and gates are attached to all MOS inputs and outputs—"fully buffered inputs and outputs"—on a circuit board. This provides a measure of static protection, since the TTL devices provide low-impedance discharge paths for static to ground. Fully buffered boards are still not static-proof, however, and should be handled with the same precautions as all-MOS boards, just for your peace of mind.

15-7 BUS-ORIENTED INPUT/OUTPUT

The I/O section of the computer is its link with the outside world. Although the computer uses its memory section to store and recover data and instructions, the data and instructions had to get into the computer from somewhere in the first place. The input devices are the source of information put into the computer. When the computer processes the data and comes up with a result of some kind, it's the output devices that are the destination of these results, used to display them in some way to the user.

An input device is one that puts data into the data bus, and an output device is one that takes data out of the data bus. The data on the bus is used by the CPU. You can picture data flow in the system like traffic on an expressway. For instance, suppose that a news report says "Outbound traffic is heavy on the expressway, but inbound traffic is light" and you're going from O'Connor's house to the downtown area. Does "outbound" mean outward from O'Connor's house or outward from downtown? Usually, news reports are based on what everybody in the city agrees is the most important part. O'Connor's house is not the most important part of the city to its citizens; the downtown district is. If you're going outbound from O'Connor's house and inbound toward downtown, the inbound lane (the one with the lighter traffic) is the one you'll use.

Similarly with the computer, the input and output are named according to the important "downtown" part of the computer system. The CPU (central processing unit) is the "important" part of the system; everything inbound or outbound goes into or out of the CPU. To decide what's an input or an output, you only have to ask yourself the question: Does it provide data that's for the CPU (input) or display data that's from the CPU (output)?

15-7.1 Decoding of Device Number

We've already described this. All we'll add here is that the job of the device decoder is to identify whether the number on the address bus is the device number of the I/O device, and whether the code on the control bus is for an input or an output operation. This can be done by a system of gates, a prefabricated decoder chip, or by the decoder that is a built-in part of a multiplexer.

15-7.2 Relevant Control Signals

We've discussed this before, too. The only things we need to recap here are that two forms of I/O exist, the memory-mapped I/O design and the true I/O.

In the memory-mapped scheme, it's how we design the hardware that determines what addesses are I/O devices and which are really memory. The control signals that distinguish input from output are the same ones that distinguish memory read from memory write. The memory read signal is used for input and the memory write signal for output when the memory happens to be an I/O device. The only way to know what's a memory operation and what's I/O is to know which addresses are mapped as memory locations and which are used for I/O.

In the true I/O scheme, all addresses for I/O devices are also legitimate addresses for memory locations. To tell them apart, I/O operations cause input (I/O read) and output (I/O write) signals to appear on the control bus while memory operations produce memory read and memory write signals. There's no chance of confusing memory operations with I/O operations because they produce completely different signals on the control bus.

15-7.3 Input Ports (Buffers)

We're introducing a new vocabulary term here, the **input port,** which has a different meaning from input device. An input device might be something like a typewriter keyboard equipped so that it generates a 8-bit ASCII code whenever a key is pressed. The input device creates the data in a digital form. An input port, on the other hand, is the device that interfaces (communicates) the data to the data bus. There may be many kinds of input devices converting data from real-world signals into digital code, but once converted, the code is brought into the data bus through the same kind of input port in every case.

Figure 15-12 shows a system with four in-

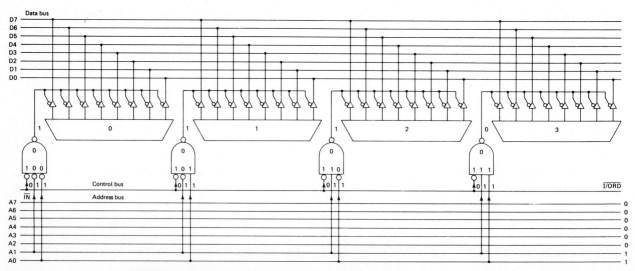

FIGURE 15-12 Selection of input devices using buffers as input ports.

put devices and four input ports connecting them to the data bus. All four input devices have a data word for the bus, but the four data words of information can't be put on the bus at the same time. Instead, the devices must take turns so that one device "talks" to the CPU at a time. The CPU controls this. It is the CPU that originates the addresses and control signals used in this diagram. A program in the CPU that contains input instructions is responsible for enabling each input port, letting the port "drop" its data word onto the data bus. In this diagram, we imagine that the instruction "input a data word from port 3" is being carried out by the CPU at this moment. The results are fourfold. The CPU:

1. Originates an address (00000011).
2. Puts out an active (LOW) "IN" signal on the control bus.
3. The port 3 decoder is a gate with the logic: "If IN is LOW and address lines (1) and (0) are HIGH, with all other address lines LOW, make the output of this gate LOW.
4. This gate enables the Tri-State buffers between the input device and the data bus and these buffers place a data word on the bus.

The other decoders only enable their port buffers for other conditions. They are not enabled right now, and their floating condition does not interfere with the HIGHs and LOWs put on the bus by port 3.

Every input port in this diagram is a Tri-State buffer enabled by a DSP from a device decoder. It doesn't matter if the data word for the CPU is generated by a keyboard, a punch tape reader, or a cassette player: once the data word is in digital code, a buffer must stand between the word and the bus. We already discussed the mutually exclusive nature of each input—and we can see that the decoders in this diagram work according to this rule; only one port at a time is ever enabled.

15-7.4 Output Ports (Latches)

An **output port** is the interface between the data bus and an output device. This means, as you might guess, that an output port enables data to pass to a particular output device from the data bus. When an output operation is taking place in the computer, data are placed on the data bus by the CPU for pickup by an output device. The task

of the output device is to display the digital code in some form that's useful in the outside world.

You might be led to believe that an output port is just an input port turned around. That's not so. A buffer can't be used for an output port for a very important reason. The display devices normally attached to computers (printers, numeric displays, video monitors) work much more slowly than the computer itself. An output command in a program might put a number on the data bus for a millionth of a second or so before the computer moves on to the next operation. This is not enough time for a printer to print a letter, or a person looking at a numeric readout or video monitor to read the number displayed there (even with help from Evelyn Wood!). The number sent to the output device must stay a lot longer than a millionth of a second to get anything done.

Do we slow down the computer to the speed of the output devices? This doesn't work, because even if we did slow the computer down, only one data word can be on the data bus at a time, and a system with hundreds of output devices would display a number at each device for a second or so once every few minutes. This is not a good way to work with computers. What we need instead is an output port that will remember the word sent to it while the computer goes ahead to other tasks. Memory in the output port would let it keep applying the data word to its output device until it gets its job done, even as the data bus carries other numbers to other ports.

The ideal device to use in this case is a latch, shown in Figure 15-13. Each of the four output ports in the picture is an 8-bit latch strobed by a clock pulse. In the diagram, we imagine that the computer is doing an "output a data word to port 1" operation. The port 1 decoder has the job of recognizing when a 00000001 appears on the address bus at the same time as the OUT signal is active. When the decoder sees a 1, it sends a LOW signal to strobe the latch. The latch then picks up the word on the data bus, and keeps it active at its outputs until another "output to port 1" strobes it again at some later time.

Since a latch has a memory capability, it will keep the data word at its Q outputs until a new data word is sent to the same port. The outputs of the latch can keep pushing the word at the output device while the computer proceeds to use the data bus for other tasks. The word in the port will not be affected by these changes on the data bus because the strobe (clock pulse) is not active.

As before, it's the job of the device decoder to make sure that this port gets data for only one

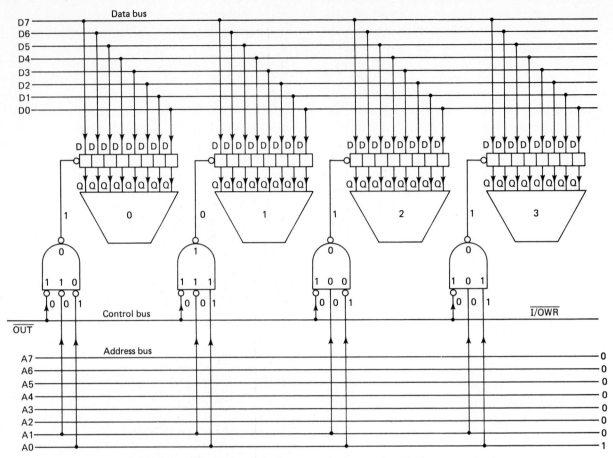

FIGURE 15-13 Selection of output devices using latches as output ports. (See Appendix 4 for 74LS373 octal latch datasheet.)

set of conditions, and all the other device decoders should be mutually exclusive and strobe their ports for different conditions.

Unlike the input situation, however, there could be more than one output device/output port combination with the same code—if multiple displays of the same data at different locations were desired. Since the output port doesn't place any data on the common bus, there's no way two different sets of bits would appear on the same conductor to "fight" with one another. It's only input devices that must be enabled one at a time only— but output devices usually are organized along the same lines, since the multiple-display arrangement mentioned above is rarely needed.

15-7.5 An Input Port for 8-Bit Data

The input ports seen in Figure 15-12 are suitable for interfacing input devices to an 8-bit microcomputer (the most common word size for micros). It's typical of the interfacing scheme used in computers like the Altair 8800 or the Radio Shack TRS-80. Another circuit that does the same job with multiplexers is shown in Figure 15-14. It's like the scheme used in the Apple II, although the Apple uses a memory-mapped arrangement instead of having a separate I/O read signal generated by its CPU.

In Figure 15-14, four 74LS253 dual 1-of-4 MUX circuits are used to select which data word will reach the data bus. The select inputs A and B are attached to address lines (1) and (0). By themselves, A and B can only "see" what address lines (1) and (0) are. This is fine for decoding device codes 00, 01, 10, or 11 (0, 1, 2, or 3), but cannot tell 00000111 (7) or 00001111 (15) from 00000011 (3) if only two of the eight address lines are connected. The remaining six address lines and "IN" control line are attached to an OR gate with seven inputs. If all seven inputs are LOW, the output of the OR is LOW. Since it's used as the enable for the MUXes, the only time the MUX circuits can work correctly is when the addresses are really 0, 1, 2, or 3, and the device is an input.

In this example, each multiplexer is tri-state logic with the capacity to float when dis-

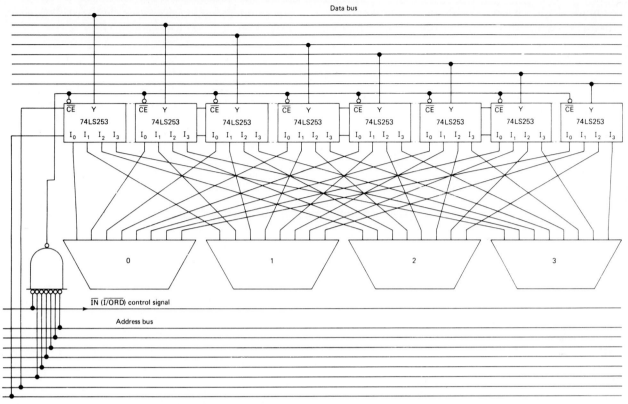

Data bus

FIGURE 15-14 Selection of input devices using multiplexers.

abled. We can assume other multiplexers are available on the bus for other combinations of address and I/O signals.

15-7.6   An Output Port for 8-Bit Data

The output ports in Figure 15-12 are suitable for any output device that converts digital code into a display or some sort of action. The four 8-bit ports shown are operated as a true I/O circuit. Data is sent to these ports using "OUT" output instructions to port 0, 1, 2, or 3.

The same four ports are operated in memory-mapped I/O systems using instructions that write in memory location 0, 1, 2, or 3. To make Figure 15-12 into a memory-mapped circuit, change the control bus signal in the diagram from "out" to "memory write."

EXAMPLE 15-2

Suppose that a certain microprocessor has an 8-bit data bus, an address bus of which only the first 8 bits, A_0 through A_7, are used for selecting I/O devices, and a pair of control bus signals, I/O Request and Write, which both become active LOW when an output operation is being done.

Design a circuit that will allow this microprocessor to select any one of four output ports (and thus any of four output devices). Each output device (they do not have to be shown in detail in the diagram) will use an 8-bit number latched from the data bus by its respective **output port.**

Solution: A decoder is needed which will have a different active output for each of four addresses, and which will be enabled only when the I/O Request and Write signals are both active at the same time. The 74138 decoder (shown on page 278) can be used for this. It actually has eight outputs, so we won't be utilizing it fully. One of its three inputs (C) will be wired to ground, so that only outputs Y_0 through Y_3 will ever be activated. The 74138 also has three chip-enable lines; one is active HIGH, and the other two are active LOW. Again, we won't be needing all of them, so the active-HIGH one will be wired to V_{cc} (enabled all the time), so that the remaining two—active LOW—lines will control whether the chip is enabled. (All three chip-enables have to be in their active states for the chip to deliver an output.)

Only two of the eight address lines are needed to select one of four output devices; the rest will be unconnected. This makes it possible to select these four output devices, but no others.

A more sophisticated design will be needed if there are more than four output devices in the entire microcomputer system, but for this case we will assume there are no others.

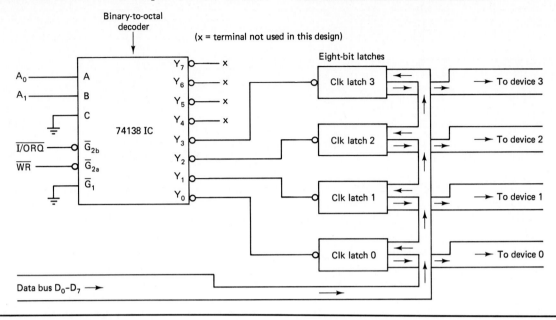

15-7.7 Parallel I/O Port ICs

The I/O ports shown in the preceding sections of this chapter are **parallel I/O ports.** The data they input and output is transferred 8 bits at a time at eight different points, so they are **parallel data.** To make the input port, for example, we need an 8-bit buffer and a decoder to enable it. An output port needs an 8-bit latch and a decoder to strobe it.

The Intel 8212.   One single-package device that has parts of both an input and an output port is the Intel 8212 buffer/latch. It contains eight tri-state output flip-flops, strobed by a gate network. Two modes of operation permit use of the 8212 as an input or output device. Two inputs called device-select permit strobing of the latch or enabling of the tri-state buffers with either active-HIGH or active-LOW logic. A strobe input strobes the latch in input mode, while the buffers are controlled separately. A reset input is also present for clearing the bits in all latch flip-flops. (The 74LS373 is a similar TTL product but without the reset.)

Since this is a TRANSPARENT latch, the flip-flops of the latch can be "clocked" so that data simply pass through the latch and are then stopped or passed by each flip-flop's tri-state output. In this mode, the 8212 can be used as an 8-bit buffer in an input port. If the latch clock is "strobed"—turned ON and OFF again after a short time—the latch will store the bits at its in-

puts during the ON time. In this mode, the 8212 can be used as an output port. In addition to these functions, the 8212 also contains an input and output for use with interrupts, which we'll discuss in a subsequent chapter.

One-chip parallel port devices.   The manufacturers of most microprocessors support their customers by providing **PIA** (parallel interface adapter) chips, also called **parallel port** chips. These devices are designed to work hand-in-hand with the microprocessor to provide interface links to input and output devices. They simplify the work of the hardware designer, who does not have to design the device decoder and port circuits. For the software designer, however, this design approach means that extra steps must be added to the programming. Instructions must be added to programs to "set up" the PIA device, so that it is in the **configuration** needed to operate the input and output devices attached to the ports. This software "overhead" is called **initializing the PIA.** Once a PIA device has been built into the microprocessor system, its parts occupy several I/O port locations in the computer system. We'll begin by looking at the function performed by some of these "devices within a device"; then we will look at some specific PIA devices that are available.

Control registers: The **control register** is sent a binary pattern of 1s and 0s at the beginning of initialization. This is called a **control word.** Some PIAs will have several control registers used for different purposes. These registers can be reached by doing *output* operations. For instance, a computer with four control words may need to have outputs done to output devices 0, 1, 2, and 3 before the PIA can actually be used.

Data direction: Since most PIA chips contain more than one 8-bit port, one function of the control word(s) is to identify which ports will be used for input and which will be used for output. Sometimes, the individual bits of each 8-bit port can be separately identified as to whether they are inputs or outputs. In these cases there will be a separate direction-control register for each port. The Motorola 6821 PIA, for instance, contains two 8-bit ports called A and B, and has two control registers called DDRA and DDRB (**data direction register** A and B). In DDRA, each bit is used to identify whether the corresponding bit in port A is an Input (if the bit is a '0') or Output (if the bit is a '1'). For port B, DDRB performs the identical function.

Another function of the control register(s) is to control each port's response to **interrupt** signals. Interrupts are described further in Chapters 21 and 22, but all you need to understand here is that an interrupt is a signal generated by an external I/O device which wants the computer's attention. By sending certain '1' or '0' levels to the control register (in a control word), we choose which interrupt signals will get the computer's attention, and in some cases, what type of signal (low, high, falling edge, rising edge, etc.) will work as an interrupt.

Additional functions are included in some PIA chips and not in others. Since they are not standard, we will not try to describe them here but will cover them as they come up in the discussion of some specific PIA devices in the following paragraphs.

The Motorola 6821 PIA. This is the granddaddy of all PIA devices. It was developed as a support chip for the Motorola 6800 microprocessor, Motorola's first 8-bit microprocessor. Although it is one of the earliest of I/O port chips,

its operation is quite complicated. It was designed at a time when no one was sure what internal functions you might want to control, so its designers made almost *everything* controllable. Many who have worked with it feel that too much flexibility was built in. As a result, you have to do a lot of setup before you can get a system going that contains 6821 PIAs.

Setup: Begin by identifying where the microcomputer system has located the PIA. Observe how the address bus is used to enable the RS0, RS1, E, CS_1, $\overline{CS_2}$, and CS_0 inputs of the 6821 IC (see Figure 15-15). Since Motorola microprocessors used with the 6821 usually have memory-mapped I/O, this means that the PIA will occupy some addresses in memory. Although its ports and control registers occupy locations at memory addresses, they are not memory devices and will have to be handled specially before the PIA can be used. Here's an example.

Let's suppose that we have a system with PIA, located at memory address 8000. Our PIA actually occupies four addresses in the memory. Its RS1 and RS0 inputs are wired to the two least-significant bits (A_1 and A_0) of the address bus, so we can use addresses 8000, 8001, 8002, and 8003 to reach it.

Next, find the registers inside the PIA. There are two peripheral registers (ORA and ORB), two data direction registers (DDRA and DDRB), and two control registers (CRA and CRB). If you're counting, you've probably just noticed that the four addresses, 8000, 8001, 8002, and 8003, are not enough for these six registers.

In the example we used above, 8001 and 8003 are the control registers, CRA and CRB. If these registers are sent a control word, XXXXX0XX, where bit 2 (the 0) is LOW, then addresses 8000 and 8002 are the data direction registers, DDRA and DDRB. But if the control word sent to 8001 and 8003 (CRA and CRB) is XXXXX1XX, where bit 2 (the 1) is HIGH, then addresses 8000 and 8002 are the peripheral registers (ORA and ORB). Whoever designed this must have thought, at the time, that it was a very

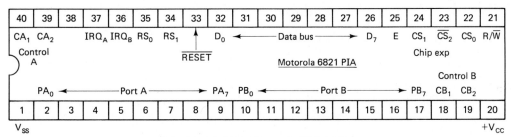

FIGURE 15-15 Motorola 6821 PIA.

clever way to cut down the number of addresses used up by the PIA.

Suppose that you want to set up the PIA at address 8000, so that its port A is an 8-bit input port and its port B is an 8-bit output port. Here's what you have to do:

Load the control registers. Send a number to addresses 8001 and 8003 that contains a 0 in bit 2. You might as well use 00000000 here, since you aren't going to keep this number in the control registers, and won't do anything else with it here except load the data direction registers.

Load the data direction registers. Send 00000000 to address 8000 (DDRA) and 11111111 to address 8002 (DDRB). This will make all 8 bits of port A inputs and all 8 bits of port B outputs. Since each bit is individually controllable, you could set half the bits of port A as inputs, and the other half as outputs, or whatever. Usually, all eight are the same.

Load the control registers again. Send a number, XXXXX1XX, to addresses 8001 and 8003 that contains a 1 in bit 2. The rest of the bits control the interrupt status of each port, and we're not going to concern ourselves with them at this time.

The ports are now configured, and addresses 8000 and 8002 are peripheral interface registers ORA and ORB. If we want to read an input from port A, we can go to address 8000 and see what's there. To write an output to port B, send it to address 8002.

Of course, we don't know when it is time to read port A, because we don't know when the input is going to change. That is what the interrupt inputs (CA1/2 and CB1/2) are for on the PIA chip. If a signal (an **interrupt request**) is generated by the input device attached to port A every time a new input appears, this signal can be used to interrupt the processor, and have it run software that will read port A's input.

Now, let's look at all those other X's in the control word that we gave the control register. We know what bit 2 is for, but what about the rest?

The control word for port A is shown in Figure 15-16. The rest of the bits in CRA (control register A) are all for interrupts. The ↑ symbol indicates the rising edge of a signal, and the ↓ symbol, the falling edge. Bits D_7 and D_6 are set by interrupt requests to the 6821 through input lines CA_1 and CA_2, and the items with a star (*) indicate when an interrupt signal to the 6821 will be passed on to the CPU interrupt request input.

The control word for port B is shown in Figure 15-17; it is identical to the word for CRA (control register A), except that all its functions apply to port B, of course. This will all make more sense when we get to interrupts in a later chapter.

The Intel 8155 RAM/IO/Timer IC. In the Intel microprocessor family, support chips have been designed to operate with the multiplexed address/data bus of the 8085 processor. The 8085-family equivalent of the 6821 is the 8155, which is a PIA, plus a programmable timer module and a RAM memory block in one integrated circuit. We could assemble a system similar to the 6800 system described previously, but Intel's processors allow I/O to be addressed separately from memory, and don't require memory mapping. We enable the 8155's pin 8 (see Figure 15-18) for the I/O port addresses we want. Then we send the control words to set up the ports—for instance: In the 6821 example just completed, we located the PIA at address 8000. Let's suppose that we have a system with the 8155 located at I/O device 08. Our PIA actually occupies six locations: I/O ports 08, 09, 0A, 0B, 0C, and 0D. These are the locations of the control register (08), ports A, B, and C (09, 0A, and 0B), and the timer count and timer-control word (0C and 0D).

6821 Control Word
Control Register A

Interrupt status (active high)		Trigger of CA$_2$ Interrupt input			CPU IRQ*	DDRA access	Trigger of CA$_1$ Interrupt input		
D_7	D_6	D_5	D_4	D_3	* = Active	D_2	D_1	D_0	IRQ*
0	0 = None	0	0	0 = ↓ trigger		0 = DDRA	0	0 = ↓	
0	1 = IRQA$_1$	0	0	1 = ↓*trigger		1 = ORA	0	1 = ↓*	
1	0 = IRQA$_2$	0	1	0 = ↑ trigger			1	0 = ↑	
1	1 = Both	0	1	1 = ↑*trigger			1	1 = ↑*	

Control, CA$_2$ as output

1	0	0 = Write to ORA resets; set if D_7 = 1
1	0	1 = Write to ORA resets; E pulse sets
1	1	0 = D_3 = 0 resets; D_3 = 1 sets
1	1	1 = D_3 = 0 resets; D_3 = 1 sets

FIGURE 15-16 Motorola 6821 Control Word (Port A).

6821 Control Word
Control Register B

Interrupt status (active high)		Trigger of CB_2 Interrupt input			CPU IRQ*		DDRB access	Trigger of CB_1 Interrupt input		
D_7	D_6	D_5	D_4	D_3	* = Active		D_2	D_1	D_0	IRQ*
0	0 = None	0	0	0 = ↓ trigger			0 = DDRB	0	0 = ↓	
0	1 = $IRQB_1$	0	0	1 = ↓*trigger			1 = ORB	0	1 = ↓*	
1	0 = $IRQB_2$	0	1	0 = ↑ trigger				1	0 = ↑	
1	1 = Both	0	1	1 = ↑*trigger				1	1 = ↑*	

Control, CB_2 as output

1	0	0	= Write to ORB resets; set if D_7 = 1
1	0	1	= Write to ORB resets; E pulse sets
1	1	0	= D_3 = 0 resets; D_3 = 1 sets
1	1	1	= D_3 = 0 resets; D_3 = 1 sets

FIGURE 15-17 Motorola 6821 Control Word (Part B).

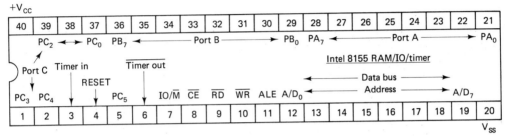

FIGURE 15-18 Intel 8155 RAM/IO/timer module. (See Appendix 4, 8155 datasheet.)

The Control register: To use the 8155, its control register (I/O port 08) must be loaded with an 8-bit **control word** whose bits specify the condition of the timer and I/O ports. In this case we have timer, interrupt, and data direction information. To configure I/O port A and port B, the bits that control their data direction are the least-significant 4 bits (see Figure 15-19). For example, if we send port 08 a hexadecimal 00, the timer will be stopped, and all I/O ports will be input ports with interrupts disabled.

Suppose that you want to set up the PIA at address 08, so that its port A is an 8-bit input port and its port B is an 8-bit output port. Here's what you have to do: Send port 08 a hexadecimal 20: the timer will be stopped, and the PIA configuration will be port A = Input and port B = Output. That's all you have to do with the 8155! If we want to read an input from port A, we can go to I/O device 09 and see what's there. To write an output to port B, send it to I/O device 0A.

Of course, you can't control the individual bits' direction. There is only one bit for each port's direction, instead of an 8-bit data direction register (2 bits for register C), but 8-bit ports generally use all 8 bits in the same direction, anyway. This clearly simplifies the software needed to initialize the PIA. The D_4 and D_5 control word bits control the interrupt status of each port, and are again simpler than the 6821 Motorola PIA, but we're not going to concern ourselves with them at this time. The D_6 and D_7 bits control the timer, about which more will be said presently.

The Timer count: A 14-bit number sent to I/O ports 0C and 0D is down-counted toward zero when the timer is started. At the end of the count, suppose that we want to send a single trig-

8155 Control Word

Timer mode		Interrupt status (Port A) (Port B)		Port C mode		Port B direction	Port A direction
D_7	D_6	D_5	D_4	D_3	D_2	D_1	D_0
0	0 = NOP	0 Disable 0		0	0 = Input	0 = Input	0 = Input
0	1 = Stop	1 Enable 1		0	1 = Mode 3	1 = Output	1 = Output
1	0 = Stop (at end of count)			1	0 = Mode 4		
1	1 = Start timer			1	1 = Output		

FIGURE 15-19 Intel 8155 Control Word

ger pulse out at the TIMER OUTPUT. Here's how we would do that. The 15th and 16th bits of the two-byte **timer count** are actually a mode selection used to decide what type of waveform the counter is sending out. The various options are shown below.

		Port 0D (MSByte)	Port 0C (LSByte)
M_2	M_1	C_{13} C_{12} C_{11} C_{10} C_9 C_8	C_7 C_6 C_5 C_4 C_3 C_2 C_1 C_0
0	0	= single square wave	(Bits C_{13} through C_0 are the)
0	1	= continuous square wave	(count that's down-counted)
1	0	= single pulse	(to zero to time-out a pulse)
1	1	= continuous pulse	

From this we can see that the mode we want (a single trigger pulse at the end of the time-out) is **mode 2** ($M_2 = 1$; $M_1 = 0$). Before we can trigger the timer, we must load it with a number that contains bits 10 in the most-significant bits of port 0D. Then we have to send a new control word to the control register (08) to turn on the timer. Changing 00000010 to 10000010 would do this without changing ports A, B, and C.

The Intel 8255. The Intel 8255 Programmable Peripheral Interface (PPI) is similar in nature to the 8155, but without the multiplexed A/D bus needed for the 8085 processor. As with the 6821, it contains two bidirectional, software-definable registers, but also contains an additional "split" register permitting two 4-bit interfaces.

All of the devices described in this section are designed to run from a single 5-V TTL-type supply, and generate TTL-compatible signals suitable for driving TTL or TTL/Schottky devices. The 8212 is itself a TTL/Schottky device; the others are MOS/LSI.

15-7.8 I/O Interface Adapters (Serial I/O Ports, UARTs, Intel 8251, Motorola 6850)

In Chapter 11 we discussed a use of shift registers and parallel registers, the serial-to-parallel and parallel-to-serial converter, called a UART (universal asynchronous receiver/transmitter). In that chapter we mentioned how a transmitter (parallel-to-serial) and a receiver (serial-to-parallel) could be found in single-package UART devices that could be made to do either job by control inputs.

Two such devices are the Intel 8251 USART (called a PCI—a Programmable Communication Interface—by Intel) and the Motorola 6850 UART (called an ACIA—an Asynchronous Communications Interface Adapter—by Motorola). They are very similar in structure and control, although the 6850 is somewhat simpler, having only an asynchronous mode of operation, whereas the 8251 has both synchronous and asynchronous modes of operation.

One job of both circuits is to take parallel data from the data bus and transmit it serially (bit by bit on one line). In this mode, the UART is being used as an output port with 8-bit SDM (space-division-multiplexed) data converted to TDM (time-division-multiplexed) data. The output device most commonly connected to this type of port is a **modem** (modulator/demodulator), which is used as a modulator when it's doing output things.

The other job of both circuits is to receive serial data arriving from another location and convert them to parallel data for the data bus. In this mode, the UART is being used as an input port with 8-bit TDM data converted to SDM. The input device that provides the serial bit stream to the UART in this mode is usually a modem, operated as a demodulator.

These LSI packages contain all the circuitry necessary to do both of these jobs simultaneously (full duplex) through separate paths, and also contain check circuitry for detection of transmission errors through parity detection and other tests. Circuits inside the UART also respond to signals that tell whether the circuit at the other end of the link is ready to receive/transmit, and generate signals that say whether *this* circuit is ready to transmit/receive. This sort of cross-communication, called "handshaking," is discussed in detail in a subsequent chapter.

EXAMPLE 15-3

In Example 15-1 we designed a 256K × 8 memory using 64K × 8 SRAM chips (schematic on page 283). Suppose that block 3 won't write—in other words, we go through the motions of writing into addresses in block 3 between 196K and 256K − 1—but all that comes back when we read block 3 is FF, hexadecimal, at all locations in the

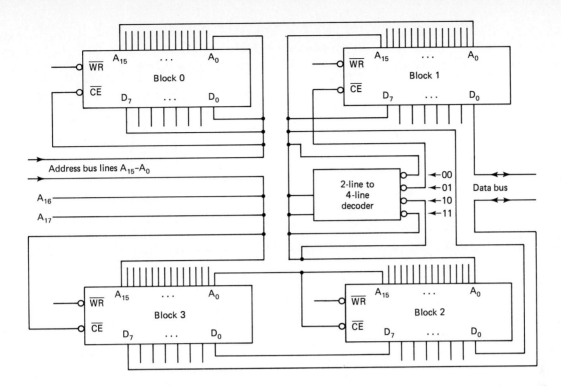

block. This indicates that all bits are floating HIGH or tri-stated. What is probably wrong?

Solution: Block 3 is supposed to be enabled by decoder output 11, and in the floating-HIGH or Tri-Stated condition otherwise. We suspect that the signal from the decoder is not being delivered to the chip-enable of block 3. To test this, we attach a signal injector to the 11 output pin of the decoder IC, that sends pulses to block 3 repeatedly, then attach an oscilloscope probe to the CE input of the block 3 RAM chip, looking for a recurrent waveform consisting of active-LOW chip-enable pulses. If no waveform is present on the CE pin of block 3's RAM chip, there is a break or open in the line from the decoder to block 3. If there is a pulse and the data in the RAM appear to change, perhaps the decoder is not working. Backtracking from the outputs toward the inputs, we check if applying code 11 to the decoder's inputs results in output 11 becoming active LOW. If the other outputs become active for inputs 00, 01, and 10, but output 11 doesn't become active for input 11, the output of the decoder chip is damaged, and the chip should be replaced. If signals are present all the way from the inputs of the decoder to the RAM chip's CE pin, but it doesn't respond, then the RAM chip input may be damaged, and it will be necessary to replace RAM-3.

If the break isn't obvious upon visual inspection, place an ohmmeter probe tip on the 11 output pin of the decoder, and start to slide the other probe tip along the trace traveling to block 3's CE pin. At some point, the resistance should jump from near-zero to high (open) resistance. The most common place for this open circuit is where the pin of the IC chip is soldered into the circuit board. This can generally be repaired by resoldering the connection. A break in midtrace (in the copper foil of the circuit board) is easily fixed by soldering a small piece of wire across the break.

EXAMPLE 15-4

In Example 15-2, we designed a four-output-port interface circuit. Suppose that in the design (shown on page 284), we find that port 2 won't deliver data to device 2—device 0 gets the data instead—and port 3 won't deliver data to device 3—device 1 gets the data instead. What is probably wrong?

Solution: The decoder is strobing the wrong latches. In binary, we would say that latch 01 is being strobed instead of 11, and 00 instead of 10. That indicates that the problem is in a specific binary bit. Input B of the decoder would have this effect if it were held LOW. Ports 0 and 1 (00 and 01) would be addressed properly, but ports 2 and 3 (10 and 11) with B grounded, would become ports 0 and 1 (00 and 01). This is exactly what happens, so next we have to take some measurements to confirm our suspicions. First, a fact: On the 74138 chip, input C is right next to input B

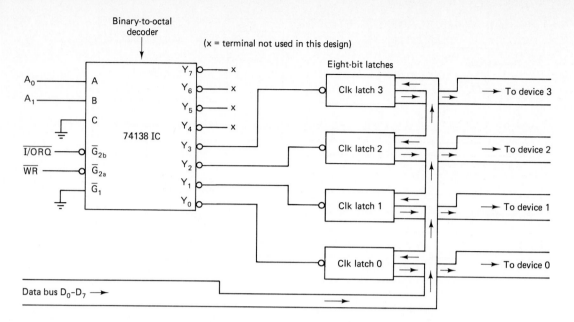

Binary-to-octal decoder

(x = terminal not used in this design)

Eight-bit latches

(adjacent pins). We have tied input C to ground in this design—could input B be shorted to its next-door neighbor, input C? We disconnect power from the circuit and measure the resistance between B and C. If there is a short circuit between the two adjacent pins on the IC, it can probably be corrected by resoldering the connections, or by using a knife to cut through the solder between the two pins, if that is the reason they are shorted together.

QUESTIONS

15-1. In Figure 15-1, which part is more like a data bus, the segment drivers or the digit drivers?

15-2. Are the displays in Figure 15-1 common cathode or common anode? (Try to determine which lines are active HIGH and which are active LOW.)

15-3. Describe how a multiplexed display with 10 digits is used to show 10 numbers.

15-4. Figures 15-2 and 15-10 show two ways to multiplex data onto a data bus. The multiplexed display receives time-division-multiplexed (TDM) information in both cases. (True or false?)

15-5. In Figure 15-2, the A and B outputs from the MOD-4 counter constitute the address bus. The A, B, C, and D inputs to the seven-segment decoder (and its a, b, c, d, e, f, and g outputs) are the data bus. What is the control bus signal needed to synchronize the transfer of data and addresses? (*Hint:* It is the signal that makes the counter advance from one count to the next.)

15-6. Which bus of a computer makes it possible to transfer inputs (like 5, +, 2, and =) from the input device to the memory?

15-7. Describe how the control bus signals I/OWR, I/ORD, MEMWR, and MEMRD control transfer of data to I/O or memory devices in a bus-oriented computer.

15-8. What is the difference between true I/O and memory-mapped I/O?

15-9. Describe how a device decoder makes it possible to make a 64K RAM out of 16K RAM chips.

15-10. Describe the difference between memory devices with common I/O and separate I/O (see Figures 15-7 and 15-8).

15-11. Draw a schematic showing how you would convert a TTL 7402 NOR logic gate to Tri-State logic, and O. C. logic.

15-12. Why is Tri-State logic preferable to open-collector logic?

15-13. Figure 15-10 is an example of time-division-multiplexing, even though there isn't a multiplexer anywhere in the picture. (True or False?)

15-14. Describe what "decoupling" is and why it's needed.

15-15. Since most bus-oriented systems contain one or more MOS devices, handling precautions for MOS circuit boards were included in this chapter. What problem does MOS logic have that makes it necessary to take these precautions?

15-16. Describe a way to determine what parts of a computer system do inputs and what parts do outputs.

15-17. Describe the difference between an input port and input device.

15-18. Describe the difference between an output port and an output device.

15-19. Draw the schematic of an input port that will enable input device 13 to put its data on the data bus (use a four-line address bus to the device decoder).

15-20. Draw the schematic of an output port that will latch data for delivery to output device 10 (use a four-line address bus in this schematic).

15-21. Why are latches used in output port design? Why isn't an output port just an input port "turned around"?

15-22. Input devices must be enabled onto the data bus through their input ports one at a time only. Does this rule apply to output devices as well?

15-23. The Motorola 6821 PIA is useful where rapid redefinition of port structure is required. Describe two situations where this might be necessary.

15-24. Describe, briefly, what "serial I/O" means.

DRILL PROBLEMS

Questions 15-1 through 15-10 refer to the following block diagram (whose function is described in Section 15-1):

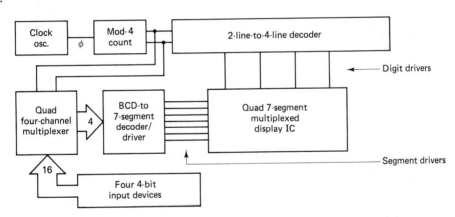

15-1. The address bus in the system above contains _____ (1, 2, 4, 8) bits.

15-2. The data bus in the system above contains _____ (1, 2, 4, 8) bits.

15-3. The segment drivers receive decoded _____. (address, data)

15-4. The digit drivers receive decoded _____. (address, data)

15-5. The displays can show _____ (1, 2, 3, 4) digits at any moment.

15-6. When operated at clock speeds of 100 Hz or more, the displays *appear* to show _____ (1, 2, 3, 4) digits at a time.

15-7. In a computer system, the clock signal, ϕ, is normally considered as part of the _____ (address, data, control) bus.

15-8. The system above is made of circuits and wiring, which are its _____. (hardware, software)

15-9. The system above holds and transfers binary 1s and 0s, which are its _____. (hardware, software)

15-10. If the data bus in the system above contains code for the number 5, and the address bus contains code for the number 3, the segments for a _____ (3, 5) will light up in digit _____. (3, 5)

Questions 15-11 through 15-20 refer to the block
diagram of the following bus-oriented computer:

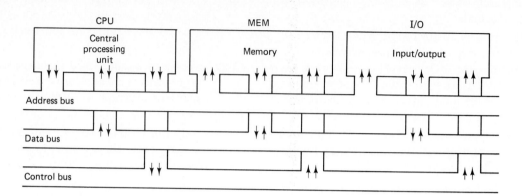

15-11. The _____ block does the actual computing.

15-12. The _____ block stores numbers and instructions.

15-13. The _____ block contains devices like printers, video screens, keyboards, and tape recorders.

If this computer carries out the computation 5 + 2 = 7, by fetching the 5 and 2 from memory locations 1020 and 1021, and delivering the 7 to output device 1, which buses carry what?

15-14. The _____ bus would carry the numbers 5, 2, and 7.

15-15. The _____ bus would carry the numbers 1020, 1021, and 1.

15-16. The _____ bus would carry signals that activate the memory to read out the 5 and 2 from locations 1020 and 1021, and the I/O system to deliver 7 to output device 1.

15-17. The _____ bus would fetch from memory the instructions that make the CPU add (+) the numbers together.

15-18. The _____ bus is bidirectional, using the same lines alternately to carry binary codes into and out of the CPU and other blocks.

Name the block:

15-19. Input data travel into the _____ from the data bus.

15-20. Output data travel out of the _____ from the data bus.

Questions 15-21 through 15-25 refer to the semiconductor RAM depicted in the following pin diagram:

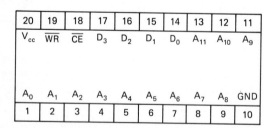

15-21. What is the total number of memory locations in the RAM? _____

15-22. To make a 16K × 8 RAM with these chips, _____ (1, 2, 4, etc.) chips would have to be enabled together in each block.

15-23. To make a 16K × 8 RAM with these chips, _____ (1, 2, 4, etc.) blocks of these chips would be enabled by the block decoder.

15-24. The "extra" address lines for the 16K × 8 memory described would have to be decoded by a _____ -output decoder.

15-25. The outputs of the decoder used to enable the memory blocks should be active _____ . (LOW, HIGH)

Questions 15-26 through 15-30 refer to the semiconductor RAM depicted in the following pin diagram:

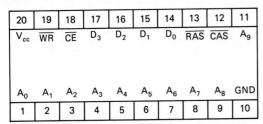

20	19	18	17	16	15	14	13	12	11
V_{cc}	$\overline{WR}$	$\overline{CE}$	D_3	D_2	D_1	D_0	$\overline{RAS}$	$\overline{CAS}$	A_9

A_0	A_1	A_2	A_3	A_4	A_5	A_6	A_7	A_8	GND
1	2	3	4	5	6	7	8	9	10

15-26. What is the total number of memory locations in the RAM? _____

15-27. How many bits are there in each memory location? _____

15-28. To make a 512K × 8 RAM with these chips, _____ (1, 2, 4, etc.) chips would have to be enabled together in each block.

15-29. To make a 512K × 8 RAM with these chips, _____ (1, 2, 4, etc.) blocks of these chips would be enabled by the block decoder.

15-30. The "extra" address lines for the 512K × 8 memory described would have to be decoded by a _____ -output decoder.

16

PERIPHERALS: I/O DEVICES (INPUT)

By the time you finish this chapter, you will be able to:

1. Identify the basic function of *input devices* attached as peripheral devices to a computer via its input ports.

2. Identify basic operating principles of *keyboards* and *keypads*, and see the two basic ways in which their input is encoded for use in the computer systems (hardware encoding and software encoding).

3. Understand the principles of *keyboard scanning*.

4. Describe the basic types of *magnetic storage devices*, including *magnetic tape*, *magnetic disk* (hard disk and floppy disk), *magneto-optical disk*, and *bubble memory*.

5. Explain how information is stored on machine-readable documents like *Hollerith cards*, and how this information is read into a computer via a *card reader*.

6. Explain how information is stored and retrieved on *punched paper tape*, and the advantages and disadvantages of this type of machine-readable document compared to punchcards.

7. Explain how *OCR characters* and *barcode* (Universal Product Code) are optically read into computer systems, and their advantages over other machine-readable documents mentioned in this chapter.

8. Describe some basic types of graphics input devices, such as:
   a. Paddle or game controller
   b. Joystick
   c. Trackball or mouse
   d. Light pen
   e. Tablet
   f. Video digitizer

9. Describe how a *transducer* combined with an *A/D converter* can constitute an input device which is an automatic measuring instrument.

10. Describe what a DMA (direct memory access) device is and how it "streamlines" the transfer of input into a computer system.

11. Describe how the system of item 8 can be applied to voice input for a computer.

In Chapter 15 we saw the layout, in broad terms, of a bus-oriented digital computer. We divided our computer block diagram into three main divisions: the CPU, memory, and I/O (input/output). Since memory devices were discussed in an earlier chapter, the memory part of Chapter 15 already had a foundation to build on when we described how the memory was interfaced to the buses of the computer system. The I/O section, however, had no background, and we simply dealt with the I/O ports that connected these devices to the computer. From what you read in Chapter 15, you could find out very little about the devices themselves.

In the next two chapters, we plan to go outward from the buses and ports in the computer system to the peripheral devices. The word *peripheral* means "around the outside". In a data processing center, the processor is usually a box in the middle of the room with other boxes arranged around it, connected to it by cables, but outside the housing of the CPU itself. These boxes are the computer's eyes and ears, and also its muscles and power of speech. We've already discussed what input and output mean in terms of the computer system. This chapter will deal with input devices.

The task of an input device is to convert an "outside" signal—which is not digital and may not even be electrical—into a binary digital word. An input port will place this data word on the data bus if its number comes up on the address and control buses.

We have already seen the details of how input ports accomplish their job. Now we will see how the input devices create the data words that pass through the ports when the computer calls for them.

Three categories of input devices will be discussed: manual switch devices, document reading devices, and sensors.

16-1 BINARY INPUT (BIT SWITCHES)

Let's say you want a simple way to enter numbers into a computer. It may be that the numbers are commands for the computer or merely numbers the computer uses in some calculation. Whatever they are for, the input device's task is to make each number available to the computer in a code its port can put directly on its data bus. Since many microcomputers use an 8-bit binary code, the input device in Figure 16-1 has eight bit switches.

What it does isn't complicated. In one position (up) the switch is open, and its data wire is pulled up to a HIGH logic level by a pull-up resistor. In the other position (down) the switch is closed, and connects its data wire directly to a LOW logic level. Since the path to the 0 level (ground) is a much better conductor than the 1-k-Ω resistor, the pull-down is stronger, and the logic level is LOW. There are eight switches, one for each bit of the data bus. If each of these data wires is connected to one bit of an input port, it

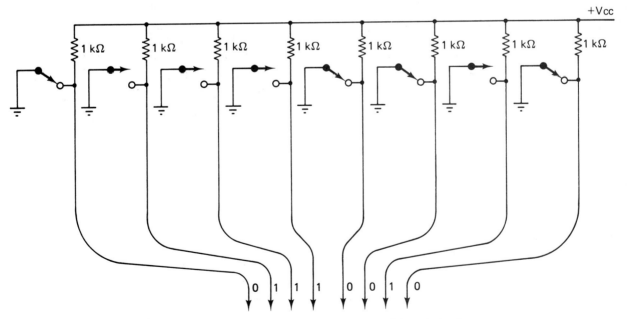

FIGURE 16-1 Eight-bit binary input device (bit switches).

will input any data word you put on the switches. Of course, to use this input device, you need to put data bit by bit onto the switch panel, and know the exact binary code for every word you want to input to the machine.

This input device is not a very good one from the human factors standpoint. The hardware is easy enough to design, but a mess to operate. You must make eight switch entries for each new word you want to enter. You have to know the bit code for each word, even if all you want to enter is the letter A, or the binary code for the number 4. It's the slowest, most error-prone method of entry there is.

The earliest models of microcomputers used this method of entry. Bit-switch panels are cheap, and don't take much hardware or power to interface. The human factors problem make this a case of "penny wise and pound foolish," however. Later designs included numeric and alphabetic keyboards as a built-in part of the unit, and companies that continued to make nothing but boxes with bit switches went *bankrupt*.

16-2 NUMERICAL INPUT (KEYBOARDS)

For human convenience, keyboards are a big improvement over bit-switch panels. They need more hardware, true—but have advantages for speedy, error-free operations that far outweigh their greater complexity and higher cost.

16-2.1 Octal Keyboard with Hardware Encoding

This is a keyboard somewhat like the panel used for code entry to the Heathkit H-8. We had the basics of this keyboard back in Chapter 5 when we discussed encoders. The idea of an encoded keyboard is that each key is a switch, and each switch closure produces not one, but a whole group of binary digits (a code). In our example (Figure 16-2), the eight keys are attached to a three-wire bus (that has binary 4, 2, and 1 place values) by diodes. Each set of diode connections passes a code from the key to the bus. In Figure 16-2(a), each switch puts a logic 0 on the diodes,

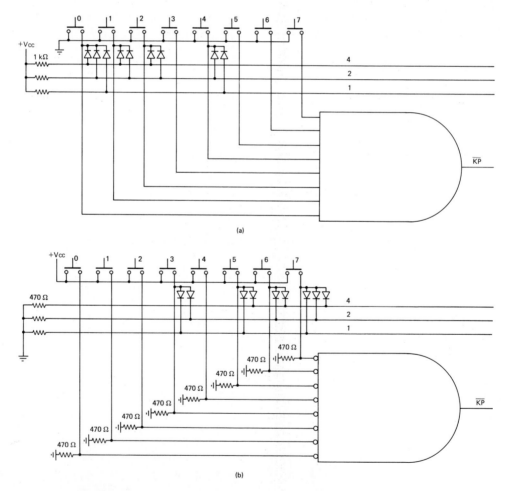

FIGURE 16-2   (a) Active-LOW and (b) active-HIGH octal keyboards.

and they conduct the 0s to the places where a 0 appears in the 3-bit code for that key's octal number. For instance, the 0 key's diodes conduct 0s to all three wires, giving a code of 000 binary for this key, while the 5 key conducts a 0 only to the 2-wire, producing a 101 code. The wires that don't receive a 0 through a diode are 1s because there are pull-up resistors on each wire. No diodes are needed on the 5 key, or on any key that makes only one connection. It is when two or more bus wires are connected to a key, like the 0, that diodes are necessary, so that the key can conduct to the bus wires, but they don't conduct to one another.

In Figure 16-2(b), each switch puts a logic 1 on the diodes, and they conduct the 1s to the places where a 1 appears in the code for the key's octal number. In this example, imagine that the 7 key is pressed. The diodes are pointed in a direction that will conduct the 1 to all three bus wires, making 111. If the 5 key is pressed, 1s will be conducted to the 4 and 1 wires, giving 101 as the code for a 5. The wires that don't receive a 1 through a diode are 0s because there are pull-down resistors on every wire. The conduction path for the 1s is a better conduction path than the pull-downs, so when a diode *does* conduct, the voltage will be HIGH, but otherwise, the voltage is LOW.

The keyboard (usually called a **keypad**) in Figure 16-2(a) uses an active-LOW design, because each key produces its code by applying LOW levels to the bus. Figure 16-2(b) is an active-HIGH design, since each key produces its code by applying HIGH levels to the bus.

Look at the 7 on the active-LOW keypad. There is nothing connected to the bus to make a 7 in this case. There is a 7 on the bus of the active LOW keypad if no key is pressed, because the pull-up resistors put 1s on all the bus wires that aren't pulled down. How does the computer know if a 7 is real or if it's just a number on the bus between key presses?

All the keys produce a signal that goes to a gate. This gate produces the $\overline{KP}$ (key pressed) signal. In both Figure 16-2(a) and (b), the keypressed gate produces a LOW output if any one of the keys is pressed. When the signal is active, it's the keyboard's way of saying to the computer "Hey! Over here! I've got a live one for you—this is the real thing!"

Let's assume that our keyboard is used as an input for a microcomputer with 8-bit input ports. You might ask: "How is this keyboard, which produces 3-bit numbers, suppose to be con-

nected to an 8-bit data bus?" There are two answers. One solution is to add a shift register to the circuit that can save up each 3-bit key code, until an 8-bit number is built up with several keystrokes. This is called the hardware solution to the problem. Another way to save up the keystrokes until an 8-bit word is entered is by program. The computer receiving the data can take in 3 bits at a time, shifting over the bits every time another key is struck.

The hardware solution is shown in Figure 16-3. If a key is pressed, the keypressed signal shifts the last 3 bits over one level in the shift register. In our example, we imagine that the key sequence 315 has been pressed on the keypad, and the 5 is still on the bus, although the keypressed signal has clocked it into the front end of the shift register already. The 8-bit number 315 converts into (from octal to binary) 11,001,101. This number is available on an 8-bit bus which can be attached directly to an input port.

Another circuit for the hardware solution to the octal keyboard is shown in Figure 16-4. This circuit uses keyboard scanning, which is the most common way of identifying and encoding keys on larger keyboards. Even on this very small keyboard, you'll see that the scanning approach reduces the number of diodes in the matrix encoder and cuts down the number of inputs needed for the keypressed gate.

The eight keys of our keyboard have been organized into two rows of four keys each. On the bottom row, the codes for the 0123 keys are shown in binary. On the top row, the codes for 4567 in binary are shown. Notice that for each pair of keys on the same column, both key codes end in the same two digits. For instance, 1 and 5 are on the same column, and their codes 001 and 101 both end in 01. This fact is critical in the design of the encoder used here.

Let's see how this encoder works by imagining what happens when the 5 key is pressed. The 5 key connects a row wire to one of the four column wires. The J-K flip-flop is toggling a 0 back and forth from Q to not-Q. When the 0 reaches Q, the 5 key connects that 0 to the column wire that 5 and 1 are on. That connection passes the 0 to one of the wires on the 421 bus (the 2 wire). While this is happening, there is a 1 on the not-Q output of the keyboard strobe flip-flop, which is also the 4 output on the bus. The remaining wire on the bus is floating, and its pull-up resistor makes its logic level a 1. This means that at this instant, the number on the 421 bus is 101—a 5 in binary code—but will this number

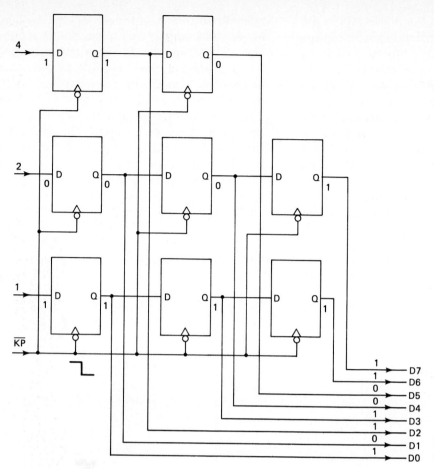

FIGURE 16-3 Octal-to-8-bit shift register.

stay there if the flip-flop keeps toggling? The answer is "no," but the flip-flop won't keep toggling. While all this is happening, the *keypressed decoder* is receiving the same 0 on the column line that made the 2 bus line LOW. This 0 will make the output signal, $\overline{KP}$, go LOW after a small delay while the C capacitor charges up. The keypressed signal does two things. It goes on to the shift register or input port of the computer, and it goes to the J and K inputs of the flip-flop, making it stop toggling and "freeze" at its present state. When the key is released, and the 0 on the

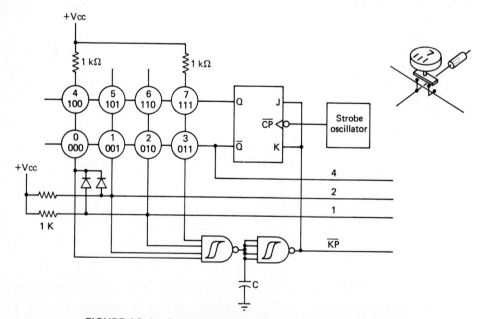

FIGURE 16-4 Octal keyboard with synchronous scanning.

keypressed line becomes a 1 again, the toggling—which is used to scan the two rows of the keyboard—will resume.

Key debounce. There is a reason for the capacitive time delay put into the keypressed signal. At first, when the key is pressed and the contacts meet, they will bounce for a while. The code on the 421 bus will switch from 101 to 111 and back again as the contacts bounce. The KP signal doesn't become active until the capacitor charges and the voltage rises above the upper trip point of the Schmitt trigger logic gates. If the capacitor is large enough, this time delay is long enough for the bouncing key to stop and the 101 to stabilize on the bus. At a few milliseconds after the 5 key is pressed, the KP signal switches active, and by that time, the 101 on the 421 bus is steady.

You can see that all the other numbers on the keyboard will work also. A combination of two actions makes this possible. The row information—what row the key is on—contributes the 4-bit of the 421 data word. The column information—the 0 on one of the column lines—provides the encoder with enough information to generate the remaining two bits of the data word.

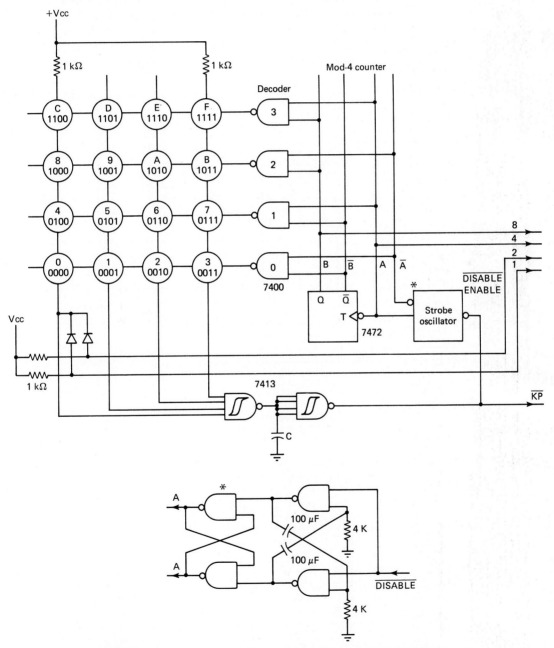

FIGURE 16-5 Hexadecimal keyboard with synchronous scanning.

16-2.2 Hexadecimal Keyboard with Hardware Encoding

In Figure 16-5 you can see the method used for an octal keyboard in Section 16-2.1 can be extended to a hexadecimal keyboard with twice as many keys. The only differences in this circuit and the octal keypad in Figure 16-4 is the line decoder used to scan the row, and the "freezable" clock oscillator for the flip-flop. Because the oscillator can be frozen at a particular place, the count can be stopped when the KP signal appears. Using the output of the oscillator as one bit of a 2-bit counter with double-rail outputs, we can get this combination of oscillator plus flip-flop to count in modulus 4, with the sequence 0, 1, 2, 3, 0, 1, 2, 3, ... Since we can stop the oscillator at any moment when the KP strobe goes LOW, the scan stops just like the scan for the octal keyboard in Figure 16-4, as soon as the key has been found and the debounce delay is completed.

16-2.3. Decimal Keyboard with ROM Encoding

We have just seen several ways to scan and encode a keyboard. In particular, we have a lot of versions of the octal keyboard we started with. Although we discussed four different methods of keyboard encoding in preceding sections, there is yet another important way to convert a key closure into a binary code. In Figure 16-6 there are two versions of a decimal keypad that takes the closure of one key and converts it into a 4-bit BCD code for the number on the key. The first keypad is organized as a row of 10 keys which provide 10 inputs to a ROM. The second is arranged in two rows of five keys, providing five inputs to the ROM, and with a scanner switching a 0 from one row of keys to the other. The scanner provides an additional bit of input for the ROM, making 6 bits in all.

What do they do? In the first ROM, the 10

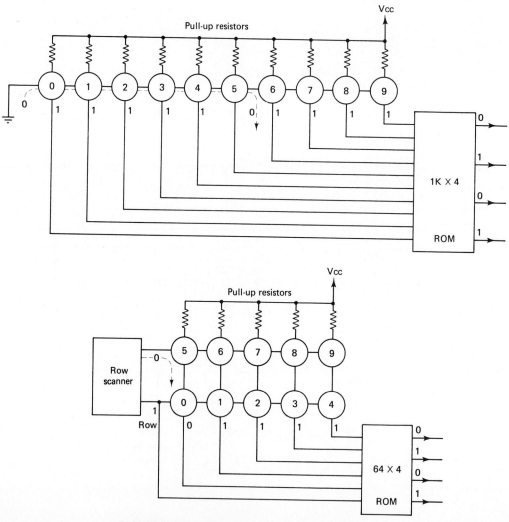

FIGURE 16-6 Keyboards with encoder ROMs.

inputs are the address lines for a 1K × 4 ROM. There are 1024 places inside the ROM with a permanently burned-in number in each place. Assuming that we only press one key at a time, there are only 10 of these 1024 places that we'll actually use. This is a very wasteful method of keyboard encoding. We use less than 1 of every 100 addresses in this design.

When a 5 is pressed on the keypad, for instance, the address 1111101111 is put on the address lines. We've burned in the code 0101 at that location in the ROM, so the four outputs that appear on the data lines of the ROM are 0101. Although the ROM is a memory device, we're using it here as an encoder. Since the contents of each location in the ROM are nonvolatile and permanent, the ROM will remember a 0101 every time a 5 is pressed, even though the power has been turned off for a while. In this regard, the keyboard ROM is used as hardware more than software.

We said that using a 1024-location ROM is wasteful when we only need to encode 10 keys. The second design is more efficient. The ROM being used has six address lines, and 64 places where 4-bit numbers are stored. Each key that's pressed generates row and column information. The row number is a 0 or a 1, and the column number is a LOW on one of the five column wires. In this design, if the 5 is pressed on the keypad, the column wires will be 01111 and the row wire will be a 1. Address 101111 is address 47 and our 64 × 4 ROM had better have a 0101 stored there, or our keyboard won't work. Similarly, the other

nine keys of the keyboard will address nine other locations, each with the key's code stored in it. This ROM encoder is still wasteful. From 64 locations, we get 10 useful numbers. That's 5.4 useless locations to every useful one.

For larger keyboards, the ROM method is more efficient, and is often used. A 10-digit decimal keypad is just too small to take advantage of this method of encoding. Keyboards with 64 and 96 keys are frequently designed with ROM encoders built in. For smaller keypads, a diode matrix encoder is more efficient than a ROM.

You may have wondered why all our keyboard designs have used active-LOW logic when we introduced both active-HIGH and active-LOW types at the beginning. This is no accident. Since most computer logic is designed to be TTL compatible, its inputs regard a disconnected line the same as a 1. We don't want a disconnected line or plug to act as an *active* signal (or a whole bunch of active signals). There could be major trouble if HIGH was the active state of logic devices in a computer. A plug accidentally kicked loose would turn on all the devices it was supposed to control—and there might be several fighting for control of the data bus at the same time. When this happens, some gates usually get burned out somewhere, and that's not a good design. Instead, computer circuits should be designed so that a disconnected plug makes everything play dead and float. That's why active-LOW logic and active-LOW enables are used in as many places as possible where TTL or TTL-compatible devices are used.

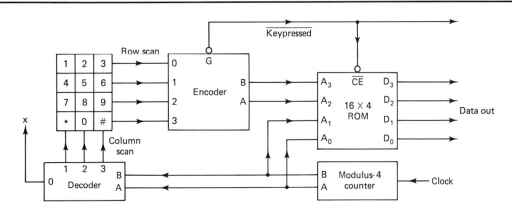

EXAMPLE 16-1

Suppose that we want to make a keypad from a telephone (a 12-key keypad) into a "smart" input device. The keyboard interface should produce codes 0–9 = 0000 through 1001 (binary), and encode * as 1010 and # as 1011. What numbers

should be "burned in" to each cell of the ROM in the following schematic?

Solution: First, it is necessary to determine what code appears at the ROM's *address inputs* when each key is pressed and scanned (producing a Keypress signal that enables the ROM).

We mark the keyboard columns with the

number in the counter that comes out of the encoder when it receives an active input from each row, as shown below:

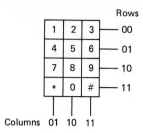

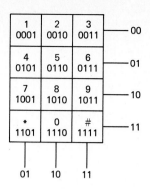

Then we put a 4-bit number into each space of the keyboard, using the row number as the most-significant 2 bits of the 4-bit number, and the column number as the least-significant 2 bits.

Each number is the ROM address accessed by the key. Now, based on what is marked on each key, we can fill data into the ROM addresses in the diagram below.

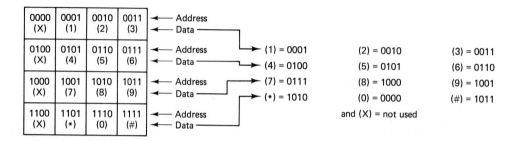

16-3 ALPHAMERIC INPUT

"Alphameric" is a contraction of two words, "*al-ph*abetic" and "nu*meric*." An **alphameric** (sometimes, **alphanumeric**) **keyboard** is one that has both numbers and letters on it. This is the type of keyboard usually found on data terminals. It is normally laid out like a typewriter keyboard, and the key operations usually produce some code with both letter and number capability, such as ASCII or EBCDIC code. The EBCDIC keyboard is largely obsolete; even IBM is s-l-o-w-l-y adapting to ASCII code. There are two basic kinds of alpameric keyboard: hardware encoded and software encoded. Let's briefly review what is meant by the terms **hardware** and **software**. A flip-flop is hardware. The resistors, transistors, and wires the circuit is made of are its hardware. To change any part of it requires a rewiring job. In the case of integrated circuits this is virtually impossible. The flip-flop circuit is "graven in stone" and is not going to change easily—hence the name hardware. If the flip-flop is set so that its Q output is HIGH, the 1 at Q is software. It is stored in the hardware, but is not a part of it. The wiring is not changed by altering the 1 to a 0, and the alteration doesn't require a major rewiring job. If power is turned off, the 1 or 0 will be lost. It's

temporary, "soft" information, easily changed or lost—hence the name "software." If we think of the circuit and the wiring as hardware, and the 1s and 0s as software, there shouldn't be much confusion about which is which.

What's a ROM? Well, the 1s and 0s are permanently wired in, so they seem to be hardware, but they're also logic states (HIGH and LOW), so they seem to be software. The ROM falls into a gray area—not bird and not beast—and is usually called **firmware**. ROM programs stored inside a computer are sometimes called software instead of firmware, but we've never seen them called hardware. Are you still confused? So are we.

16-3.1 ASCII Keyboard (Fully Encoded)

In Figure 16-7, we're cheating a little. The box in the diagram marked "magic" might be a diode matrix, a ROM, or a gate array. This hardware encoder converts the row and column information from the keyboard into 8-bit ASCII code. The ASCII code is passed on to the computer through an (input) I/O port.

Even if this port is memory mapped, the keyboard data will only occupy one byte of memory where a memory address has been replaced by the buffers of the keyboard input port.

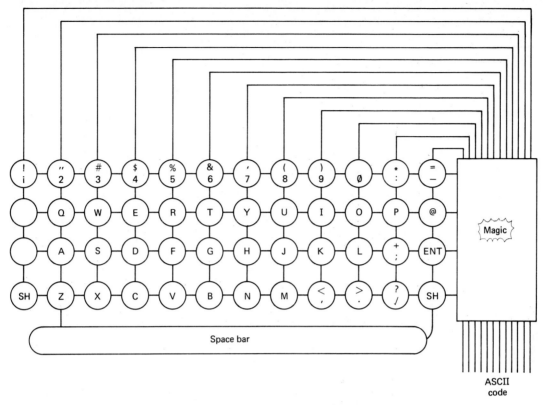

FIGURE 16-7 "Smart" ASCII keyboard.

The advantage of this type of keyboard is that it takes up very little of the computer's memory space for ports or a keyboard-scanning program, and provides very fast conversion of keystrokes into code, without taking any of the computer's time for the process. The disadvantage is that the keyboard encoder circuitry is expensive and takes up extra power and board space as compared to the software-encoded type. This type of keyboard is used on the Apple II computer.

16-3.2 ASCII Keyboard (Software Encoded)

In Figure 16-8 we see a keyboard with no encoding of its own. The computer scans a row on the keyboard by reading a certain address in memory. The address bus provides eight signals which are used by the program (the software) in the computer to scan the keyboard row by row. A single 1 is put on one of the eight address lines to see if a key is pressed on that row. The 1 is converted to a 0 by the inverting buffers used to pass the address bits A(7) through A(0) to the keyboard rows. If a key is pressed, the 0 on the row wire (notice that the keyboard is strobed with active-LOW signals) appears on one of the column wires. These are passed to the data bus through

inverting buffers—and the lone 0 becomes a single 1. The computer program identifies from the 1 which key has been pressed, and extracts a suitable code for that key from a table stored in the computer's memory. Then it reads another address that is the next row down on the keyboard, and continues until all rows on the keyboard have been read. This is repeated every time the computer "looks" at the keyboard.

This software-encoded keyboard design has one advantage. The hardware is simpler and cheaper than a hardware-encoded keyboard. The software that runs the keyboard-scan program takes up little space in the computer's operating system, and costs almost nothing. Data the computer gets from this keyboard can only tell it the row and column the key is on. The rest is done by the computer program.

This is a memory-mapped keyboard. Addresses where the eight rows of the keyboard are located (256 of them) are off-limits for use as ordinary memory by the computer. The keyboard is wired in at these locations. Part of the memory has been sacrificed so that the keyboard requires less hardware to make it operate.

The advantages of software encoding are the reduced cost of the simpler hardware (all that's needed is two buffers with gate logic for enables), and the lesser power requirement and smaller

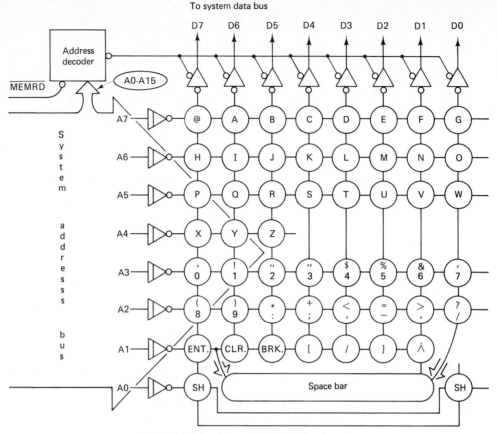

FIGURE 16-8 ''Dumb'' ASCII keyboard.

board space this provides. The disadvantages are the extra time it takes for the computer to run a keyboard-scan program every time it wants to see what key's been pressed, the keyboard-scan program that takes up memory space we could have used for other programming, and the large amount of memory space that must be "lost" to fit the undecoded keyboard into the memory map.

This is the kind of keyboard used in the Radio Shack TRS-80 Model III computer.

In the trade-off of hardware costs versus software costs these two styles of keyboard come out about even. The added expenses of hardware decoding (extra parts, more board space, more power consumption) are balanced by the inconveniences of software decoding (less available memory, slower conversion time). In the marketplace, where dollars and cents is most important, the software-encoding scheme gives the vendor a slight advantage in price (and few people can hit keys fast enough to notice the speed difference). A fully encoded (hardware encoded) keyboard attached to a true I/O port, however, would give the programmer more room to work in, and this is a selling point for the serious business or scientific

user. It's hard to say which of these schemes will win out in the long run—if either one does—and it appears we'll have both of them around in different machines for some time to come.

16-3.3 Difference between ''Smart'' and ''Dumb'' Input

For the two types of keyboard described in the preceding two sections—and for input devices in general—the names "smart" and "dumb" are used. A **smart keyboard** has its own "smarts"— that is, it has its own on-board decoding—and doesn't need to use any of the computer's smarts to convert the keystrokes into ASCII. A **dumb keyboard** has the bare minimum of logic ("smarts") on the board with the keys, and depends on the computer program to have the smarts to convert each key (row and column) into an ASCII code.

In a larger sense, smarts come from the encoding and logic circuitry that's on board the input device. If all the code conversion work is done at the input device before data are put into the input port—if the data are all digested before

EXAMPLE 16-2

Given the keyboard design from Example 16-1 (shown below), what is probably wrong with the keyboard if the only keys that work properly are 1, 2, 3 and *, 0, #? Keys 4, 5, 6 and 7, 8, 9 either produce the outputs you should get for 1, 2, 3 and *, 0, #, or else they produce nothing at all.

Solution: We suspect that the address delivered to the ROM has something faulty in the "row" portion of the address, because the columns of the 1, 2, 3 row and the *, 0, # row work OK, but the output seems "confused" about rows other than row 0 and row 3. Since the columns within those rows work appropriately, we suspect that the column decoder is scanning the columns properly.

The second thing we notice is that for the two rows that work properly, the binary codes out of the encoder are 00 and 11. In both of these cases, A = B. The other rows of the keyboard, we are told, produce outputs appropriate to 1, 2, 3 or *, 0, # if they produce anything at all. Those are outputs for which the address in the ROM has A_2 = A_3. (These are attached to the encoder's A and B outputs.) Since the only outputs we ever get in this case are outputs appropriate to ROM addresses where A_2 = A_3 (Encoder output A = B), we suspect that *the A and B outputs of the encoder are shorted together*. If they are shorted, A cannot be different from B, and the only legitimate outputs we can get from the ROM are from addresses where 11 or 00 are the first 2 bits.

To test this suspicion, we turn off system power and measure the resistance between points A_2 and A_3 with an ohmmeter or continuity tester. We will probably see zero ohms or a very small resistance between terminals A_2 and A_3. A visual inspection may reveal the location of the shorted connection if the resistance check results in a low-resistance reading. Also, using a sharp knife to scratch a line between the A_2 and A_3 traces from the encoder to the ROM may cut through a copper-bridge or solder-whisker short too small to be readily seen. Check the resistance after doing this to see if it has eliminated the low-resistance connection.

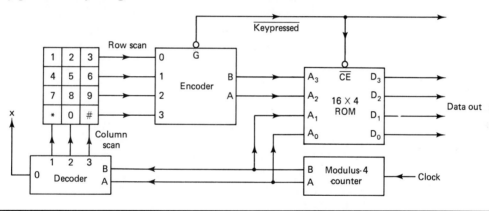

they're sent to the computer—the input device has the smarts. A dumb device lets the computer do more (or all) of the converting work, and doesn't have any capability on board to do anything the computer could be doing instead. Since the smarts are already a built-in part of the computer, building input devices with smarts that duplicate a lot of the functions of the computer adds extra cost. The cost of smart peripherals is not always easy for the buyer to live with. A dumb peripheral that utilizes the computer's smarts can be purchased for less money, and unless the computer system it attaches to is "I/O-bound," it will appear to work the same way the smart one does. It may be that the smart designer is the one who uses the dumb peripheral.

16-4 MAGNETIC (I/O) STORAGE DEVICES

At the beginning of this chapter, we split the I/O device world into three parts. Keyboards, keypads, and switch panels fell into the category of direct input devices. They allowed generation and use of the input in "real time." The rest of the input world we called document-reading devices and sensors. We'll begin our discussion of **document-reading devices** with magnetic storage (tape, disk, magnetic card, magnetic bubble, etc). We might also consider these as forms of auxiliary memory, since they are used to store information that takes up too much space for the computer's main memory. It is possible, through the use of these mass-storage devices, to transfer chunks of information into and out of the real computer memory so that the virtual memory (the size the memory appears to be) is hundreds of times larger than the main memory actually is.

16-4.1   Tape

Several different schemes for recording 1s and 0s on tape exist. One simple system uses a north magnetic polarity on the top surface of the tape for a 1, and a south magnetic polarity for a 0. More complex methods better adapted to the response of magnetic r/p (record/playback) heads are used, but in all cases, the reading (this is a chapter on input devices, after all!) takes place like this:

A layer of plastic coated with a ferromagnetic oxide like ferrite has magnetic fields on its ferrite surface. This surface passes by the r/p head, which is made of a good magnetic-field conductor. A coil of wire is wrapped around this magnetic-field conductor, and as the field in the head changes, the changing field induces a voltage in the coil, which is amplified and waveshaped by the electronics in the tape-drive mechanism to digital logic levels before being input to the computer.

In some machines, **saturation recording** is used. The 1s and 0s are recorded as magnetic fields that completely saturate the ferrite magnetically. Each bit is usually represented by one region of magnetization, although the borders between magnetized regions may be used as bits instead.

In other machines, two tones or frequencies are used to represent the HIGH and LOW logic levels. As you might expect, the high frequency usually stands for the HIGH logic level, and the lower frequency stands for the LOW logic level. This system of recording is called **audio recording**. The frequencies used are generally in the middle (a few kilohertz) audio range, and are recorded as sinusoidal, nonsaturated magnetic-field variations, instead of saturated pulses.

Figure 16-9 shows three methods of recording on tape. If we imagine that magnetic fields are visible as dark spots on the tape, Figure 16-9 illustrates the types of recording discussed in the next three sections.

Reel-to-reel seven-track and nine-track formats.   The most popular mass-storage medium in large data processing centers are **saturation-recording tape drives** that record data in multiple tracks simultaneously. Early IBM machines that used BCDIC code (a six-level code) were able

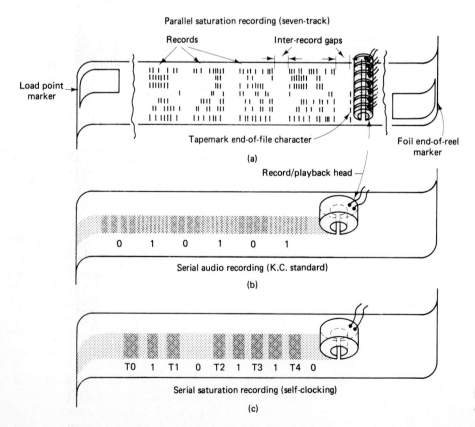

FIGURE 16-9 Magnetic tape data formats.

to store each byte on a seven-track strip (six for the code plus a parity bit). A seven-track tape drive has a record/playback head which records 7 bits at once to store a byte, then as the tape moves by, the next byte and the next are stored in strips of bits on the tape.

Later machines that use 8-bit EBCDIC or ASCII code record nine tracks on a strip 1 bit wide and 9 bits deep (8 bits of code plus a parity bit). All of these machines have certain things in common.

A standard-size reel of tape is about 25 cm in diameter and holds 800 meters of tape. Bytes are read from the tape at a few hundred to a few thousand bits per centimeter. U.S. tape recording densities are described in bits per inch (bpi). At 6400 bpi (2520 bits per centimeter) a reel of tape like this holds 192 million bytes of data.

Blocks of bytes recorded as a group are called **records**. A data processor recording data from a box of punchcards onto tape might record each card as a separate record, separated from the last card and the next card by an **inter-record gap** (IRG). The end of the box might be the end of all the records in that file of data. On tape, this point would be identified by recording a special end-of-file character called a **tapemark**. At the beginning (**load point**) and end of the reel, permanent foil reflectors—also called **tapemarks**—identify the points at which rewinding or further recording must stop. These reflective foil markers are sensed by photocells that "see" light from a small lamp bounced off the foil strips.

The tape-transport mechanism can move the tape in both forward and reverse directions at speeds up to several feet per second, yet can stop the tape in a fraction of an inch at the end of a record, file, or reel. It is possible to start and stop the tape within these short distances because the whole reel of tape isn't started and stopped as every record is read. Two loops with several decimeters of slack tape are held (on either side of the r/p head) under tension by a partial vacuum behind vacuum doors. When the tape is moved, one loop is shortened, and the other loop gets longer. Only the slack tape has to move, and the small inertia of a few decimeters of tape (compared to the whole reel) makes it easy to start and stop the tape in short intervals of time. Eventually, when one loop gets too short and the other gets too long, pressure sensors activate the drive and advance or rewind the reels until the loops are equalized. The IRGs between tape records provide startup time and braking time for the tape as each new record is read. When a tape drive is reading tape, even though the tape appears to move smoothly through the heads (it isn't; it's jumping ahead a fraction of a centimeter—a record—at a time), the reels appear to jump forward or back in steps of several decimeters.

Tape is much cheaper per byte of storage space than other magnetic media, and a reel of tape stores data in a smaller volume than any other type of machine-readable document. A 25-cm reel of tape, for instance, holds as much data as (at 6400 bpi) you could fit into 1200 boxes (about 6000 kg!) of punchcards.

Cassette: Serial KC standard (audio) recording. Microcomputers that cost a few hundred to a thousand dollars aren't likely to use parallel multitrack digital recorders like those just discussed. Those cost tens of thousands of dollars. You're more likely to find a small computer using a cassette recorder that sells for under a hundred dollars. These recorders have only one r/p head that records one track. Most of them aren't suited to overdriving the heads for saturation recording. They also can't handle the bodacious data rates computer tape is normally expected to handle (100K to millions of bits per second). Cassette tape recorders used with microcomputers record data one bit after another—serially—instead of 7 or 9 bits side by side.

In 1976, a convention of microcomputer and small-computer users devised a standard recording scheme used by several manufacturers. Since the conference was convened by *BYTE* magazine in Kansas City, the standard developed there is called the **BYTE standard** or **Kansas City Standard** (KC standard). Although it is by no means the only technique used for recording on small-computer cassette recorders, KC standard is representative of a number of other recording schemes. Here's how it's organized:

1. A logic 0 is recorded as four cycles of 1200-Hz sine-wave voltage. (a **space**)
2. A logic 1 is recorded as eight cycles of 2400-Hz sine-wave voltage. (a **mark**)

You can see that both a 1 and a 0 take 1/300 of a second, permitting data recording at a rate of 300 baud (300 bps).

This is an audio-recording standard. The 1s and 0s are identified by tuned circuits or a phase-locked loop (PLL), and converted to logic levels in the input device. The serial-to-parallel conversion can be done either by software or by a UART.

Cassette: TRS-80 tape (saturation) recording. Cassette tape can be saturation recorded. One method of doing this was used by Tandy in their first TRS-80. It is, in fact, the same method IBM uses to record on disks. A train of magnetic pulses is recorded on a single track of tape, serially. Two kinds of pulses are recorded, clock pulses and data pulses. A data pulse position is found between every two clock pulses. If there's no data pulse between two clock pulses, the data bit is a 0. If there is a data pulse between two clock pulses, the data bit is a 1. The recorded magnetic information is waveshaped by an op amp circuit in the computer and then input through a buffer. The computer program sorts clock from data pulses and converts the data from serial to parallel form.

Saturation recording has the advantage that bits can be recorded at a faster rate than KC standard—only one pulse has to be recorded per bit instead of 4 or 8—but the disadvantage that only one pulse has to be lost to lose the whole bit of data. The KC standard audio will be more reliable—but much slower—than saturation recording.

Cassette recorders used with small computers don't have the sophisticated direction-control and end-of-reel detection capability of the commercial drives for the big mainframes. The record, playback, and rewind/fast forward functions are all manual. About the only automatic thing you can do with these recorders is turn them on and off through the remote jack. Many computers don't even have provisions for this much control (or the control doesn't work very well).

The amount of data that can be recorded on a cassette is not as large as on a reel, partly because the data are recorded serially instead of one byte at a time, and partly because lower frequencies and data rates are used on audio tape. Even so, a 60-minute cassette recorded at 500 baud can hold 225,000 bytes.

16-4.2 Disk

"Hard" disk (disk packs and Winchester disk). Disks and drums were invented as high-speed mass-storage devices for data processors in the late 1950s. Their main advantage over tape is that they are random access as opposed to the serial access of magnetic tape. An example of random access is a phonograph record. If you want to listen to the first cut on a record, then the last, then one in the middle, it's easy to do this on a phonograph by just picking up the tone arm and setting it somewhere else. To do the same thing with a tape recording—you can't do the same thing with tape. Since the tape re-

cording is all in a straight line you have to fast forward and rewind through all the intervening tape to get to the parts you want to hear. On a phonograph disk, since it's two-dimensional, the needle doesn't have to fast-forward through the groove to get where you want it. You just jump over the part you want to skip.

This feature of phonograph disks made a group of engineers at IBM think of doing the same thing with a magnetic recording disk for mass data storage. Unlike photograph records, disks can be erased, recorded, and reused. Like phonograph records, both the top and bottom surfaces are used. In large data-processing systems, groups of platters called ''packs'' are spun on a common shaft and read by a comblike arrangement of r/p heads in a holder called an **access arm**. On a drive with six platters, there will be 10 usable surfaces (the top and bottom surfaces are protective covers) so the access arm will have 10 r/p heads. Figure 16-10 shows the original IBM-1311 RAMAC disk pack developed for the IBM 1401 system in the late 1950s. It had 10 surfaces, 100 tracks, and 20 sectors per track. These three dimensions define a system of *cylindrical coordinates* that allows access to any record on the disk. The surface number is selected by enabling just one of the 10 r/p heads on the access arm. The track number is selected by moving the access arm in toward the center of the disk or out toward the edge. Once the r/p head is positioned at a specific track, all we do to get a particular sector is just wait for it to come around on the disk, since the disk is spinning.

In some ways a **disk pack** is like a 3D memory. It has three dimensions, and is accessible at almost any location in (nearly) the same amount of time. The time it takes to find any piece of data on a disk is a combination of the time it takes the access arm to reach the desired track, and the time it takes the disk to rotate to the desired sector. Enabling the r/p head to select the surface takes no time on this scale. For instance, suppose that the disk pack in Figure 16-10 is rotating at 1500 rpm and the access arm is moved from track to track by a stepper motor that takes 100 steps per second. In a worst-case situation, the access arm would have to seek across all 100 tracks to find the right one (1 second), but in a typical case, the access arm only has to move one track at a time (10 ms). Again in a worst-case situation, the disk would have to complete one full revolution to reach the sector desired (40 ms), but a typical seek would find its target in half a revolution (20 ms) around the disk. Enabling the r/p head would take less than a microsecond (0.001 ms). On the

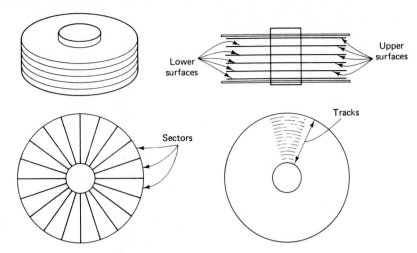

FIGURE 16-10 Magnetic disk architecture (hard disk).

whole, in our hypothetical example, it would take an average 30 ms to reach any record on the disk, but could take as long as 1.04 seconds.

Is this significantly faster than tape? Suppose that we have a 25-cm reel (800 meters) of tape moving at 1 meter per second while a specific record is being searched for. If the tape mechanism can read data as it flies by at this speed, looking for a specific record number, how long would it take to find one of the records on tape? In the worst case, there are 800 meters of tape, and the next record might be 800 meters away from the last. That would make the worst-case time to find a record 800 seconds (13 minutes 20 seconds). An average record might be the next record over, but if tape is used for random access—not trying to read the records in the order they were recorded—the next record wanted would average 400 meters (6 minutes 40 seconds) away. For the worst case, tape is almost 800 times slower than the worst-case disk-seek operation (random access), and in the typical random-record disk seek, tape is 13,000 times slower.

The name **hard disk** or **rigid disk** comes from the fact that these platters are made of metal (usually aluminum) coated with ferrite—the same magnetic material on tapes—and recorded/played back with r/p heads very similar to those in a tape drive. Disk packs are fairly easy to change, but are more expensive, and have less capacity, than tape. Our IBM-1311 disk pack had 100 tracks × 10 surfaces × 20 sectors. That makes a total of 20,000 records on the pack (records are sometimes called sectors, too—just to confuse things). Suppose that each record contains 512 bytes. (On the innermost track, this is just slightly above the 6400 BPI we used on our tape example earlier.) The total content of the disk pack would be 10,240,000 bytes. This is

larger than the main memory of many computers, but a lot less than the 192 million bytes on a 25-cm reel of tape recorded at the same BPI density.

Disk packs are precision-machined devices. The r/p heads float above the surface of the disk on a cushion of compressed air at a height of a few millionths of a centimeter. If the disks are warped or have irregularities in the surface that bump up more than this amount, the heads will "crash" and damage the disk. A head crash will usually gouge some of the ferrite off the disk surface (making it unrecordable) and damage or destroy the heads themselves. Dust (and even the particles of cigarette smoke) is made of particles big enough to get caught between the heads and the disk surface, causing scratches and damage to the disks. Most disk drives blow filtered air across the surface of the disks to keep them free of dust. A "clean room" environment is not absolutely necessary for disk drives, but is recommended.

As you could no doubt guess, disk packs cost a lot more than tape reels. A new disk pack would cost in the hundreds of dollars, whereas a reel of computer-grade tape would be in the tens of dollars.

Winchester disk drives are hard-disk drives for small/microcomputers, that usually have just one platter, which may not be removable. Winchester drives have higher recording densities than the disk pack discussed in our hypothetical example. Early units of this type had 30 megabytes on a surface with 30 sectors on a track. The designation 30-30 for this unit reminded someone of a 30-30 hunting rifle (a Winchester), and this name was coined for the disk and its drive.

A 10-megabyte 5-inch "Winchester" disk drive used on an IBM-PC might have 306 tracks

× 4 surfaces × 9 sectors. That makes a total of 11,016 records on the two-platter disk. For a 10-MB drive, each record contains 1024 bytes. The total content of this Winchester disk is 11,280,384 bytes (which is actually larger than 10 MB). This Winchester disk does not use a removable pack, so all four surfaces of the two platters are used, since they're protected. Since the disks aren't taken out of the drive and swapped with other disks as floppies are, drives of this type are sometimes called **fixed drives**.

Floppy disk. Popular in the microcomputer/small computer area, **floppies** are smaller, cheaper, slower, and have less precision (so they require less careful handling) than hard disks. A typical floppy is either $5\frac{1}{4}$ inches (about 13 cm) or 8 inches (about 20 cm) in diameter. The name "floppy" comes from the material (flexible plastic) on which the ferrite is bonded. Figure 16-11 shows a typical ($5\frac{1}{4}$-inch) floppy. The basic features are indicated on the diagram. Unlike the hard disk, floppies are enclosed in an envelope with a low-friction lining, and only one disk is placed on the drive at a time. Disks are removable and easily changed. The drive spindle in the disk drive is a tapered plug with a shoulder that grabs the disk and holds it by friction (when the disk is put into the drive and the door closed) so that it can turn without slipping. **Diskettes**—another name for floppies—are spun at 360 rpm. An indexing hole in the disk is read by photocell to identify its rotational position. There may be a hole at the beginning of each sector on the diskette (hard-sectored) or just one hole at the beginning of the first sector, with consecutive sectors identified by magnetic information recorded on the disk (soft-sectored). A single r/p head is usually all there is in a floppy drive, but both sides of each disk are recordable, so double-sided drives with two heads do exist.

The slower speed and lower bit density of floppies makes head crash less of a problem. Machine tolerances are less critical on floppies, and the "clean room" problem is less of a difficulty. The disk in our example has 35 tracks with 10 sectors on each track. A typical disk like the one in Figure 16-11 might have 9 sectors and 40 tracks, using both surfaces of the floppy (called "double-sided, double density"). With 512 bytes per disk record, this provides 368,640 bytes, usually called "360 K" by computerists. This format is the standard established for the IBM-PC, and copied by many "clones," for compatibility.

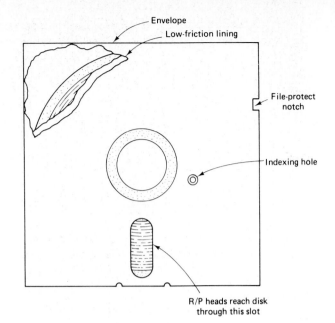

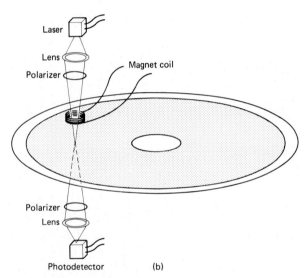

FIGURE 16-11 (a) Magnetic disk architecture (floppy disk); (b) optical read-write (magneto-thermo-optical) memory (Verbatim beam-addressable memory).

File protect. Magnetic media like reel-to-reel tape, cassette tape, and floppy disk can be **file protected**. That is, the recording on the disk or tape can be protected against being overwritten by a later recording. Cassette tapes have two plastic tabs on the back end of the cassette which, when broken out, prevent recording on the cassette. Floppy disks have a square notch on the side of the disk that can be covered with tape to inhibit recording on the disk. Data-processing reels have a rubber ring which can be removed to inhibit recording. In all of these cases, the reason for doing this is to keep important programs or data safe from accidental destruction by being recorded over.

EXAMPLE 16-3

Suppose that a certain Winchester drive has two platters (four surfaces) with 308 tracks (308 cylinders), nine sectors, and 2048 bytes per disk record. What is its total capacity in bytes?

Solution: Each disk record contains 2048 bytes, and if we travel entirely around one track, we find nine sectors, thus nine records per track. This means that one track (on one surface) contains

$$2048 \times 9 = 18,432$$

bytes per track. Since the Winchester has four surfaces, a cylinder contains four times as much as a track, or

$$18,432 \times 4 = 73,728$$

bytes per cylinder. There are 308 cylinders if the r/p head can be positioned into any one of 308 tracks, so the total capacity of the disk is

$$73,728 \times 308 = 22,708,224$$

which is over 20 megabytes. With part of the disk used for directories and addressing, the hard-disk unit will probably be sold as a "20-megabyte hard disk."

16-4.3 Bubble Memory

We discussed bubble memory in Chapter 14. We're mentioning it here again as a possible replacement for disks in computer mass-memory systems. Unlike disks, bubble memory chips are all solid-state, no-moving-parts technology, so there's none of the wear associated with drive spindles access arms and heads. The bubble can be read faster than disk, because no mechanical transport is necessary to move the bits past the reading mechanism. Single one-wafer devices (about 1 inch square) can now store 8 million bits, (1 million bytes), compared to the floppy in the preceding section, which stores less than one-fifth of that.

The compactness, speed, and mechanical simplicity (no motors, heads, or drive needed) of bubble memory suggests that these integrated circuits will replace floppies and fixed-hard disk in many applications. They will probably be configured (designed) to be transparent to the disk user—they'll look just like the disk they replaced to the programmer—but will be built into the system design in a much simpler way. Perhaps the

bubble memories will interface into the computer as plug-in cartridges, looking something like tape cassettes as they are snapped into and out of a holder.

16-4.4 Optical and Magneto-Thermo-Optical Disk Drives

Optical disk. Within recent years, technology developed for recording and playback of sound and video has been adapted to data-transfer use. Like the computer's floppy-diskette, **laser disk** is a removable medium that contains binary codes. Although the binary codes were originally intended for conversion into an analog video (television) picture, the original binary data could actually represent binary data, which could be read into a computer without A/D conversion. A smaller version of the video disk, called a **compact disk (CD)**, was developed for storing audio (sound) recording. Again, its intended use was for digital recording of analog sound, to be converted back into an analog wave upon playback. Both of these media store information with a much higher density than magnetic diskettes. Several hours of television recording, such as a video version of a full-length movie, could be stored on a single video disk. Converted into digital form, the movie becomes several *billion* bits of information. The compact-disk medium, while smaller in scale, provides storage space for several hundred million bits in a $3\frac{1}{2}$-inch plastic disk.

The technology of video disk and CD is optical. A laser beam is reflected off the surface of the medium and reflected back into a sensor (from the flat parts of the disk) or scattered off the surface (where a pit, or irregularity, in the smooth surface, has been formed). These two states are picked up by the optical sensor as the 0 and 1 of binary logic. This has several advantages and disadvantages compared to magnetic disk media:

Magnetic (Floppy) Disk	Optical CD Disk
1. May be recorded on and played back as many times as desired. (*Advantage: Floppy*)	1. Bits are burned in or formed at the time of production. Once written, a bit cannot be erased. (*Disadvantage: CD*)
2. Capacity = a few hundred thousand bytes on both sides of the disk. (*Disadvantage: floppy*)	2. Capacity = a few hundred million bytes on both surfaces of the disk. (*Advantage: CD*)

Magnetic (Floppy) Disk

3. Record-playback heads contact the surface of the disk; both disk media and heads are subject to wear. *(Disadvantage: floppy)*

4. Can be erased totally by an external magnetic field. Data can be erased accidentally. *(Disadvantage: floppy)*

Optical CD Disk

3. Noncontact laser reading principle ensures no wear on the medium or its optical sensors. *(Advantage: CD)*

4. Cannot be erased accidentally by external magnetism. Safe around magnets. *(Advantage: CD)*

Clearly, the use of video disk or compact disk for data storage and retrieval (sometimes called **CD-ROM**) is not for everybody. Its most practical use would be in storing large quantities of information that does not change but needs to be on-line for instant access (such as a dictionary). A system that takes a "blank" optical disk and burns pits into the surface to store data (called a **WORM,** for **write once–read many**) could be used like a floppy or hard-disk system, wherein updating information in a file simply means rewritting the file onto another part of the disk, then changing the directory to indicate that former versions of the file are inactivated. When billions of bytes of data space are involved, a single, removable WORM medium could be used like a floppy, and the fact that the disk is becoming gradually cluttered up with inactivated versions of updated files would be "transparent" (not noticed) to the user.

Magneto-thermal-optical disk. In 1987, Eastman Kodak's Verbatim division announced the release of a *read-write* optical medium similar to CD-ROM, but capable of being reused indefinitely. Their **BAM** (**beam-addressable memory**) device employs laser and magnetic technology to write data with the density of a laser disk.

Instead of burning a pit or rough spot onto the surface of the disk, the BAM uses the laser to heat a small portion of a magnetic medium, making it easier to magnetize. An external magnetic field, too weak to affect the rest of the (cool) disk, magnetizes the "hot spot." Later, a laser beam of polarized light is transmitted through the disk. Where the disk is magnetized, the polarized light is rotated. This rotation can be detected by an op-

tical sensor on the other side of the disk, which can tell the magnetized spots from the unmagnetized parts of the disk; thus it can distinguish 1s from 0s.

The fact that a permanent-magnet material (such as ferrite) is easier to magnetize (and demagnetize) when heated is called the **Curie effect.** It was discovered by Pierre and Marie Curie, who are also noted for the discovery of radium. The rotation of polarized light by a magnetic field occurs in many substances, and is called the **Faraday effect,** after Michael Faraday, who also invented the electromagnet and described magnetic lines of force for the first time. The magnetic layer of the BAM disk is coated with a plastic that has a strong Faraday effect.

Unlike the action of ordinary magnetic disk media, BAM magnetizes a space as small as the spot heated by the laser. This is the same size as the "pits" on CD-ROM, so a BAM disk $3\frac{1}{2}$ inches in diameter can also hold several hundred million bytes of data. Apparently, this medium has all the advantages of an optical disk and none of its disadvantages, except that it can be erased by a magnet. Don't use a refrigerator magnet to stick a BAM (or a floppy) up on the refrigerator with a note saying you found the lost data that slid under the sofa last week. If you do, the data will be *lost again. . . .*

16-5 HOLLERITH/PUNCHED PAPER TAPE, ETC.

Disk drives and tape players read magnetic documents. These are simple and readily available media for data storage and retrieval, but the first machine-readable documents were not magnetic. The earliest machine-readable documents used by data processing machines were punchcards, and they are still in use a century after Herman Hollerith developed them to tabulate the census of 1890. Hollerith punchcard readers and punchedtape readers work in about the same way, so the working principles described in the card section will not be repeated in the tape section.

16-5.1 Punchcards

Hollerith cards. The Hollerith code and Hollerith card layout were described in Chapter 7, so we won't repeat that information here. A **card reader** is shown in Figure 16-12. As the card passes between the *brushes* (electrical conductors actually made in the form of brushes with steel bristles) and the *contact roll,* the cardboard keeps the brushes and roller apart. When a hole

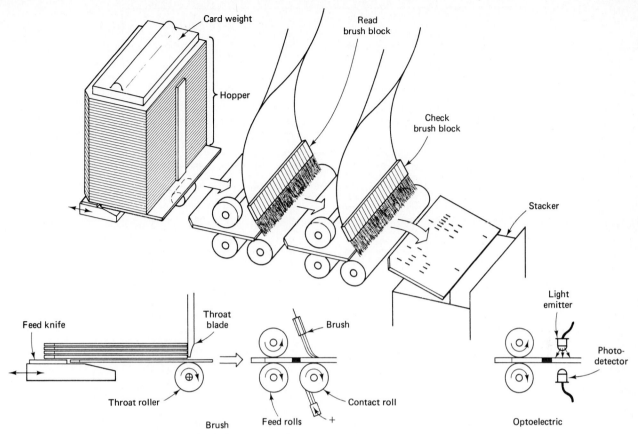

FIGURE 16-12 Punchcard reader mechanisms.

punched through the card passes between the brush and roller, the brush drops through the hole and makes an electrical contact with the voltage on the contact roll. This is a logic 1 (on early data-processing equipment, +40 V was used), which can be input into the computer through an input port and processed into whatever code the computer needs to use.

Hollerith made his cards the size of a dollar bill (1890 issue)—$3\frac{1}{4}$ by $7\frac{3}{8}$ inches. Dollar bills have gotten smaller in all dimensions—especially value—since then, but the old "computer card" is soaked in so many years of tradition that its size didn't change when the currency did. The other dimension of a punchcard is 0.0065 inch—to a very high degree of precision.

Since punchcards are usually fed to the computer's card reader in stacks (decks), a mechanism is required for separating cards from the stack and feeding them into the brush block one at a time. This card-feeding mechanism depends on the exact physical dimensions of the card. The bottom card on the stack is pushed from the *hopper* into the *throat* of the card reader by a *feed knife*, which catches the card with a squared-off metal blade thinner than one card (0.0040 inch). The throat is a space between the *feed roller* and

the *throat knife* that is just a bit wider than one card, but not wide enough to admit two cards (0.0080 inch). The card slides through the throat while the card above it is held by the throat knife, and gets fed into the brushes, where it is read. Other rollers continue to move the card until it ends up in a *stacker*.

The same job can be done in a noncontact way using photodetectors to see flashes of light that pass through the holes in the card from a light underneath the card. Both types of reader are shown in the diagram.

Problems with card readers usually involve "card jams" that occur when cards are stacked poorly (all the edges don't line up) or when the edges of the cards get worn—dog-eared—to the degree that their edges are frayed and become wider than 0.0080 inch. Hot, damp weather will also cause cards to swell up with moisture beyond this critical threshold and jam in the reader. This is one reason why computer rooms are air-conditioned and dehumidified in large data-processing installations.

96-Column cards. Based on the Hollerith code, but actually encoded in BCDIC instead of Hollerith, 96-column cards make six holes do the

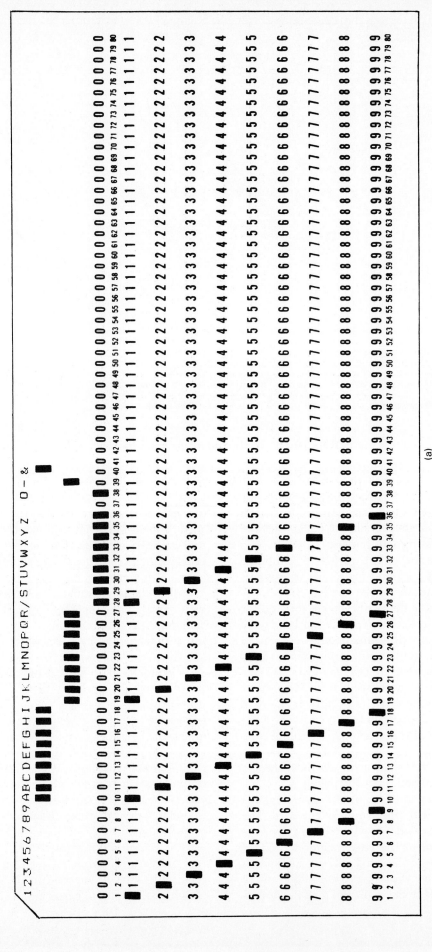

FIGURE 16-13   (a) 80-column punchcard format, (b) 96-column punchcard format.

(a)

308

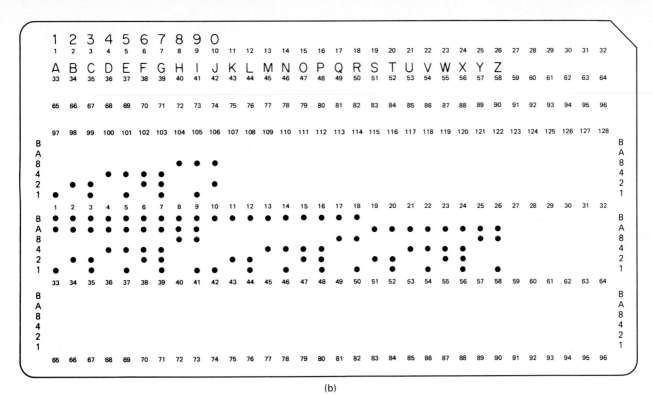

(b)

FIGURE 16-13 (*Continued*)

same job as the 12 in a Hollerith column. Although there are more data bits on each card, 96-column cards are smaller than Hollerith cards, because they are organized as three rows of 32 columns on a single card [Figure 16-13(b)]. The holes are round and smaller than those on a Hollerith card, and presumably a higher-precision type of reader is needed for them, but their compactness and higher storage density suggest that they should replace Hollerith cards in all applications. This doesn't seem to be happening—perhaps because, as I said before, the old Hollerith card is soaked in tradition, and data-processing people aren't going to change established systems just for the sake of compactness and higher storage density.

Like the 80-column Hollerith card [Figure 16-13(a)], 96-column cards are the size and shape of a common twentieth-century currency—the plastic charge card.

16-5.2 Punched Paper Tape

Figure 16-14 shows two versions of code punched on paper tape. As with punchcards, a hole stands for a 1 and everything else is a 0. The code in Figure 16-14(a) is Baudot code, a five-level code discussed in Chapter 7. In Figure 16-14(b), a seven-level version is shown which uses BCDIC

with added odd-parity bits and an end-of-line (EL) character.

The punching and reading equipment for paper tape is often found on Teletype machines and terminals. Paper tape was used in telegraph communications before there was data processing at all, which accounts for the use of Baudot code on some older machines, and the association of punched tape with Teletypes. It is also the most popular medium for input to industrial numerical-control systems, and computer-numerically controlled machines.

The basic principle of reading punched tape is the same as that of the punchcard reader. Brushes, pins, or starwheels may be used to make electrical contact when they fall into the holes, and optical sensors are used in current-generation noncontact high-speed readers. The only important difference between a reel of punched tape and a deck of punchcards is that the tape is one continuous document, whereas the punchcard deck is broken up into unit records none of which is more than 80 (96) characters long. If you make a mistake in a deck of punchcards, you've only got to duplicate one card with the mistake corrected, and throw out the old card in the deck, replacing it with the new one. Changing the order of instructions in one-command-per-card programming languages like FORTRAN or BASIC

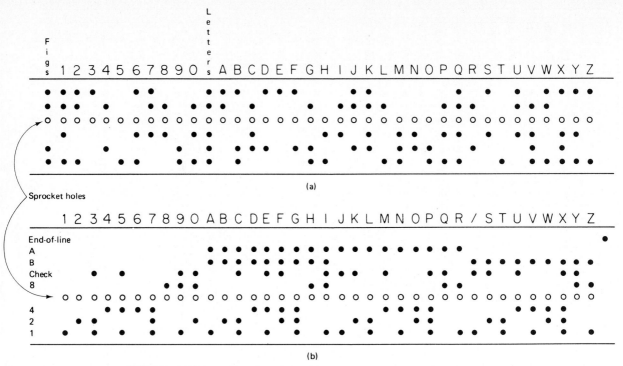

FIGURE 16-14 (a) Five-level and (b) seven-level punched paper tape format.

is as easy as shuffling the order of cards. With punched tape, however, if you make a mistake after 500 meters of tape, too bad! You have to duplicate the entire 500 meters, correct the error, and throw out the old tape, or try to splice in a corrected section of tape to fix the error, a messy and unreliable procedure. As for shuffling the order of commands by swapping cards, there's no way to do this with a reel of tape, except by cut and splice, or redoing the whole reel!

The good thing about tape is that being one continuous form, tape doesn't require a complicated feeding system like a punchcard reader's, and thickness or humidity aren't critical factors. You'll also never have the sad job of trying to put a 1000-card program back in order after knocking the deck off the top of a table. Tape readers are cheap, simple, and fast compared to card readers. They were even found on early microcomputer systems. A really simple optical punchtape reader can be gotten for under a hundred dollars, whereas card readers with deck-feeding capability cost many thousands.

16-6 OCR/BARCODE READER

These are becoming very popular as data-entry input devices for *point-of-sale* (POS) *terminals* (what we used to call cash registers) at retail stores. Both **OCR—optical character recognition**—wands and *barcode readers* identify a product at the checkout counter, simplifying the clerk's job. OCR wands contain a grid of photodetectors arranged to identify the shapes of alphameric characters on labels. Most of this work is done by the smarts of the main computer or a microprocessor built in to the POS terminal. OCR readers usually can read only one type of type font (character set), called OCR characters, but some programs are now smart enough to read several standard typewriter fonts as well as special OCR labels. The idea of making machine-readable print on paper that is also people-readable began with magnetic-ink OCR characters on bank checks, read by magnetic sensors. As the type of sensors that could read ordinary ink became cheaper and more reliable, the bank check reader moved out to department stores as the POS terminal.

At the same time, barcode readers using only one photosensor to read light and dark bars on a special label (UPC or *Universal Product Code*) began to appear in grocery stores, and a wide variety of products were required to carry these labels.

In both cases, the sensor picks up a pattern of changing light and dark spots as 1s and 0s, which are buffered into the computer from an input port. After that, some really hairy program-

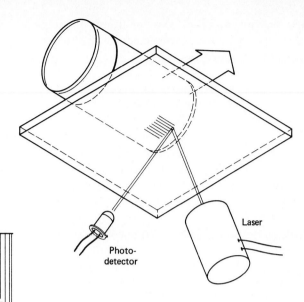

FIGURE 16-15 Barcode reader mechanism.

ming is required to digest these into useful alphameric information inside the computer.

Ultimately, printed labels are a lot simpler to make than punchcards or punchtape, and the reader/sensors are much simpler than those needed for either punched document. One design for a barcode reader is shown in Figure 16-15 together with the barcode itself. OCR fonts and sensors are not standardized and we've made no attempt to cover the field with a chart or diagram.

16-7 A/D CONVERTERS USED WITH SENSORS

16-7.1 Graphics Input: Light Pens, Tablets, Joysticks, and Video Digitizers

These input devices are used to enter position information into computers. Their position usually controls the position of something else, like the position of a spot of light on the video screen or the place where a dot will print on a sheet of paper.

A **light pen** in use appears to be operating or controlling the video screen. It is usually used to draw onto the video screen or to move objects around on the screen. Although it appears that the light pen is some sort of output controller associated with the video, it is in fact the simplest of input devices. All the light pen tells the CPU is "light on" or "light off." Clever programming uses this one-bit data buffered in through an in-

put port to locate the position of points of light on the screen. More clever programming is used to hunt, track, and follow the light pen around the screen with a point of light. A picture can be drawn on the video display, and its points digitized directly into the memory of the computer while this program is running. Without a suitable software driver program, the light pen does nothing.

A **tablet** is used like a pad of paper you draw on with a *stylus*—a pencil-like instrument. The stylus may—or may not—be an electrical contact. As it is moved across the tablet, a grid of wires under its surface detect the position of the stylus and send two numbers X and Y to the input ports of the computer. A drawing can be placed on the tablet and traced with the stylus to put its lines and curves into the computer as a set of numbers that describe the position of the points on the picture.

A **joystick** is a control handle used to steer things around on a display. It consists of variable resistors or switches that vary with position in two degrees of rotation. The variations in resistance are converted by an A/D converter into numbers that indicate two dimensions of position. Originally, the joystick was a steering handle in front of the pilot in a fighter plane that stuck up from between his legs. We'll let you guess why the pilots gave it that name.

Video digitizers take the signal from a television camera and sample the voltage at intervals of time. The voltages are converted from analog to digital, and each digital number—called a *pixel* or picture element—is stored in memory for future recovery and redisplay, printout, or *image processing*, in which certain aspects of the picture are sought out and enhanced.

16-7.2 Digitized Instruments: Sensors for Measurement and Data Logging

Any measurable quantity that can be converted into an electrical signal by a transducer can be input to a computer. An A/D converter will take the electrical voltage and turn it into a digital-coded number. When the number is input to the computer, it can be buffered in through an input port and used by the computer program, provided that the programmer knows what quantity is being measured and what the number means.

In Chapter 8 we discussed a digital thermometer. We are using a digital thermometer as an example of an input device in Figure 16-16. The *thermistor* is a resistor whose resistance goes down as its temperature goes up. The oscillator's frequency increases as the resistance goes down. The counter will count pulses from the oscillator. That's the whole input device. To use it, the computer program must reset the counter, let it count for a specified amount of time, then input the number in the counter and subtract a number from it to adjust for zero Celsius or zero Fahrenheit.

"So why," you ask, "should I use a computer to take temperatures?" "I can take a thermometer out with me and record temperatures on a pad of paper perfectly well myself."

By way of an answer, we'll ask another question; what would you do if you needed a temperature reading every thousandth of a second from 10 thermometers spaced every 10 meters outward from an underground nuclear weapons test until the sensors vaporize? Read the thermometers yourself and write them down on a pad? Not likely! This is a job for a computer.

Any time you need a lot of readings taken in a very short time, and remotely if necessary,

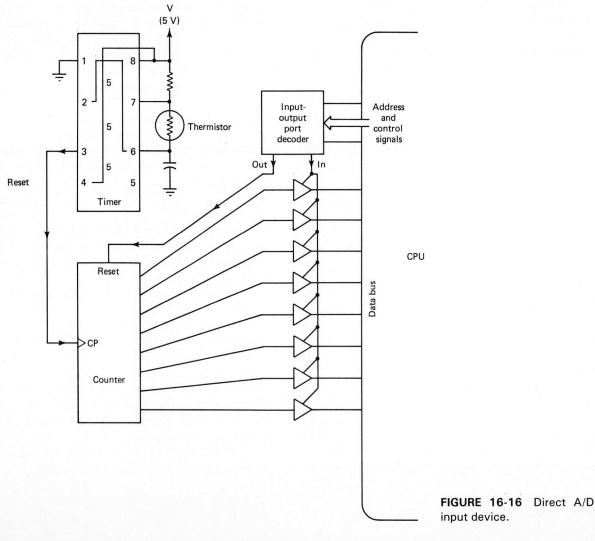

FIGURE 16-16 Direct A/D input device.

digital data transmission and automatic (computer) data logging are called for.

16-8 DMA CONTROLLER

DMA stands for **direct memory access.** Although we've included DMA with other input devices, a DMA peripheral is not an ordinary input port at all. It's an input device that puts its data directly into the memory of the computer. The **DMA controller** takes over the functions of a CPU and generates data, address, and control signals itself. One of the control signals is called a *DMA request* (*bus request* and *hold* are other names for the same signal). When the peripheral has a data word ready for the computer, it generates a DMA request that causes the address, data, and some control signals from the CPU to float at the end of the next machine cycle. As soon as the CPU has gone "on hold" and everything's floating, a DMA granted signal is sent back to the requesting device from the CPU (this signal is also called *hold acknowledged* or *bus acknowledged*). The requesting device now has the "go-ahead" signal and can enable all its buffers that were *tri-stated* and holding back on address,

data, and a memory write signal. Since these signals are enabled onto a bus that is floating already, they take over the system and the data are written into the memory at whatever address is placed on the address bus by the DMA controller.

In our example (Figure 16-17), a keypad generates a data word and a keypressed signal every time a key is pressed. The keypressed signal creates a DMA request and advances the address counter in the keyboard assembly. When a DMA granted signal returns from the CPU, the (memory) control signals, address, and data buses are floating, and when the buffers in the DMA keyboard are enabled, a memory write signal appears on the control bus, the address in the keyboard's address counter appears on the address bus, and the encoded character from the keyboard appears on the data bus. In our example, the address counter has counted to 0000000000001100 (12) and a 01000001 (the letter A) is pressed on the keyboard. The DMA request is the output of the keypressed flip-flop, set by the keypressed signal. When the DMA granted signal appears, in addition to enabling the 12 onto the address bus and the A onto the data bus, the DMA granted also resets the keypressed flip-flop so that the DMA request goes away after a

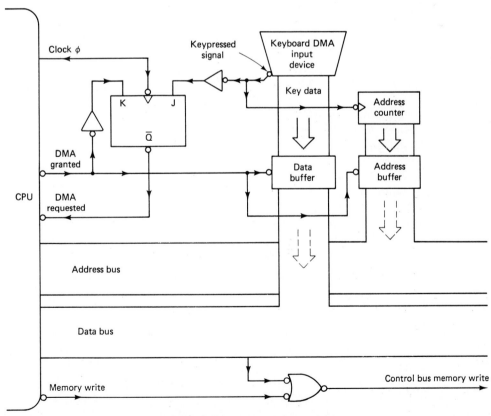

FIGURE 16-17 DMA input device.

delay of one clock pulse. The key will have to be released and another key pressed before the DMA request signal becomes active again.

The memory receives the 12 and the A and the memory write signal on the buses. This immediately causes the memory to write an A in memory location 12.

Meanwhile, the computer has been "playing dead." It's stuck at the end of its last machine cycle, ready to begin the next machine cycle, but unable to do so until the next clock pulse delay time is over. When the delay time passes, the A has been written into address 12, and the computer picks up where it left off as though nothing had happened. As far as the computer is concerned, there was a sudden puff of smoke and an A appeared! This is called a **cycle steal,** and the DMA controller that does it has done a cycle stealing DMA operation.

16-9 UARTs, Modems, Serial I/O for Telecommunications

We have discussed UARTs and modems in several previous chapters. The important thing to remember about these serial I/O devices is that they are "just another I/O device" where they connect to the I/O ports of the computer. As far as the computer is concerned, it comes to a port attached to a modem and picks up a byte of data just like it was from a keyboard. The fact that this byte of data might have come from the other end of the continent doesn't impress the computer at all!

Obviously, telecommunications is not a one-way process, and the dataset formed of acoustic coupler, modem, and UART is not just an input device. Just as obviously, there must be an input device among the things inside the dataset. To recap from previous chapters, the input part of the dataset takes serial FSK audio data from the acoustic coupler, the modem demodulates the FSK frequencies into two voltage levels, logic 1 and logic 0, and the UART collects the data bits one by one, converting from serial to parallel. The parallel data can now be gated onto the data bus by the input port.

16-10 AUDIO INTERFACE (MICROPHONE)

Finally, in the realm of audio interfaces, we'll discuss a slightly "blue sky" idea that is catching on. A microphone can be used as a detector (as it is in the acoustic coupler in the preceding section) for digital data, but it can also input music, sounds, and even human voice. With suitable analog-to-digital conversion, codes identifiable to the computer can be generated by spoken words.

With a suitably interfaced (and this is the complicated part) microphone, a music-digitizer or voice-to-typewritten output converter could be imagined. The actual hardware and software reality is a bit short in these departments, but computers—even very small computers—that understand the spoken word are already available.

EXAMPLE 16-4

How much data (how many bits) can be recorded on a C-90 (90-minute) cassette, using the Kansas City Standard recording technique?

Solution: We start out by noting that Kansas City Standard defines a mark—a logic 1—as eight cycles of 2400 Hz and a space—a logic 0—as four cycles of 1200 Hz. Both have a duration of 3.333 ms, so the data rate of Kansas City Standard is 300 bps, for either marks or spaces. Since we have the number of bits recorded in a second, we have to determine how many seconds long a C-90 cassette is. This is

$$\frac{90 \text{ minutes} \times 60 \text{ seconds}}{\text{minute}} = 5400 \text{ seconds}$$

Since 300 bits are recorded in a second, how many bits are recorded on a C-90 cassette (in 5400 seconds)?

$$300 \text{ bits/second} \times 5400 \text{ seconds/cassette}$$
$$= 1,620,000 \text{ bits/cassette}$$

The answer is 1,620,000 bits.

QUESTIONS

16-1. What is the task of an input device?

16-2. Redraw the bit-switch input device in Figure 16-1 to show how the switches would be set for the number 01000001.

16-3. Redraw either Figure 16-2(a) or (b) to

handle a 10-key keyboard (with keys numbered 0 through 9).

16-4. What is the purpose of the Schmitt trigger logic in Figure 16-4?

16-5. Could a ROM be used in the place of the

diode encoder for the eight-key octal keyboard in Figure 16-2? Would the ROM be used efficiently?

16-6. Why do we prefer active-LOW signals in TTL keyboards?

16-7. What are hardware, software, and firmware?

16-8. Which is faster, a hardware-encoded input device, or a software-encoded input device?

16-9. What is the difference between a smart I/O device and a dumb I/O device?

16-10. Describe the difference between saturation recording and audio (linear) recording.

16-11. Write definitions for the terms "inter-record gap" and "tapemark." What is the difference, if any?

16-12. What is the major reason microcomputer manufacturers use serial audio or saturation (single track) recording instead of multitrack reel-to-reel recording?

16-13. KC standard tape data are recorded at a lower density than multitrack (saturation) recording. (True or False?)

16-14. Magnetic tape is like a one-dimensional recording (on a straight line). To access information, you must travel up and down the line, past all the bits, to find the one you want. You have also heard of 2D (two-dimensional) and 3D (three-dimensional) RAMs and core memory. Which kind is most similar to a magnetic disk pack?

16-15. Which provides faster access to randomly distributed records on a magnetic recording, tape or disk?

16-16. A floppy disk is made of flexible material. (True or False?)

16-17. What would you do to file-protect a cassette tape?

16-18. How many characters are needed to fill a Hollerith punchcard?

16-19. Why are there check brushes in a punchcard reader?

16-20. Describe three reasons why punchcards might jam in a punchcard reader.

16-21. How many bits of data encode a character in 96-column, 80-column, and 90-column cards?

16-22. List one advantage, and one disadvantage, of punched paper tape compared to punchcards.

16-23. What sort of sensors are used to read OCR symbols and UPC code?

16-24. Which of these would be the easiest way to digitize a drawing that's already done, to store it in a computer's memory?
(a) Light pen
(b) Tablet
(c) Joystick
(d) Video digitizer
(e) OCR reader.

16-25. Describe how a DMA input device stores numbers in a computer's memory.

16-26. A digital tape recording is played back from the tape into a computer. If it's a monaural tape recorder, the input port connecting the recorder to the computer is probably a:
(a) serial I/O port
(b) parallel I/O port

16-27. Voice-recognition input equipment is a form of:
(a) D/A input device
(b) A/D input device

DRILL PROBLEMS

Drill problems for Chapter 16 have been combined with those for Chapter 17 and appear at the end of Chapter 17.

17

PERIPHERALS: I/O DEVICES (OUTPUT)

━━━━━━━━━━━━━CHAPTER OBJECTIVES━━━━━━━━━━━━━

By the time you finish this chapter, you will be able to:

1. Identify the basic function of *output devices* connected to a computer via its output ports.

2. Describe how *LED readouts* are interfaced to the computer, and how decoding is provided for complex LED readout devices such as *seven-segment* and *dot-matrix* displays.

3. How to use software to replace hardware decoding for output devices.

4. Describe how an *LCD segmented display* works, and its advantages for battery-operated equipment compared to LEDs.

5. Describe the functions of *fluorescent* and *incandescent* displays and their characteristics.

6. Describe the two ways in which *CRT displays* are used to display graphics and alphanumeric information (random-scan and raster-scan).

7. Describe how a typical raster-graphics text display system works.

8. Describe the basic types of *hardcopy devices* (*printers* and *graphic plotters*).

9. Explain how the commonly used printer mechanisms work.

10. Explain how graphic plotters work.

11. Explain the basic types of output devices for generating *machine-readable documents*, and the types of recording used.

12. Describe how *actuators* and *electromechanical devices* are controlled.

13. Discuss, briefly, the application of devices like those in item 12 in *computer numerical control* and *robotics*.

14. Describe how an *audio synthesizer* operates.

15. Describe how the input and output devices in the preceding two chapters are classified into:
    a. The *human–computer interface*
    b. The *document–computer interface*
    c. The *machine–computer interface*

In Chapter 15 we discussed output ports. What is done with the digital code once it is latched in the output port is the job of the I/O device. Whether it is a display, a document-generating machine (even one that makes documents for another computer to read), or a controller—that activates some sort of machine on computer command—the output device is the computer's way of talking to the outside world. In this chapter we split output devices into **displays**, which produce some visual effect to indicate the digital state of the output port, and **electromechanical devices**, which use the digital signal to control some sort of motion.

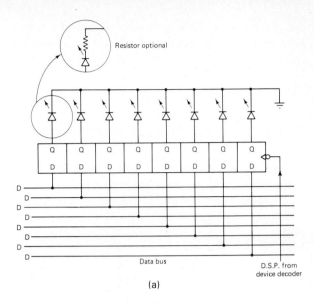

(a)

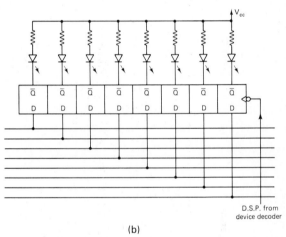

(b)

FIGURE 17-1 (a) Active-HIGH and (b) active-LOW binary output devices.

17-1 BINARY READOUTS (LED PANEL)

This is the output-device equivalent of a bit-switch panel. There is one LED lamp attached to each bit output of the output port's latch. Figure 17-1 shows that all there is to this sort of device is a light, a resistor (optional in configuration a), and a connection to the power supply ground ($-$) or V_{cc} ($+$) terminal for each bit of data. The LEDs which are the output device merely make the bits latched in the output port visible. What is displayed is the same information taken from the data bus, without any modification other than being converted from voltage into brightness.

In Figure 17-1(a), the HIGH states of the latch's Q outputs make the anodes of the lamps positive. The cathodes are grounded, so they are more negative than a HIGH anode, and the LEDs attached to HIGH Qs light. LEDs attached to LOW Qs have the same LOW voltage on anode and cathode, and don't light.

In the case of Figure 17-1(b), the $\overline{Q}$ output of each bit is used to drive the LED's cathode, while the anode is attached through a resistor to a voltage that's always HIGH. When a HIGH bit is taken from the data bus, the not-Q output (and the cathode) becomes LOW. This LOW makes the LED light, since its anode is more positive than its cathode. When a LOW bit is taken from the data bus, the not-Q output is HIGH. This HIGH voltage at the LED's cathode is just as HIGH as the anode and the LED's are dark, since the cathode must be at least 2 V less positive than the anode before the LED will light.

The result, either way, is that the binary readout shows a snapshot of what was on the bus at the instant an output operation was done. As with all other output devices, the bit panel continues to show the same data until its output port is strobed again.

This type of display is simple, but it requires the user to read directly the binary code used inside the machine. In Chapter 16 we described smart and dumb peripherals, with the smart peripheral having its own smarts and the dumb peripheral having the smart work done for it by the CPU. The binary LED display, and its input counterpart, the bit-switch panel, have no smarts of their own, but do not depend on the smarts of the CPU either; they must be used by a smart user.

17-2 SEGMENTED DISPLAYS

To display numbers and letters from binary output, rather than lights in ON and OFF states, a smarter output device than the binary LED readout is necessary. This method of displaying numbers and letters (segmented displays) is used in calculators and digital clocks. We've discussed it before, so we'll simply detail two ways these devices can be used.

17-2.1 LED Seven-Segment (Numeric)

Figure 17-2(a) shows a familiar seven-segment display and a **character set** that can be formed by these seven segments for the decimal numbers 0 through 9. One method of using this display is shown in Figure 17-3, where a decoder is combined with the display to make the actual output

device. The combination of a binary-to-seven-segment decoder/driver with the seven-segment readout gives a device that can be operated directly from the binary bus yet display two digits of decimal, provided that BCD numbers are placed on the bus. All current microprocessors include an instruction in their **instruction set** that transforms binary codes into BCD numbers.

The other method of using the seven-segment display is to attach its LEDs directly to the bit outputs of the output port latch. Then, special codes are sent on the bus that display in the shapes of the character set. For example, the code 01111111 can be put on the data bus for the number 8. To do this instead of using the normal bi-

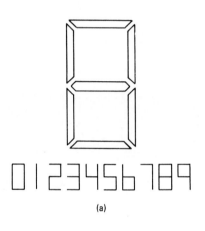

(a)

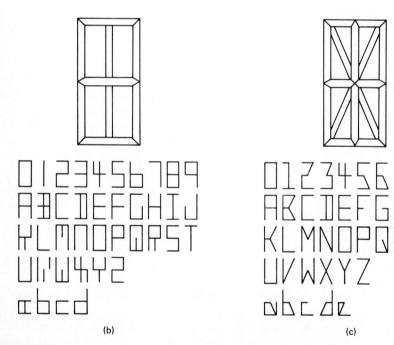

(b)

(c)

FIGURE 17-2 Segmented readouts: (a) 7 segments; (b) 9 segments; (c) 16 segments.

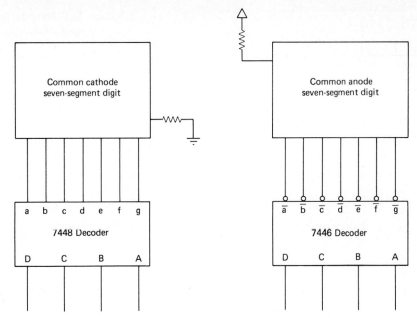

FIGURE 17-3 Decoder/drivers for segmented readouts.

nary codes, a special program (called a *display driver*) has to be running in the computer when this display is used. The combination of seven-segment readout plus display-driver program works like the combination of the decoder/driver circuit and display. We can say in this case that the software (driver program) replaces the hardware (decoder/driver). The circuit for this "dumb" version of the output device is shown in Figure 17-4, along with a software block-diagram (a **flowchart**) of what the display-driver program has to do.

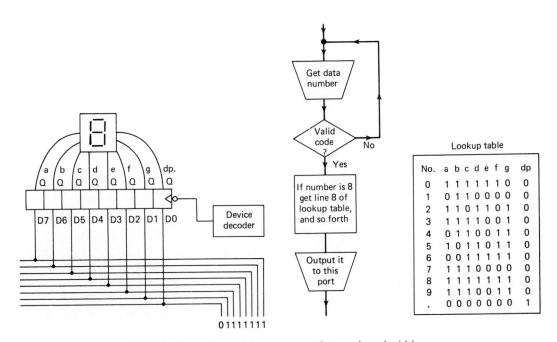

FIGURE 17-4 Software used to replace a decoder/driver.

EXAMPLE 17-1

Examine the circuit of Figure 17-4. If the pattern of segments on the seven-segment readout is:

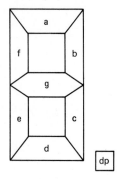

What code should we send to the port in order to have the seven-segment readout display the letter "F"?

Solution: In the lookup table of Figure 17-4, the data structure of bytes for the seven-segment readout is

$$\begin{array}{cccccccc} a & b & c & d & e & f & g & dp. \\ D_7 & D_6 & D_5 & D_4 & D_3 & D_2 & D_1 & D_0 \end{array}$$

We need these segments to make an "F." They are segments a, e, f, and g. To turn on these segments, and not any of the others, the data word we will have to construct should contain 1s in the bits whose segments should be lit, and 0s in all the other segments.

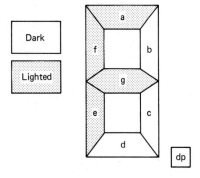

The data word should appear like this (in binary):

$$\begin{array}{cccccccc} a & b & c & d & e & f & g & dp. \\ D_7 & D_6 & D_5 & D_4 & D_3 & D_2 & D_1 & D_0 \\ 1 & 0 & 0 & 0 & 1 & 1 & 1 & 0 \end{array}$$

If the computer requires data to be entered in hexadecimal, this would be 8E, and if data are entered in decimal, 142.

17-2.2 LED Nine-Segment and 16-Segment (Alphameric)

Figure 17-2(b) and (c) show displays with more segments and more characters in their character set. As more segments are added, the number of types of characters it is possible to make with a segmented display goes up. With nine or more segments, it's possible to define a whole alphabet as well as the numbers of the seven-segment display, and these can be called **alphameric displays** for that reason. There are drivers, both hardware and software, for these types of display. These devices, and their interface to the output port, differ from those in Figures 17-3 and 17-4 only in having more LEDs and more wires.

17-2.3 Dot-Matrix Displays

We said that as more segments are added, the number of different characters that can be formed out of them gets larger and larger. This is true not only for the absolute number of characters (with 16 segments, over 65,000 combinations of lit and unlit segments are possible), but also for the precision with which familiar shapes can be represented using these segments. For instance, a 2 represented in seven-segment displays looks like a backwards, square S, but with 16 segments, it is a lot more recognizable as a 2. The 5 × 7 **dot-matrix display** in Figure 17-5 makes very recognizable shapes for every alphabetic or numeric symbol, and handles a number of punctuation marks as well. There are a total of 35 LED dots in this display. Although it's possible to produce over 34 billion different patterns with this device, that's not the point. The shapes you want to make—no matter what they are—can probably be found among these 34 billion patterns, and a pretty good fit can be made to any alphameric or punctuation character.

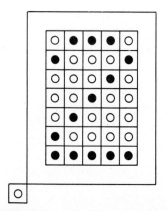

FIGURE 17-5 A 5 × 7 matrix display.

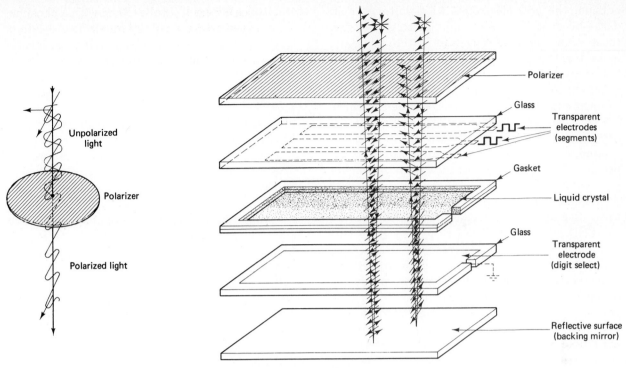

Unpolarized
light

Polarizer

Polarized light

Polarizer

Glass

Transparent
electrodes
(segments)

Gasket

Liquid crystal

Glass

Transparent
electrode
(digit select)

Reflective surface
(backing mirror)

FIGURE 17-6   LCD display.

17-3   LCD SEGMENTED DISPLAYS

LCDs (**liquid-crystal displays**) are passive light transmitters or light reflectors that do not emit any energy. Since they do not emit light energy—in fact, they're visible only if illuminated from elsewhere—they don't consume much energy to run. In fact, they consume almost no energy, which is what makes LCDs desirable in applications like digital wristwatches, where the unit must run from small batteries for a year or more.

As regards character set and character formation, the LCD is just like the LED display. The working theory, however, is much different. To understand how LCDs work, you must first understand what polarized light is.

Normal light, such as daylight or electric light, is *unpolarized*. That means that the waves of electromagnetism that make up the light vibrate any way they please, and a beam of **unpolarized light** is made of waves vibrating in all sorts of different directions at once. Now imagine a sort of filter with slots in it that only let through light that vibrates in one direction. Waves vibrating in the other directions don't get through the slots. This is shown in Figure 17-6. The filter is called a *polarizer*, and is actually made from crystals whose molecular lattice structure forms the slots in sizes small enough to be important to a light wave. Once you pass normal light through this filter, you have only the part of it that was

polarized in the direction the filter was aligned. The **polarized light** can pass through another polarizer only if it is aligned in the same direction. If the polarized light or the second polarizer is rotated, less and less light gets through the second filter (called an *analyzer*) until, when the light waves and the polarizer are exactly crossed, no light at all gets through.

In a digital wristwatch display (reflective type), as shown in cross section by Figure 17-6, the light coming into the top of the picture is polarized by the polarizing filter (the top layer in the diagram). The polarized light is reflected from a mirror layer on the bottom back up and out through the polarizer again, to the outside world. If nothing else happened to the light, it would always get back out because the reflected light is still polarized in line with the polarizer. The space between the polarizer and mirror is filled with a nematic liquid crystal which does nothing to the polarized light passing through it, except when an electric field lines up the molecules of the *LC material* (which is why it's called a liquid crystal). Transparent electrodes placed just under the polarizer can produce an electric field in the LC material when a small voltage is applied to them. Three electrodes are shown in the picture, two with voltages, and one without. The electric field rotates the polarized light so that on the bounce it's crossed to the polarizer, and can't get out through the same filter it got in. From the outside, the two electrodes with voltages look dark—because there's no light getting out from those areas—and the electrode without voltage looks transparent—since the light bouncing off the shiny bottom layer can be seen. The whole effect of the charged and uncharged electrodes is to form characters out of dark segments where the voltages are, as compared to the LEDs, which are lighted where the voltages are. The background, with no electrodes, is shiny all the time, because there's no rotation of the polarization angle to block the light.

The seven-segment LCD display is typical of these displays. Its electrodes are shaped just like the LED displays. Unlike LEDs, very little current flows between the electrodes in an LCD, because the LC material is an insulator, something like oil or grease chemically. The electrodes form a capacitor, which consumes virtually no true power, except for a small leakage current. The currents used for LEDs are a few milliamperes per diode, but the current for LCDs is less than a thousandth of that. Of course, this means that battery lifetimes, largely determined by the current LEDs draw in devices like calculators, can be extended (thousands of times!) by using LCD readouts instead. Like capacitors, though, LCDs are slow to react, and the switching of segments is slow enough to drag visibly. Also, the LC materials used in a lot of these applications don't work well at all temperatures. In a wristwatch, where the device is strapped to a human wrist, it's kept warm, and functions OK. When the LC material gets cold, its responses slow down and the contrast between lighted and dark segments changes. Also, you need some source of light to see the LCD by, whereas LEDs are self-illuminated. Finally LCDs have short life spans compared to LEDs, because their complex organic chemistry tends to break down—but the life span is several years anyway, and is likely to be long enough to match the obsolescence life span of most digital devices. In the next few years, look for LCDs to take over in many of the digital applications where LEDs are presently used.

17-4 GAS DISCHARGE/INCANDESCENT/ MECHANICAL DISPLAYS

Displays that are similar to the LED/LCD segmented types but use older technology include fluorescent/gas plasma discharge tubes, incandescent displays, and mechanical or electrochemical displays.

Fluorescent displays are basically electron tubes or vacuum tubes lined with a phosphor that shines when struck by an electron beam or beam of energized secondary particles generated in a rarefied gas. The name "fluorescent" indicates that these are similar in nature to a fluorescent lamp, although the beams of charged particles that excite the phosphor can be steered or switched by electrodes you won't find in your kitchen fluorescent tube. Some pocket calculators use these instead of LED segmented displays. They are blue or green, and usually brighter than LEDs, but they eat up batteries at an even faster rate.

Gas plasma discharge tubes come in a variety of forms, some organized like segmented LED displays and others quite different. One type popular before LEDs came on the market was called a *Nixie* tube. This device was basically a neon glow-lamp with a grid-shaped positive electrode and 10 negative electrodes made of wires in the shape of the decimal numerals. Since a neon lamp's negative electrode glows when a dc voltage is applied to it, the number of your choice could be displayed by connecting its negative electrode to the negative voltage and leaving all the other number electrodes floating. The shaped-wire electrodes were thin, and 10 of them could be stacked one behind the other in a glass envelope the size of a miniature vacuum tube. Voltage between the positive and negative electrodes was about 80 V. As you can imagine, controlling 80-V displays with 5-V logic devices took quite special integrated circuits, and a separate 80-V supply for the readouts. Being quite complicated to make, Nixies cost several dollars per digit, and being glass vacuum tubes (low-pressure neon, really), they break when they're dropped. LEDs are fairly indestructible in this regard, and cost less, and run from the same 5-V supply as everything else—so they've replaced Nixies for all display applications in present-generation equipment.

If you're servicing older units with funny orange-shaped digits that walk back and forth as they change, we suggest caution when you're testing the live circuits. Circuits that run on a single 5-V supply can't do much to you, but you can catch a nasty tickle from an 80-V Nixie-driver power supply.

Incandescent displays are basically Edison light bulbs with filaments shaped like numbers or segments in a segmented display. The filaments carry currents large enough to get them hot, and they glow. Being less efficient than either LEDs or fluorescent displays, these really eat up batteries. Nobody uses them in pocket calculators, but they're found in cash registers (point-of-sale terminals) and other plug-in applications where low power consumption is not essential. They are bright and easy to read. Since they emit white light, they can be covered by colored filters and be viewed in any color. Incandescent displays are available in a variety of voltages, and being filament devices, they aren't very particular whether the voltage they get is dc, ac, smooth, or noisy.

Fluorescent displays and incandescent displays contain filaments that burn out. The lifetime of a filament device is limited by this factor. The fluorescent devices are additionally limited by charged-particle etching of the phosphor until it is burned away in high-current, high-brightness devices.

Mechanical/electromechanical displays use the bits of the output port (or a decoder) to switch on currents to electromagnets. The electromagnets move displays mechanically (for example, flipping light-colored segments over to their dark side). Interfacing electromagnets and electromechanical devices is described in detail later in the chapter.

17-5 CRT DISPLAYS/VIDEO DISPLAYS

These are what almost everybody calls *TV* or *monitor displays*. There are really two basic types, *raster scan* and *random scan*. Both are called **CRT (cathode ray tube) displays** because they use the picture-tube display device familiar from TV sets and oscilloscopes. Cathode rays are a quaint Victorian term for electrons, the negative charge carriers in all circuits. In a CRT, a focused beam of electrons is formed and aimed at the front of the tube from a section in the back called the *electron gun*. Where the beam strikes the front of the tube, it produces a bright spot because the face of the tube is coated with a phosphor. The beam is deflected or steered to various places on the face of the tube by either electrostatic or magnetic fields (oscilloscopes usually use electrostatic deflection, but televisions invariably use magnetic deflection). Both types of tube and their deflection circuits are shown in Figure 17-7.

17-5.1 Vector-Graphics Display (Random Scan)

Random scan is used by CRTs in oscilloscopes. The electron beam is moved in two dimensions by forces that steer the spot of light so that it traces

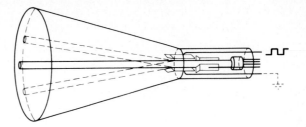
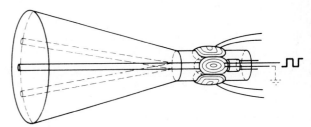

FIGURE 17-7 Two methods of CRT beam deflection.

the outline of a figure on the phosphor. For every two-dimensional shape, there is a combination of two waveforms (one in the X direction and one in the Y direction) that move the electron beam around in the outline of the shape. The figure formed by the beam when it is steered by these two waveforms is called a *Lissajous pattern*. Figure 17-8(a) shows how a Lissajous pattern for the number 8 forms the character on the screen. To produce the waveforms, an output port attached to an A/D converter must be fed numbers very quickly to move the beam in an 8. The 8 must be retraced often enough in a second to give the impression of a continuous, stable figure 8 on the screen—about 30 times a second—which is very quickly in human terms, but no big deal to digital logic circuitry.

In Figure 17-8(b), the Lissajous pattern for an A is shown. A third waveform for *spot brightness* (called *Z-axis modulation*) is responsible for "blanking" the beam (dashed line) while it is traced up to the crossbar of the A from the lower right-hand corner. A combination of X-, Y-, and Z-axis modulation signals can form any character, whether it is continuous or made of segments like a dotted line. Blanking is also used as the beam moves from one letter to another when it is tracing out a line of letters.

17-5.2 Raster-Graphics Display (Raster Scan)

Raster scan is the method used in television picture generation. It is sometimes called "video" for this reason, although many raster displays are not standard video at all.

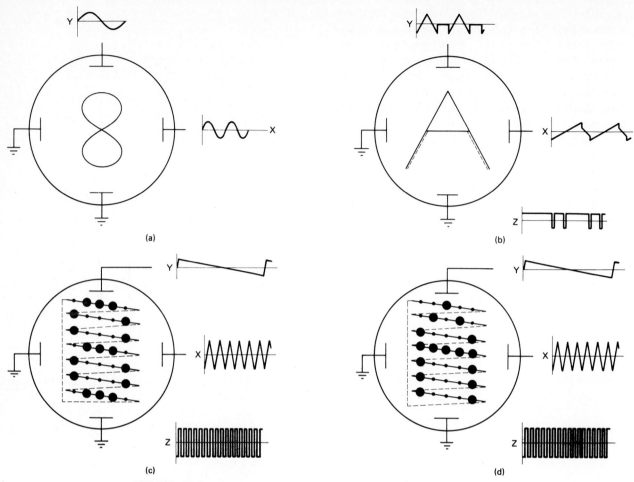

FIGURE 17-8 Random-scan and raster-scan displays: (a) random-scan 8; (b) random-scan A; (c) raster-scan 8; (d) raster-scan A.

Figure 17-8(c) and (d) are raster displays of the number 8 and the letter A. Notice that the figures are all made of dots like a dot-matrix readout. The X and Y waves are the same for both figures; only the Z-axis modulation is different. Basically, each character is formed by blanking the beam until the raster (the pattern scanned by the X and Y waveforms) reaches the right point on the screen, then turning the beam ON at that instant. The raster scans every point on the screen, so it's just a matter of time before the scanning action reaches every point in an 8 or an A where a dot should be. The character set is the same one used in an earlier section for the 5 × 7 LED matrix. Timing generation and synchronization between the raster and the Z-axis modulation is accomplished by triggering the scanning action of each line in the raster with a pulse from a counter (called a *horizontal sync pulse*). A chain of counters (called a *video divider chain*) is responsible for synchronizing just how many dot times appear before a new line is scanned. Each dot time is a moment at which a bit can be trans-

ferred to the Z-axis modulation control to make a *dot* (ON) or *undot* (OFF). The video divider chain also increases the downward position of each scan line below the last until a certain count (for TV, 525 lines, but for monitors, it's often 512 or 256 lines) of lines is reached. Then the counter responsible for vertical position resets and the next scan line appears again at the top of the screen.

Who transfers each dot or undot to the Z-axis modulator? The answer varies. It could be the computer's output port feeding serial data directly from the data bus—but the computer has to run horrendously fast to do this and have time for anything else. More generally, the character generator is a ROM programmed with dots addressed by the count in the video divider chain and an ASCII from a memory device's outputs (the memory is called *video RAM*) or an output port of the computer. The raster display that gets its ASCII codes from a part of the memory (video RAM) is a memory-mapped I/O device, because, to see the characters on the video, one has merely to place them in the video RAM part of the memory. The TRUE I/O approach to video is inconvenient in raster display for the same reason that feeding dots one by one from a serial output port is inconvenient—the CPU just wouldn't have time for anything else, running at microprocessor speeds. Most microprocessors are too slow to keep up with this task at all unless a limited number of characters per line (such as 32) is used. Limiting the number of characters per line reduces the number of dots and thus the frequency of dot transmission.

Random scan and raster scan each have advantages and disadvantages for certain applications. Random-scan graphics permits very high resolution line drawing, but as the number of lines on the screen gets larger, the scan gets slower and slower. Raster-scan graphics uses familiar and inexpensive television technology and is ideally suited to text (lots of letters and numbers on the screen) display, but is not capable of producing the high-precision quality line drawings of random scan. The number of figures on a raster-scan display has no effect on the speed

(frame rate) of the display. The hardware used for raster scan—a television circuit with a video divider chain and character-generator ROM—is cheaper than the hardware for a vector graphics (random scan) display—a set of D/A converters and decoders for the X, Y, and Z signals.

17-5.3 A Typical Raster-Scan Text Display System

Let's look at the way a screen display that uses raster-scan to display text might be constructed. Figure 17-9 illustrates how the video memory is accessed to display the characters on the screen. For the moment we will not concern ourselves with the question of how the characters got into the video RAM memory. We will simply assume that 24 rows of 80 characters are already in the RAM and will be displayed in the 5×7 dot-matrix format shown at the bottom of Figure 17-9.

First, the dividers in the video divider chain are all counter chips, which count in binary code, having the modulus indicated in each block of the

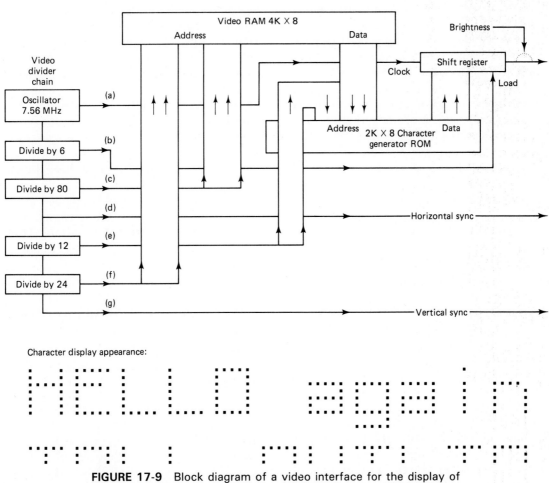

Character display appearance:

FIGURE 17-9 Block diagram of a video interface for the display of 24 lines of 80 characters each. Characters are 5×7 with descenders.

diagram. Output (a) in the diagram is the **dot clock,** which completes one cycle every time another dot in the matrix character is to be sent to the brightness, or Z-axis modulation, of the CRT. Since the characters being sent to the CRT are 5 × 7 dot-matrix characters, and one dot space is being left between characters (an **undot**), every character is six dots wide, and six binary states will be sent to the CRT brightness control during the time it would scan one character.

The shift register gets the patterns, six dots at a time, from the **character generator ROM,** and a new pattern is loaded into the shift register from the **video RAM** every six cycles of the dot clock. Signal (b) from the divide-by-6 circuit does this. It is called the **character clock,** since it completes one cycle every time a new character begins to be scanned on the CRT.

Every cycle of the character clock, the next character must be fetched into the character generator ROM's address inputs from the video RAM. Output (c) of the divide-by-80 counter goes to the least-significant bits of the video RAM's address bus, and selects a different ASCII character from among those stored in video RAM every time the character clock completes a cycle. Output (c) contains a number between 0 and 79, which is the position of the character on the 80-character line.

At the end of each 80-character side-scan line, a **horizontal sync** pulse must be sent to the CRT to return the electron beam to the left side of the screen and start another scan line. The most-significant bit, output (d), of the divide-by-80 chip, is used for this purpose.

Output (e) is the 4-bit code from the divide-by-12 chip, and has a binary value between 0 and 11. It identifies which scan line in a line of characters is currently being sent to the CRT. This number is combined with the ASCII code of the video RAM output (the character currently being scanned onto the CRT) to make a unique address which identifies the storage location of a six-dot pattern that represents one of 12 "slices" through the matrix character (the fourth row of dots in the letter A, for example, has a different address from the fifth row of dots in an A). There are 12 lines used for each line of characters because the 5 × 7 matrix characters take seven lines plus those that lowercase characters with descenders, such as the letter "g" in the matrix-character example at the bottom of Figure 17-9, need. The descender on "g" uses two additional lines in the example, and three lines are left between lines of text so the descenders don't "bang into" the next line of characters down on the screen.

Output (f) is a binary number between 0 and 23, which is advanced every time the most-significant bit of output (e) completes a cycle. It identifies which of the 24 lines is being scanned on to the screen at the moment. Since there is a divide-by-12 counter between the two parts of the address sent to the video RAM, the same 80 addresses are accessed in the video RAM 12 times, before a new set of 80 addresses is accessed.

This allows each of the 12 scan lines of a row of characters to be scanned. The characters gotten from the video RAM are in ASCII code. Their code is combined with a number from output (e) that makes up an address for the character-generator ROM. Although the same 80 characters from the video RAM appear each of 12 times, they are combined with a different **scan-line number** every time, thus fetching data from a different place in the ROM for different scan lines of the same character.

When all 24 text lines of 12-scan-line data have been read from the video RAM, converted into dot patterns by the character generator ROM, and clocked out to the CRT brightness control by the shift register, it is time to start the whole thing over from the top. This is the function of output (g), the **vertical sync** pulse, which restores the location of the electron beam to the upper-left-hand portion of the CRT. The electron beam will then scan out the picture again, unless the data have changed in the video RAM, and in that case, it will scan out a new picture showing the changes.

With the 7.56-MHz dot-clock frequency shown, a single scan line of 80 characters (six dots each) takes place 15,750 times per second. The entire screen is 288 scan lines from top to bottom (24 text lines of 12 scan lines each) and would be repeated about 55 times per second. That would be often enough to eliminate objectionable flicker as the screen is rescanned.

Eighty-character text lines and 24-line screens are commonly used throughout the computer industry as a standard CRT screen display format.

EXAMPLE 17-2

Assume that a video raster interface is being used to produce 7 × 9 dot-matrix characters from binary numbers stored in a character-generator ROM. What are the hexadecimal numbers that should be stored in nine ROM addresses to produce the character shown below? (Assume that a shift-left register is being used in the design, so that the leftmost bit in the binary word makes the leftmost dot on a scanned character.)

Solution: If we assume that illuminated pixels (dots) are 1s and the rest are 0s, the pattern translates into

```
  1 1 1
 1     1
1   1 1   1
1    1    1
1   1     1
1   1 1 1 1
1   1 1 1 1
  1 1 1
```

(displaying 1s only) or

```
0011100
0100010
1010101
1000001
1001001
1100011
1011101
0100010
0011100
```

(displaying both 1s and 0s). Assuming that every line needs an "undot" to separate the character from the next character to its left, we add a zero to the most-significant bit of each line above, and convert to hex:

		Hexadecimal
byte 1	00011100	1C
byte 2	00100010	22
byte 3	01010101	55
byte 4	01000001	41
byte 5	01001001	49
byte 6	01100011	63
byte 7	01011101	5D
byte 8	00100010	22
byte 9	00011100	1C

17-6 HARDCOPY (PEOPLE-READABLE DOCUMENTS)

Hardcopy is a term that refers to devices that produce some sort of permanent document you can take with you as their output. Two types of documents (hardcopy) are possible; there are documents that can be read by people and documents that can be read by machines. The people-readable documents fall into two classes; *printouts*—which are alphameric documents printed on paper—and *graphic plotting*, which is also printed on paper—but is in the form of pictures rather than words. The **machine-readable documents** come in a variety of types we've discussed before. We'll devote a little space to how these documents are written, but not to how they're read (that's in Chapter 16).

17-6.1 Impact Printers

These are the most familiar types of ink-on-paper printing devices. Many of them are available as office typewriters, but when they are used as output devices for computers, they become printers (or sometimes, *lineprinters*). Some are only used as printers in computer systems—these are found in large data-processing systems. We'll begin with these printers. They handle a high volume of output and are popular peripherals for business data processors.

Formed-Character Printers. **Formed characters** are made by striking the paper with an embossed character (a raised piece of metal in the shape of a character) through an inked ribbon. Each character is written by a separate metal or plastic shape.

Drum printers. This is one of several types of *speed printers* that print a line at a time, rather than a character at a time, and are appropriately called **lineprinters.**

In Figure 17-10, the working mechanism of the **drum printer** is shown. Part d is the *drum*, a metal cylinder embossed with raised letters. It is rotated by the motor, and as it rotates, a portion of it arranged as a shaft encoder is read by a group of photodetectors, p (each of these has a self-contained light source and photocell for detecting reflected light). The photodetector array provides a position signal that identifies what row of letters is under the *print hammers*, h. In this case, it happens that the letter G is lined up with the hammers.

When a voltage 1 is applied to one of the *print magnets* m, it *energizes*, as shown for magnet 5. The armature of the electromagnet pulls on a *print wire* w, which makes a *hammer* strike the paper. There are two hammers shown in the picture striking the paper. A hammer drives the paper against an inked ribbon, which strikes a letter on the drum as it flies by. In this example, there are two places on the current line where a

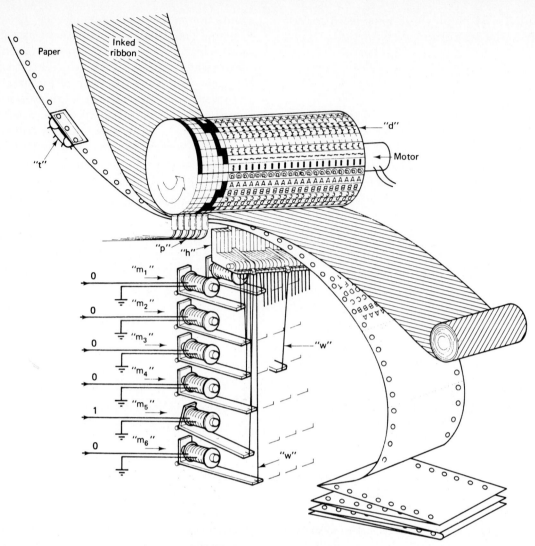

FIGURE 17-10 Drum printer.

G is being printed. When the drum has made a complete revolution, the whole character set on the drum has passed by the line, and hammers have been triggered to print every character on the line. The paper can be advanced to the next line by the *tractor* t, which is like a drive sprocket.

The drum printer can print whole lines every time the drum revolves. Maximum speed is a few thousand lines per minute. The limit on how fast this can be done is the speed with which the electromagnet-driven hammers can strike the paper and bounce off the drum. If the drum is moving too fast, the printed characters will smear in the direction of drum rotation. Improper hammer adjustment can also result in lines of print where the letters stagger up and down along the line (vertical misregistration).

If the drum becomes worn, the whole drum must be replaced, rather than just the characters that are worn, because the drum is a solid unit.

Cost for drum printers is in the ten-thousand- to hundred-thousand-dollar price range. The standard number of characters printed across a line for printers in the business/data processing area is 132. Drums with 132 characters on a line cost around $3000 to replace. The expense of drum replacement and constant electronic/timing adjustments for hammer firing make the drum less desirable than some other types of speed printers. Over the next few years, drum printers will be gradually replaced by the horizontal moving-font printers described in the following section. The drum printer was used in early printing telegraph mechanisms designed in the 1850s.

Chain (Train, Band, and Belt) Printers. The drum printer discussed in the preceding section had one alphabet of embossed characters for each *print position* (each place where a character would print on a line). The characters moved in the direction of paper motion, which made the drum a *vertical moving-font* printer. Figure 17-11 shows the mechanism of a **chain printer.** It is typical of **horizontal moving-font printers,** which also include *train printers*, *belt printers*, and *band printers*. All of these types have several full alphabets of embossed characters moving around a track across the line being printed.

In the example (Figure 17-11), a hammer behind the paper at each print position is fired as the character for that print position passes it. Exact timing is needed to hit the character on the fly. The *chain* c carries letters past the *hammers* h, which are fired as magnets m pull on wires w connected to the hammers. A *drum encoder* (shaft encoder) d keeps track of the position of the chain. Except for the chain itself, this mechanism is identical to the drum printer.

In the chain printer, type slugs carrying one or two letters each are held together (in a chain). Train printers also have type slugs carrying embossed characters, but the characters aren't connected together. The type slugs push each other around a track (like a train). Belt and band print-ers have a ribbon of embossed characters which are not made of separate type slugs. Since all of these pictures move type past the paper horizontally, they're called horizontal moving-font print-ers.

Like drum printers, chain printers are speed printers. They print a complete line in the time it takes for a single character set to pass by a print position. The speed of chain printers is in range from a few hundred lines per minute to a few thousand. Belt and band printers tend to be at the low end of the speed range; chain and train are somewhat faster.

Advantages of Chain, Train, Belt, and Band Printers. If type characters become worn, single slugs can be replaced in chain or train printers, and a belt or band with replaceable pet-als carrying embossed characters has been designed. This ability to replace single slugs or characters is a big advantage of these printers over the drum. In word-processing applications, for instance, the e and t become worn before the q. With a chain, train, belt, or band, you replace the e instead of the whole mechanism (as you

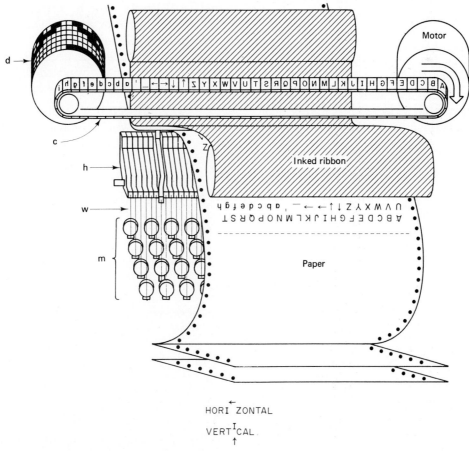

FIGURE 17-11 Chain printer.

would with a drum). Even if the entire band has to be replaced, as is the case where the band is not made of separate petals, the cost of a band with four or five character sets embossed on it is much less than a drum with 132 complete character sets.

If the timing adjustment of the print hammers is slightly off, the resulting horizontal misregistration is much less objectionable than the vertical misregistration that drum printers develop. Look at the two lines of letters shown at the bottom of Figure 17-11. They both have the same degree of timing error, yet the error in horizontal registration can hardly be seen, whereas the error vertical registration is clearly visible.

The price of chain and train printers is in the same range as drum printers; belt and band printers are lower in cost. The important cost savings with these printers is in the maintenance/adjustment cost, and replacement of worn parts.

Typewriter (Traditional Type-Basket). Invented by Sholes in the late 1860s, the **typewriter** mechanism can be used as a serial printer (one letter at a time rather than a whole line at a time). Solenoids (electromagnets) under each key mechanism would pull down on the key or push on the key from above [Figure 17-12(a)]. The average typewriter keyboard has about 52 keys, including control keys and space bars. Using 52 solenoids to push or pull on the keys sounds awkward, but there are interface units out there that do just that: in fact, we've seen some that just fit over the keyboard of the typewriter like a hood. This sort of mechanism is simple, but not partic-

ularly fast. Manual typewriters can type about five characters a second without jamming; electric typewriters are slightly faster.

Figure 17-11 shows the mechanisms of the standard typewriter, cylinder, golfball, and daisywheel printers.

Cylinder (teletype). This is similar to a standard typewriter, but instead of moving the entire carriage, platen, and paper every time another character is struck, the print head moves. The print head of the cylinder printer is shown in Figure 17-11(b). It has several rings of characters embossed around it. The head-carrier mechanism lines a particular character up with the ribbon and paper by rotating the print head until the correct column is aligned with the paper, then lifting or dropping the cylinder until the right row is aligned with the paper. Once it has been aligned, the head is driven toward the ribbon and paper, strikes the ribbon, and presses it against the paper, making an impression of the character.

The basic mechanism described here is used in Teletype machines, and is called a Teletype printer by everyone except companies competing with Teletype Corp. It is faster than a manual typewriter, printing about 10 characters per second—still pretty slow by lineprinter standards. The teletype printer is an old work horse that's been around for many decades. The mechanism is noisy but rugged, and used/reconditioned teletypes, costing a few hundred dollars, are very popular with small-computer enthusiasts.

Golfball (IBM Selectric). This type of mechanism [shown in Figure 17-12(c)] is called an

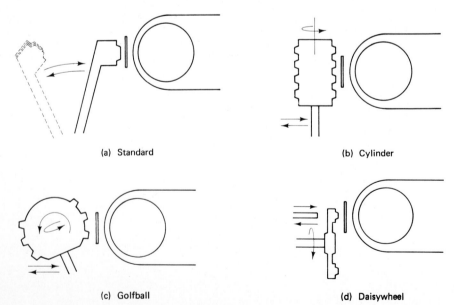

(a) Standard

(b) Cylinder

(c) Golfball

(d) Daisywheel

FIGURE 17-12 "Typewriter"-type mechanisms.

IBM Selectric by everyone except IBM's competition (sound familiar?). It is a variation on the cylinder (Teletype) printer that permits faster printing. The characters are embossed in rings around a sphere, and are aligned with the paper by a combination of a tilt and a rotate motion. This aligns the proper latitude and longitude on the sphere with the paper, and then the head is driven toward the paper, striking the inked ribbon and then the paper. The combination of tilt and rotate accomplishes the same type of positioning as the rotate and lift of the Teletype without having to lift and drop the entire printhead. This makes the golfball faster than the cylinder, because it's easier to swivel the head than lift it. Golfball printers print up to 15 characters per second, but cost more than Teletypes. There are fewer used or surplus units available, since the Selectric is a more recent development than the Teletype. Over the next few years, as used and surplus Selectric teleprinters become available, prices should come down from a thousand to a few hundred dollars.

Daisywheel/Thimble (Diablo/Spinterm).
In these designs [Figure 17-12(d)] a windmill-shaped wheel (**daisywheel**) carries petals with an embossed character on each petal. A hammer strikes a petal on the fly, driving it against the inked ribbon and paper, to print each character. Timing of the hammer-firing determines which character is printed. As with the chain printer, horizontal misregistration is possible if hammer timing is off, but vertical misregistration cannot happen.

Since only two motions—continuous rotation and hammer firing—are needed to print, the daisywheel is faster than either the golfball or cylinder printer. It prints at speeds up to 60 characters per second.

Daisywheel printers provide the highest speed and print quality available for printers in the under-$10,000 range. They are more expensive than golfball or cylinder printers—over a thousand dollars—but their high speed and inexpensive print element (daisywheels can be replaced at lower cost than golfballs or cylinders) are attractive features. Daisywheel/thimble printers seem to be capturing a large part of the golfball/cylinder market despite their higher initial cost.

Matrix printers: 5 × 7 matrix wire printer.
Matrix printers eliminate the need to rotate or swivel the embossed characters on a print head by eliminating the print head. Instead of a separate embossed character for each letter, the **matrix wire printer** uses seven wires driven by electromagnets to print dots by striking an inked ribbon against the paper. As the seven-wire print head is moved across the paper, letters are formed in the same character set used for an LED matrix.

In this design (Figure 17-13), the print head is also its own print hammer. As it moves across the paper, it prints part of a character on each stroke. For a 5 × 7 head, five strokes are used for each character with a sixth "dead" stroke between characters (an "undot"). The limiting factor on speed is how long it takes a wire to strike the ribbon/paper, bounce back, and reposition to be fired again. Top speed is about 200 characters per second. (Florida Data Corp. has one that runs at 900 cps!)

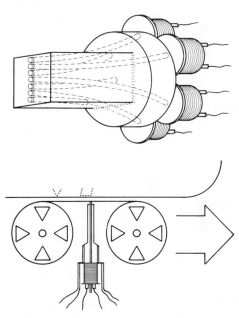

FIGURE 17-13 Matrix or "needle" printer mechanism.

Designs that use overlapping dots can produce dot matrix characters that look almost as good as formed-character print. These use more dots vertically (11 or 13 as opposed to 7 or 9) and more strokes horizontally, so they are slower than 5 × 7 printers, but still faster than daisywheel/thimble machines. As print quality and speed improve, matrix printers may replace formed-character printers. At the present time, fast matrix printers produce characters of such low quality (compared to "letter quality" formed-characters printers) that they are not acceptable for word-processing (office) applications. For data-logging and calculational applications, even low-quality print is good enough, so the 5 × 7 matrix has already replaced formed-character heads in such applications as cash-register-tape printers and printing calculators.

17-6.2 Nonimpact Printers

In the foregoing sections we have discussed printers that have multicopy capability—that is, they can be used to make carbon copies at the same time the original is being printed. In this section we'll see printers that are faster or cheaper, but do not have multicopy capacity.

Formed-character printers. The only formed-character nonimpact printer we've seen is the **oscillographic printer,** which uses photo-sensitized paper to make a copy of an image on an oscilloscope. If the scope is used as a vector display, the characters on the paper are formed, but if raster display is used, the characters on the paper are matrix characters.

A variation on this is a printer that uses a row of LEDs or a moving LED to scan a picture on photosensitive paper. Photosensitive paper has a tendency to fade with time, especially if it's left out exposed to sunlight.

Both of these are used more as plotters than as printers and do not represent an important fraction of any market as printers.

Matrix printers (Nonimpact)

Thermal. **Thermal printers** [Figure 17-14(a)] use a print head made of small electric heating elements with a small thermal lag (they heat up and cool down quickly). A head with seven elements can print 5 × 7 dot-matrix characters by moving across a sheet of thermal paper and printing one stroke at a time, like the wire matrix printer. Thermal paper is coated with a chemical that changes from light to dark when it is heated. The printer is inexpensive. It has the same transport mechanism (to move it across the paper) as a wire-matrix printer, but doesn't have the expensive electromagnets in the print head. The paper is more expensive than plain paper, but not outrageously so. The only quibbles we have with thermal printouts are the lack of ability to make carbon copies as the original is being printed, and the thermal paper's tendency to go gray over long periods of time. There is one other problem with thermal paper that makes it "soft" hardcopy. One of our friends discovered this when he left a five-page printout from a thermal printer lying on the radiator for a couple of hours. He came back to find a beautiful thermal photograph of the radiator top, with *some* of his printout still visible in the spaces.

Ink-Jet. **Ink-jet printers** [Figure 17-14(b)] have no print head at all. A *nozzle* squirts tiny drops of ink at the paper. The drops of ink are charged electrically, then deflected on their way to the paper like the electron beam in a CRT. The dots (actually small splatters) formed as each drop hits the paper are formed into matrix characters. With overlapping dots, the print quality approaches that of formed-character printers.

Since there are no moving parts in an ink-jet printer except the ink, these printers should be faster than anything we've seen so far. Their speed should be limited only by the time it takes the ink to dry. If the ink isn't dry when the paper starts to fold up on the floor, you end up with a Rorschach inkblot instead of a printout.

We can't think of any reason why this type of printer hasn't caught on a lot better than it has. It seems to have great promise, but hasn't made much of an impact on the market. Its price has remained rather high compared to other types.

Electrographic Printer. The **electrographic printer** [Figure 17-14(c)] writes on aluminum-coated paper by vaporizing the aluminum covering with an electric current. Two steel needles trail along the paper making marks whenever current is passed between them. A group of seven pairs of styli can produce reasonable 5 × 7 matrix characters. Underneath the aluminum coating, the paper has a black layer, so that the print has a black-on-silver appearance.

Advantages: Although this is a serial matrix printer, we've seen an experimental model cranking out paper at 4000 characters per second. That's faster than some lineprinters with drums and chains.

Disadvantages: No multiple-copy capability; the paper looks like aluminum foil and crinkles like aluminum foil. Fingerprints and black creases where the paper has been wrinkled make it unreadable after it has been subjected to a large amount of handling.

Xerographic (Laser) Printer. This is a xerox copier without an "original" to copy. It works in four steps:

1. A laser beam scans (raster) the image of letters onto a charged selenium drum [Figure 17-14(d)].

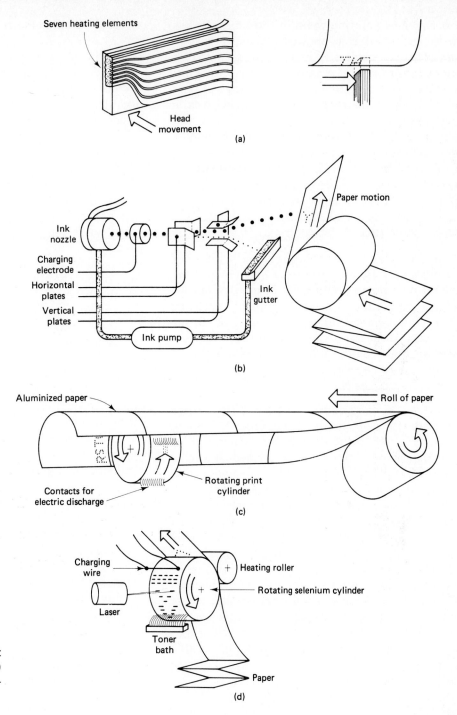

FIGURE 17-14 Nonimpact printing mechanisms: (a) thermal; (b) ink jet; (c) electrostatic; (d) xerographic.

2. The charges are dissipated by the photoelectric effect where the laser beam strikes the drum, so that a charge image of the dark spaces on the drum remains.

3. The charged areas on the drum pick up black powder toner, which sticks where the drum is charged.

4. A heated roller transfers the toner to a sheet of paper by melting it from the drum onto the paper. It soaks into the paper and bonds to the fibers when it cools off.

Since this type of printer prints a whole page at a time from a raster scan, it's faster than any other type of printer. Like electrostatic copiers in general, this printer is expensive, and not for the microcomputer. Although it doesn't have carbon-copy capability, this printer is useful in high-volume business applications because it can run off five copies by printing them in the time it takes an impact printer to print one five-ply carbon copy. This is a modest estimate. Speeds of 10,000 lines per minute can be handled easily by an electrostatic printer. Prices are in the range of tens to hundreds of thousands of dollars. Service calls and repairs/adjustments to machines of this type are quite frequent compared to other printer types. The main application of this type of printer is in very high volume printing applications, where thousands of pages per hour may be required (also called a *laser printer).*

EXAMPLE 17-3

Your boss asks you to recommend a printer for his office secretary's word processor. The printer should be able to produce letter-quality output with a capability to produce multiple carbons, since your business handles government forms and must have everything in triplicate. The cost should be less than $1500 and the volume that it produces each day would not have to be more than a few dozen pages of print.

Solution: A low-volume, letter-quality output with multiple-copy (carbon) capability will have to be an impact printer to get the carbon copies. It will probably have to be a formed-character printer to get true letter quality (although matrix printers that come *very close* are available). If low volume is needed (not much more than an office typewriter), one of the typewriter-like printer mechanisms seems appropriate. This means that you will be looking at an electric typewriter, cylinder printer, golfball printer, or daisywheel printer for this application. The laws of supply and demand have brought the cost of daisywheel printers with computer interface capability down below that of other typewriter-type mechanisms, and a variety of easily changeable type fonts (roman, italic, etc.) are available for daisywheel printers.

If you have decided on a daisywheel, the next place to look is the professional trade magazines that cater to the data processing profession. You could check with your public or company librarian, to help you locate the appropriate *Index to Periodical Literature* for computer products. These indexes list what reviews of printers are available in recent issues of trade magazines. *Datamation,* for example, runs a printer review annually that reviews printers by manufacturer, quality, cost, and other factors. From this, you could evaluate the best buy for your application from among the printers currently available.

17-6.3 Graphic Plotters; X-Y Drives

We split the world of human-readable documents into two parts, printouts and graphic plotting. Devices that produce pictorial rather than alphanumeric output are called plotters. Some types of printers (especially matrix types) can be used as plotters, but in this case we'll confine our discussion to devices that move a pen around on a sheet of paper.

An **X-Y drive** (Figure 17-15) is the mechanism that moves the pen around on the surface of the paper. It's basically a digital-to-analog (motion) converter whose output is a position specified by a binary number input. There are several methods for using a binary-code number to select a position; the simplest is to use the same kind of feedback we used with the printers to "see" what letter was being printed (Figure 17-15). An input device detects the position of the pen, the computer (or the plotter, if it's a smart plotter) compares it with the desired position, and activates an actuator to move the pen. The pen's moved forward (if it's too far back), backward (if it's too far forward), or stays where it is (if it's already where it belongs).

What's an actuator? Actuator is the word for something that moves things in response to a digital signal. A good candidate for the job is a *stepper motor,* a device that moves in equal-size steps when its controller gets pulses of digital code. Since steppers can be directly controlled digitally, they need no feedback to determine if they are positioned correctly.

Dc motors are more inexpensive than steppers, and can move either forward or reverse depending on the (+) or (−) polarity of the voltage. A rotation counter or shaft encoder is needed to feed the pen's position back to the computer so that it can properly control the motor.

The X and Y positioning motors are used by the plotter to draw pictures in the following way:

1. Two numbers (X, Y) are sent to the controller.
2. The controller moves the pen to a position where (X, Y) matches the desired position.
3. The controller lowers the pen point to make a dot on the paper—or it's been drawing a line on the paper all the time.
4. The next numbers (X, Y) are sent to the controller. (The next step is step 2.)

Using the plotter, all pictures are just a list

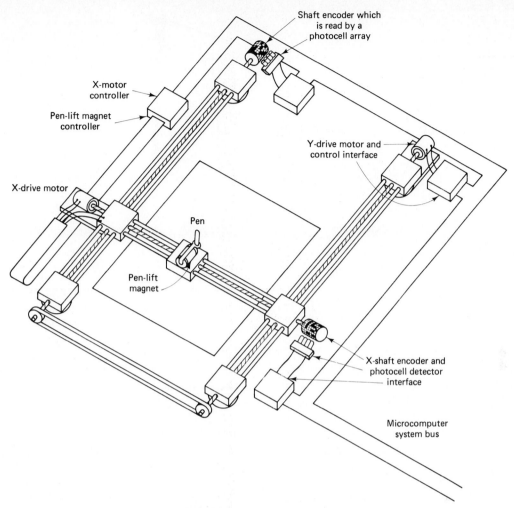

Shaft encoder which
is read by a
photocell array

X-motor
controller

Pen-lift magnet
controller

Y-drive motor and
control interface

X-drive motor

Pen

Pen-lift
magnet

X-shaft encoder and
photocell detector
interface

Microcomputer
system bus

FIGURE 17-15   X-Y plotter.

of points described by pairs of binary numbers (X, Y) and joined by lines connected to the last and next point.

17-7   HARDCOPY (MACHINE-READABLE DOCUMENTS)

Some people-readable documents are machine-readable documents (see OCR, Section 16-6) and some people can read the documents intended for machines (if they know Hollerith or BCDIC code). What we mean by "machine-readable" in this section is "documents generated by a machine that are uniquely intended for machine reading." For the most part, we've discussed the documents and their use in Chapter 16. Details that are discussed here will concern only the output devices that generate the documents; the details of those documents themselves will not be repeated here.

17-7.1   Hollerith/Punched Paper Tape, etc.

Figure 17-16 shows a general-purpose **punch** mechanism for punching holes in cards, tape, or whatever. Its actuator is an electromagnet, as with many of the other output devices we've discussed (the interface for switching on and controlling electromagnets is discussed at the end of this chapter). Unlike some of the other designs we've looked at, this punch derives its power mainly from a motor, and the magnets only control the action; they don't provide the driving force.

The *motor* (not shown) rotates the shaft of the *eccentric roller* e, which makes the *bail* b rock up and down. If the *punch magnet* m is energized, its armature pulls on *punch wire* w, and moves the *interposer* i so that it gets caught by the moving bail and pushed down. The *punch blade* p is attached to the interposer, and is pushed through

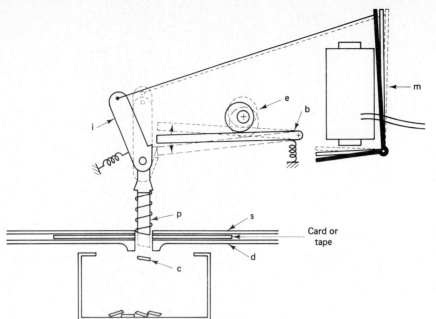

FIGURE 17-16 Punch mechanism for paper tape or punchcards.

the *stripper* s into the card, where it punches a hole. The rest of the card is held back by the *die* d as the blade cuts through the card, and the cardboard hole which is called *chad* c falls out the bottom of the die into the *chad bucket*. When the bail rocks back up, the spring withdraws the punch blade from the card, which is held down by the stripper. If the *magnet* deenergizes, the interposer will return to a position where the bail does not touch it, and no hole will be punched on the next stroke of the bail unless the magnet is energized again.

Since the energy used to punch the cards comes from the motor and not the magnets, small magnets are used which can be interfaced directly with the outputs of an output port. The exact nature of this interface is discussed later in this chapter.

Hollerith cards. For Hollerith cards with 80 columns and 12 rows of punches, two card-punch designs are possible. A punch assembly with 80 punch blades, which completes a card in 12 punch cycles, gives maximum speed, with cards moving through the punch assembly 9-edge (bottom) or 12-edge (top) forward.

A punch assembly with 12 punch blades, which completes a card in 80 punch cycles, is slower but less expensive. It would be used to punch cards traveling through the punch assembly 1-edge (left) or 80-edge (right) forward.

IBM automatic card punches (such as the model 1402) are constructed along the "80-blade,

12-cycle" design. They punch 250 cards per minute, with cards fed into the machine with their 12-edge forward.

IBM keypunches (such as the model 29 and 129) are normally manual machines, but can be interfaced as output devices for a computer, have 12 punches, and are constructed along the "12-blade, 80-cycle" design. They punch about 60 cards per minute running automatically, with cards fed into the machine with their 1-edge forward.

96-Column cards. The IBM 96-column card is punched as three rows of 32 columns of six-punch holes. The punch mechanism for these cards still looks like Figure 17-16, but the punch blade is round instead of rectangular, and smaller than the hole for an 80-column card. The arrangement of punches could be either 18 punch blades operated in 32 punch cycles, or 32 punch blades operated in 18 punch cycles

Punched paper tape. The punch mechanism in Figure 17-16 can be used for paper tape. The holes will be round, but larger than those on the 96-column card. Only one arrangement of blades is possible since tape is a continuous form. The five-, seven-, or eight-track tapes require five, seven, or eight punch blades. These could be operated from a single 8-bit output port. A drive sprocket uses the sprocket holes down the middle of unpunched tape to synchronize punch cycles with tape motion.

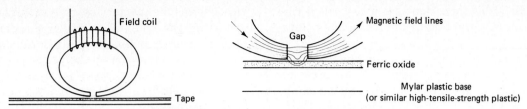

FIGURE 17-17   Writing on magnetic tape.

17-8   MAGNETIC (I/O) STORAGE DEVICES

17-8.1   Tape

Figure 17-17 shows the basic construction of a magnetic r/p head. When it's used to record, current pulses run through the coil producing a magnetic field in the iron magnet core. At the gap in the core, the magnetic field jumps to the ferrite recording surface—because it's a better conductor of magnetism than the air in the gap. The ferrite is magnetized by the field from the head. Some designs employ an erase head to clear the ferrite of magnetic fields before the record head magnetizes new data onto the surface.

In Chapter 16 we discussed two types of data recording, saturation and audio. Although the TRS-80 standard and Kansas City standard methods of cassette recording were described, we said nothing about the multitrack and disk recording schemes (although we did mention that the TRS-80 recording scheme was similar to IBM disk recording).

Reel-to-reel seven-track and nine-track formats.   Figure 17-18 shows five schemes used for saturation recording on tape and other magnetic media.

The first [Figure 17-18(a)] is called **RZ (return-to-zero) recording,** because the current in the head has a HIGH direction and a LOW direction, but returns to zero amperes of current (neither a HIGH nor a LOW pulse) between data pulses.

The second [Figure 17-18(b)] is called **RB (return-to-bias) recording,** and has current flowing in either a HIGH or LOW direction during data pulse times, and is in the LOW direction between data pulses. The name derives from the fact that the head is magnetized (it has a bias) at all times.

The third and fourth methods [Figure 17-18(c) and (d)] are called **NRZ (non-return-to-zero)** and **NRZI (modified non-return-to-zero) recording.** There is no space between data pulses. One data level follows immediately after the previous data level. In NRZ recording, current flows through the write head in the HIGH or LOW direction through the entire bit time. In the NRZI method, a toggle in the direction of write-head current during the middle of bit time identifies a 1, while a steady level identifies a 0.

The fifth method [Figure 17-18(e)] is called **phase-encoded recording** (also called the Manchester standard, or "biphase"). If the other methods of recording were level-triggered, this one is edge-triggered logic; a falling edge during

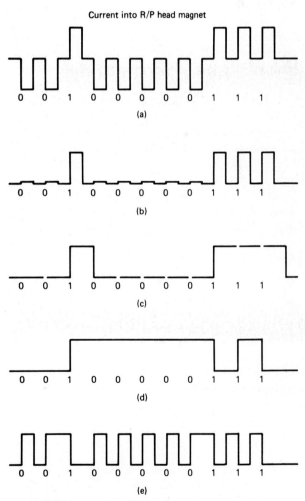

Current into R/P head magnet

(a)

(b)

(c)

(d)

(e)

FIGURE 17-18   Magnetic tape formats: (a) RZ; (b) RB; (c) NRZ; (d) NRZI; (e) Manchester Standard or Biphase.

the middle of bit time identifies a 1, while a rising edge during the middle of bit time identifies a 0.

Although this section is concerned primarily with output (i.e., how to put the recording onto the tape), we've included waveforms that show how the magnetic field looks on the tape and how the playback head will "see" the field when the tape moves past it. If you've had a course in the branch of mathematics called calculus, or if you've studied differentiator and integrator circuits in an electronics course, you'll recognize that the read-head waveform is a differentiated version of the write-head waveform.

Cassette: serial KC standard and TRS-80 tape. Both types of cassette, and most types of floppy disk and rigid disk as well, record on ferrite-coated surfaces with r/p heads the same as the one in Figure 17-17. Details of the recording formats were discussed in Chapter 16.

17-8.2 Disk

Both rigid disk and floppy disk recording methods have been discussed in Chapter 16. Some disk and drum recording schemes involve recording a timing pulse on a separate alongside data. For the five recording methods described for tape in the last section, a timing pulse would be essential to determining when bit time was for each new data bit. The self-clocking code used to record TRS-80 cassette tapes (Figure 17-19) simplifies this by putting the timing pulses on the same track with the data, between data pulses. This does, of course, reduce the density with which data can be recorded on the track by half, but simplifies the hardware requirements from a dual-track to a single-track head. For floppy diskette recording, where one-track data recording saves cost on the diskette drive, this same format is used.

17-8.3 Bubble Memory

The basic details of bubble memory operation were described in Chapter 14. As an output device, the bubble memory storage records data by distorting the domain structure of a ferromagnetic crystal. A bubble is created when current passing through a loop on the surface of the YIG (yttrium-iron garnet) crystal forms a strong local magnetic field in a direction opposite to the external bias on the crystal. A compact mobile *domain* is formed (a bubble) when current in this *generate loop* is strong enough to saturate the magnetic surface under it. The bubble is a 1. It can be destroyed by pulsing a reversed current through the generate loop as the bubble passes beneath it.

17-9 ELECTROMECHANICAL OUTPUT (CONTROLLERS)

In previous sections of this chapter, we've discussed output devices where the actual interface between the computer's output port and the final document (printers, etc.) was an electromagnet. Using the logic outputs of the latches in an output port to switch on magnets, relays, and motors is termed control, and the circuits that do the switching are termed *controllers*. The mechanisms (usually electromagnetic) that convert the control into actual work are called *actuators*. They produce motion from electricity, and do the work we want them to by producing controlled motion at the command of signals from an output port.

17-9.1 Relays/Electromagnets

Figure 17-20 shows an electromagnet interfaced to one bit of an output port. The action of the electromagnet (also called a *solenoid*) is binary in nature. It's magnetized, and pulling on something, or it's not. In the case of a *relay*, the magnetic field is used to close (or open) switch contacts. It may seem that a switch closed or opened by a digital signal should be solid-state, to eliminate moving parts and reduce switching time. For most purposes, this is true, but if you want to use a +5-V digital signal to switch on a 220-V (ac) motor that uses 10 A of current, the solid-state

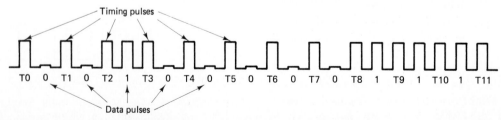

FIGURE 17-19 TRS-80 tape or IBM disk format.

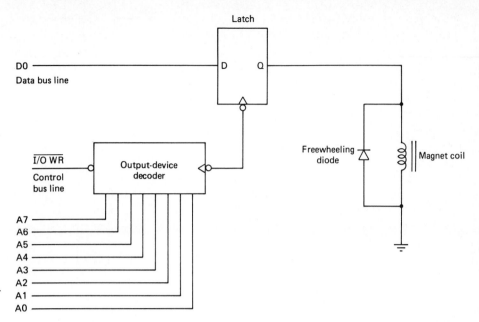

FIGURE 17-20 Actuator for electromagnets.

control gets complicated—especially if you don't want any of that motor's horrendous noise output to get back into the 5-V logic system. With a relay, the switch contacts closed by the magnet can be electrically isolated (not connected in any way) from the logic circuits. Those contacts can then carry any kind of electric energy—whether it's dc, ac, high-frequency RF, or whatever.

Since the relay's contacts are a metallic switch, they can connect things together with very low resistance, but like other switches, they bounce. In most control applications, this doesn't matter. Few loads attached to a relay are affected by bounce.

A more serious problem with all kinds of solenoids is the "back-EMF" problem. When the current in a magnet's coil is switched off, the collapsing field of the magnet generates a voltage that tries to keep the current going. This voltage is called a *back EMF*, and can be hundreds of times as large as the steady-state voltage on the coil. To protect the driver transistor and logic from these "EMF spikes," a diode called a *freewheeling diode* is added in parallel with the coil. In normal operation, the diode does not conduct, because it has the wrong polarity. When the back EMF develops, its polarity is backwards to the normal operation, and the diode conducts the back current. This clips off the high voltage of the diode and protects the rest of the circuit.

The punch magnets, print magnets, and so forth, that we saw in earlier parts of this chapter are all interfaced like the one in Figure 17-20, at one magnet per bit of output. For a gismo like the 80-column card punch, the 80 magnets would re-

quire 10 output ports, or 10 addresses of memory in a memory-mapped 8-bit system.

17-9.2   SCR/Four-Layer Devices

A family of solid-state devices called thyristors has the control characteristics of a relay without moving parts. The SCR can be used to switch on power to dc devices and the Triac can be used to switch on (and off) ac devices. Although these devices don't provide as much isolation as a relay, they can switch faster, consume less power, and take up less space than relays. Figure 17-21 shows how an SCR and Triac could be switched on and off by a bit from an output port. Like the magnet interface, these four-layer devices are basically binary in nature, and operate off one bit of digital data. The SCR is, unlike the relay, a latching device, and will keep conducting once its gate has received a pulse. To turn off the SCR, the current in the load must be interrupted.

17-9.3   Solid-State Relay (Optocoupler)

In Figure 17-22 a circuit called an **optocoupler** provides the link between the digital control signal and the load. One bit of an output port is used to turn an LED on and off. The light from the LED controls the current flow through a *photodetector*, and the photodetector's current switches on a power-handling device like a power transistor or Triac. This type of circuit provides isolation as good as a relay's. It has better speed, doesn't bounce, and doesn't generate back-EMF voltage spikes when the driver shuts off. Its only

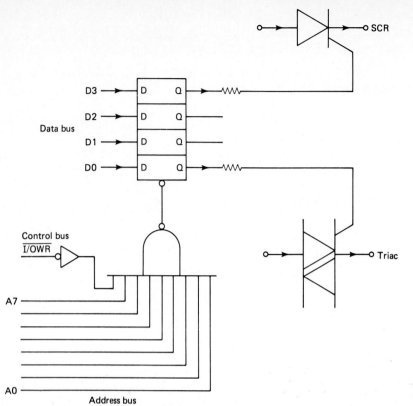

FIGURE 17-21 Dc and ac actuators.

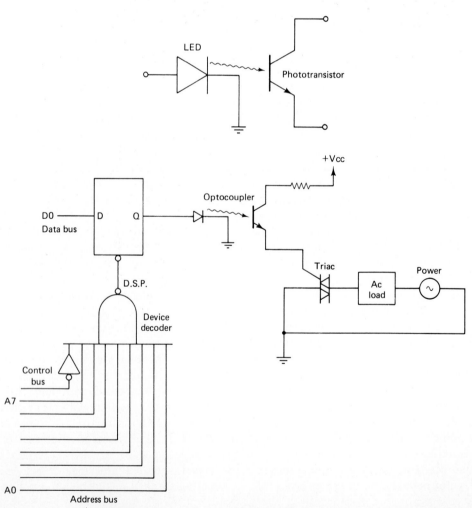

FIGURE 17-22 Optocouplers for output actuators.

disadvantage compared to relay control is slightly higher resistance in the circuit used to switch the load.

17-9.4 A Motor Drive with Directional Control

Figure 17-23 shows a circuit that uses SCRs and diodes to control a dc motor. The SCRs are the control devices and the motor is the actuator. The final user of the mechanical power is whatever the motor's attached to. Two motors like this could be used to make an X-Y drive, provided that some sensor tells the controlling computer when the actuator has moved everything to the right place.

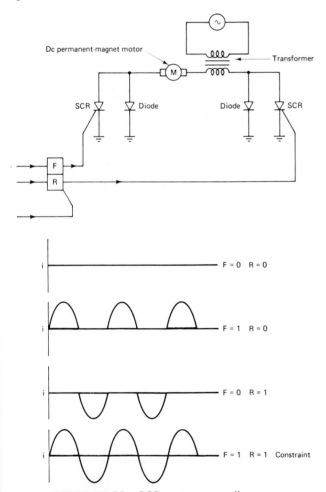

FIGURE 17-23 SCR motor controller.

One bit of the output port is used to switch on an SCR that lets current flow around the circuit in the forward direction. The other SCR permits current in the reverse direction. There is ac voltage available at the transformer secondary, but the SCR permits current in only one direction. When the forward bit of the output port is

HIGH, the current passing through the motor will make it rotate in one direction. When the reverse bit is HIGH, the current passing through the motor will make it rotate in the opposite direction. If both forward and reverse bits are LOW, there will be no current either way, and the motor will stand still. If both forward and reverse bits are HIGH, you're trying to do a silly thing—run the motor both ways at once. What actually happens is that an ac current flows in the dc motor, making it vibrate back and forth at 60 Hz. It's doing its level best at trying to go both ways. What will happen? Probably the motor will burn out, since the electrical energy is producing heat rather than motion.

All these conditions are shown on the diagram—with the last one (both forward and reverse HIGH) indicated as a constraint. Remember from Chapter 10 that a constraint is a "Don't do it!" condition?

17-9.5 Computer Numerical Control (Discussion)

Motors controlled like the one in the preceding section are used in industrial settings for things like smart machine tools. Combined with disk or drum encoders, motors like this can be controlled by computer programs to perform automatic drilling, milling, cutting . . . any industrial machining operation. With enough sensors, actuators, and smart enough programming, the smart machine tool becomes a robot machinist. The operation of such machines changes hands from a skilled manual laborer to a skilled computer programmer.

Such systems are called **numerical control systems,** because the motions of all parts of the mechanism are controlled by numbers from the output ports. In the overall view of system operation, the system is a digital-to-analog converter that makes binary numbers into mechanical action.

17-9.6 Audio Output

By attaching a speaker [Figure 17-24(a)] to one bit of an output port, the computer can generate sounds of various frequencies ("beeps"). Using a D/A converter [Figure 17-24(b)], waveforms of more complexity can be formed by the computer up to, and including, human speech and music synthesis. In the future, the computer that "talks back" may become more popular than the computer that "prints back."

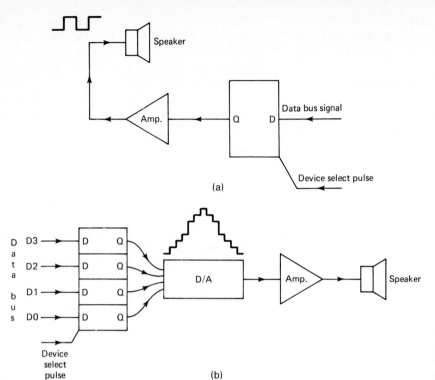

(a)

Device
select
pulse

(b)

FIGURE 17-24 Audio output devices: (a) simple tone generator; (b) waveform synthesizer.

EXAMPLE 17-4

How many bytes would it take to store 1 second of hi-fi (monaural or single-channel) audio in digital form? Assume that each digitized sample of the audio waveform is an 8-bit number.

Solution: High-fidelity (but not stereo) audio contains frequencies up to 20,000 cycles per second (20,000 Hz). Each wave of the audio waveform must be sampled several times during the waveform, and each sample converted into binary code, for digital storage. The minimum number of samples required to reconstruct a wave at all is two (Nyquist's sampling limit), one for the positive alternation and one for the negative alternation. That means that the absolute minimum number of samples per second, to handle frequencies up to 20,000 Hz, is 40,000 samples. If each sample is 8 bits (a byte), then

$$40,000 \text{ bytes} = 1 \text{ second}$$

is the absolute minimum needed to reproduce high-fidelity audio.

Audio sampling rates are often 56,000 or 64,000 samples per second for good sound quality, and for stereo recording, where two channels of audio are separately recorded, twice as many bytes are needed. Is it any surprise that CD (compact disk) recording media contain hundreds of millions of bytes of binary information?

17-10 SUMMARY: INPUT AND OUTPUT DEVICES

In the preceding two chapters, we have divided I/O devices into three main categories:

1. The *human–computer interface*

Input:	Output:
Bit switches	Binary readouts
Keyboards	Segmented/matrix
Light pens	readouts
Joysticks	CRT displays
Tablets	
Audio	Audio (speaker)
(microphone)	

2. The *document–computer interface*

Input:	Output:
Card readers	Card punches
Paper-tape	Paper-tape
readers	punches
OCR readers	Printers
Video digitizers	Plotters
Magnetic tape	Magnetic tape
(player)	(recorder)
Magnetic disk	Magnetic disk
(read)	(write)

3. The *machine–computer interface* (instrumentation and control)

Input:

Sensors with:
A/D converters
  Digital
    measuring
    instruments
  Code wheels/
    shaft encoders

Output:

Actuators with:
Controllers
  Lights
  Heaters
  Motors
  Solenoids, etc.

QUESTIONS

17-1. What is the task of an output device?

17-2. The $\overline{Q}$ outputs of Figure 17-1(b) drive the LEDs. Why?

17-3. Describe how a software lookup table replaces a 7446.

17-4. As more segments are added to a segmented display, the range of numeric and alphabetic characters it can display is extended. (True or False?)

17-5. What is the main advantage of LCD displays for calculator and wristwatch digital displays?

17-6. Describe how the electric field in an LCD digit darkens and lightens each segment.

17-7. What hazard is present in circuits with Nixie displays that is not generally found in circuits with LED displays?

17-8. Briefly describe each of the numeric/alphameric displays (described in Sections 17-1 through 17-4) in terms of brightness, readability, power consumption, life span, and versatility.

17-9. Name the two types of CRT deflection techniques.

17-10. Develop the waveforms for a letter T done with random scan and raster scan CRT displays. Show X, Y, and Z waveforms as in Figure 17-8.

17-11. Describe what a character-generator ROM does.

17-12. Name two different kinds of hardcopy.

17-13. In the following 10 parts, identify whether the type of printer listed is impact or nonimpact, formed character or matrix character, serial printing or parallel printing.
  (a) Drum printer
  (b) Chain printer
  (c) Cylinder printer
  (d) Golfball printer
  (e) Daisywheel printer
  (f) Wire matrix printer
  (g) Thermal printer
  (h) Ink-jet printer
  (i) Laser (xerographic) printer
  (j) Electrographic printer

17-14. Suppose that a graphic plotter is used to draw a giant letter A on a piece of paper. Would its action be more like a raster-scan display or a vector-scan display (random scan)?

17-15. The mechanism in Figure 17-16 derives all its "punch power" from the punch magnet. (True or False?)

17-16. Two different types of Hollerith card punches could be designed, one with 12 punch blades and another with 80 punch blades. What's different about the way each one feeds cards?

17-17. Describe briefly the following magtape recording schemes.
  (a) RZ
  (b) RB
  (c) NRZ
  (d) NRZI
  (e) Manchester standard
  (f) KC standard

17-18. The IBM disk recording format is used by Tandy for TRS-80 tape recording. (True or False?)

17-19. Define "actuator."

17-20. Why is a "freewheeling diode" used in a relay or electromagnet interfaced to an output port?

17-21. What solid-state switching device could take the place of the electromechanical relay for power switching?

17-22. What type of coupling is preferred between the output port and the actuator, if the load it controls is noisy?

17-23. CNC (computer numerical control) cannot be done as only output devices. Some feedback must be provided through input devices as well. (True or False?)

17-24. Describe how you would make a voice output device for a computer combining the circuit in Figure 17-24 with a ROM and a counter.

DRILL PROBLEMS
(Chapters 16 and 17)

Which devices listed are primarily *input*, primarily *output*, or is used about equally for *input and output*?

17-1.	Matrix printer	_____
17-2.	Floppy disk drive	_____
17-3.	Graphic plotter	_____
17-4.	Winchester disk	_____
17-5.	Cylinder printer	_____
17-6.	Joystick	_____
17-7.	ASCII keyboard	_____
17-8.	Raster video display	_____
17-9.	Graphics tablet	_____
17-10.	Shaft encoder	_____
17-11.	Motor speed control	_____
17-12.	Drum printer	_____
17-13.	Pressure sensor	_____
17-14.	Numerical keypad	_____
17-15.	Cassette tape	_____
17-16.	Chain printer	_____
17-17.	Barcode reader	_____
17-18.	Voice synthesizer	_____
17-19.	Golfball printer	_____
17-20.	Random (vector) scan display	_____
17-21.	Music synthesizer	_____
17-22.	Bit switch panel	_____
17-23.	Segmented LED display	_____
17-24.	Reel-to-reel tape drive	_____
17-25.	Punchcard reader	_____
17-26.	Matrix LED display	_____
17-27.	Punchcard punch	_____
17-28.	Paper-tape reader	_____
17-29.	Matrix LCD display	_____
17-30.	Daisywheel printer	_____
17-31.	Disk encoder	_____
17-32.	Digital light meter/sensor	_____
17-33.	Video digitizer	_____
17-34.	Laser disk drive	_____
17-35.	Nixie tubes	_____
17-36.	Thermal printer	_____
17-37.	Fluorescent display	_____

17-38.	Motor on-off controller	_____
17-39.	X-Y plotter	_____
17-40.	Voice-recognition system	_____
17-41.	Electric typewriter mechanism	_____
17-42.	Electrostatic printer	_____
17-43.	Incandescent display	_____
17-44.	OCR reader	_____
17-45.	Laser xerographic printer	_____
17-46.	Digital temperature sensor	
17-47.	Modem	_____
17-48.	Teletype (TTY)	_____
17-49.	Light pen	_____
17-50.	Code wheel	_____

Which devices provide a *human–machine interface*, a *document–machine interface*, or a *machine–machine interface*?

17-51.	Matrix printer	_____
17-52.	Floppy disk drive	_____
17-53.	Graphic plotter	_____
17-54.	Winchester disk	_____
17-55.	Cylinder printer	_____
17-56.	Joystick	_____
17-57.	ASCII keyboard	_____
17-58.	Raster video display	_____
17-59.	Graphics tablet	_____
17-60.	Shaft encoder	_____
17-61.	Motor speed control	_____
17-62.	Drum printer	_____
17-63.	Pressure sensor	_____
17-64.	Numerical keypad	_____
17-65.	Cassette tape	_____
17-66.	Chain printer	_____
17-67.	Barcode reader	_____
17-68.	Voice synthesizer	_____
17-69.	Golfball printer	_____
17-70.	Random (vector) scan display	_____
17-71.	Music synthesizer	_____
17-72.	Bit switch panel	_____
17-73.	Segmented LED display	_____

17-74. Reel-to-reel tape drive _____

17-75. Punchcard reader _____

17-76. Matrix LED display _____

17-77. Punchcard punch _____

17-78. Paper-tape reader _____

17-79. Matrix LCD display _____

17-80. Daisywheel printer _____

17-81. Disk encoder _____

17-82. Digital light meter/sensor _____

17-83. Video digitizer _____

17-84. Laser disk drive _____

17-85. Nixie tubes _____

17-86. Thermal printer _____

17-87. Fluorescent display _____

17-88. Motor on-off controller _____

17-89. X-Y plotter _____

17-90. Voice-recognition system _____

17-91. Electric typewriter mechanism _____

17-92. Electrostatic printer _____

17-93. Incandescent display _____

17-94. OCR reader _____

17-95. Laser xerographic printer _____

17-96. Digital temperature sensor _____

17-97. Modem _____

17-98. Teletype (TTY) _____

17-99. Light pen _____

17-100. Code wheel _____

18

INTRODUCTION TO MICROPROCESSORS

CHAPTER OBJECTIVES

By the time you finish this chapter, you will be able to:

1. Describe how computers use a *program* to perform automatic calculation, by fetching and decoding *instructions* from a memory device.

2. Describe the difference between *opcodes* and *data*.

3. Describe how the *buses* of 8-bit microcomputer systems are organized, and what signals are common to all of them.

4. Identify the characteristics of commonly used 8-bit CPUs.

5. Describe the function of *registers*, and identify what registers are found in commonly used 8-bit CPUs.

6. Describe the function of the ALU, and the significance of the register called the *accumulator*.

7. Read the microprocessor *timing diagrams* to identify the sequence of actions in a microcomputer system.

8. Describe the characteristics of some commonly used 16-bit CPUs, and explore their functions through the example of the Intel 8086 family of CPUs.

Even before the first integrated circuits were made, engineers engaged in a sort of wishful thinking that asked the question: "Someday, couldn't we put complicated circuits together in a single piece of semiconductor as nature has "built" the complicated circuits of the brain?" Always in their minds, as more and more complicated circuits were developed on single chips, the engineers imagined the ultimate digital circuit—perhaps the ultimate electronic circuit—being built in a single package as a single device. They were imagining the digital computer—the "thinking machine" of 1950s science fiction—as a single throwaway component like a resistor or a light bulb. Today, that wish is a reality. The **microprocessors,** "computers-on-a-chip," which first appeared in the early 1970s, contain all the essentials of a computer's central processing unit (thus the name "microprocessor").

In the realm of digital electronic circuits, the microprocessor falls into the same position the TV set does in the realm of analog electronics. To the analog electronics engineer, the TV is the one circuit that contains "a little of everything else." In the word of digital systems, the microprocessor is the one circuit that has almost everything digital in it.

Because the microprocessor is programmable, the many circuits inside it can be switched on in any sequence the programmer desires, and the numbers each circuit produces can be passed to another circuit for further processing. Since there's a circuit inside the micro for virtually any digital function, the microprocessor, suitably programmed, can imitate any other digital system—or combination of them—that you want. A clever and knowledgeable programmer can make the microprocessor do anything that another digital system does, replacing boards and boards of hardware (custom wiring between gates) with a microprocessor and suitable software.

In the beginning, microprocessors were CPUs for computers of very limited size. Early microprocessors handled numbers of 4 or 8 bits at a time. Later models had word sizes of 16 bits, and one-chip CPUs with 32-bit word size are now in common use.

Many of the currently popular microprocessors are still 8-bit machines (byte machines), and it is to these that we will devote most of this chapter. A wide variety of types and plans exist—but they share certain features in common. We will look first at those features found in all 8-bit micros, and later at what makes each of the three main families unique. At the end of the chapter, we explore 16-bit machines and look at those features that make them similar to, and different from, the 8-bit micros.

18-1 HOW COMPUTERS DECODE INSTRUCTIONS

We'll begin our discussion of microprocessors by seeing how digital computers, in general, do what they do. It's easier to understand new ideas when they relate to something familiar, so we'll begin with an analogy. Let's imagine that we want to use a common pocket calculator to do the automatic computing of a number. Our starting point will be an "el cheapo" pocket calculator with 16 keys [Figure 18-1(a)]. The keypad in our picture permits you to key the 10 decimal digits; a decimal point; the add, subtract, multiply, and divide operations; and an "equals" key. A simple calculator like this is called a "four-function" or "four-banger" because it has only four math operations.

This calculator does only one thing at a time—the thing on the key you are pressing at the moment. To add 8 plus 6, you must perform four keystrokes, 8, +, 6, and =. This is a manual operation—it is not automatic. If, after you have struck the 8, +, and 6 keys, you do not strike the = key, you won't be able to see the answer—14—on its display. Even though 8 + 6 has been calculated, and we're pretty sure that there's a 14 hiding in the calculator, the answer isn't displayed because we haven't struck the =. Actually, we can't even be sure that there's a 14 inside the calculator. Suppose that your key sequence proceeds like this: 8, +, 6, 0. You're actually entering the calculation of 8 + 60 (68), and it won't do to have a 14 popping out at you before you finish keying in the 60. You're not ready for an answer, yet, and besides, the 6 belongs in the tens' place.

It seems that this pocket calculator business is more complicated than we thought. The order in which you strike the keys is as important as the actual keys you strike. If we strike the 8, +, and 6 keys in a different order (+ followed by 8, followed by 6, for instance) we get a different result (86).

In the sequence 8, +, 6, =, we see two different kinds of key strokes—the 8 and 6 are data. These are the numbers that are the victims of our mathematical operation (another name for these data we're operating on is **operands**). The + and = keys are the operations that will be performed on the data. The + key, we suspect, will send the 8 to the full-adder circuit in the calculator's in-

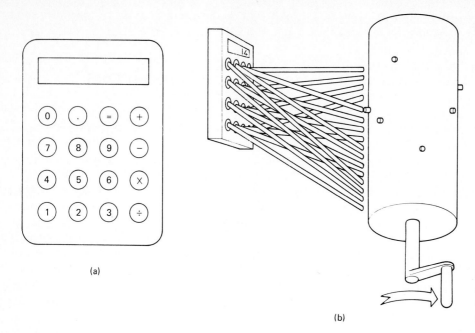

(a)

(b)

FIGURE 18-1 Automatic calculation—the hard way: (a) simple "4-banger" calculator; (b) program unit.

nards. The = key has a more complex task—letting the calculator know that the second number (6) is finished (it's not part of a 60), and this number should be added to what's already waiting in the adder (the 8). Further, the = key directs the calculator to put the adder's result on the display.

Now, a series of operations that adds 8 to 6 and displays the 14 on a display is hardly worthy of a digital computer. Suppose that we are using our calculator for a more challenging task, such as computing sales tax on items in a store. If the sales tax on merchandise is 6%, every price must be multiplied by 1.06 (the whole value, plus 6% more) to determine the cost after taxes. This can be done on our calculator for any number. First we key in the number; we'll represent this keywork by N, even if the number has a lot of digits and a decimal point. Then we multiply it by 1.06. The whole key operation is N, ×, 1, ., 0, 6, =. That's seven keystrokes even if N is only a one-digit number. Wouldn't it be nice to make this 6% tax calculation an automatic operation? How do we go about doing it?

Since we push on the buttons of our calculator to operate it manually, the first thing that comes to mind is an automatic button-pushing mechanism like the one in Figure 18-1(b). The drum with pins sticking out of it is used in music-box mechanisms to plink the notes. We've just scaled it up to a size that pushes calculator keys. If we forget about how this thing gets in the way of our fingers, we can imagine entering N by hand, then turning the crank and having "N +

6% tax" displayed when it's all over. A great labor-saving device!

This is, in fact, exactly what a digital computer does when it carries out automatic computation. True, there aren't any push rods or a drum with pins, but the idea is what's important. An "automatic" job like 6% tax computation has the same operation, ×, and the same second operand, 1.06, every time. The number 1.06 is a constant that doesn't need to be reentered by hand every time, if we can just manage to make the calculator remember it. This is the function of the drum-with-pins. It is a memory device that remembers the sequence of keystrokes mechanically.

It would make a lot more sense to use an integrated circuit memory to do this, wouldn't it? The memory can hold a list of keystrokes in a lot less space than our drum, and retrieve them a lot faster than fingers can move. We can do away with the push rods, too, by using a decoder to close each set of key contacts in the following way.

Suppose that each key on the keyboard has a four-digit binary identification number associated with it, as shown in Figure 18-2. The numbers 0000 through 1001 are the codes for the keys with the decimal numbers 0 through 9. That makes sense, because these are the binary codes for these decimal numbers. The rest of the keys are identified by the remaining 4-bit codes (1010 = +; 1011 = −; 1100 = ×; 1101 = ÷; 1110 = . and 1111 = =). The first 10 codes are data codes and the last six are **opcodes** (operation codes). Now, we want to push each key when its four-

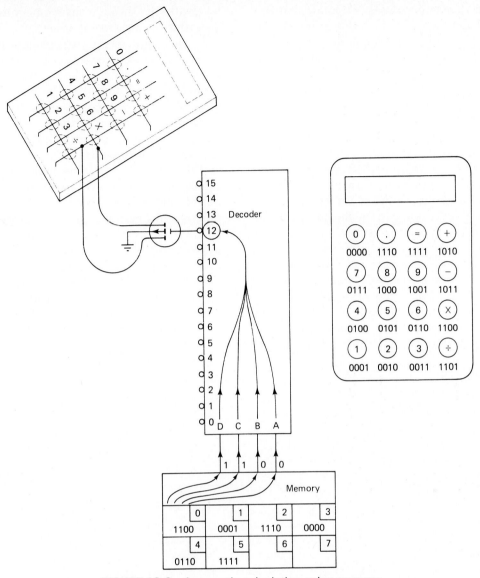

FIGURE 18-2 Automatic calculation using memory.

digit code appears at the memory device's outputs. What are we doing when we push a key? The key is just a switch connecting two wires beneath the keyboard. Instead of pushing on the key, we can get the same effect by switching a transistor ON, if the transistor is connected to the same two points as the key. A 4-line-to-16-line decoder can take any 4-bit code from the memory, and switch on an FET transistor wired across the key contacts (Figure 18-2). The 6% tax problem is now a string of opcodes and data codes. For our example, we need six machine codes to operate our "automatic machine" (1100, 0001, 1110, 0000, 0110, 1111).

We now have a *digital computer*. It's still very crude, and has a ridiculously small instruction set (+, −, ×, ÷, ., and =), but it works the same way the "big guys" do. An *instruction decoder* in "real" computers identifies the current opcode or data from the memory and one of its outputs switches on circuit(s) that carry out the operation or transfer the data.

We have a name for the list of instructions to the keyboard (1100, 0001, 1110, 0000, 0110, 1111). We call such a list a **program.** When the list works the way this one does, on a circuit wired to respond directly and uniquely to each code, the codes being used are called **machine codes** and the program is a *machine language program.* **Machine language** is the language the computer is wired for. For any real machine, the machine language it is wired for is the only language it knows.

Our pocket calculator "computer" is wired

to recognize 16 machine codes. We know that the decoder is important, but the order in which the codes operate the circuits is equally important. How do we get the memory to send the codes to the decoder in the right order?

In Figure 18-3, our computer has a *counter* attached to the memory device. We assume that the memory already contains the program stored in its memory cells. In the figure, the 1100 is stored in address 0, the 0001 in address 1, the 1110 in address 2, and so forth. The instructions of our program are stored in the same address order as the time order we want the computer to see them. To get the items in the memory to come out in sequence, a binary counter is attached to the address lines of the memory. In our example, this is a very small memory with eight storage locations. The addresses are all 3-bit numbers, so we have a 3-bit binary counter (the **program counter**) that counts 0, 1, 2, 3, 4, 5, 6, 7. This program counter makes the computer do the instructions in our program in the right sequence, by "walking" through the memory addresses one at a time, in increasing order.

To do just one 6% tax calculation, we have to be able to start and stop the program. When we do the computing, we want to run the program once, and stop when the result is displayed (after the =). For this reason, we've added a connection to the output of the decoder that goes on when there's an = (1111) code. Since the = is the last step in most calculations, we've arranged it so that the counter stops counting when an = comes along. The = (1111) code makes the 15 output of the decoder become LOW. A LOW level switches on the FET transistor across the = key on the calculator, but it also disables the clock pulses going to the program counter. The counter then freezes at the last step and won't go any further (which is a good thing, because there aren't any more steps!). To get the computer to run another 6% tax program, we have to get the program back to the instruction in address 0. This is done by using the reset input of the counter.

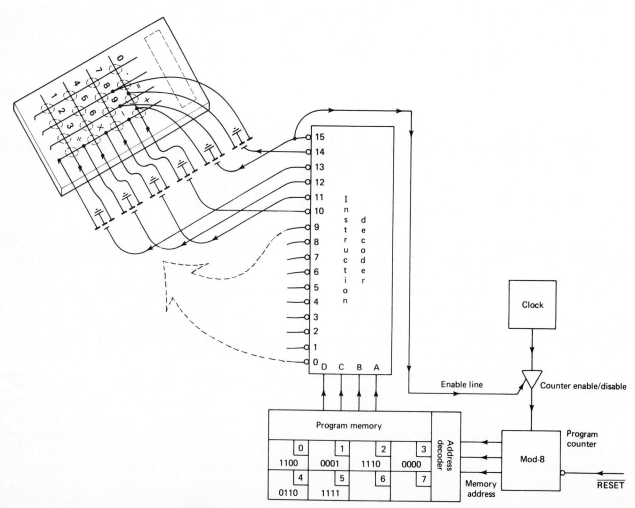

FIGURE 18-3 Addressing and the program counter.

We've added a pushbutton to the counter that will reset it (to 000) when pushed. This reset button is our way of making the computer run the program starting with "step 0" after we've keyed in the data N, which we want to add tax to.

Clumsy though this example is, it has all the important parts of a computer. We'll see items like the program counter, reset control, and clock in the following sections where we discuss some real microprocessors.

18-2   8-BIT MICROPROCESSORS

Since the microprocessor was designed for, and is intended to be, the CPU of a digital computer—all circuits that use microprocessors are really *microcomputers*, even if the label on the package says "football" or "calculator." The architecture of any circuit with a microprocessor in it is the architecture of a microcomputer. We'll begin by looking at what any 8-bit computer must contain.

All microcomputers are bus-oriented (rather than monolithic). This means that, for example, if you're adding an 8 to a 6 the 8 *or* the 6 can be on the bus at one time, but both are not simultaneously brought into the adder. Your computer has to fetch the 8, then store it somewhere, then get the 6, before the adder can be used. Why is this necessary?

A *monolithic* computer has separate wiring for each of the numbers. A two-address machine, for instance, fetches the 8 and the 6 into the adder at the same time. The monolithic computer also has more circuits, more wiring, costs more, and takes up more space than the bus-oriented computer. It also runs faster, but the other factors make bus-oriented architecture the "only way to go" in one-chip computers. In IC chip design, where the designer is trying to push the greatest number of advanced features into the smallest space (smallest number of gates and interconnects), monolithic design would be nonsense. Bus-organized architecture also cuts down the number of external connections to the IC package, an important consideration where additional pins add cost to packaging. (For 20% more pins, you pay a lot more than 20% more money.)

18-2.1   Bus Organization in 8-Bit Systems

The address and data buses.   Three buses are necessary for any microcomputer or minicomputer; these are the address, data, and control buses described in Chapter 15. The name "8-bit computer" comes from the size of the data being handled by the machine—8-bit words—and the data bus has eight wires. These are called D_0, D_1, D_2, D_3, D_4, D_5, D_6, and D_7.

The address bus (of most 8-bit microprocessors) has 16 wires called A_0, A_1, . . . , A_{15}. They permit control of 2^{16} (65,536) locations. In some systems, all 65,536 addresses may be used for memory devices; in others, some of the locations are used for I/O devices.

The control bus: signals found on all micros.   The control bus is the most variable part of the microprocessor. Each type of microprocessor has a different mix of signals used to control the parts of the microcomputer outside the CPU (output control signals) and to operate and control the CPU itself (input control signals). Some signals which all microprocessors share in common are:

Input   **1.** Clock (ϕ): synchronizes and drives all actions in the synchronous logic of the microcomputer—which is practically all the logic of the microcomputer

Input   **2.** Reset: resets all bits in the address contained in the program counter to zero (forces the computer to start running a program that begins with instruction 0)

Input   **3.** Interrupt request (INT): asks the CPU to accept commands from a peripheral device and run a special subroutine directed by that peripheral device (details in Chapter 20)

Output   **4.** Interrupt granted: shows that the CPU has accepted the interrupt request and is used by the requesting peripheral as a "go-ahead" signal for it to proceed with its interrupt

Input   **5.** DMA request: asks the CPU to go "on hold" and float the data, address, and part of the control bus

Output   **6.** DMA granted: shows that the CPU has floated the buses, and the DMA controller can take control of the memory. Data can be transferred directly from a peripheral to the memory

Output   **7.** Read (RD): says when CPU is using the data bus for incoming data (picking data up off the bus)

Output	**8.** Write (WR): says CPU is using the data bus for outgoing data (placing data on the data bus)
Input	**9.** Wait: makes CPU freeze data, address, and control signals on the buses to wait for slow memory or peripheral devices

The control bus: special design concepts. The control bus of practically every microprocessor has the nine signals listed above. For this chapter we have chosen the Intel 8080, Zilog Z-80, Intel 8085, Motorola 6800, and MOS Technology 6502 as representative 8-bit microprocessors. Additional signals described below are found only on some of these micros, depending on how the designer wants to handle certain important concepts.

Multiplexing: To limit the number of pins on the microprocessor's IC package, some designers use the same pins for multiple purposes.

The Intel 8080 uses the data bus to output eight control signals called a status byte. The rest of the time, the data bus pins carry instructions and data between the CPU and the memory or I/O systems, but during status time, the data are latched and decoded into control signals. A special signal called the status strobe identifies when it's status byte time on the data bus.

The Intel 8085 has separate pins for all its control signals, but its data bus is also the "low" half of its address bus. When the data on the A/D bus are really address, a signal called *Address Latch Enable* is used to catch the low half of the address in a latch, for later use by the memory.

These are basically things designers do when they run out of pins. Going from a 40-pin package to a 42- or 44-pin package costs a lot. The 40-pin package is so standard that designers will do some pretty extreme things to keep it.

True and Memory-Mapped I/O: Microprocessors in the 8080 family generate signals called I/O Read and I/O Write (input and output) as well as Memory Read and Memory Write signals for normal memory operations. With a 16-bit address bus, this allows the CPU to read and write at 65,536 memory locations, and to input and output from 65,536 I/O devices as well (the Intel machines use only half the address bus, allowing 256 I/O devices, but the Zilog Z-80 can actually address 64K of I/O ports).

Other microprocessors that have only the Read and Write signals cannot tell memory from I/O. There's no way they can use all 65,536 ad-

EXAMPLE 18-1

The Intel 8086 microprocessor has a 20-bit address bus. If it is used in a microcomputer system designed for **memory-mapped I/O,** how many memory locations, and how many I/O ports, can the system contain? How many memory locations, and how many I/O ports, would be possible if the microcomputer design used **true I/O** (also known as **separate I/O**)?

Solution: In *memory-mapped I/O,* an address is either a memory location or an I/O port (it can't be used for both). There isn't any one right answer, except that the total number of memory addresses plus I/O ports must add up to

$$2^A = 2^{20} = 1,048,576$$

which is also generally known as a "meg" or megabyte. For any memory-mapped I/O computer, the number of possible memory locations is

$$2^A - (\text{number of ports})$$

and the number of possible ports is

$$2^A - (\text{number of memory locations})$$

In a system with *true I/O,* the same address can be used for a memory location, and used, separately, for a port, without any conflict. In such a microcomputer system, the maximum possible number of ports would be

$$2^A = 2^{20} = 1,048,576$$

and the number of memory locations would be

$$2^A = 2^{20} = 1,048,576$$

which is 1 meg of memory and one megaport.

dresses. Some of the memory must be sacrificed to acquire I/O ports. Part of the I/O system must be mapped into memory locations.

Nonmaskable Interrupt (NMI): Interrupts normally break into a running program. If there is a command in the program to disable interrupts, it causes the CPU to ignore an interrupt request. The Disable Interrupts command has masked the INT pin (like masking tape is used to cover up the chrome and glass on a car before a spray-paint job). An NMI is an interrupt demand, not a request. The CPU, regardless of what software it is running, cannot ignore a nonmaskable interrupt. This input is used for some "can't-wait" routine, like one that kicks in the emergency battery pack when the power fails. Details of interrupt action are given in Chapter 20.

Polyphase and Single-Phase Clocks:

Some microprocessors require two clock inputs. The phase 1 clock, $\phi(1)$, goes from LOW to HIGH, then LOW again. When the $\phi(1)$ is LOW again, the phase 2 clock, $\phi(2)$, goes HIGH. When $\phi(2)$ goes LOW again $\phi(1)$ can go through its cycle again. This is called a *two-phase, nonoverlapping clock* [Figure 18-4(a)].

A microprocessor that only requires a single-phase square-wave input for its clock uses a *single-phase clock* [Figure 18-4(b)].

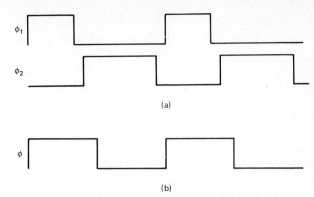

(a)

(b)

FIGURE 18-4 (a) Two-phase clock; (b) single-phase clock.

18-2.2   8-Bit CPUs

Figure 18-5 shows the CPUs of the Intel 8080, Zilog Z-80, Intel 8085, Motorola 6800, and MOS Technology 6502. The 8080 and 8085 have multichip CPUs.

The 8080 has a three-chip CPU. In addition to the 8080, two other chips are necessary for complete CPU performance. The 8228 systems controller has two jobs. One is to "catch" the status byte when the STSTB (status strobe) is active

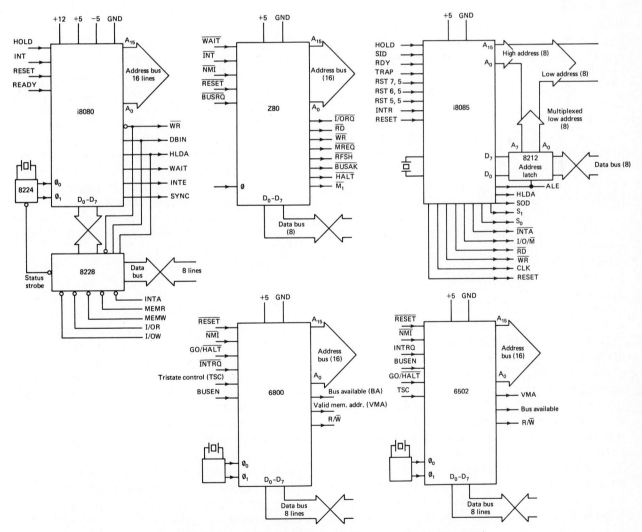

FIGURE 18-5 Popular 8-bit microprocessors (logic diagrams).

and decode the control signals multiplexed onto the data bus. The other job is to buffer and boost the signal power of the data bus signals when authentic data are traveling in and out on the bus. In this function, it is a bidirectional buffer.

The 8080 also requires a clock-generator chip (the 8224) to generate the two-phase nonoverlapping clock it needs. There is also logic in the 8224 for decoding the status strobe signal and boosting the power of one of the clock phases.

The 8085 needs a latch to catch the low half of the address bus when the ALE signal is active. No elaborate clock generator is needed. To clock the 8085, a crystal is attached to two pins on the IC, or a single-phase square wave is input to pin 1 (pin 2 is left floating). For full implementation of the address bus, the 8085 is really a two-chip CPU.

The 6800, like the 8080, needs a two-phase, nonoverlapping clock. Unlike the 8080 system's 8224, the Motorola clock generator is simply an oscillator, and does not contain any logic-decoding functions. Although generation of this type of clock signal is a fairly complex function, we'll call the 6800 a one-chip CPU. Clock pulses, like voltage from the power supply, are essential to the operation of the CPU, but the circuits that provide them aren't usually considered a part of the CPU itself.

The Zilog Z-80, hardware-wise, is like the 8080, 8224, and 8228 rolled into one chip. A single-phase square-wave clock is still needed to keep it going. Like the 6800, we don't count the simple square-wave oscillator as a part of the CPU, so the Z-80 qualifies as a single-chip CPU.

The MOS Technology 6502 attempted to go the 8085 one step better, having its two-phase crystal oscillator built in and needing only an external crystal to operate, while functioning without the 8085's multiplexed-bus latches. Unfortunately, the internal clock design lacked reliability in early 6502 models and has been replaced by a model that has an external clock.

All of these devices, with the sole exception of the 8080, operate from a single 5-V power supply, already used for the TTL latches, buffers, or multiplexers normally attached to the buses in full-buffered designs. The 8080, being the earliest, was designed at a time when 12 V was still needed to obtain optimum operation of the MOS gates in its CPU. The 8080 has +5 V, −5 V, +12 V, and ground connections. With 10% of the pins on a 40-pin package tied up in power connections, it's no wonder some of the control bus signals had to be multiplexed onto the data bus.

18-2.3   Working Registers and Their Use

Within every microprocessor there are a limited number of memory locations called **registers.** Some of them are connected with the calculating circuitry in the processor, whereas others are not, but are used to hold numbers while the calculating registers are in use (*scratchpad* registers). There are registers that count addresses for use by the address bus, registers that hold instructions fetched in from the data bus, and registers made of **flags** (which indicate whether the results of arithmetic came out positive, negative, or zero, overflowed the available space, and so on).

Figure 18-6 shows what registers are available in each of the 8-bit processors we described in earlier sections of this chapter. Just because a processor has more registers, it is not automatically better than another processor with fewer registers. The number of registers in a processor is only one measure of how powerful it is—there are many others.

In each processor, the register called the **accumulator** is identified as register A. The accumulator is the one register that participates in all arithmetic operations. Some of the processors have more than one accumulator, but every micro has at least one. In micros with true I/O, the accumulator is also the recipient of all input and the source of all output data. It has a "privileged" relation to the data bus. When memory is read or written, the data are transferred between a memory cell and a register in the CPU. Some processors permit transfer between memory and a large number of registers—others, only one or two—but the accumulator is always one of the registers used for Memory Read and Write operations.

Registers in 8-bit computers are 8 bits in size. Most micros permit grouping of register pairs which are then treated like 16-bit registers—but not all micros allow this. Some 16-bit registers exist—especially for addressing—which can't be used 8-bits at a time. On the illustrations, registers that can be paired are shown as rectangles joined at the edge, single registers are shown as detached rectangles, and 16-bit registers are shown as double-wide rectangles without a line in the middle. Flags, which are really independent one-bit registers, are grouped together in a byte (8-bit register) called the *flag byte* or *CCR.* On the diagrams, the flags in the byte look like a register broken into eight smaller parts.

The *program counter* and *stack pointer* are used to keep track of addresses in two areas of the

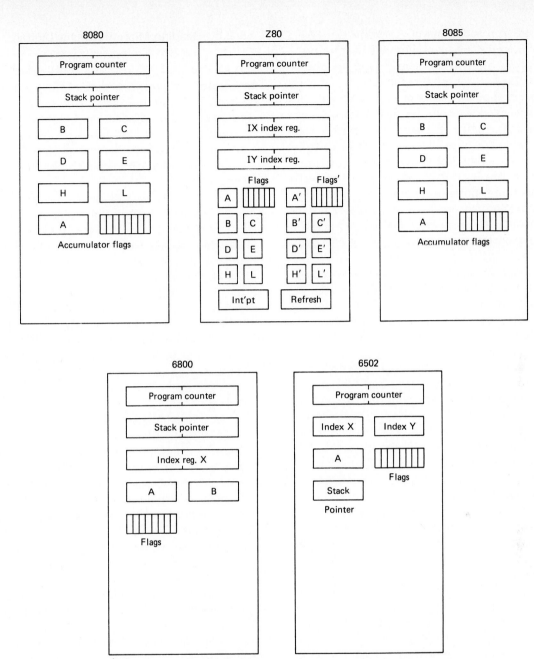

FIGURE 18-6 Popular 8-bit microprocessors (register architecture).

memory. The program counter contains a memory address which points at the next instruction in the program. The stack pointer also contains a memory address, which points at a part of memory called the *stack*. This is described in greater detail in Chapter 20.

All registers in microprocessors aren't available from outside to the programmer. Some of them are "for internal consumption only." We aren't including them on the diagrams, because they're of absolutely no use to the microcomputer technician—if you can't do anything about them . . . why worry about them?

18-2.4 The ALU

The **arithmetic-logic unit** (ALU) of a computer is the part that gives it its name. It is the section that computes or calculates with numbers. Most microprocessors have no arithmetic circuits beyond an *adder/subtracter* (see Section 6-1). The logic part of the ALU includes AND, OR, and NOT circuitry, as well as exclusive-OR.

Register A is used by the ALU for all arithmetic operations. A limited number of operations may be done in other registers. In the Motorola 6800, register B is a second accumulator, and duplicates all the functions of A, while the 8080, Z-80, 8085, and 6502 have only one accumulator that performs all arithmetic and logic functions.

The *flag byte* contains independent 1-bit registers which show the state of numbers after the ALU has worked on them. In the 8-bit processors we've described, all five have a plus/minus *sign* flag, a *zero* flag (that tells if the result is all 0s in A), a *half carry* flag (used only for BCD arithmetic), and a *carry* flag (which acts as a sort of ninth bit on the accumulator when numbers get too big for 8 bits). (You may recall from Chapter 6 that the carry output of a multibit adder/subtracter is also its borrow bit when it's subtracting.) An odd–even parity flag is provided on the 8080/Z-80/8085, but not the 6800. The 6800 has an *overflow* flag, that tells if the number is in true positive or twos'-complement negative form. In the 6502, a separate BCD mode of operation can be invoked by setting the *BCD/binary* flag—not found in any of the other processors.

All five processors contain, as part of their ALU, a "10-or-bigger" detector and "fudge factor" adder that convert binary to BCD when that conversion is desired. Binary-to-BCD conversion is not specifically used in the 6502, since it can operate in an all-BCD mode (if you skipped to this section without reading Chapters 6 and 7, or if you've forgotten the details of BCD arithmetic, this is a good time to go back).

A limited amount of 16-bit arithmetic may be done within the register pairs of the 8080/Z-80/8085 family (addition and subtraction in the Z-80; addition only in the 8080).

For the technician, the ALU and registers of the CPU are a "black box." If everything doesn't work, there's not a thing you can do to replace just a piece of the CPU; the whole part has to be replaced with a new one. That's the whole point to telling you what's inside the CPU. If you know it's one of the things locked inside the box, you know that there's nothing further to do but replace the box.

If, for instance, you find that something's wrong with register A, or that the ALU doesn't do a certain piece of arithmetic properly, but everything else works, you need to know that these parts are "locked in the box," and the only thing to do is replace the whole CPU (no matter how much of the rest of it still works). On the other hand, if you see a problem involving an address or data, such as writing to memory, the problem can be inside or outside the box and exploring the buses first (looking for shorts and opens) is a good place to start.

18-2.5 Reading Microprocessor/Digital System Timing Diagrams

In earlier chapters on gates and counters, we looked at timing diagrams. We noted that these diagrams are an important diagnostic tool for analyzing faults in a digital system. It is important to know what the waveforms for a part *should* look like, in order to find out what's wrong.

Some cases are simple. Counters generally have a predictable 1 or 0 at each output, which can be predicted if we know what the last state of all the outputs was. In some cases, however, even a simple digital system does not have a predictable 1 or 0 level in every timing state. For example, consider what happens in a parallel register when it is clocked. In Figure 18-7, an 8-bit latch is clocked by the falling edge of its CP input. The timing diagram shows what happens during three clock cycles. It is easy enough to see what the CLR (clear) and CP (clock pulse) inputs are doing, but what is happening to the Q outputs?

At first, the Q outputs appear to be 0. That is the result of CLR being active at the beginning of the diagram, and it is exactly what we would expect an active clear input to do. Then CLR becomes inactive at the beginning of the first clock (CP) cycle. At the falling edge of the clock, the Q outputs appear to become *both high and low*. What is that all about?

The first thing that bothers most people about this is the "sausage link" appearance of the wave, which appears to show each Q as high and low at the same time. The explanation is simple. After the first falling edge of the clock, each of the Q outputs can take on either a high *or* a low state, depending on the D inputs (which are not shown on this diagram). Since we have *no idea* what might be arriving at the D inputs during each clock cycle, and no way to show eight outputs at once in any case, the "crossovers" shown between the "sausage links" tell us where the output *can change*. Between crossovers, where the high and low lines are straight lines, the Q outputs, whatever they are, stay the same. Every crossover represents a moment in time when the last state of Q goes to the next state, without saying what that state might be.

Examine Figure 18-7. There are three falling edges on the CP (clock) wave. Following each falling edge, there is a **transition** at the Q outputs. This is a possible change from high to low, from low to high, or simply from the state of the

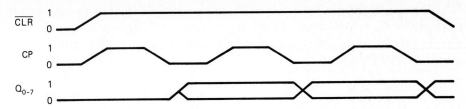

FIGURE 18-7 Eight-bit latch timing diagram.

last D input to that of the next D input. It could even be a change from an old low to a new low, or from an old high to a new high, which would never show up on a *real* multichannel oscilloscope trace at all.

If you used the CLR (clear) input to retrigger an actual oscilloscope trace every three cycles, however, while random data entered the D inputs, and looked at the pattern visible on the oscilloscope, any one of the Q outputs would actually produce a pattern like Figure 18-7. Why? Because what you see on the 'scope is produced by several sweeps and what you see is an "overlap" of all these traces. Because the wave starts out just after the CLR (clear), all Q outputs are low. Every time the pattern is traced out, the time between CLR and the first falling edge of the clock will have a 0 at Q. After that, however, things are unpredictable. Sometimes we are clocking a 1 into Q, and at other times, a 0 is clocked into Q. Overlapping traces with each possibility (from several consecutive sweeps of the oscilloscope), your eye sees both the high and the low states following the first falling edge. At the next two falling edges, sometimes the waveform rises, sometimes it falls, and you will see both outcomes as the repeated sweep of the scope produces both patterns. The crossovers in both cases will be visible, so the actual appearance of a repeated sweep on the oscilloscope will resemble Figure 18-7.

That's where this diagram came from. It is an idealized attempt to depict what an oscilloscope trace of an 8-bit latch (our example) would actually look like if we repeated the trace every three cycles (clearing the register at the beginning of each trace), and provided the D input of each latch bit with a random pattern of 1s and 0s (different every time the trace was redone).

The second thing that bothers most people when they see a diagram like this is the fact that all eight Q outputs are represented by a single waveform. Shouldn't there be eight?

All eight Q outputs would produce a similar-appearing trace in these circumstances. Every Q output would begin at 0, and then show transitions at the next three falling edges. All eight

patterns (overlapped from multiple traces) would look the same. There would be no point to showing eight identical patterns, so one pattern is labeled with the names of all eight outputs.

Example 18-2

From the computer timing diagram given below, identify whether the HOLD and HLDA signals are active HIGH or active LOW. HOLD is a request for the CPU to go Tri-State on its address and data buses, useful for allowing peripheral devices to obtain direct memory access (DMA). HLDA is a "hold acknowledge," a signal that responds to the request for a HOLD state, notifying the outside world that the DMA can be done, because now the address and data buses are Tri-Stated.

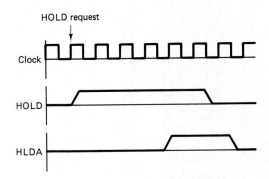

Solution: The state of the HLDA signal is active HIGH. Since it spends most of the time LOW, and is LOW before a request to go into the HOLD state arrives, then goes HIGH, it cannot be active low. It would make no sense for it to be active before a HOLD was requested, and in fact, it would take some time to become active after a HOLD was requested, because the microprocessor must complete whatever it's doing in its current machine cycle. Interpreting the HLDA signal as active high makes perfect sense in this situation.

The HOLD signal must be active HIGH also, since (1) it went HIGH right after the peripheral device's HOLD request, and (2) the HLDA signal (which responds to HOLD) doesn't go active until

HOLD goes HIGH, and then goes inactive one clock cycle after HOLD goes LOW. It is reasonable to assume that HOLD is active HIGH, since HLDA responds to it in this way.

18-3 16-BIT MICROPROCESSORS

The main difference between 8- and 16-bit micros is the word size of the data they handle. Of course, a 16-bit microprocessor handles 16-bit words of data, and has 16-bit arithmetic for all operations instead of just a few.

Our target processors are the Zilog Z-8000, Intel 8086, and Motorola 68000. Figure 18-8 shows the package layout of these three processors.

18-3.1 Zilog Z-8000

The Zilog Z-8000 processor is available in two packages called the 8001 and 8002.

The 8001 is a maximum system chip which can control more memory and has expanded capabilities compared to the 8002. It also has a 48-pin package.

The 8002 is a simplified, 40-pin version of the 8001. We will concentrate on it here, because it has all the essential features of a 16-bit machine.

The 8002 multiplexes its data bus and address bus on the same 16 pins (called A/D in the figure). Control signals AS and DS (Address Strobe and Data Strobe) identify whether the numbers on this bus are address or data bits.

Special signals called M_o and M_i are provided to permit the Z-8000's use in a network of processors. *Networking* or *multiprocessing* is an important concept (also called *distributed processing*) that lets a bunch of independent CPUs share a resource, such as a printer. The M_o and M_i are signals that allow the designer of such a network to daisy-chain the processors together so that they share the resource evenly, and take turns using it.

Three interrupts, VI (vectored interrupt), NVI (nonvectored interrupt), and NMI (nonmaskable interrupt) are available on the Z-8000 (more than on a Z-80, but not as many as an 8085).

An S/N control signal identifies whether the processor is in system or normal mode. In the system mode the CPU can do special privileged instructions that it can't do in normal mode. If this sounds obscure, consider that you, as a technician, will probably not have to worry about it.

You will seek in vain for interrupt granted, I/O, or memory strobe signals. These and other

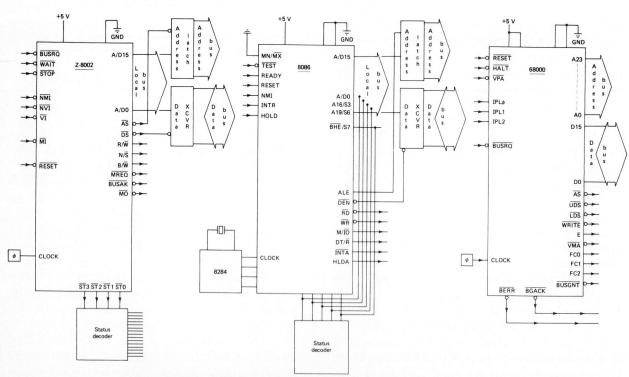

FIGURE 18-8 Some 16-bit microprocessors. (see Appendix 4 for 8086 data sheet)

conditions are encoded in a group of 4 bits called ST (*status*) signals. They can be decoded to identify any one of 16 conditions. The conditions chosen are mutually exclusive (only one of them at a time is ever active), so the designers decided that four pins (decoded externally) were more economical than 16.

The rest of the control signals on the Z-8000 are the standard ones you would expect on any microprocessor (see Section 18-2.1). It has a single-phase clock and a single 5-V power supply.

From the standpoint of external circuitry, the Z-8000 is fairly similar to 8-bit micros we've seen. It can handle a memory of 65,536 addresses through its bus (a 48-in package called the 8001 can handle more). Although there's a 16-bit data bus, each byte (8 bits) of memory is separately addressable; thus 65,536 bytes of memory exist in a segment. The 8001 can have up to 128 segments of 65,536 bytes, while the 8002 stays within one segment.

The Z-8000 has a two-chip CPU, even in the minimum system (the 8002), because an address latch is needed to handle the multiplexed A/D bus. The maximum system (an 8001) needs a special memory-management chip as well.

18-3.2 Intel 8086

The Intel 8086 has features similar to the Z-8000. Like the Z-8000, it comes in a maximum version and a minimum version, with the minimum one requiring far simpler support hardware. Unlike the Z-8000, both versions are available in the one chip, selectable by wiring an input (MAX/MIN) to a HIGH or LOW logic state. In keeping with the Z-8000 description above, the 8086 description that follows is for the simple system.

The 8086's 16-bit data bus is multiplexed onto the same pins as the address bus. Two signals, ALE (Address Latch Enable) and DEN (Data EN-able), identify what's on the A/D bus (they correspond to the Z-8000 signals AS and DS). An additional five address-expansion signals are provided for maximum system memory of 1,048,576, while minimum system operation uses only 16 bits of address for 32,768 addresses. These four address bits are also multiplexed as status bits, used like the Z-8000 ST signals.

Like the Z-8000, the 8086 addresses memory one byte at a time, but can address either the low, high, or both halves of the data bus, through a combination of the address-bus bits and a signal called BHE (Bus High Enable). BHE is multiplexed with a fifth status bit. The rest of the control signals are the same ones found on 8-bit microprocessors, although a special wait request input exists (called TEST), that can be software enabled and disabled.

Like the 8080, the 8086 requires a special clock chip (the 8284), and like the 8085, it needs a latch for the address. The maximum mode of operation requires additional decoding and buffers, but at a minimum, the 8086 is a processor with a three-chip CPU.

18-3.3 Motorola 68000

The Motorola 68000 is an upward enhancement of the 6800/6809 family of 8-bit processors. This 16-bit processor has a 24-bit address bus and 16-bit data bus, but unlike the '80 family, these two buses are not multiplexed on the same pins. This makes the 68000 a real centipede (it's in a 64-pin package). With a 24-bit program counter, the 68000 can address 16M bytes of memory. Like the other machines, this memory is byte addressable, and can be controlled 8 bits at a time rather than in 16-bit words only.

The control signals on the 68000 are pretty standard 6800-series stuff. One "exotic" feature of the 68000 is its ability to pull a software interrupt when certain illegal operations (called *bus errors*) are done. If, for instance, a word operation, which is supposed to take place at an even address (because words start at even addresses) is done at an odd address instead—that's a bus error. The 68000 will pull a *software interrupt* to an *interrupt service routine* (we'll see what that is later) when one of these errors ocurs. The 6800 had software-interrupt capability, but the 68000 is vastly more powerful.

The 68000 has a number of long word operations, that use 32-bit words—much like the 8080 and Z80 have limited 16-bit operations although they are 8-bit machines.

If you have an understanding of the operation of 8-bit machines and their hardware, the 16-bit machines are just more complicated—not different. Troubleshooting 16-bit systems will be just like troubleshooting 8-bit systems, except everything's bigger (This reminds me of Tom Lehrer's statement that octal is just like decimal—if you're missing two fingers!)

18-3.4 The Intel 8086/186/286/386 Family of Microprocessors

The Intel family of 16- and 32-bit microprocessors developed from the basic architecture of the 8086 have been used in several of IBM's personal computer models. As a result of this, and the widespread development of IBM-PC "clones"—processors that mimic the IBM personal computers—these microprocessors, and the 8/16 version of the 8086 (the 8088), have become dominant in the personal-computer market. IBM-compatibility has become an important selling point in personal computer design. Fortunately, the software of all these 16- and 32-bit machines contains the same instructions, the same addressing modes, and the same arrangement of registers, with slight modifications from one model to the next.

Register architecture. Let's look first at what the Intel **8086-family register architecture** is like. In the 8086 family, its 16-bit registers are used for two purposes, data manipulation and address formation. The registers shown in Figure 18-9 (a) are called general registers. They may all be used for arithmetic and logic operations, but the **data group** (AX, BX, CX, and DX) is divisible into 8-bit subregisters while the **pointer and index group** are not. The pointer and index group are, on the other hand, involved in addressing memory. The pointer and index group consist of registers **SP** (the **stack pointer**), **BP** (the **base pointer**), **SI** (the **source index**), and **DI** (the **destination index**). They are combined with the **segment registers** (CS, SS, DS, and ES) shown in Figure 18-9(b) to make addresses delivered to the address bus. Another pointer besides the ones listed may be combined with one of the segment registers to make an address. Instructions are executed, as in the 8-bit processors described earlier, by an up-counter

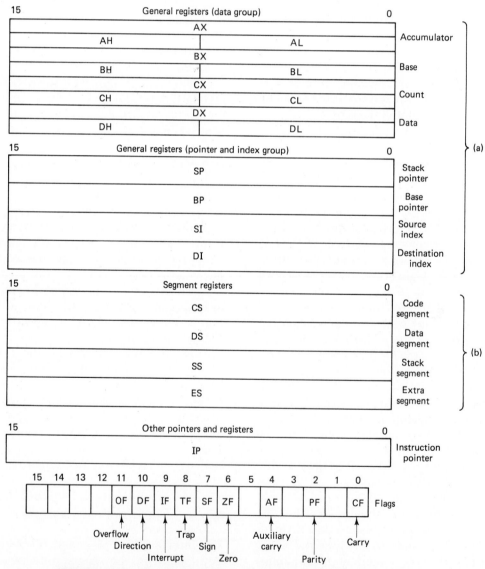

FIGURE 18-9 The 8086 family: (a) general registers; (b) segment registers.

that fetches consecutive bytes (or 16-bit words, in the more advanced models of this family). In the 8-bit processors, this was called the program counter, but in the more elaborate addressing scheme used in the 8086 family, a segment address in the CS (code segment) register is combined with this up-counter. Intel calls the up-counter an **instruction pointer** and confines the executable code in a program to a single segment of memory (the code segment) limited to 64K of memory. The total available size of memory depends on which member of the family we are talking about. (More on this later.)

Memory and addressing. Memory in the 8086 family is "segmented" into 64K **segments** within a total space of 1 megabyte (or more). Segments may overlap, if there is no reason to expect all 64K of space to be used. When the programmer wants to use *more* than 64K of space, the program or its data must be broken up into modules of 64K or less. This encourages programmers to use modular, or structured, programming. Everybody says that structured programming is a Good Thing, so this is (according to Intel) the reason for employing this system of addressing.

The names of the segments are **CS (code segment), SS (stack segment), DS (data segment),** and **ES (extra segment).**

Addressing memory is done by a two-part procedure. The address is held partially in a pointer-and-index-group register and partially in a segment register. The address actually delivered to the address bus is the number in the pointer (or index register, plus 16 times the value in the segment register. The segments of memory can begin only at exact multiples of 16, so theoretically it isn't possible to set up a segment smaller than 16 bytes. For most purposes, 16 bytes is too small a block of memory for a program or its data, so this is not a serious limitation.

This means that the location of each segment used must be defined at the beginning of a program. The instructions that assign initial values to the segment registers, and any other instructions to set up devices like PIAs, are usually put at the very beginning of the programmer's code. Programmers refer to this obligatory code as **boilerplate.**

Once the segment registers have the addresses where the segments begin, addresses used within the program work within each segment, according to what's happening. For example, the code segment contains the program, and the instruction pointer walks around inside the code

segment picking up instructions in the program. In this case, the number inside the instruction pointer is the **logical address,** while the total, combined address formed by adding

$$\text{instruction pointer} + 16 \times (\text{code segment})$$

is called the **physical address.**

Just as the code segment (which holds the program code) goes together with the instruction pointer, some of the other segment registers have a specific pointer or index that they "go together" with. In the case of the SP (stack pointer) and SS (stack segment), this relationship can be seen from the names. Together, the SP and SS registers address a portion of memory called **stack memory.** Its function can be more clearly understood after reading the following Chapter 19, especially the material on PUSH, POP, CALL, and RETURN instructions. In this chapter, however, all that we need to know is that SP and SS "go together." That means that the address in stack memory is

$$\text{stack pointer} + 16 \times (\text{stack segment})$$

which is the same pattern as that seen with the code segment and instruction pointer earlier. The base pointer (BP) and index registers (SI and DI) usually "go together" with the data segment (DS) register, or the extra segment (ES) register.

Example 18-3

The **logical address** of an 8086 instruction is C5D8 (hexadecimal) in the instruction pointer, but the contents of the code segment register is 4000 (hexadecimal). What is the actual **physical address** of the instruction in memory?

Solution: In 8086 architecture, the physical address is

$$16 \times (\text{code segment}) + \text{instruction pointer}$$

This is the same (in hexadecimal) as shifting the number in the code segment register over one place to the left, and adding on the number in the instruction pointer. In this case

```
  4000    code segment boundary
+ C5D8    logical address
  4C5D8   physical address
```

The instruction is at location 4C5D8 (hexadecimal) in memory.

The rest of the '86 family. If all the registers and instructions are (fundamentally) the same, you might well ask: Then what is the difference between all these numbers? Among the features that distinguish among the 8086, 80186, and 80286 are higher clock speed and more memory addressing power. The 8088, 80188, and 80288 are 8/16-bit processors. The 8088, for instance, is like an 8086 cut down to an 8-bit bus. Information cannot be fetched 16 bits at a time, as is the case with the 8086. In exchange for using the more common 8-bit data path, the **throughput** of the processor (how fast data are processed through the system) is slowed down. The 80386 is a 32-bit processor. It has CMOS technology, the ability to grab data in chunks twice as wide as the smaller 8086-family processors, and a higher clock speed (despite the CMOS technology utilized). Although it is compatible structurally with the processors/instructions/register set of the 8086, it is a freight train compared to a hay wagon.

The trend toward more microprocessor processing power and larger data size seems to be holding at a "ceiling" of 32 bits, although the early predominance of 8-bit processors in personal computers has been supplanted by 8/16- and 16/32-bit processors in the personal computers of the "power users." In the next chapter we look at the software of 8-bit processors, ending with a summary of what software features the 8086 family of processors add to their earlier 8080/8085/Z80 predecessors.

Example 18-4

An 8086 contains the following numbers in its segment registers:

$$CS = 0400; DS = 0C00;$$

Segment register: Code   Data

$$SS = 1800; ES = 3000$$

Stack   Extra

(a) What is the largest program that could safely be placed in the code segment to avoid data operations destroying code (instructions)?

(b) How much data can be loaded into the data segment without affecting the stack (or vice versa)?

Solution: To make this easier to see, we'll turn the numbers in the registers into a memory map, showing the addresses in memory occupied by each type of segment according to the numbers in the segment registers.

(a) We can see that the physical addresses of the Code, Data, and Stack segments overlap. That means that data could destroy code written beyond address 0C000, because that is where the data segment begins. Writing program code into addresses 0C000 to 14000 − 1 is

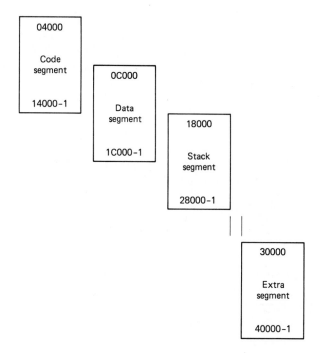

risky because data writing could overwrite program code instructions in this area. So the largest program that could be safely written into the code segment must fit between addresses 04000 and 0C000 − 1. The longest safe program is 0C000 − 04000 = 08000 bytes.

(b) We can see that the Stack segment overlaps the Data segment at 18000. Data start at 0C000. Between 0C000 and 18000 is 08000 bytes, so the contents of the data segment should not exceed 08000 (hex) bytes in length.

18-1. Write a program for the calculator in Figure 18-2 that will calculate 10% sales-tax. Show all the machine codes in a memory like the one at the bottom of the figure.

18-2. Where is the data bus in Figure 18-3?

18-3. Describe briefly the difference between bus-oriented and monolithic computer design. Use anecdotes or examples to aid you in your description.

18-4. How many memory addresses can a microprocessor with a 24-wire address bus control?

18-5. Describe how Intel uses multiplexing to reduce the number of control bus pins on its 8080 microprocessor.

sor's ALU do not work properly. What do you, as a technician, replace?

18-11. Every program run on a microprocessor "messes up" when the program counter reaches address 16 in the memory. Would you replace the CPU? If that didn't work, what would you replace?

18-12. The Motorola 68000 has 64 pins. What's the advantage of using a design like this that doesn't multiplex address and data (or address and control) onto the same pins?

The timing diagram below shows how a microprocessor carries out an OUTPUT operation. Answer the following four questions about the diagram.

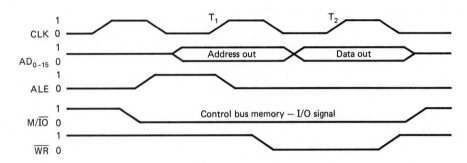

18-6. What's a polyphase clock?

18-7. Identify which of the microprocessors below have a one-chip, two-chip, or three-chip CPU.
   (a) Intel 8080
   (b) Zilog Z-80
   (c) Motorola 6800
   (d) Mos Technology 6502
   (e) Intel 8085

18-8. Of the five 8-bit microprocessors discussed in this chapter, which has the largest number of registers?

18-9. Of the five 8-bit microprocessors discussed in this chapter, which has the largest number of addressing modes?

18-10. Part of the functions of a microproces-

18-13. Is the address valid before or after the "IO" control bus signal becomes active?

18-14. Is the "IO" control bus signal active high or active low?

18-15. Does the "write" (WR) signal become active before or after the data become valid?

18-16. From the signals in the timing diagram, would you expect an edge-triggered address latch to be rising-edge triggered or falling-edge triggered?

18-17. In the 8086 family of processors, when stack memory is being addressed via the SP (stack pointer) register and the SS (stack segment) register, which register contains the *logical address*, and which register contains the *physical address*?

Suppose that a programmable calculator with 32 number and function keys uses a 5-bit code for each key, and can recall a "program" of up to 256 keystrokes:

If the calculator is limited to using decimal numbers:

18-1. How many of the 5-bit codes are used for *opcodes*? _____

18-2. How many of the 5-bit codes are used only for data? _____

18-3. Does the calculator contain a memory? _____

18-4. Does the calculator contain I/O devices? _____

18-5. Does the calculator contain an *address bus*? _____

18-6. Does the calculator contain a *data bus*? _____

18-7. Does the calculator contain a control bus? _____

18-8. Does the calculator contain an instruction decoder? _____

18-9. Does the calculator contain a clock oscillator? _____

18-10. Does the calculator contain a program counter? _____

18-11. Does the calculator contain *software*? _____

18-13. Name nine control signals that are found on any microprocessor:

18-12. Can this calculator be considered a computer? _____

18-14. Name four types of registers that are found in any microprocessor:

18-15. Name the four segment registers found in one of the Intel 8086 family of microprocessors.

18-16. How many bits are contained in the instruction pointer of the Intel 8086 processor? _____

19

MACHINE-LANGUAGE PROGRAMMING FOR THE 8080 AND Z-80

CHAPTER OBJECTIVES

By the time you finish this chapter, you will be able to:

1. Describe how the *machine language* of 8-bit and 16-bit processors works, in general terms.
2. Explain why a knowledge of software is essential to troubleshooting problems in digital systems.
3. Describe the function of Intel 8080/8085 and Zilog Z-80 8-bit machine-language instructions in the:
   a. Data transfer group of instructions
   b. Arithmetic-logic group of instructions
   c. Control group of instructions
4. Describe the instruction set of the Intel 8086 family of 16-bit processors by comparison to the 8-bit processors already described.
5. Convert a list of instructions in *assembly language* into a binary or hexadecimal listing of *machine code* using a programmer's reference card, or the chart in Appendix 1 at the back of the book.

The **machine language** of a computer was described in general terms in our calculator example in Chapter 18. We found out that opcodes and data were used—like keystrokes on the keyboard of a calculator—to activate circuits within the CPU. You may have heard the names of high-level languages like FORTRAN and COBOL and wondered how many languages a microprocessor, like an 8080 or Z-80, really knows. The answer may surprise you. Any computer you see knows just one language—its own machine code—and all the other languages are translators that "cook down" FORTRAN or COBOL (or whatever) into the machine's own machine code. If the system breaks down, all those other languages will be gone, and there will be only one language in which you can talk to the machine, really—the stuff that's wired into the machine.

We hope that this chapter can answer the question: What is a "typical" microprocessor like (in terms of the commands it is wired to do)?

As of this writing, the majority of microprocessors being used are still 8-bit machines. We chose the Z-80 as a model, because it has the best features of most 8-bit microprocessors.

In some places, a command for the Z-80 will have a different mnemonic on the 6800, 6502, or 8080. Too bad that they can't all agree on what to call a *move* command.

We're not going to attempt to describe the alternate mnemonics and commands of every processor, but we will use Z-80 and 8080 mnemonics side by side, and most manufacturers use either Intel's or Zilog's name for a mnemonic.

The Z-80 has more registers than any other 8-bit processor in this group. It also has at least as many addressing modes as any of the others. Seeing how programs are written for the Z-80 should provide good background for learning (and using) the language of any 8-bit processor, and a good headstart for learning 16-bit machine programming.

19-1 WHY A TECHNICIAN SHOULD KNOW ABOUT PROGRAMMING

Recently, we tested 61,440 flip-flops in packages of 1024 each. It took us about half an hour. How long would it take a technician with a breadboard, a meter, an oscilloscope, and a logic probe to test the same 60K flip-flops? Odds are, the poor soul would die of boredom or old age (or both!)

before the job got done. This is a job for a computer!

Computers have an infinite capacity for boredom. The more boring, tedious, and repetitive a job is, the better a computer does it. In fact, repetition and sameness are the things computers do best. Consider testing RAM integrated circuits. You'd have gone nuts if you tried to test 60 (1K × 1) RAM chips by hand. Not only would it take too long to be worth it, by the time you got even one chip tested, you'd be so thoroughly bored that you'd be getting really sloppy about the last half of the testing.

Also, computers can do the boring, repetitive stuff faster than any human being can—or would want to. Our test takes less than a minute to set up, run, and interpret, and most of that's the human time delay. The computer program tests all 1024 locations on the chip in a fraction of a second.

We have no intention of attempting to teach you the complete art of programming in this book. Our purpose is to explain what operations a typical microprocessor can do on its own, and show how some of these can be put together to solve some small digital design problems. We expect that after reading the next few chapters, you should be able to pick up a manufacturer's documentation of any microprocessor and understand enough of it to pick up the details you need to read and understand a specific program.

There are two reasons why this is necessary. The first has to do with the "my computer doesn't work—fix it" problem. If the "doesn't work" is hardware, you can fix it with chips, wires, and so on—but if the problem is bad software, the computer itself isn't really broken. It is necessary for the technician to know the difference. Few technical experts can "debug" both hardware and software, but you need to know something about both to know if it is the computer or the program that needs fixing.

The technician who comes to fix a "sick" computer will often be equipped with a set of diagnostic programs in a package. If these programs work, then the hardware is OK. Occasionally, the technician will have to make up the software test procedure on the spot to look at a specific problem. Knowing how to use the machine's instruction set for this purpose is as important as knowing how to operate a hardware test instrument like a scope or meter.

The second reason concerns digital design. The engineer of the future will not design a unique random-logic board for each digital application. Instead, the design will use a standard microprocessor development board programmed to do the tasks of the random logic it replaces. This means that the technician who services the units will often see the same board doing vastly different tasks. You, as the technician, will have to know—or have some understanding of—the way the engineer used the program to solve each design problem, in order to troubleshoot the board.

It seems likely that the technician's job in the future will polarize into two categories. There will be the techs who mash the keys and watch the lights on a high-technology tester, and throw the boards into two piles—good and bad. And then there will be the techs who program that same high-technology tester with the test procedures for each model of the board being tested. The first group of techs will be doing "chimp work" and getting paid accordingly. All the "smarts" will be in the board tester. If you want to be in the second group of technicians, *be patient.* You'll have to put up with what's in this chapter, which is pretty dry, to get to the real technician stuff (troubleshooting with machine language programs) in the next chapter.

19-2   THE DATA TRANSFER GROUP OF INSTRUCTIONS

The simplest and most typical operation you can do within a computer system is to transfer data from one place to another and make copies of them. Inside the CPU, there is a local data bus that is connected to all the registers in the CPU. Each register is an 8-bit latch whose outputs and inputs are attached to this bus. The data on the bus can be clocked into the inputs or enabled onto the bus from the outputs. Suppose that you want to transfer a number from register A to register B. The outputs of register A are enabled to the bus and register B is clocked. This makes a clone of register A's contents in register B. Nothing happens to the contents of register A. The same is true for all transfers from anyplace in a computer to anyplace else. The closest analogy we can think of is copying a tape cassette. The "source" cassette is played back on one cassette deck; its output is recorded on another cassette deck onto a "destination" cassette. If there is already something recorded on this destination cassette, it is overwritten and replaced by the copy of the source cassette's contents.

19-2.1   Register to Register

In the 8080, there are eight 8-bit registers, called B, C, D, E, H, L, M, and A, that can be addressed. Data can be transferred from any one register to any of the others. The Z-80 has these eight registers, and eight more called the *primed registers* (B', C', D', E', H', L', M', and A'). A single command in Z-80 code permits swapping the primed register set with the others, and the commands that transfer data between them are the same as the 8080 commands.

The Intel name for the instructions that transfer data from one register to another is MOV. The Zilog name for instructions that transfer data between anywhere and a register is LD (LoaD). An example would be:

(Intel) MOV B,A   (Zilog) LD B,A

It might look like this says, "Move B to A" but the MOV described here actually moves "A" to "B"! Why Intel chose to do it this way is a mystery somehow connected with an earlier 8-bit chip, we're told. Zilog chose to keep the destination before the source, but changed their mnemonic to read, "Load B with A". We load B with the contents of A when we do this LD.

The actual binary code for a MOV or LD instruction has 8 bits. Each register can be identified by a 3-bit identification number since there are eight 3-bit numbers and eight registers. The actual opcode for one of these MOV instructions contains a register code for the destination register, a register code for the source register, and two more bits. If these bits are 01, that identifies the instruction as a MOV.

The arrangement of the bits in the MOV instructions is shown in Appendix 1. It shows

MOV r,r'   01rr'

as the code conversion rule. Let's see how this is used for the instruction MOV B,A. First, we see that the letters r and r' (r-prime) are used in place of the actual destination (r) and source (r') registers. At the top of Appendix 1 is a list of 3-bit register codes for the eight registers:

B = 000   C = 001   D = 010   E = 011
  H = 100   L = 101   M = 110
  A = 111

To find the binary code for MOV B,A we "plug in" the missing letters for registers r and r':

MOV B,A
    ↑ ↑
MOV r,r'   01rr'
ld r,r'   binary

and we find the register code for each letter, namely:

B = 000   A = 111

Where we plugged in the missing letters in the **mnemonic** (the name of the instruction), we plug in the missing numbers in the corresponding parts of the binary number:

MOV B,A   01 000 111
    ↑ ↑
MOV r,r'   01rr'
ld r,r'   binary

Now that we've plugged in the missing letters for the registers into the mnemonic, and the register codes into the binary byte, we have MOV B,A = 01000111, the binary **opcode** for the instruction. To convert it into hexadecimal (since most microprocessor keyboards only let you enter code into memory in hexadecimal code), we cluster the bits in groups of four and identify the corresponding hexadecimal character for each group:

0100 0111 = 47 (hex)

Most of the mnemonics in Appendix 1 are already converted into hexadecimal opcodes. Wherever this is *not* the case, the instruction is given partially in binary, with the rest of the binary bits to be filled in using register-code substitution, as shown in this example.

19-2.2 Register to/from Memory

The same instruction that transfers data from register to register within the 8080 can also transfer information from the registers (B, C, D, E, H, L, A) to a memory (M) location—provided that the address of the memory location is already in registers H and L. Since an address in the 8080/Z80 memory is 16 bits long, only half an address can fit in one register. The top half (most significant 8 bits) of the address we want to use

is held in the 8-bit register H (the HIGH byte). The bottom half of the address is held in register L (the LOW byte). Together, H and L hold the entire 16-bit address of memory location M. Any 8080 instruction that comes along with an M in it will end up using HL to find where M is.

Zilog uses the *infix notation* representation of the M register—(HL)—when they represent their register-memory MOV instruction. We'll see how infix represents numbers, registers, and addresses in a little while, but you'll get some idea of how it works from Zilog's mnemonic in this example:

(Intel) MOV A, M   (Zilog) LD A, (HL)

Suppose that you want to move the contents of the memory location at address 5000 (decimal) to register A. The instruction you need is the one above, but you must first prepare the memory pointer (register H and L) by putting the number 5000 in it. This is done by finding the binary value of 5000 (Table 19-1). (If you haven't done this for a while, here's a good chance to practice decimal-to-binary conversion.) We see that the binary code for 5000 is 1001110001000, a 13-bit number. This is too large for one byte. We expected that, and planned to put half the address in register H and the other half in register L. First we'll pad our number out to 16 bits: 0001001110001000. The HIGH byte of this word is 00010011, and its LOW byte is 10001000.

TABLE 19-1

Decimal-to-binary conversion

Numbers	Odd/even	Place value
1	1	4096
2	0	2048
4	0	1024
9	1	512
19	1	256
39	1	128
78	0	64
156	0	32
312	0	16
625	1	8
1250	0	4
2500	0	2
5000	0	1

Before we can use the MOV A, M instruction, we need to load 00010011 (which is a decimal 19) into H and 10001000 (which is a decimal 136) into L. We don't know how to do that yet.

EXAMPLE 19-1

Show what happens to the numbers in the registers below when an Intel 8085 processor executes the instruction shown.

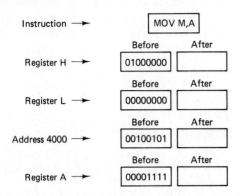

Instruction → MOV M,A

	Before	After
Register H →	01000000	
Register L →	00000000	
Address 4000 →	00100101	
Register A →	00001111	

Solution: MOV M,A transfers data from A into whatever memory location is at the address to which the 16-bit number in the H,L register points.

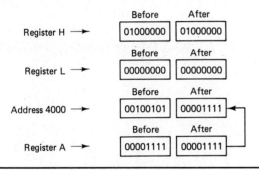

	Before	After
Register H →	01000000	01000000
Register L →	00000000	00000000
Address 4000 →	00100101	00001111
Register A →	00001111	00001111

19-2.3 Constant to/from Register

The Intel MOV commands (Zilog LD commands) can move a number from one register to another like this:

Mnemonic		Result
(Intel) MOV H, B	(Zilog) LD H, B	H ← B

We would like to have a command that can move a 19 directly into H and a 136 into L without getting the numbers from another register. There is such an instruction for the 8080 and Z-80; to load a 19 into H, for example, the mnemonics are:

Mnemonic		Result
(Intel) MVI H, 19	(Zilog) LD H, 19	H ← 19

Since these instructions load H with a number that's *immediately* available (it's right in the code for the instruction itself!) they're called MoVe Immediate (MVI) instructions by the Intel peo-

ple. Zilog just calls them LoaD (LD) like all other data-transfer instructions.

The difference between these load instructions and the ones in Section 19-2.1 is that B is a variable and 19 is a constant. With the instruction MOV H, B, the number put into H can vary. If the computer executes this instruction a number of times in a program, the value in B can be different every time, depending on the history of what has just happened to B (since it can vary, it's a variable).

The number 19 cannot change (it is constant). It will always be put into H when a MVI H, 19 instruction is done. The computer may execute this same instruction at different times in a program, and H will have a value of 19 afterward every time.

So, to summarize, instructions with the immediate addressing mode use a constant (which is usually found right in the body of the instruction itself). Instructions that use only register addressing use variable quantities whose value cannot be guessed just by looking at the instruction.

Now, in case you forgot, we needed the immediate instructions because we wanted to put the contents of memory location 5000 in register A using a MOV A, M instruction. The 5000 must be put into registers H and L as page 19 and line 136 (decimal) of memory. In the HL register pair (the memory pointer), these two numbers will address memory location 5000 as M whenever an 8080 instruction containing an M, or a Z-80 instruction containing (HL), is executed. Using the following program segment, we can get the number in memory location 5000 into the CPU's register A:

Mnemonic		Result
(Intel)	(Zilog)	
MVI H, 19	LD H, 19	H ← 19
MVI L, 136	LD L, 136	L ← 136
(the number 5000 is now in register pair HL)		
MOV A, M	LD A, (HL)	A ← (5000)

where the infix notation A ← (5000) says: "Put the number found in memory location (address) 5000 into register A. This does not put a 5000 into A. 5000 *won't fit*! Register A is like any 8-bit microprocessor register; the largest 8-bit binary number it can hold is 11111111 (decimal 255).

If we choose to put a number into A from a different memory address, all that we need to change is the numbers in H and L. Address 5000 was at page 19 and line 136 in the memory. Another address, like 3000, would be at a different location (line 11 and page 184). We could get a number into A from address 3000 this way:

Mnemonic		Result
(Intel)	(Zilog)	
MVI H, 11	LD H, 11	H ← 11
MVI L, 184	LD L, 184	L ← 184

(the number 3000 is now in register-pair HL)

MOV A, M	LD A, (HL)	A ← (3000)

(since HL is 3000 now, there's a different result)

There's another way to get a number loaded into register A from memory location 3000. This one is more direct than the MOV command, and doesn't use the HL registers to point the way to the address:

Mnemonic		Result
(Intel)	(Zilog)	
LDA (3000)	LD A, (3000)	A ← (3000)

Goodness, that looks easy! If you want to load register A with the number stored at memory address 3000, you tell the computer the address 3000 right in the instruction. This is called **direct addressing.** The address is available to the CPU directly in the instruction. That's certainly simpler than looking for it in the register HL. Also, Intel and Zilog almost agreed on the mnemonic for this one.

Unfortunately, this instruction is quite limited. It's a great way to load something into register A, but it's only available for register A, and you can't load register B or C, or any other register with it. The MOV instruction with register M—the Zilog LD instruction with (HL)—could be used to load any register with the number stored at 3000 (or anyplace else, for that matter).

Another thing that makes LDA a bit awkward is its size. Although it appears there would be only two parts to this instruction—the opcode, which says that it's a LoaD, and the address 3000—there are really three bytes of code in the "final" machine language. The opcode is 58 in decimal and the address is made of the bytes 184 and 11. Notice that the line number (184) comes before the page number (11). This is always true of Intel machines. When a 16-bit number is broken into a page and a line, the line comes first, even though it's the LOW part of the address. The whole instruction LDA 3000 would be stored in consecutive memory addresses as 58, 184, 11.

Since information is usually put into memory in hexadecimal, this instruction would be entered in through the keyboard of a microprocessor trainer, or the monitor program of a microcomputer, as 3A, B8, 0B.

Also, in Intel jargon, a LD instruction always transfers data out of memory to a register. If you want to transfer data into memory from a register, Intel calls everything a STore (see below):

Mnemonic		Result
(Intel)	(Zilog)	
STA (3000)	LD (3000), A	(3000) ← A

The number in register A is stored in memory location 3000 by this instruction.

The memory pointer—register pair HL—is not the only place where the CPU can go to find an address. The indirect mode of addressing allows BC and DE to be used as memory pointers too. Intel calls these instructions LDAX (LoaD A indirect—don't ask us where X comes from) and STAX (STore A indirect). Like other Intel LD and ST instructions, these only work between memory and register A:

Mnemonic		Result
(Intel)	(Zilog)	
LDAX B	LD A, (BC)	A ← (BC)
LDAX D	LD A, (DE)	A ← (DE)
STAX B	LD (BC), A	(BC) ← A
STAX D	LD (DE), A	(DE) ← A

Let's use one of these instructions to solve the same problem we've solved with other instructions in this group—moving a number to register A from the memory:

Mnemonic		Result
(Intel)	(Zilog)	
MVI B, 19	LD B, 19	B ← 19
MVI C, 136	LD C, 136	C ← 136

(now the BC register pair contains the number 5000)

LDAX B	LD A, (BC)	A ← (5000)

The Zilog Z-80 has two additional 16-bit registers that may be used as a memory pointer. These are registers IX and IY. They can be used like BC, DE, or HL, but have an additional feature. In the instruction, you can tell the computer to find a memory location forward or backward from the number in the pointer. A one-byte signed binary number (this will be discussed in detail later) is added to the contents of the 16-bit register. If it is a positive number, the address formed by this addition points ahead of the address in the pointer. If the number (called a **displacement**) is negative, the address formed by this addition points behind the address in the pointer. This method of reaching a place in memory in the neighborhood of a specific address is called **indexed addressing.** We'll use the same example we've used in the rest of this section to illustrate this point:

Mnemonic	Result
LD IX, 5000	IX ← 5000
LD A, (IX + 1)	A ← (5001)

Now this is something new! We loaded A with the number stored at address 5000 + 1. The number 1 could be replaced by any positive number up to 127, or any negative number down to (−128).

The same operation can be done in the reverse direction. A register can be put into memory using

LD (IX + 1), A − LD (IX + 2),
B − LD (IX + 3), C

and so forth. Every register is usable, and all addresses can be reached up to half a page on either side of the address in IX or IY.

There are two other registers Zilog has that Intel does not. The R (refresh) register and I (interrupt) register can be loaded to or from register A (and only register A). The function of these registers is outside the scope of this chapter. For now, think of them as "just another" 8-bit place in the CPU where we can get or put numbers.

19-2.4 Extended Precision

Although the 8080 and Z-80 are 8-bit computers, they can do some 16-bit operations as well. Such operations are called double precision when they deal in double bytes. The number of 16-bit "things" we can do in the 8080 and Z-80 are more limited than the number of 8-bit things. This is partially due to the fact that 8-bit registers must be grouped together in **register pairs** to do double-precision operations. Since there's only one register pair for every two registers, there aren't as many 16-bit places to work in as there were 8-bit places.

We hinted at 16-bit load capability in the last section (did you spot it?). In the example of a

LD A, (IX + 1) instruction, we showed the instruction LD IX, 5000. The IX register is a 16-bit register. You can load a 5000 into it, so it must be bigger than 8 bits! There are five 16-bit registers (four in the 8080) that can be used for double-precision work. They are:

AF   BC   DE   HL   SP   (8080 or Z-80)
         IX   IY   (Z-80 ONLY)

(the Z-80 also has a duplicate set of registers)

AF'   BC'   DE'   HL'   SP'   IX'   IY'

Four methods of addressing exist for these data transfers: register, immediate, direct, and indirect.

Register transfers let you move a 16-bit number from one register pair to another:

Mnemonic		Result
(Intel)	(Zilog)	
XCHG	EX DE, HL	HL swapped with DE
SPHL	LD SP, HL	SP ← HL
	LD SP, IX	SP ← IX
	LD SP, IY	SP ← IY
Z-80 only	EX AF, AF'	AF swapped with AF'
	EXX	BC ↔ BC' DE ↔ DE' HL ↔ HL'

We used an **immediate** load instruction when we put the number 5000 into register IX. Other instructions of this kind are:

Mnemonic		Result
(Intel)	(Zilog)	
LXI B, 5000	LD BC, 5000	BC ← 5000
LXI D, 5000	LD DE, 5000	DE ← 5000
LXI H, 5000	LD HL, 5000	HL ← 5000
	LD IX, 5000	IX ← 5000
	LD IY, 5000	IY ← 5000

Of course, any 16-bit number could be used; it doesn't have to be 5000. Numbers loaded into one of the register pairs can have values from 0 to 65535 (unsigned arithmetic) or −32768 to +32767 (signed arithmetic).

Direct addressing permits you to load the register pairs from memory addresses. Each address can be either the source or the destination of a data transfer. In our examples below, we'll use the address 3000, but any 16-bit address will do equally well:

Mnemonic		Result
(Intel)	(Zilog)	
LHLD 3000	LD HL, (3000)	HL ← (3000)
SHLD 3000	LD (3000), HL	(3000) ← HL
	LD BC, (3000)	BC ← (3000)
	LD (3000), BC	(3000) ← BC
	LD DE, (3000)	DE ← (3000)
	LD (3000), DE	(3000) ← DE
	LD SP, (3000)	SP ← (3000)
	LD (3000), SP	(3000) ← SP
	LD IX, (3000)	IX ← (3000)
	LD (3000), IX	(3000) ← IX
	LD IY, (3000)	IY ← (3000)
	LD (3000), IY	(3000) ← IY

When these are done, the bytes in two consecutive addresses (in this case, 3000 and 3001) are transferred between memory and the registers. You might expect that the bytes from 3000 and 3001 would be put into the registers (in a register pair) in alphabetic order. Not so. If we're using register pair HL, the 3000 byte goes in L, and the 3001 byte goes in H. Since the transfer instructions that put HL back into memory work the same way, there's no reason to notice what order this happens, unless you want to use just one half of the 16-bit word that was transferred. In that case, you'll need to know this to find the register that has the half you want.

Indirect addressing permits you to load a register pair to/from an address in memory, but the memory address doesn't have to be a part of the instruction. Instead, the address is taken from another register pair. In the case of 16-bit numbers, the memory pointer is almost always "SP" (the stack pointer). The memory the stack pointer points to is called the stack. The data-transfer operations that transfer 16-bit numbers between the stack and the register pairs are called PUSH and POP instructions instead of LD (even by Zilog!).

A **PUSH** puts a number onto the stack and a **POP** takes it off the stack. Unlike the 8-bit instructions that use indirect addressing, stack instructions change the pointer after they've been done. The stack pointer is an up/down-counter. It will decrease by 2 every time a PUSH has been done. When a POP is done, it will increase by 2. This makes the stack a last-in, first-out (LIFO)

shift-register. The most recent entry PUSHed on the stack will be the first one POPped back off the stack if a POP is done.

Here are the PUSH and POP instructions available in the 8080 and Z-80 instruction sets:

Mnemonic		Result
(Intel)	(Zilog)	
PUSH B	PUSH BC	(SP) ← BC
PUSH D	PUSH DE	(SP) ← DE
PUSH H	PUSH HL	(SP) ← HL
PUSH PSW	PUSH AF	(SP) ← AF
	PUSH IX	(SP) ← IX
	PUSH IY	(SP) ← IY
POP B	POP BC	BC ← (SP)
POP D	POP DE	DE ← (SP)
POP H	POP HL	HL ← (SP)
POP PSW	POP AF	AF ← (SP)
	POP IX	IX ← (SP)
	POP IY	IY ← (SP)

There are even some instructions that do both a PUSH and a POP at the same time. These are the **indirect exchange** instructions:

Mnemonic		Result
(Intel)	(Zilog)	
XTHL	EX (SP), HL	(SP) ↔ HL
	EX (SP), IX	(SP) ↔ IX
	EX (SP), IY	(SP) ↔ IY

Each time an EX instruction is done, two 16-bit numbers change places. In this case, one of the numbers is stored on the stack, and the other is in the CPU. Afterward, they have changed places.

We can say that stack operations are like other memory operations that use a memory pointer, except that the numbers we are storing and retrieving are 16-bit numbers, and the pointer follows the numbers as they're stacked and unstacked.

19-2.5 Block Moves

In the Z-80, a group of **block transfer** instructions allow you to move a block of memory as large as you want from one place to another. This is a characteristic of *variable-word-length computers* found in no other microprocessor. To use these instructions, three numbers 16 bits long must be put into the BC, DE, and HL register pairs:

HL ← starting address of the source memory block

DE ← starting address of the destination

BC ← byte counter for the length of the block

There are up-counting and down-counting versions of the block move instruction, and even ones that "bump" the counter but stop after one byte of data has been moved. Here they are:

Mnemonic		Result
LDI	(DE) ← (HL);	bump BC down and HL, DE up
(Single byte move)		just one time
LDIR	(DE) ← (HL);	bump BC down and HL, DE up
(Block move)		and repeat until BC is 0
LDD	(DE) ← (HL);	bump BC, DE, and HL down
(Single byte move)		just one time
LDDR	(DE) ← (HL);	bump BC, DE, and HL down
(Block move)		and repeat until BC is 0

These instructions may seem a bit mysterious to you. That's understandable, especially for the single-byte memory-to-memory transfers. Unless you've had the experience of programming an IBM 1400-series data processor or a Honeywell 200-series processor, it's unlikely that you'll see the usefulness of these instructions. We've had the experience of those machines. Take our word for it, an experienced programmer can find these very useful.

19-2.6 Register to/from I/O

There are three possible places for numbers to live in a microcomputer system, CPU registers, memory, and I/O devices. The 8080 has only one method of I/O operation. A number transferred between the I/O port and CPU of an 8080 must use register A. Each I/O port has a number called its **device code.** In the 8080, this code is included in the instruction directly. If data are being transferred in to the CPU, the instruction is "IN" and the port an input port. If data are being transferred out of the CPU, the instruction is "OUT" and the port an output port.

The Z-80 has a lot more ways of handling I/O. The device code can be direct or indirect. When indirect addressing is used, the pointer holding the device code is register C. Instead of A, an I/O port can transfer data to or from any one of the 8-bit registers. There are also single-byte and block moves between a port and a block of memory. In our examples, we'll use I/O port 7 and register B in the IN and OUT instructions, but others are certainly possible.

Mnemonic (Intel)	(Zilog)	Result
IN 007	IN A, (7)	A ← PORT 7
OUT 007	OUT (7), A	PORT 7 ← A
	IN B, (C)	B ← Port (C)
	OUT (C), B	Port (C) ← B
	INI	(HL) ← PORT (BC)

(BC will count down and HL up, but only a single byte will be moved)

	INIR	(HL) PORT (BC)

(this instruction can be used to load data from 256 I/O devices to 256 memory locations by counting down at B and up at HL until B is 0)

	IND	(same as INI except HL counts down by 1)
	INDR	(same as INIR except HL counts down until B is 0)

19-3 ARITHMETIC-LOGIC INSTRUCTIONS

The 8080. There are eight arithmetic/ Boolean operations that the 8080 microprocessor can use to combine or compare two numbers. They are:

(ADD) Add (A plus Register)
(ADC) Add with Carry (A plus Register)
(SUB) Subtract (A minus Register)
(SBB) Subtract with Borrow (A minus Register minus the Borrow)
(ANA) (logical) AND (Register) with A
(XRA) (logical) XOR (Register) with A
(ORA) (logical) OR (Register) with A
(CMP) Compare (A minus Register) (look-ahead subtraction)

There are also a number of operations that the 8080 can do on single numbers. They are:

(INR)   Increase (by 1)
(DCR)   Decrease (by 1)
(CMA)   Complement (invert all bits)
(DAA)   Decimal Adjust

19-3.1 Register-to/from-Register Arithmetic

The eight operations that combine two numbers can be used to combine a number in register A with a number in any other register. A typical instruction of this type is:

Mnemonic		Result
(Intel)	(Zilog)	
ADD B	ADD A, B	A ← A plus B

All the instructions in this group of 8 do something to the A register. In Intel mnemonics, this is implicit. The A register is not shown as a part of the instruction, but any time numbers in two registers are combined, one of the registers is always A. Zilog chooses to state register A explicitly (sometimes) in some of its mnemonics. We're not sure why the Boolean functions have an implicit A and the arithmetic ones have an explicit A in Zilog assembly language. Intel's code, although it doesn't show that register A is used, is at least consistent.

Add with Carry and Subtract with Borrow allow "chaining" together numbers in several registers to do multibyte arithmetic. Examples of this will be discussed in a subsequent chapter. For now, we'll just say that Add and Add with Carry are software that does the same thing as a half-adder and full-adder do in hardware. Subtract and Subtract with Borrow are the software equivalents of the half-subtracter and full-subtracter in hardware.

As with Intel's other register instructions, there is a register M that may be used in any of these instructions, but is really a memory location addressed by HL. This means that a number may be taken out of memory to be combined with A by the ALU. The results will be in A, of course, after the arithmetic/logic is completed—so if you want to store the results in memory, you'll still need a Data Transfer instruction to do the job (see the preceding section).

We should note at this point that the *Increase* (INR) and *Decrease* (DCR) instructions defy the general rule that arithmetic is only done in the accumulator (register A) and permit you to "bump up" or "bump down" a number in any of the 8-bit registers (including M).

The Complement, the Accumulator (CMA) and Decimal Adjust Accumulator (DAA) instructions, of course, only exist for register A.

EXAMPLE 19-2

Show what happens to the numbers in the registers below when a Zilog Z-80 processor executes the instruction shown.

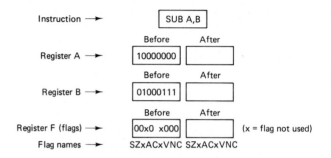

Solution: SUB A, B subtracts B from A (or A minus B), placing the result of the subtraction in register A. All flags are affected. This subtraction results in a positive (S = 0), nonzero (Z = 0) number without any overflow (V = 0), carry, or borrow (C = 0). Since a 1 was borrowed across from the high nibble of the byte to the low nibble, the auxiliary carry (borrow) flag is set (AC = 1). Because this is a subtraction and may need to be BCD-adjusted, the N flag (for subtraction) is set (N = 1).

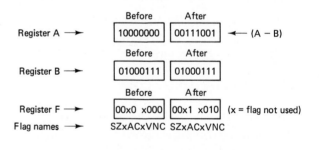

The Z-80. The Z-80 has two additional memory pointers (IX and IY) that are used for indexed addressing. Arithmetic in the Z-80 has all the instructions of the 8080 but has slightly different mnemonics, and allows use of (IX + d) and (IY + d) as well as (HL) to do arithmetic with numbers in the memory.

Zilog's mnemonics for these instructions are:

(Dyadic Arithmetic Operations)

(ADD) Add (A plus Register)

(ADC) Add with Carry (A plus Register plus the Carry)

(SUB) Subtract (A minus Register)

(SBC) Subtract with Carry (A minus Register minus the Carry)

(Dyadic Logical Operations)

(AND) (logical) AND (Register) with A

(XOR) (logical) XOR (Register) with A

(OR) (logical) OR (Register) with A

(CP) Compare (A minus Register) (look-ahead subtraction)

(Unary Arithmetic Operations)

(INC) Increase (Register) by 1

(DEC) Decrease (Register) by 1

(Unary Logical Operations)

(CPL) Complement (invert all bits)

(NEG) Negate (form two's complement of A)

(DAA) Decimal Adjust (register A)

19-3.2 Constant-to-Register Arithmetic

The 8080. The same eight operations can be done with a register and a **constant.** In the 8080, the constant is always combined with register A. Instructions have slightly different mnemonics in Intel code when one of the numbers is a constant. Like the MVI instructions, these instructions are immediate (the constant is immediately available in the instruction itself). Because of the immediate mode of addressing, these mnemonics all contain an I:

(Arithmetic Operations)

(ADI) Add Immediate (constant to A)

(ACI) Add with Carry Immediate (constant to A)

(SUI) Subtract Immediate (constant from A)

(SBI) Subtract with Borrow Immediate (from A)

(Logical Operations)

(ANI) AND Immediate (constant with A)

(XRI) EXCLUSIVE OR Immediate (constant with A)

(ORI) OR Immediate (constant with A)

(CMP) Compare Immediate (constant with A)

We'll go into the uses of these instructions in a future chapter, especially the Boolean ones, since it may not be obvious what they're good for.

The Z-80. The Z-80 permits the same immediate mode addressing operations as the 8080. The mnemonics used are the same ones Zilog uses for its register-to-register arithmetic. The only difference is the use of a constant in the operand—represented by a number—instead of a register (a variable)—represented by a letter.

19-3.3 Extended-Precision Arithmetic

The 8080. When you plan to use numbers larger than 255 (decimal), you need more than 8-bit spaces to work in. We've already mentioned that a way exists to use the Carry and Borrow flags with the 8-bit arithmetic instructions to extend the results of a one-register ADD or SUB into another register. The 8080 has one 16-bit ADD instruction, Double Add (DAD). It uses the HL register pair as a 16-bit ACCUMULATOR. The remaining register pairs can be added to HL. The register pairs can also be increased and decreased. Extended-precision mnemonics for the 8080 always contain an X (except double add):

(DAD) Double Add (register pair to HL)

(INX) Increase (register pair) by 1

(DCX) Decrease (register pair) by 1

These operations allow you to build up numbers as large as 65,535 in register pairs as easily as single-byte arithmetic allows you to build numbers up to 255.

The Z-80. Zilog permits 16-bit subtraction as well as addition. There are ADD, ADC, and SBC mnemonics for these 16-bit operands (the numbers in the register pairs), but no SUB (subtract without using the Borrow from the last subtraction). Since two more register pairs, IX and IY, exist in the Z-80, some 16-bit operations exist which use them as accumulators, like the HL. There are ADD, INC, and DEC operations for these two registers. You'll recall that Zilog uses the same name for all ADD instructions—including DAD—whether they're 8-bit or 16-bit arithmetic.

19-3.4 Boolean Operations

You may have wondered how a register could be ANDed or XORed with another register or an 8-bit constant. All the AND and XOR gates you've seen combine 1-bit numbers, not 8-bit numbers. Actually, the AND, XOR, and OR instructions have eight parallel gates that combine the 1-bit of one operand with the 1-bit of A, the 2-bit of the operand with 2-bit of the A, the 4-bit of the operand with the 4-bit of A, and so forth, combining each of the operand's 8 bits with the corresponding bit of A.

For example, let's suppose that the number in A is 10101010, and we're going to AND it with 11110000. The operand and A are written like this:

$$
\begin{array}{r}
10101010 \\
\text{AND} \quad \underline{11110000}
\end{array}
$$

as though we were going to ADD them. Then, starting at the right, we can AND each column of 2 bits (as though we were doing addition or subtraction). On the rightmost column 0 AND 0 equals 0. Moving to the left, we have 1 AND 0 equals 0. We continue this column by column, until we reach the leftmost column—1 AND 1 equals 1. The result looks like this:

$$
\begin{array}{r}
10101010 \\
\text{AND} \quad \underline{11110000} \\
10100000
\end{array}
$$

If you know what the output of a two-input gate is for any kind of logic, you know the output for a combination of two 8-bit numbers (or any other size), by following the rule above. Since each Boolean gate has a single output, each column of two 1-bit numbers results in a 1-bit answer, and two 8-bit operands give an 8-bit answer without generating any Carry or Borrow output.

There is also one Boolean (sort of . . .) operation the Z-80 does that the 8080 does not. This is the Negate (NEG) operation, which forms the *two's complement* of the number in register A. It's a two-byte mnemonic. 8080 users could use two one-byte instructions CMA (invert all bits of A) and INR A (A + 1) to get the same results.

Retracted precision(?). There a few 1-bit arithmetic-logic operations in the 8080 and Z-80.

They set, reset, complement, or test a single bit. Two operations affect the Carry flag:

(Intel)	(Zilog)	
STC	SCF	Set the Carry (flag)
CMC	CCF	Complement Carry flag

In the Z-80, any bit in any register can be set or reset using one of the following:

(SET)   Set a bit
(RES)   Reset a bit

These registers include any memory location that HL, IX + d, or IY + d points to.

A 1-bit Compare instruction, called a bit test (BIT), is available in the Z-80 instruction set to test whether any bit in a register (includes memory) is ON or OFF. The Z (zero) flag in the ALU is set or reset to indicate if the bit is 0 or not. For example, the instruction

$$\text{BIT} \quad 7,\text{A}$$

tests whether bit 7 (the most significant bit) of A is a 0 or a 1. After the instruction, the Z flag is ON if the number is 0 and OFF otherwise.

19-4 CONTROL INSTRUCTIONS

There are three types of control operations, called jump, call, and return. These operations affect the program counter, which is the register that points at the next instruction. A jump puts a new number into the PC, so that the computer "jumps" to a new section of the memory to get its next instruction. Imagine that the program counter is "fetching" instructions from memory in this order: 23, 24, 25, 26, 27, ... If a jump occurs on line 26 of memory, the count could look like 23, 24, 25, 26, 12, ... The next address in the program counter, which would have been 27, has been overwritten by a 12. From here on, the program counter will count 13, 14, 15, . . . , until it reaches 26 again. At that point, it would go back to line 12 again, and continue to repeat the instructions between 12 and 26 over and over.

Why should anyone want to do that?

As an answer, think of how you would make the computer test 1024 memory locations by writing and reading data. You could write the same test routine 1024 times, or write it once and go back to the beginning of the routine 1024 times. The second procedure is much simpler than the first, if you can make the jump that goes to the beginning of the routine stop after 1024 times.

This requires a **conditional** type of instruction—in this case, one that checks a counter to "see" if the count is 1024 yet, and jumps as long as it is not. There are several conditions that can be tested that will enable or disable a conditional jump instruction. They are all related to the flags in the arithmetic/logic unit. We will begin by studying these instructions.

19-4.1   Unconditional/Conditional Jump

The term "unconditional" means "always, no matter what." There is an instruction, JMP (Zilog JP), that loads a number into the program counter every time. This number is gotten directly from the instruction itself, something like a MVI, except that when the program counter is loaded with a number (a 16-digit number) the number is an address, not a data. In the program counter, this number controls the program itself.

The form of JMP instructions is shown by an example:

Mnemonic		Result
(Intel)	(Zilog)	
JMP 5000	JP 5000	PC←5000

We used the address 5000 in an earlier example. When the number 5000 is stored in H and L, it is broken up into two bytes, the page number and line number. You might remember that address 5000 was the same as page 19, line 136 (decimal) in our example. We don't have to deal with two separate bytes of code in the JMP instruction, because the PC is large enough to hold 5000 as a single number. In reality, however, the 5000 is part of a three-byte instruction in machine code. The first byte has the decimal value 195. This is the opcode, and it tells the computer that the next two bytes will be "jammed" into the PC (and will clobber anything that's already there). The next two bytes of the code are 136 and 19, which are there because, despite the size of the PC, the program itself is stored in a memory with words of only 8 bits' length. The complete instruction for JMP 5000 is stored in three consecutive bytes of memory as 195, 136, 19, although it's unlikely they would be expressed as decimal numbers in any computer printout. The general standard is to use hexadecimal, which represents the same instruction as C3, 88, 13. You'll notice that the line number comes before the page number just as it did in the LDA instruction in Section 19.2.

The term "conditional" means "sometimes—when something else is true." There is a group of instructions for the 8080 and Z-80 that jump only if a condition is met in one of the flags. There are four flags that are tested by conditional jump instructions: the Zero, Carry, Parity, and Sign flags. These are flip-flops that are set or reset by the ALU when it does arithmetic or Boolean operations. 8080/Z80 conditional jump instructions are called "IF" statements or "conditional BRANCH" in other languages. They all have pattern "Jump IF (condition) TRUE." The four flags tested by the 8080/Z80 conditional jumps give rise to eight jumps:

Mnemonic		Action
(Intel)	(Zilog)	
JNZ	JP NP	Jump if Not Zero
JZ	JP Z	Jump if Zero
JNC	JP NC	Jump if No Carry
JC	JP C	Jump if Carry
JPE	JP PO*	Jump if Parity Reset*
JPO	JP PE*	Jump if Parity Set*
JP	JP P	Jump if Plus Sign
JM	JP M	Jump if Minus Sign

Each of these instructions is a three-byte code, an opcode followed by a two-byte address that's put into the PC if the condition is true. A typical instruction of this type is:

Mnemonic		Result
(Intel)	(Zilog)	
JM 5000	JP M, 5000	PC←5000 IF sign is (−)

When the Sign flag is set (the ALU has just calculated a number with a negative result) the address in the PC is replaced by 5000. What happens if the Sign flag is reset? If the number in the PC is not replaced, it's still there! When this happens, the CPU fetches its next instruction from the place it was going to, anyway. The program flows on from the last instruction to the next as though nothing has happened. (This action is sometimes called a "fall through" in programming.)

Example 19-3

Show what happens to the numbers in the registers below when an Intel 8085 processor executes the instruction shown.

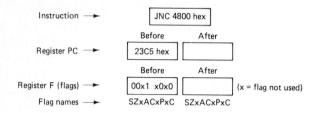

	Before	After
Instruction →	JNC 4800 hex	
Register PC →	23C5 hex	
Register F (flags) →	00x1 x0x0	(x = flag not used)
Flag names →	SZxACxPxC	SZxACxPxC

Solution: JNC 4800 jumps to address 4800 if the carry flag is equal to 0 (no carry). In this case, C = 0, so the address 4800 replaces the address in the PC. The flags are not affected.

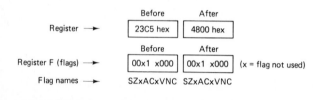

	Before	After
Register →	23C5 hex	4800 hex
Register F (flags) →	00x1 x000	00x1 x000 (x = flag not used)
Flag names →	SZxACxVNC	SZxACxVNC

19-4.2 Relative Addressing versus Absolute Addressing

An address "points to" a place in the memory where the CPU wants to fetch or put a byte. In the case of the program counter, the address in a Jump instruction is fetched out of the second and third bytes of the instruction itself and put into the PC. The only difference between a Jump and a Load Immediate is the register where the number is put; there are no LXI instructions that load the PC. Just as the Load Immediate is not the only way to move a number into a 16-bit register, the Jump instructions we've seen are only one of several ways to move a new number into the PC.

The Jump instructions we've just seen load a 5000 into the PC. This is a type of immediate load called *absolute* addressing; because the address is right there in the instruction, it's absolute and unvarying. It (the address) is just loaded directly into the PC. There are other ways of changing an address. You can increment, decrement, add to, or subtract from numbers in a register. Since the PC automatically increments as the CPU moves from one instruction to the next (which leads us to suspect that we can ADD to the PC), we might ask if numbers other than one can be added to the PC.

The answer is yes, and the numbers can be either positive or negative. A Jump relative in-

struction is used to tell the CPU things like "Jump ahead five lines" or "Jump back three lines," instead of "Jump to 2351" or "Jump (if No Carry) to 6315." The relative addressing implied in the JR instructions (found on the Z-80, but not the 8080) is useful for two things:

1. Relocatable code. If you want to take the same instructions that work in the 5000 area of memory and transplant them to the 8000 area, there will be no need to rewrite the Jump instructions for the new addresses. The software will be perfectly transplantable if all the jumps are relative.

2. The instructions themselves will be shorter, since the "jump forward" and "jump backward" are done with only one byte of address. The one byte (in signed binary that can be positive or negative) permits the programmer to "jump" 127 bytes forward or 128 bytes backward.

Examples of the JR (Jump Relative) instructions include both unconditional and conditional types. An explanation of each one is given with the mnemonic:

Mnemonic	Result
JR FROG	PC←PC + displacement
JR NC, FROG	PC←PC + displacement, IF No Carry (Carry flag OFF)

In both cases, "displacement" refers to the number you would have to add to the present address in the PC to get to the address called FROG. Only the Carry and Zero flags are used by JR instructions to do conditional, relative branching.

There are also instructions that a programmer can use to jump the computer somewhere indirectly. An address in HL, IX, or IY can be used as the address of a jump. In this case, the instruction itself doesn't even need to contain a part of the address. The PC is simply loaded from the HL, IX, or IY (16-bit) registers. The 8080, of course, has no IX and IY registers, and can only load the PC with HL.

Mnemonic (Intel)	(Zilog)	Result
PCHL	JP (HL)	PC←HL
	JP (IX)	PC←IX
	JP (IY)	PC←IY

19-4.3 Subroutine Calls and Returns

You use the Jump instructions to tell the computer where to go (to fetch its next instruction). Usually, when you tell someone where to go, you don't bother telling them how to get back. In this case, there's a polite instruction for the jump that tells the CPU how to get back.

This instruction is called a **CALL**. When a CALL takes place, the computer doesn't just transfer a new number into the PC, it also saves the old PC. The old PC is saved by being written into the memory. A special memory pointer is used to keep track of where the return address is.

Later, another instruction—Return (RET)—is used to get back to the place where the PC was at the time of the CALL.

At the place where the CALL jumps to, we do a number of instructions terminated by a RET. This group of instructions is called a **subroutine**. The program that contains the CALL is referred to as the **main program**. Any time a CALL to a subroutine takes place, the RET at the end of the subroutine brings the computer back to the main program. The difference between a subroutine and any other routine a computer might jump to is the reusability of the subroutine from any place in the program. If you want to do the same thing seven or eight times in a program, for instance, there are two ways to do it; either write the same identical group of instructions in the program seven or eight times, once for each place where it's wanted, or make the group of instructions into a subroutine.

Writing the same identical routine eight times is wasteful. Writing it once is elegant. Subroutines are the only way to go in cases like this. Now, you might ask: "Why would you want to use the same routine eight times in a single program?" One example we can think of is *division*. None of the popular 8-bit microprocessors that we have listed have built-in circuits that can divide. To do a division, a small program (an *algorithm*) must be written that can divide two numbers using the *add*, *subtract*, and *move* circuitry. While the algorithm isn't terribly complicated, if you divide something at eight different points in the program, the code for the algorithm will have to be written eight times in the program. If you use a division subroutine, it's only necessary to write the algorithm one time. It can then be used by a CALL from any point in the program, or any other program that needs it. With the CALL instruction, a single command can take the place of the whole algorithm every time it's needed.

Like the Jump instructions, CALL and RE-

Turns can be found in unconditional and conditional forms. Here are some examples:

Mnemonic		Result
(Intel)	(Zilog)	
CALL MULT	CALL MULT	STACK←PC
		PC←MULT
CNZ DIVIDE	CALL NZ, DIVIDE	STACK←PC
		PC←DIVIDE
		IF Not Zero
		(Zero Flag OFF)
CALL 5000	CALL 5000	STACK←PC
		PC←5000
RET	RET	PC←STACK
RNZ	RET NZ	PC←STACK IF
		Not Zero
		(Zero Flag OFF)

In the third example, we really did the same thing as the first example—an unconditional CALL to a place in the memory. In the first example, there's a label (MULT) that identifies a place in the memory that the assembler keeps track of, but which we don't know. In the third example, the CALL uses a "real" address (5000) which isn't represented by a label. Instead, it is shown as an absolute address. The machine code that the assembler turns these mnemonics into has a number in either case; it's just easier to remember what the CALL does when the address of the subroutine is represented by a suitable label with some mnemonic value.

The RET instructions take back the number stored in the stack by the last CALL that was done. This number was the old PC that was pointed at the next instruction the main program was going to do after the CALL.

19-4.4 Restart (Trap) Instructions

There is a one-byte CALL instruction called a **Restart** (RST) in the 8080 and Z-80 instruction sets. An ordinary CALL instruction contains three bytes, an opcode, and two bytes of address. This address allows the subroutine being CALLed to be at any one of 65,536 places in the memory. Since the RST instruction is only one byte long, there isn't enough room in the instruction for a 16-bit address. Instead, there are eight different opcodes for the RST instruction, each of which CALLs a subroutine at a different place. The "destination" of each RST subroutine call is contained inside the opcode itself, as shown below:

Mnemonic	Action
RST 1	Same as CALL to address 010 (octal)
RST 2	Same as CALL to address 020 (octal)
RST 3	Same as CALL to address 030 (octal)
⋮	⋮
RST 7	Same as CALL to address 070 (octal)

All the addresses are on page 0 of the memory. There isn't enough room between each of these subroutines and the next for a decent subroutine. The instruction that's usually found at the destination of a RST is a Jump to a place where the rest of the subroutine is written. The rest of the subroutine is then located where there's plenty of room for all the instructions that are needed. When a Jump instruction is used this way, it's called a vector. Restart instructions are usually used in a hardware-driven subroutine call known as an *interrupt*. When the interrupt uses a vector as we just described, it's called a *vectored interrupt*. We'll study interrupts in more detail later.

Unlike subroutine CALL and RETurn instructions, the RST instruction has no conditional form. There isn't any such instruction as "RESTART, IF . . ." in the 8080 or Z-80.

19-5 MISCELLANEOUS OPERATIONS

These are operations which don't necessarily have much in common with each other, but differ in special ways from most of the instructions in the other groups.

19-5.1 Stack Operations

We discussed the stack for the first time in Section 19-2.4, in conjunction with 16-bit data-transfer operations. The stack is a special part of the memory, not because other memory operations can't reach it, but because it has its own special memory pointer—the stack pointer (SP)—which works differently from the index registers or program counter in the way it operates. The PUSH and POP instruction, described in Section 19-2.4, are 16-bit operations that permit the user to save or recover 16-bit words one at a time between memory and register pairs in the CPU. These are used in much the same way that Move, Store, and Load instructions are used for 8-bit bytes. The

stack pointer, however, differs from the (HL, IX, or IY) memory pointer by being a counter. Unlike the program counter, which marches forward through memory as it picks up instructions, the stack pointer marches backward in double steps after each word is PUSHed onto the stack. When a word is taken off the stack (POPped), the SP marches forward. The *stack* thus operates like a shift register that can reverse directions. Such a register is called a LIFO (Last In, First Out).

The CALL, RST, and RET instructions, described in Sections 19-4.3 and 19-4.4, use the stack in the same way as PUSH and POP, except that the words being stored and recovered are addresses used by the program counter.

19-5.2 Rotates and Shifts

The 8080 has a large number of *rotate* instructions, and the Z-80 has even more. *Shift* instructions use a register (usually register A—the accumulator) as a shift register for the bits it contains. The instructions called *rotates* use the register as a ring counter (see Section 11-4) in which the Carry flag is part of the ring. In either case, these shifts and rotates include the ability to operate as a *left-shift register* or a *right-shift register*. In the Z-80, most registers can be rotated or shifted, while the 8080 allows these operations only on register A. The Z-80 also allows BCD shifts between the accumulator and memory. Whenever a register is used in ring counter mode, the Carry flag is added onto the front (most significant bit) of the register and its contents are rotated along with the bits in the register, or else it copies some bit inside the register. Here are examples of the rotate instructions for both processors:

Mnemonics		Action
(Intel)	(Zilog)	
RAL	RLA	Use the Carry flag and A as a ring counter, rotating to the *left*.
RAR	RRA	Use the Carry flag and A as a ring counter, rotating to the *right*.
RRC	RRCA	Use register A (only) as a ring counter, rotating to the *left*. The Carry flag "copies" bit 7 of A.
None	RL*	Same as RAL, but can use "*"

*Any register or memory location M that (HL), (IX + d), or (IY + d) points to.

Mnemonics		Action
(Intel)	(Zilog)	
None	RR*	Same as RAR, but can use "*"
None	RLC*	Same as RLCA, but can use "*."
None	RRC*	Same as RRCA, but can use "*."
None	SLA* (Shift Left Arithmetic)	Acts like a left-shift register which uses the Carry as an MSB.
None	SRL* (Shift Right Logical)	Acts like a right-shift register which uses the Carry as an LSB.
None	SRA* (Shift Right Arithmetic)	Acts like a right-shift register which uses the Carry as an LSB and copies the MSB into the next lower bit as the shift happens.
None	RLD (Rotate Left Decimal)	Rotates a BCD digit from memory into the accumulator.
None	RRD (Rotate Right Decimal)	Rotates a BCD digit into memory from the accumulator.

Rotate and Shift instructions are used for various mathematical purposes which we will not attempt to explain here. The usefulness of these instructions will become clear later.

19-5.3 NOP (Dummy) Instruction

This is an instruction that does nothing. It's as though there is no circuit connected to the instruction decoder that this instruction can turn on. The CPU fetches this instruction, puts it into the instruction decoder, and . . . nothing happens. The code for a NOP (No OPeration) is reasonable, considering it does nothing—its binary code is 00000000 (nothing!). NOP instructions are one byte long, and are often used to fill unused memory space, or to cover up an instruction with nonactive bytes during software writing and debugging. NOP is also a useful instruction when a short time delay—that does nothing else—is desired.

19-5.4 Halt Instruction

This is just what you would expect it to be. At the end of a program when you want to stop the computer from going further in the memory, the Halt instruction stops the PC from counting ahead and fetching the next instruction. This is important

because there will always be "something" in the memory locations that follow the program—even if the programmer didn't write anything there—and in most cases, the "something" is "garbage." If the computer should go on and fetch these bytes after the last instruction in the program, it will attempt to do what the garbage says it should do. The computerese expression "garbage in, garbage out" is all too true in this case; the instructions in the garbage that follows the program will probably destroy part, or all, of the data, program, and output, if the program isn't stopped. The Halt instruction also waits for an interrupt when the computer is stopped. The computer can be restarted if the interrupt input is activated. We'll see more about this later, but in this case, we'll just say that in certain microprocessor instruction sets (Motorola) the Halt instruction is called a WAI (WAit for Interrupt) instead of a HLT. It's really the same instruction with another name.

19-6 ASSEMBLERS AND ASSEMBLY LANGUAGE

In this chapter we've represented each code for the 8080 and Z-80 with a mnemonic. The actual binary number stored in the memory for each of the instructions we've discussed is shown in the Appendix. To simplify the numbers, we've used octal as "shorthand" for the binary value of each opcode. An assembly language program for the 8080 or Z-80 is a combination of the mnemonics and labels the programmer has used to solve a problem. If the computer is programmed directly in binary, the programmer will probably use a lookup table (like Appendix 1) and write a program in assembly language first. It is easier to keep track of loops and variables in assembly language, and easier for someone else to look at an assembly language program when they're trying to see how it works. Ultimately, however, the programmer must code the program into the computer in binary. This means the assembly language program will have to be translated into binary before it can be entered. This can be done by hand, laboriously writing down, line by line, the translation of each mnemonic and label, and then entering the completed translation into the memory. An alternative is to have a program that takes lines of assembly language as its input and does the translation into binary for the programmer. Such a program is called an **assembler**, because it puts together the actual opcodes and addresses that take the place of the mnemonics and labels. When it's done, the assembler has assembled a twin of the assembly language program in the absolute machine language of the microprocessor. This program can then be entered by the programmer manually, but is usually used directly by the computer, since it's in the computer's memory already when the translation is completed.

19-7 16-bit Processors (8086 Family)

Now, let's look at the 16-bit (and 32-bit) family of processors mentioned at the end of Chapter 18. The 8086 family uses instructions that are basically similar to those used in the 8080/8085/Z-80, although different opcodes are used for the instructions and the registers aren't the same. Three important areas of difference are in the addressing of memory, the arithmetic instructions, and the names (mnemonics) assigned to the conditional JMP instructions.

19-7.1 Data Transfer

The 8086 family uses the **MOV** mnemonic for most data transfers, just as the 8080/8085 do. There are a few more ways to move data in these 16-bit computers, including the block move and indexed modes provided in the Z-80. As with the 8080/8085/Z-80, the accumulator (AX) is a "privileged" register and takes part in many operations that the other registers don't.

Register-to-register MOV instructions are the same as in 8080/8085/Z-80, although it is now necessary to keep track of whether a byte (8 bits) is being moved or a word (16 bits) is being moved. If you try to move a 16-bit number into an 8-bit place, for instance, it's clear why *that* won't work, but it's not so clear why moving an 8-bit word into a 16-bit place is a mistake. For example, if byte F8 (hexadecimal) is moved into a 16-bit register formerly holding the number 1234, the outcome might be 12F8 or F834, depending on which place in the destination register the source number was sent. If you are expecting to find F8 in the destination register (or 00F8) you'll be disappointed. As a result of this problem, instructions are provided to do MOV instructions only where both the source and destination are bytes, or both are words. There is a special **CBW** (Convert Byte to Word) instruction to make 8-bit numbers into 16-bit numbers, to handle this kind of problem. It "stretches" the accumulator's least-significant byte (AL) to fit the entire AX accumulator (AH and AL together) (see Figure 18-9 for register layout).

Incidentally, the terms *bit*, *nybble*, and *byte* have specific sizes (1, 4, and 8 bits), but the word *word* can be anything, depending on whether it's a 16-bit machine, a 32-bit machine, and so forth. In this case, we're talking about 16-bit machines, and we'll assume that *word* means 16 bits, except maybe in the case of the 80386, which is 32 bits. If you hear IBM 370 programmers jabbering about words, you'll find that they often assume that 32 bits is the only *possible* size for a word. An IBM 370 is a 32-bit machine (like the 80386).

Addressing modes available for data transfer via MOV instructions are, understandably, more complex in the 8086 family than in the 8080/8085/Z-80. The 8086 family has so many more ways to put together an address that a multitude of instructions exist.

For comparison, Table 19-2 is a list of MOVs, and whether the other machines we have discussed have similar addressing modes. The really mystifying MOV commands on lines 12 through 17 of the list are used for arrays of numbers. An example would be to have one array of resistors, holding all the resistor values in a circuit, another array of capacitors, and so on. The fifth resistor might be in the array RESIS, at location RESIS+5, and the fifth capacitor in the array CAPAC, at location CAPAC+5. If you wanted to multiply the first resistor times the first capacitor, the second resistor times the second capacitor, and so on, you could use the source index register, SI, to count through the R's and C's. When the number in SI is 5, then the instructions

MOV AX, RESIS [SI]

and

MUL CAPAC [SI]

would result in a number (the *RC* time of the fifth pair of components) being computed. You can probably guess from the example above that when we look at the arithmetic commands the 8086 family has, we'll find multiplication added to the 8080/8085/Z-80 arithmetic capabilities. You can also probably guess that, as before, the accumulator (AX) is where the multiplication is done, and where the product ends up after the arithmetic is done.

TABLE 19-2

8086 mode	Example	8080/8085	Example	Z-80
1. Reg. ← Reg.	MOV BX,CX	Yes→	MOV B,C	Yes
2. Reg. ← Memory	MOV BP,STACK	Yes→	MOV B,M	Yes
3. Memory ← Reg.	MOV COUNT,CX	Yes→	MOV M,C	Yes
4. Load Direct	MOV AX,MVAR	Yes→	LDA MVAR	Yes
5. Store Direct	MOV ARRAY,AX	Yes→	STA ARRAY	Yes
6. Reg ← Immed.	MOV CL,2	Yes→	MVI C,2	Yes
7. Memory ← Immed.	MOV MASK,4	Yes→	MVI M,4	Yes
8. Indirect A←M	MOV AX,[BX]	Yes→	LDAX B	Yes
9. Indirect M←A	MOV [BX],AX	Yes→	STAX B	Yes
10. Based A←M	MOV AX,[BX]+5	No	LD A, (IX+5)	←Yes
11. Based M←A	MOV [CX]+3,AX	No	LD (IY+3),A	←Yes
12. Indexed A←M	MOV AX,COUNT [SI]	No		No
13. Indexed M←A	MOV TABLE [DI],AX	No		No
14. Based Indexed	MOV AX,[BX][SI]	No		No
15. Based Indexed	MOV [BX][DI],AX	No		No
16. Based Indexed and Displacement	MOV AX,[BP][SI]+15	No		No
17. Based Indexed and Displacement	MOV [BP][DI]+12,AX	No		No
18. Move String	MOVS/REP	No	LDIR	←Yes
19. Load String	LODS	No	LDI	←Yes
20. Store String	STOS	No	LDI	←Yes

Example 19-4

Show what happens to the numbers in the registers below when an Intel 8086 processor executes the instruction shown.

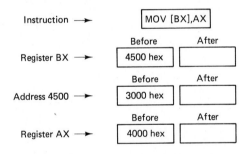

Solution: MOV [BX],AX moves data from AX into whatever memory location is at the address to which the 16-bit logical address in BX points.

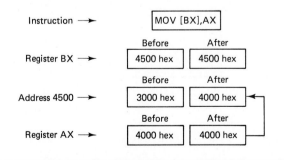

You'll notice that the multiplicity of MOV, MVI, STA, LDA, STAX, LDAX, LXI, SHLD, and LHLD commands for various 8- and 16-bit transfers in assorted addressing modes has been simplified to just one name: MOV. There are still some 8086-family data transfer commands that have separate names:

The **PUSH** and **POP** instructions found in the 8080/8085/Z-80 instruction set exist in the 8086 family, with the only difference being the names of registers used.

The **IN** and **OUT** instructions do input and output in much the same way as 8080/8085 I/O—from the accumulator. One difference is that 8-bit (AL→port and AL←port) or 16-bit (AX→port and AX←port) transfers can be specified. Using DX, ports from 0 to 65,535 may be addressed, similar to the Z-80's indirect I/O mode.

Exchange instructions, limited to a few register pairs in the 8080/8085, and the alternate register set in the Z-80, are available for swapping *any* pair of 16-bit registers in the 8086 family. An **XCHG** (exchange) between memory and a register are possible, too, but not in all addressing modes.

The memory-to-memory **MOVS** instructions listed on lines 18, 19, and 20 are block moves or string moves, but use different registers than the Z-80 instructions, and perform slightly different operations. Where the Z-80 uses register pairs HL, DE, and BC as source address, destination address, and byte counter, the 8086 family uses SI as source address, DI as destination address, CX as the counter, and requires the use of a **REP** (repeat) instruction to repeat the string operation until the entire string or block of data is moved. In addition, whereas the Z-80 has separate instructions for block moves that increment the memory address (LDIR) and decrement the address (LDDR) during a block move, the same opcodes can work either way in the 8086 family simply by changing a bit (the direction flag, DF) in the flag byte.

19-7.2 Arithmetic/Logic

For comparison, Table 19-3 is a list of the 8086-family arithmetic logic instructions that have equivalents in 8080/8085/Z-80.

EXAMPLE 19-5

Show what happens to the numbers in the registers below when an Intel 8086 processor executes the instruction shown.

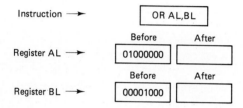

Solution: OR AL, BL combines corresponding bits of 8-bit register AL with 8-bit register BL via the logical OR operation, and places the resulting 8-bit number in register AL.

	Before	After	
Register AL →	01000000	01001000	← AL OR BL
	Before	After	
Register BL →	00001000	00001000	

As indicated previously, 16-bit **MUL** (multiply) and **DIV** (divide) operations have been added to the 8080/8085/Z-80 operations. The accumulator (AX) is multiplied times another register, and the results placed in AX and DX (in the event that the result is greater than 16 bits. If 8-bit arithmetic is being done, only AL times (byte) will be calculated. Since this will not exceed 16 bits, the result is put into AX only.

$$\text{(product) DX, AX} = \text{AX} \times \text{Register}$$

Division uses AX and DX as the dividend (for 16-bit arithmetic) and divides the 32-bit number thus formed by any 16-bit number in a reg-

ister. If 8-bit arithmetic is being done, only AX will be divided by the (8-bit) register.

$$\text{(remainder) AH, (quotient) AL} = \frac{\text{DX, AX}}{\text{Register}}$$

Multiplication and division use the opcodes

MUL (unsigned byte or word multiplication)

IMUL (signed byte or word multiplication)

DIV (unsigned byte or word division)

IDIV (signed byte or word division)

Unsigned arithmetic uses a byte or word to represent a number in which all the bits of the number are true binary code for numbers between 0 and 255 (byte) or 0 and 65,536 (word). All numbers are assumed to be positive. Signed arithmetic uses the MSB (most-significant bit) of a byte or word to represent a minus sign (if the bit = 1). The bits in the number are true binary if the number is positive (MSB = 0), and the number is in twos'-complement code if the number is negative (MSB = 1).

Logical operations found in the 8080/8085/ Z-80 have an equivalent command for every 8086 logic operation, although 16-bit logic is not available on the 8080/8085/Z-80. The **TEST** command in 8086-family code is like BIT—bit test—in Z-80, but not exactly. BIT allows testing or comparison with only one bit at a time, with a yes–no outcome in the zero flag. TEST actually does a lookahead AND of its two operands, to determine if any of the bits in the test word match bits in the target word. We have noted in the past that 8086-family machines have many instructions that the 8080/8085/Z-80 don't have. Well, here's a turnabout: The Z-80 has instructions in the rotate group—Shift-right BCD and Shift-left BCD—not found in the 8086 family.

19-7.3 Control Instructions

In the 8080/8085 instruction set, the control instructions we examined were the **JMP, CALL,** and **RET**urn instructions, in unconditional and conditional forms. (This included PCHL as a special case of JMP, and RST as a special case of CALL.) The Z-80 provided us with JR (relative jump) and DJNZ (decrement B and jump if not zero), which is listed in Appendix 1, but about which we haven't said anything yet.

The 8086 family has all the 8080/8085 instructions and all the Z-80 modes as well as some variations of the RST (restart) instruction called

TABLE 19-3

8086 function	Example	8080/8085	Example	Z-80
1. A = A + Reg	ADD AL, BL	Yes	ADD A, B	←Yes
2. A = A + Reg + Carry	ADC AL, CL	Yes	ADC A, C	←Yes
3. A = A − Reg	SUB AL, DL	Yes	SUB A, D	←Yes
4. A = A − Reg + Borrow	SBB AL, BL	Yes	SBC A, B	←Yes
5. A = A and D	AND AL, DL	Yes→	ANA, D	Yes
6. A = A ×or C	XOR AL, CL	Yes→	XRA, C	Yes
7. A = A or B	OR AL, BL	Yes→	ORA, B	Yes
8. Compare A to B	CMP AL, DL	Yes→	CMP D	Yes
9. C = C + 1	INC CL	Yes	INC C	←Yes
10. B = B − 1	DEC BL	Yes	DEC B	←Yes
11. A = $\overline{A}$	NOT AL	Yes→	CMA	Yes
12. A = (−A)	NEG AL	No	NEG	←Yes
13. BCD Adjust	DAA	Yes→	DAA	Yes
14. Test Bit	TEST AL, 1	No	BIT 1, A	→Yes
15. Rotate → A + carry	RCR AL, 1	Yes→	RAR	Yes
16. Rotate ← A + carry	RCL AL, 1	Yes→	RAL	Yes
17. Rotate → A	ROR AL, 1	Yes→	RRC	Yes
18. Rotate ← A	ROL AL, 1	Yes→	RLC	Yes
19. LogicShift ← A	SHR BL, 1	No	SRL B	←Yes
20. ArithShift → A	SAR CL, 1	No	SRA C	←Yes
21. ArithShift ← A	SAL DL, 1	No	SLA D	←Yes

INT (interrupt) instructions. The 8086-family names are mostly identical with the 8080/8085 names, except for the conditional jumps. The conditional call and return instructions, unfortunately, are not available in the 8086 family (or most other microprocessors).

Table 19-4 is a list of the 8086-family unconditional control instructions and their equivalents in 8080/8085/Z-80.

A **short jump** or call alters the IP (instruction pointer) register by adding a signed byte (−128 to +127). A **near jump** adds a signed 16-bit word (−32768 to +32767). An **indirect jump** replaces the IP (instruction pointer). An **intersegment direct** jump replaces both the IP (instruction pointer) and the CS (code segment register). Usually, the assembler program decides whether a jump or call is going far enough to need

short, near, or intersegment forms of the direct jump. (Indirect jumping is specified by the type of address written in the instruction.) If you are doing hand assembly, of course, you will have to figure this out for yourself.

All of the **conditional jump** 8086-family instructions shown in Table 19-5 can go to an address (not shown), which can be short, near, intersegment, or indirect. Conditional call and return instructions do not exist in the 8086 family.

The DJNZ (decrement, jump if not zero) instruction of the Z-80 instruction set downcounts the B (byte counter) register and jumps if B is not yet equal to zero. This is a useful instruction for setting up loops. In 8086-family instructions, the instruction that does the same thing is called **LOOP**. It decrements the CX (counter) register and jumps if it is not zero yet. An alternate name,

TABLE 19-4

Unconditional control

8086 function	Example	8080/8085	Example	Z-80
1. Jump	JMP LONG	Yes→	JMP LONG	Yes
2. Jump Relative	JMP SHORT	No	JR SHORT	←Yes
3. Jump Indirect	JMP CX	Yes→	PCHL	Yes
4. Call (address)	CALL FAR	Yes→	CALL FAR	Yes
5. Return	RET	Yes→	RET	Yes
6. Interrupt	INT 3	Yes→	RST 7	Yes
7. Interrupt RET	IRET	No	RETI	←Yes

TABLE 19-5

Conditional control

8086 condition	Example(s)	8080/8085	Example	Z-80
1. C = 0	JNB/JAE/JNC	Yes→	JNC	Yes
	Jump if not below/above or equal/no carry (unsigned, ≥0)			
2. C = 1	JB/JNAE/JC	Yes→	JC	Yes
	Jump if below/not above or equal/carry flag = 1 (unsigned, < 0)			
3. Z = 0	JNZ/JNE	Yes→	JNZ	Yes
	Jump if not equal/no zero (signed or unsigned, ≠ 0)			
4. Z = 1	JZ/JE	Yes→	JZ	Yes
	Jump if equal/zero flag = 1 (signed or unsigned, = 0)			
5. S = 0	JNS	Yes→	JP	Yes
	Jump if plus sign/no sign (signed, ≥ 0)			
6. S = 1	JS	Yes→	JM	Yes
	Jump if minus sign/sign flag = 1 (signed, < 0)			
7. P = 0	JNP/JPO	Yes→	JPO	Yes
	Jump if parity odd/no parity flag			
8. P = 1	JP/JPE	Yes→	JPE	Yes
	Jump if parity even/parity flag = 1			
9. O = 0	JNO	No	JP PO[a]	←Yes
	Jump if no overflow			
10. O = 1	JO	No	JP PE[a]	←Yes
	Jump if overflow flag = 1			
11. (C or Z) = 0	JNBE/JA	No		No
	Jump if not below or equal/above (unsigned, > 0)			
12. (C or Z) = 1	JBE/JNA	No		No
	Jump if below or equal/not above (unsigned, ≤ 0)			
16. (S xor O) = 0	JNL/JGE	No		No
	Jump if not less/greater than or equal (signed, ≥ 0)			
17. (S xor O) = 1	JL/JNGE	No		No
	Jump if less/not greater than or equal (signed, < 0)			
18. ((S xor O) or Z) = 0	JNLE/JG	No		No
	Jump if not less or equal/greater (signed, > 0)			
19. ((S xor O) or Z) = 1	JLE/JNG	No		No
	Jump if less or equal/not greater (signed, ≤ 0)			
20. ↓count · Z = 0	LOOPNZ/LOOPNE	No	DJNZ	←Yes
21. ↓count · Z = 1	LOOP/LOOPZ/LOOPE	No		No
22. CX = 0	JCXZ	No		No

[a]Zilog Z-80 jumps if either parity or overflow in this case

LOOPZ (loop until zero), is used, but the opcode is the same. In many of the conditional instructions listed, there are two, or even three, mnemonic forms for the instruction (although the opcode is the same). This is provided because the same instruction may be used for different purposes. For example, the C (carry) flag is part of a ring counter in some of the rotate instructions, and may be checked to identify if a 1 has been located when it is rotated into that bit. You would probably write the **JC** mnemonic because the contents of the C flag has no other significance. However, since the C flag is also set if a subtraction yields a result less than 0, you might write the **JB** (Jump if Below) mnemonic for this instruction if you are checking the C flag following a subtraction to see if the result is below zero.

EXAMPLE 19-6

Show what happens to the numbers in the registers below when an Intel 8086 processor executes* the instruction shown.

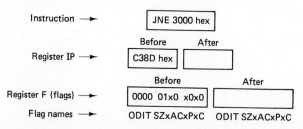

*The address in the IP register is the address at a time after all the bytes of the JNE 3000 instruction have been fetched but before it has been executed.

Solution: JNE 3000 (jump if not equal) jumps to address 3000 if the zero flag is false (if some numbers being compared are not equal, the difference between them is not zero, and the Z flag is false). In this case, Z = true (zero, thus equal), so the address 3000 does *not* replace the address in the PC. The flags are not affected.

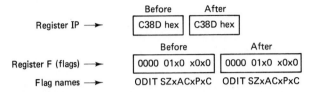

Note that the condition (not equal) was not true, so no jump occurred. We have been asked what happens to the instruction pointer after the instruction is executed. Since the instruction must be fetched to be executed, we have to ask the question: Was the instruction already fetched at the time the number in the Before box was written-down? In this case the answer is "yes" and we can assume that the IP remains unchanged after execution.

19-7.4 Miscellaneous Instructions

The Z-80 bit set, reset, and test instructions for single bits and flags are not all available for the 8086 family. The C (carry), I (interrupt enable), and D (direction) flags may be directly set and cleared; the low half of the flag byte can be loaded from AH using the **SAHF** (store AH into flags) instruction, which provides control of five flags S (sign), Z (zero), A (auxiliary [BCD] carry), P (parity), and C (carry) at once. The 8080 CMC (complement the carry flag), which toggles the bit in the carry flag, also exists in the 8086 family.

We have already noted that the **DAA** (decimal adjust accumulator) instruction exists in the 8080/8085/Z-80 and 8086 family for decimal (BCD) adjustment to do BCD addition, and the **DAS** (decimal adjust for subtraction) exists in the 8086 family to do decimal subtraction (the Z-80 DAA adjusts for both addition and subtraction in BCD). There are also a number of instructions for doing arithmetic starting with ASCII numbers. These instructions are **AAA** (ASCII adjust for addition), **AAD** (ASCII adjust for division), **AAM** (ASCII adjust for multiplication), and **AAS** (ASCII adjust for subtraction).

Like the 8080/8085/Z-80, the 8086 family has an **NOP** (no operation) instruction and an **HLT** (halt) instruction. There is also a **WAIT** instruction (which waits until a pin on the IC, called TEST, becomes active LOW). Instructions ESC (escape) and LOCK are used for external synchronization in computer systems containing multiple processors.

19-8 ASSEMBLY AND MACHINE LANGUAGES FOR 8/16-BIT PROCESSORS

8-bit processors. The instructions for 8080/8085/Z-80 processors are located in Appendix 1. To use Appendix 1 for most instructions, find its Intel (capital letters) or Zilog (lowercase letters) mnemonic. The opcode, in hexadecimal, will be located to the right of the mnemonic. For some instructions, a group of instructions has the same opcode and ending bits, with a source or destination register missing in the mnemonic. When you put the missing letter into the mnemonic, you put that register's **binary** number into the binary opcode at the right of the mnemonic. If you want a hexadecimal opcode, it will be necessary to cluster the bits by fours, and convert.

Hint 1: Don't forget that assembly language uses labels for addresses, which may look like mnemonics but *won't be found* in Appendix 1. For hand assembly, you will have to count the addresses *in your program* to find what hex bytes belong in the line and page numbers of the instruction.

Hint 2: When counting addresses, take note of the fact that Appendix 1 breaks the instructions down into one-, two-, and three-byte instructions (for Intel 8080/8085). To get the address of the next instruction after a one-byte instruction, add one to the present instruction's address; after a two-byte instruction, add two; and to find the next address following a three-byte instruction, add three to the present address. It sounds simple, but it's astounding how many people don't do this!

16-bit processors. In describing the 8086 family of 16-bit processors, we've made no attempt to explain every instruction, nor to show how to use an 8086 assembler program. The main purpose of this section and the section on 16-bit processors at the end of Chapter 18 has been, based on a more-complete description of 8-bit microprocessors and their instructions, to show how 16-bit processors differ from, and how they are similar to, the 8080/8085 and Z-80 processors.

Since hand assembly of 8086-family code is (1) very complicated and (2) not likely to be needed, no attempt has been made to provide an 8086 family assembly language-to-machine language conversion chart like the one in Appendix 1 for 8080/8085 and Z-80 opcodes. In the larger systems that use 16-bit processors, assembler programs are generally available even for microprocessor trainers (which is not the case for most 8-bit microprocessor trainers). Some good sources of information (and conversion charts) for 8086-family processors, published by Intel, are:

The 8086 Family User's Manual
(Intel Order No. 9800722-03)

iAPX 86/88, 186/188 User's Manual
(Programmer's Reference) (Intel Order No. 210911-001)

ASM86 Macro Assembler Pocket Reference
(Intel Order No. 121674-002)

The last item listed also includes a description in general terms, of how to use Intel's ASM86 assembler program. For more detailed information on how to operate the ASM86 program or similar assemblers for the Intel 8086 family, order:

ASM86 Macro Assembler Operating Instructions (for 8086-based systems)
(Intel Order No. 121628-003)

QUESTIONS

19-1. We have stated that computers are best at doing very repetitive, boring tasks. Describe one such use for the computer. (Do *not* use the one used in the text.)

19-2. How does the technician usually distinguish software from hardware "bugs" when a computer goes down during a program?

19-3. Describe the source and destination of the following Intel and Zilog mnemonics.
  (a) LD A, B   **(g)** LD A, (BC)
  (b) MOV M, A   **(h)** LD A, (IX + 4)
  (c) LXI H, 5000 **(i)** SPHL
  (d) MVI H, 19   **(j)** LD DE, 5000
  (e) LD A, (3000) **(k)** EXX
  (f) STA (3000)   **(l)** LHLD 3000

19-4. The 16-bit instructions in the 8080 and Z-80 transfer only 8 bits at a time, and do transfers to external memory in two steps. (True or False?)

19-5. What is meant by the terms "fixed word length" and "variable word length?"

19-6. In the 8080, I/O data transfers are limited to communication with only one register in the CPU. Which register?

19-7. Arithmetic and logic operations in both Z-80 and 8080 are mostly limited to one register. Which one?

19-8. Describe the difference between ADD and Add with Carry, and state where each is used.

19-9. Describe how the Compare instruction uses the ALU and flags in the 8080 or Z-80.

19-10. In the following examples, draw a diagram that shows what's in the registers and flags *after* the instruction is completed.
  (a) Instruction: ADD A, E (Zilog)
      Before   Register A   Register E
              10110100   01010111
  (b) Instruction: SUB A, E (Zilog)
      Before   Register A   Register E
              00110110   01100101
  (c) Instruction: XRA E (Intel)
      Before   Register A   Register E
              01110001   01000110
  (d) Instruction: CMP E (Intel)
      Before   Register A   Register E
              11110100   11110111

19-11. What do the instructions Jump, Call, and Return all have in common? What register do they affect?

19-12. What is the difference between a conditional and an unconditional Jump instruction?

19-13. Besides the opcode, each Jump instruction (JP-Zilog, or JMP-Intel) requires two bytes of data. Why two bytes?

19-14. Why doesn't a relative jump (JR-Zilog only) opcode require two bytes of data?

19-15. Describe what happens to the program counter if a conditional jump instruction *doesn't* jump.

19-16. What instruction is a RST (restart) most similar to in the group of Jump, Call, and Return instructions?

19-17. Are there Conditional Restart instructions in the 8080 and Z-80?

19-18. Describe what happens in a POP and a PUSH instruction. Is the POP or PUSH of single-precision or extended-precision type?

19-19. The Rotate and Shift instructions match the actions of two hardware circuits discussed in Chapter 11. What are these two circuits?

19-20. Where would you use NOP in programming?

19-21. What does an assembler program do?

19-22. How big is a *word* in an 8086 microprocessor? How big is a word for an 8085, a 68000, and a 80386?

19-23. The 8088 microprocessor uses addresses 20 bits long. If the IP (instruction pointer) register is only 16 bits long, how can the computer fetch instructions from any place in memory? How big a program can be held in the program segment of memory?

19-24. Why isn't it a good idea to move an 8-bit number to a 16-bit register?

19-25. The Z-80 has an instruction to move a block of data from one place in the memory to another; does the 8086 have such an instruction?

19-26. Name one arithmetic/logic operation that the 8086 can do, which the 8080/8085/ Z-80 can't do.

19-27. What form of the 8086-family divide mnemonics (MUL, IMUL) would you use to divide ($-355,000,000$) by ($+113$)? Remember, the dividend is a double-word 32 bits long.

19-28. Describe the difference between a short jump, a near jump, and an intersegment direct jump.

19-29. In 8086-family software, you need to check if an unsigned number is less than zero. What jump instruction mnemonic would you use to jump if this condition is true?

19-30. What is the Z-80 equivalent of a LOOP (8086 mnemonic)?

19-31. If you wished to set and reset several of the arithmetic flags with one 8086-family instruction, which would you use?

DRILL PROBLEMS

Show what happens to the contents of registers A and B after an Intel 8085 executes each instruction listed. Assume that in each case, the starting conditions are given in the illustration below.

Before

Register A → | 01000000 | | 00001000 | ← Register B

Flags: CARRY (BORROW) FLAG IS **ON**

Instruction:
↓

After

19-1. MOV A,B Register A → [ ] [ ] ←Register B

After

19-2. MOV B,A Register A → [ ] [ ] ←Register B

After

19-3. ADD B Register A → [ ] [ ] ←Register B

After

19-4. ADC B Register A → [ ] [ ]   ←Register B

After

19-5. SUB B Register A → [ ] [ ]   ←Register B

After

19-6. SBB B Register A → [ ] [ ]   ←Register B

After

19-7. ANA B Register A → [ ] [ ]   ←Register B

After

19-8. XRA B Register A → [ ] [ ]   ←Register B

After

19-9. ORA B Register A → [ ] [ ]   ←Register B

After

19-10. CMP B Register A → [ ] [ ]   ←Register B

19-11. What is the status of the sign and carry flags after the instruction above is executed? S_____ C_____

Show what happens to the contents of register PC (the program counter) after an Intel 8085 executes each instruction listed. Assume that in each case, the PC points at the first byte of the three-byte JUMP instruction shown before it is fetched and executed, and the starting conditions are those given in the illustration below.

Before

Register PC → [1000H]   (1000 Hexadecimal)

Flags: SIGN (MINUS) FLAG IS **ON**
Flags: ZERO FLAG IS **OFF**
Flags: PARITY FLAG IS **OFF** (EVEN)
Flags: CARRY (BORROW) FLAG IS **ON**

Instruction:
↓

After

19-12. JMP 3000H Register PC → [ ]

After

19-13. JNC 3000H Register PC → [ ]

After

19-14. JC 3000H Register PC → [ ]

After

19-15. JNZ 3000H Register PC → [ ]

After

19-16. JZ 3000H Register PC → [ ]

After

19-17. JPE 3000H Register PC → [ ]

After

19-18. JPO 3000H Register PC → [ ]

After

19-19. JP 3000H Register PC → [ ]

After

19-20. JM 3000H Register PC → [ ]

Show what happens to the contents of registers AL and BL after an Intel 8086 executes each instruction listed. Assume that in each case, the starting conditions are given in the illustration below.

Before

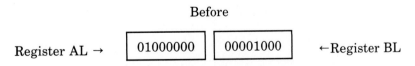

Register AL → | 01000000 | | 00001000 | ←Register BL

Flags: CARRY (BORROW) FLAG IS **ON**

Instruction:
↓

19-21. MOV AL,BL Register AL →

After

    ←Register BL

19-22. MOV BL,AL Register AL →

After

    ←Register BL

19-23. ADD AL,BL Register AL →

After

    ←Register BL

19-24. ADC AL,BL Register AL →

After

    ←Register BL

19-25. SUB AL,BL Register AL →

After

    ←Register BL

19-26. SBB AL,BL Register AL →

After

    ←Register BL

19-27. AND AL,BL Register AL →

After

    ←Register BL

19-28. XOR AL,BL Register AL →

After

    ←Register BL

19-29. OR AL,BL Register AL →

After

    ←Register BL

19-30. DEC BL Register AL →

    ←Register BL

20
SOFTWARE STRUCTURES/ FLOWCHARTING

_____CHAPTER OBJECTIVES_____

By the time you finish this chapter, you will be able to:

1. Make and read a *flowchart* for a computer program in 8080/8085 or Z-80 language.
2. Identify the basic structures in a flowchart, such as *loops*, decisions (*conditional branches*), and *subroutines*.
3. See how a practical application (a RAM tester) uses both *hardware* and *software* to diagnose system faults.
4. Explain how *interrupts* are hardware-driven subroutines.
5. Describe the various types of interrupts, and explain how the hardware of a *vectored interrupt* informs the computer where it should go to find the appropriate software.

In this chapter we look at simple computer programs for the Z-80/8080 family of microprocessors. The purpose of these programs will be two-fold:

1. The programs will show how some of the instructions in the instruction set are used, and how they are put together into larger structures (such as loops and subroutines) from which the entire program is constructed.

2. The programs will show how software may be used to check out the hardware to see if it is working properly, and how software may be used to aid in troubleshooting it when it isn't working properly.

In most cases, each program will be listed in 8080 and Z-80 assembly language and in absolute machine code (written in hex for compactness) as it would appear if you translated the instructions using Appendix 1.

Flowcharts which indicate the action of the program will be shown alongside the assembly/machine listings. New flowchart symbols will be introduced as programs that use them appear.

20-1 BRANCHING AND LOOPING

In our first example of software, we'd like you to imagine that you've built the binary output device described in Section 17-1. Figure 20-1 shows this device attached to an output port (port 0), similar to the ones we first showed you in Section 15-7.6. The instruction we use to control this output device is OUT 00. It takes any number that's in register A and places it on the data bus; at the same time, it puts address 00 on the lowest eight lines of the address bus and makes the $\overline{\text{I/O WR}}$ (output) signal active on the control bus.

20-1.1 How to Get Lost

We are going to test the output device by sending it an 8-bit number and seeing if the number lights up on the LEDs of the device. Our first attempt will be simple, but incorrect (see Figure 20-2). We have overlooked one major point in writing our test program—can you see what it is? Did you see the flaw in our reasoning? First the number hex AA (binary value: 10101010) is put in register A, then the number in A is output to port 0, then . . . ?

"Aha!" you say. "Then what?"

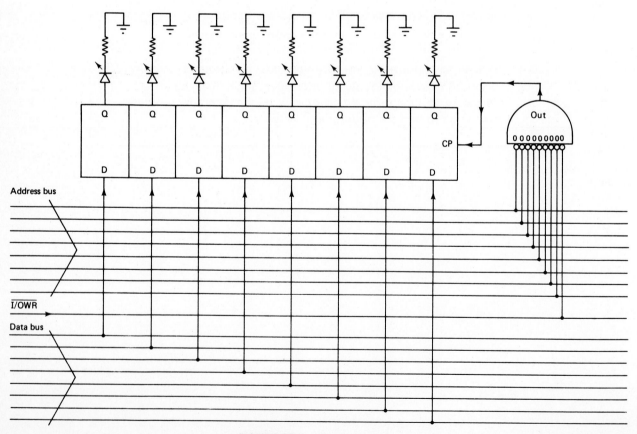

FIGURE 20-1 Output port 0.

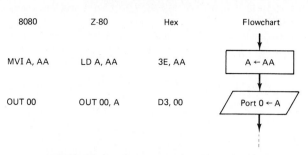

FIGURE 20-2   Output to Port 0: Program 1

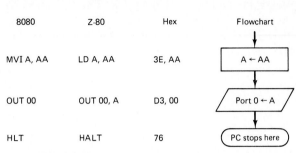

FIGURE 20-3   Output to Port 0: Program 2

The answer is unpredictable. The next thing that the computer does depends on the instruction in the next memory location. The two instructions shown occupy four addresses of memory apiece. The 3E is, let's say, at memory location 0, and the AA is at location 1. The D3 is at location 2 and the 00 is at location 3. What's at location 4? Will the computer fetch an instruction at location 4 after it finished what's at 3? To answer the last question first, the answer is yes. The computer will go ahead to 4 when it finishes 3. The question "What's at location 4?" is a tougher one. The answer must be: "Whatever the computer had left over in the location before you put in *your* instructions." This could be anything. We don't want the computer doing "garbage," but we didn't do anything to make the computer stop.

20-1.2   Staying Out of the "Garbage"

Figure 20-3 shows an improved version of program 1: Now, the program counter stops at address 4 of memory (if we started loading the program at 0). The HLT (halt) instruction makes sure that the PC won't "go ahead" to future lines of code we didn't write. This ensures that the computer won't be doing things we didn't plan just because it has finished what we wanted and kept going to parts of the memory that contain garbage.

20-1.3   Looping

There is another way to avoid the garbage past the end of our little program. Instead of stopping, we'll Jump away from the garbage in program 3.

This program goes back to the top line every time it reaches the bottom line (see Figure 20-4). The computer keeps running instead of being stopped, but still never reaches the garbage part of the memory (where you didn't write any instructions). In this case, our program avoids the garbage area by using an unconditional jump to form a loop. The name loop is obvious when we look at the flowchart at the right of the program. Whenever the computer reaches the bottom of the diagram, it finds its next step at the top of the same diagram. Thus it loops around, recycling the same instructions over and over, forever. In this case, what we have is called an **endless loop**. If we wait for this program to get finished (to halt) it will never happen. This may be a bad thing to do later, but in this example, we see it as just another way to avoid "hitting the garbage" after we do the last instruction in the program.

20-1.4   About Flowchart Symbols

A word about flowcharts. You have seen a flowchart at the right of each of the three example programs. The "shapes" in the flowchart each

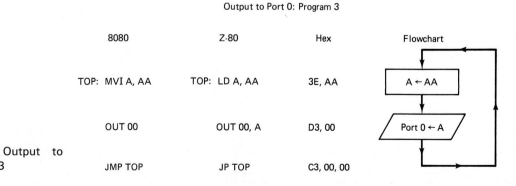

FIGURE 20-4   Output to Port 0: Program 3

have a specific meaning. In program 1 we have a rectangle and a parallelogram. The rectangle is the shape for a process, which is something that goes on in the central processor or main memory. Data-transfer and arithmetic/logic instructions are represented with a process block. The parallelogram is the shape used with any input/output operation (either input or output). In program 2, we add an *oblong* shape that represents an *end-of-routine*. In this case, the routine ends when we HALT, but there are other ways to end a routine. Connecting all the blocks in the flowchart are *flow arrows*. You have probably already guessed that these show what comes next. In each case, the computer must finish the instruction it is currently doing before it can move ahead to the next one. The flow arrows show where this next instruction is found. In program 3 we see the flow arrows change direction for the first time. This has a special meaning. When the arrows keep flowing downhill from one instruction to the next, that means that the program counter is counting normally 0, 1, 2, 3, 4, . . . and when the arrows change direction, and go back to the top of the program in this case, the program counter has "broken the rules." Where the flow arrow loops back to the beginning of the program the PC counts 0, 1, 2, 3, 4, 5, 6, 0, . . . and keeps going back to 0, instead of counting from 6 to 7. This is how a Jump looks on the flowchart, and in this case, we used an unconditional jump to close the program into a loop. As we said before, the program keeps running and doesn't HALT, but once it's in a loop like program 3, it can't escape from that loop and must repeat the same instructions forever.

20-1.5 All Good Things Must End

Program 4 (Figure 20-5) is almost the same as Programs 1, 2, and 3. It has two major differences; the number that is in A comes from a counting process—it has a different value every time the program loops—and the loop doesn't go indefinitely, because the JUMP is a CONDITIONAL BRANCH. When the number in A has counted high enough, it will carry a one into the carry flag. This will bring an end to the loop, because the loop only loops while the carry flag contains a 0 (the No Carry condition). After the carry flag is turned on, the program counter isn't JUMPed. What it does instead is to keep counting in its normal fashion, as shown below the program.

Notice the use of the Boolean XOR operation. This XOR combines A with itself. Since A always matches itself, the XOR gate's output is always LOW. This ends up zeroing the contents of the A register. It's a fairly standard thing to do this (zeroing the value in A) with the XOR instruction, so we thought we'd better show it to you in an early part of the chapter.

Notice also the new flowchart block. The diamond is the symbol for a decision operation where a CONDITIONAL BRANCH is used to direct the program onto one of two different paths

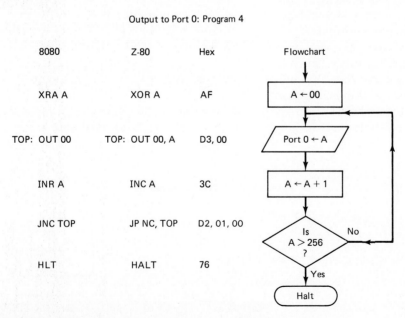

Output to Port 0: Program 4

8080	Z-80	Hex	Flowchart
XRA A	XOR A	AF	A ← 00
TOP: OUT 00	TOP: OUT 00, A	D3, 00	Port 0 ← A
INR A	INC A	3C	A ← A + 1
JNC TOP	JP NC, TOP	D2, 01, 00	Is A > 256 ?
HLT	HALT	76	Halt

PC = 0, 1, 2, 3, 4, 5, 6, 0, 1, 2 . . . (255 times) . . . 5, 6, 7 (end)

FIGURE 20-5 Output to Port 0: Program 4

TABLE 20-1

Output to port 0 program 4

| Assembly language | | Absolute machine code | | |
8080	Z-80	Address Line	Page	Instruction
XRA A	XOR A	00	00	AF
TOP: OUT 00	TOP: OUT 00, A	01	00	D3
		02	00	00
INR A	INC A	03	00	3C
JNC TOP	JP NC, TOP	04	00	D2
		05	00	01
		06	00	00
HLT	HALT	07	00	76

at a "fork in the road." The decision flowchart symbol always has one flow arrow entering into it and two (or more) coming out. In our flowchart, the two outcomes (yes or no) determine which arrow the program will "follow."

Since the A register of an 8080 or Z-80 is an 8-bit register, it will count like an 8-bit counter when it is incremented by the INC or INR instruction on the third line of the program. An 8-bit counter has a modulus of 256. This means that after the loop has run 256 times, the counter will count around to 00000000 and set the carry flag. That is the condition that lets the computer "escape" from the loop.

Table 20-1 shows another way to look at the machine code which is produced by translating program 4.

The lines marked "line" in the machine language program are one memory address each. The "address" is a combination of the line number and the page number. In this example, all the lines are located on page 00 (the first 256 locations in memory). The line called "top" is really line 01. On lines 04, 05, and 06, the Jump if No carry instruction jumps to line 01 on page 00.

This is the same place as "top." The opcode of the jump instruction (C2) says what kind of a jump it is, and the (01) and (00) are the line and page number of the place called "top" in assembly language.

Normally, all the bytes of an instruction (like the Jump, which is a three byte instruction) are put on the same line, and at the beginning of each line, the address of the opcode byte is shown (Table 20-2).

After line 01, the next line is 03 because the instruction on line 01 takes up two bytes, stored at address 01 and 02. The next available space for an instruction is line 03. In another place, the address skips from 04 to 07 because the instruction on line 04 takes up lines 04, 05, and 06. The next available space for an instruction is on line 07.

Notice that there are never any gaps between instructions. In machine language, if you skip over a space between one instruction and the next, there will be something in that space. The program counter will not skip over that space. It will do whatever is in that space, whether it's something you want or not.

TABLE 20-2

Output to port 0 program 4

| Assembly language | | Absolute machine code | | |
8080	Z-80	Address Line	Page	Instruction Opcode data
XRA A	XOR A	00	00	AF
TOP: OUT 00	TOP: OUT 00, A	01	00	D3, 00
INR A	INC A	03	00	3C
JNC TOP	JP NC, TOP	04	00	D2, 01, 00
HLT	HALT	07	00	76

EXAMPLE 20-1

Follow the flowchart below and list the number of each block executed between the START of the program and its END.

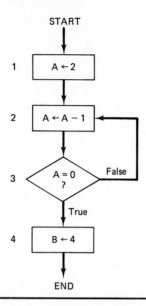

Solution: The program will repeat blocks 2 and 3 until the number in A has down-counted from 2 to 0. Since the number is checked *after* it has been counted down, the instructions in blocks 2 and 3 will be repeated twice. The list of numbers will look like this:

start, 1, 2, 3, 2, 3, 4, end

We can verify this by following the values of A and B, after each block is executed:

START,	1,	2,	3,	2,	3,	4,	END	←block
A = unknown	A = 2	A = 1	A = 1	A = 0	A = 0	A = 0	A = 0	←value of A
B = unknown	B = ?	B = ?	B = ?	B = ?	B = ?	B = 4	B = 4	←value of B

20-2 SUBROUTINES

We discussed subroutines and stack operations briefly in Chapter 19, but the only way to see how they really work is to see a subroutine in a program, and see how the program uses the subroutine.

To write a machine language program, ask yourself the following five questions:

1. Is this a problem that a computer can solve faster or with less work than if I solved the same problem "by hand"?

2. How do I want the output and input to look?

3. What steps will the computer have to do when it solves this problem?

4. How will I use this computer's instruction set to do the steps I worked out in question 3?

5. What are the actual codes and addresses that I will enter into the computer for this program?

The answer to the first question forces you to define the problem and decide if it is feasible for the computer to solve it. The answer must be yes, or there's no point going on to step 2.

The answer to the second question forces you to lay out a chart (an I/O chart) that shows exactly how you want the input to appear before it is put into the computer, and how you want the output to appear after the program is done.

To complete the remaining steps, you have to develop a flowchart (step 3), an assembly language program (step 4), and a machine language program (step 5).

20-2.1 RAM Tester Program without Subroutines

We'd like to build a RAM tester to test RAM boards as they come off the assembly line. The available hardware includes a working CPU and a working RAM TEST program written on a ROM located at page 00 of memory. The board we want to test is plugged into a socket that maps it into page 01 of memory. It is this program that we're going to develop. Before we do, we ask question 1: "Is it worth it?"

Since the memory boards contain thousands of flip-flops, and the CPU can address them to write and read them all in a fraction of a second, we decide that it's worthwhile to test these flip-flops with a microprocessor. Testing each flip-flop by hand would be time consuming (and boring) beyond description.

Since the answer to question 1 is yes, we go ahead to an I/O chart for this system (Figure 20-6).

In this I/O chart, we see that the input (plugging a board into the slot and pressing reset) is very simple. The output shows what would happen if a bad bit were encountered on line 10 of page 01. The page and line of the bad memory

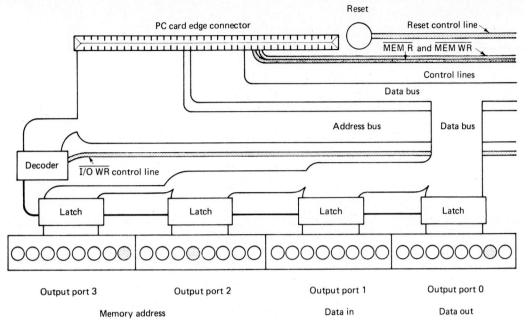

FIGURE 20-6   I/O chart for a RAM troubleshooter board.

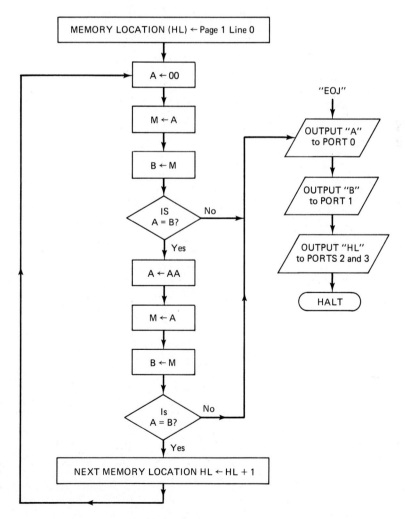

FIGURE 20-7   Software flowchart for RAM trouble-shooter

byte are shown on output ports 3 and 2. In addition to knowing where something is wrong, the numbers at output port 1 and 0 show what went wrong, by showing what we (attempted to) write into the memory and what we read back from the same memory location. This display should only be activated when the number written does not match what we read back from the same location. The output displays information the technician should find helpful in troubleshooting the board that has failed.

What should the technician look for, and what types of tests will find these things? We'll look at a very simple group of tests that only look for two things: a short on the data bus and an open on the data bus. These are typical of the type of problems that might appear on boards that have just left the flow-soldering machine at the end of a production line. Sometimes the solder doesn't hit all the points it's supposed to, and this results in an *open*; at other times, too much solder flows, connecting points that are not supposed to be connected, a *short*.

Figure 20-7 shows a flowchart for the program we're talking about (all numbers shown are hex).

We could go ahead from this flowchart to steps 3, 4, and 5, but before we do, there's a reason not to. If you look at the flowchart, you'll see this sequence of instructions repeated twice. The only thing that's different in the first and second versions of this sequence is the value of the number we called (SOMETHING) (see the flowchart in Figure 20-8). This is the number we've chosen to write and read in the memory.

Is there a way to write this routine just once and use it twice, with a different value of (SOMETHING) each time? We do this by changing the routine into a subroutine:

To test the memory, we'll write a number into the memory, then read it back and compare what's written with what's read. The number we write depends on the type of test we want to carry out.

To detect an open, we'll write 00000000 into the RAM. If an unexpected 1 appears in the number that returns, it probably indicates an open. TTL and other TTL-compatible memory devices—most MOS memories are designed to be TTL compatible—behave as though a disconnected lead is a logic 1. We say that inputs "float HIGH" when they're open, instead of being connected to something.

To detect a short, we'll write 10101010 at the RAM location being tested. This is based on the assumption that the traces on the PC board

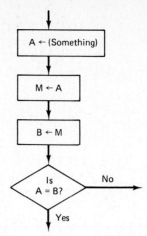

FIGURE 20-8 Section of flowchart 20-7 containing repeated coding

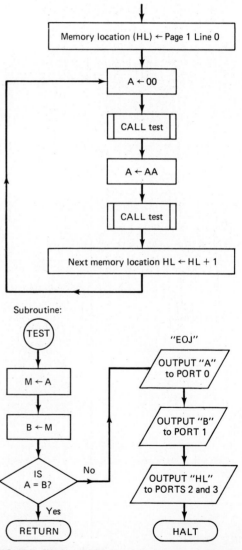

FIGURE 20-9 RAM troubleshooter software using a subroutine

are next to each other in the same order as shown above. We expect that there's a 1 next to a 0 on adjacent traces of the data bus when the number 10101010 is being written or read. TTL-compatible outputs will have strong pull-downs and weak pull-ups, so a 0 that's shorted to a trace carrying a 1 will "fight" with the 1, and the 0 will always win. This will result in unexpected 0s in the number, as a 1 is grounded out by an adjacent 0 wherever there's a short.

The instruction CALL TEST asks the computer to use the subroutine "Test" at the bottom of the diagram (see Figure 20-9). In the first instance, the value of A is 00 when Test is called. In the second, A is "AA" (10101010) when the Test subroutine is called. Each time the same test is done on the memory; we write the number into the memory location from A, then read it out again into B, then compare A with B to see if they match. If they don't, we go to the EOJ routine. The name EOJ stands for "End of Job". It's a fairly standard computerese name for the end of a program. In this case, the program halts if an error is found so that the operator has time to read the displays and correct the defect in the RAM board. If the program continued to run full speed with the errors being displayed as each one is found, the average error message would flash on the displays for a microsecond (a millionth of a second) or two. Even with a super speed-reading course, nobody could keep up with that data rate.

What if all the As and Bs match everywhere? Will the program ever stop? If you look at the flowchart, there doesn't seem to be any elegant way to get out of this program—something has to go wrong for it to end.

This program does have an end. It isn't elegant, but there is always an end—when the program runs off the end of the memory board being tested. We can assume that the PC board we plug into the slot on our tester isn't a full 65,536 bytes of RAM (if it were, we'd have no place to put the ROM with our tester program on it). If the RAM board we're testing is 32K or less, there will be a gap in the memory map of the computer which has no memory circuits in it. When the computer tries to write and read in this area, it will fail. The first test—which looks for unexpected 1s—will fail when the computer attempts to read an area where there's a nonchip (which is open everywhere) since every line on the data bus will be floating, and all the floating lines will look like 1s.

Now let's finish the translation of this program into assembly language and machine language from the flowchart. Each block in the flow-

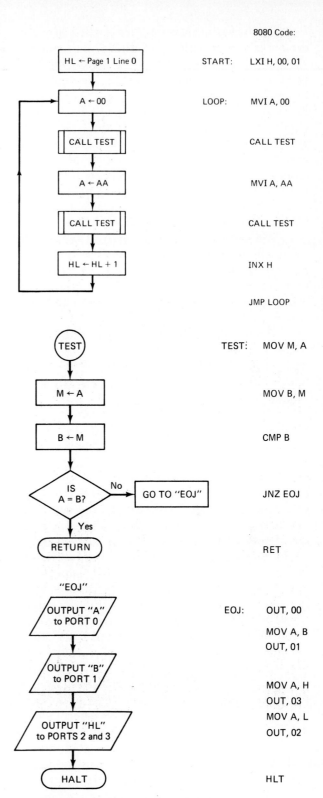

FIGURE 20-10 RAM program coded into 8080 assembly language

chart becomes a group of one or more instructions in assembly language. Each line of assembly language becomes a line of machine language (hex) code.

To answer question 4 ("How will I use this computer's instruction set to do the steps I worked out in the flowchart?"), we convert the flowchart into assembly language as shown in Figure 20-10. Two things need explaining here. First is the use of the CMP (compare) instruction. CMP B does a "look-ahead subtract" that finds out if "A minus B" would be negative, zero, or positive. After the CMP B instruction, a JNZ (Jump if Not Zero) instruction does a test that's really "Jump if Not Equal." This is so because if A and B are equal, the look-ahead subtraction (CMP) says "A minus B" would be 0.

The second thing that needs explaining is the use of MOV instructions in the last three outputs. In the 8080, only register A may source a number to the output device. If the number we want to see is in another register, we must MOV it to register A.

To go ahead to the final step of writing this machine language program, we ask question 5 ("What are the actual codes and addresses that I will enter into the computer for this program?"). This is shown in Table 20-3.

20-2.2 What Happens during a Call and How Returns Are Possible

Figure 20-11 shows what happens in our tester program as the program counter "walks" through the memory picking up instructions. Each time the CALL TEST instruction takes place, the computer goes to the same place, but returns to a different return address. The first time the CALL takes place, the program counter is pointed at line 8 (hex), which is the next instruction after the CALL. This number is saved on the stack, and when a RETURN takes place, it RETURNs to line 8 of memory. The second time the CALL takes place, the program counter is pointed at line D (hex), where the instruction after the CALL is located. This time, a D is saved on the stack, and although the same TEST subroutine is being done at the same place as before, when it RETURNs, this time it RETURNs to D.

Using a subroutine in this program didn't save much space over just writing the routine twice, but it did save some space. In programs where a subroutine is called at three or four or more places in a program, the more times the routine is called, the more space is saved by subroutining. Programs written to take advantage of "subroutinability" are called **structured programming**. Such programs run a little more slowly than programs written straight through,

TABLE 20-3

Final machine language program

| Assembly language (8080 Code) | Address | | Hex instruction |
	Line	Page	Opcode data
START: LXI H, 00, 01	00	00	21, 00, 01
LOOP: MVI A, 00	03	00	3E, 00
CALL TEST	05	00	CD, 11, 00
MVI A, AA	08	00	3E, AA
CALL TEST	0A	00	CD, 11, 00
INX H	0D	00	23
JMP LOOP	0E	00	C3, 03, 00
TEST: MOV M, A	11	00	77
MOV B, M	12	00	46
CMP B	13	00	F8
JNZ EOJ	14	00	C2, 18, 00
RET	17	00	C9
EOJ: OUT, 00	18	00	D3, 00
MOV A, B	1A	00	78
OUT, 01	1B	00	D3, 01
MOV A, H	1D	00	7C
OUT, 03	1E	00	D3, 03
MOV A, L	20	00	7D
OUT, 02	21	00	D3, 02
HLT	22	00	76

but they have advantages in taking up less memory space and being easier to recognize if a rewrite is necessary. Structured programming is sometimes also called *modular programming*.

EXAMPLE 20-2

Follow the flowchart below, and list the number of each block executed between the START of the program and its END. What is the value of A at the END?

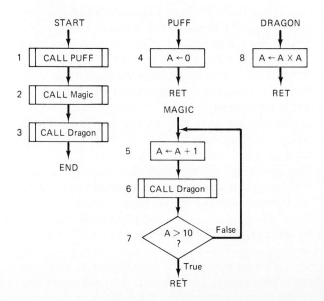

Solution: This program includes subroutines and loops, so it bounces around a bit. First, we list the block numbers from START to END, without explanation:

START, 1, 4, 2, 5, 6, 8, 7, 5,
6, 8, 7, 5, 6, 8, 7, 3, 8, END

To verify this, we follow the conditions of each of the variables as the program is running.

Block	A	Stack	Comments:
START	?	?	
1	?	2	Call PUFF
4	0	2	
2	0	3	Return — Call MAGIC
5	1	3	A + 1
6	1	7(3)	Call DRAGON
8	1	7(3)	
7	1	3	Return
5	2	3	A + 1
6	2	7(3)	Call DRAGON
8	4	7(3)	A^2
7	4	3	Return
5	5	3	A + 1
6	5	7(3)	Call DRAGON
8	5	7(3)	A^2
7	25	3	Return
3	25	END	Return — Call DRAGON
8	625	END	A^2
END	625		Return

The final value of A is 625.

20-3 STACK OPERATIONS WITH SUBROUTINES

In our RAM test program, information is passed from the main program to the subroutine in register A. The first time the memory location M is tested, the number used is 00, and the second time, it's AA. In both cases, the number to be tested is placed in register A before the CALL goes to the subroutine. When the subroutine does the test, it doesn't care what's in A; whatever it is, it gets written into and read out of the memory. After the subroutine is finished and the program counter RETURNs to the main program, the number in A is replaced with another value. If an error occurs, the number in A is displayed on output device 0.

It's important, in case of an error, that nothing happens to A. We must be able to see what number we tried to write at M after the test fails, in order to identify the type of trouble. Another number that must be untouched during the time the subroutine runs is the number in HL (which points at the address being tested). In this program, the subroutine does not do anything to HL, so it's the same number both before and after the subroutine is called. The only register that's different before and after the subroutine runs is B, which receives the number that comes back from memory M. If we need the number at B to remain untouched after the subroutine (we don't need it, in this case), our TEST subroutine can't be used. Look at what it does:

20-3.1 PUSHes and POPs

To protect the number in B, if it's an important number, the subroutine should save and recover the original value of B as shown in Figure 20-12. The operation on line (1) is called a PUSH (Zilog name: PUSH BC; Intel name: PUSH B), which is a 16-bit save-to-memory operation we discussed earlier. On line (2), we recover the number we pushed with a POP (Zilog name: POP BC; Intel name: POP B). With PUSH and POP operations, we can preserve any register or register pair that we don't want to clobber by temporarily storing it in the memory.

Which brings us to the question: "Where, in fact, is the stack in the memory?" We know that if we're running a program, the current instruction is where the program counter points. In doing stack operations, the current location of the top of the stack is the place in memory where the stack pointer (SP) points.

So where in the memory of our program is the stack pointer pointing? If there's no instruction that specifically puts a value into the SP (a LD SP, LIN, PAG instruction), it will be pointed wherever the laws of chance happened to leave it when the power was turned on. This will not be "zero." The SP could be literally anywhere, and if the "anywhere" turns out to be ROM, or a section of memory without any memory chips in it, too bad. Even if the section of memory where the SP points is populated by real RAM memory devices, it's possible that they could be defective— since the only RAM memory in the tester is the board being tested. We'll have to revise our RAM tester to provide a safe piece of RAM memory for the stack to be in. After we make sure that the hardware (some bytes of RAM memory) is there,

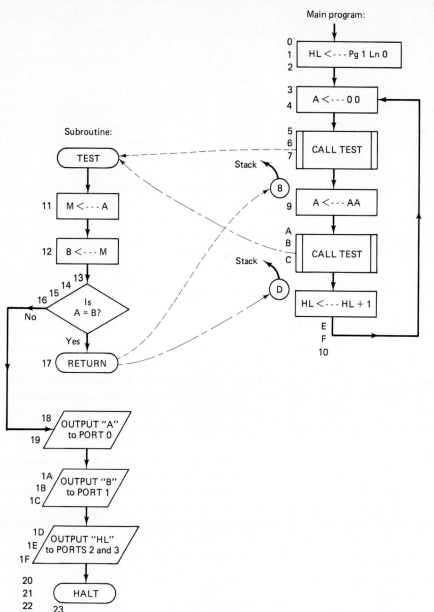

Main program:

FIGURE 20-11 Software for a RAM troubleshooter.

Subroutine:

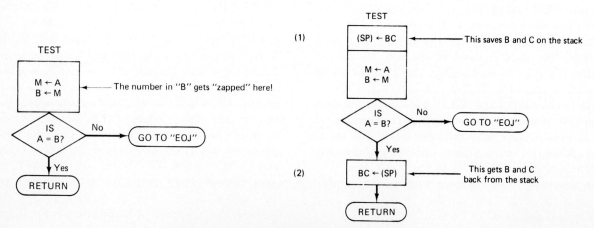

FIGURE 20-12 Using the stack to save registers.

we also need to provide the right type of software to position the stack in this piece of RAM.

But ... wait a minute! Isn't all this worry based on the idea that we've got to save register B with PUSHes and POPs? What if we don't have to save register B? (We don't!) Is there any need to worry about the stack in this case?

Yes! The stack is used every time a subroutine CALL and RETurn is done. It's the stack that holds the return address while the subroutine is running, and from which the address is recovered into the program counter when a RETurn is done. To save and recover a return address, we must use RAM memory that can record and play back. The RAM test program itself is in a ROM, the rest of the memory is empty, except the board we're testing—which may not be working.

Does that mean that the RAM tester program we looked at in the last section won't work?

It sure does! The chances that the SP would be pointed at a usable section of RAM are very small. Imagine that our RAM tester is testing boards that hold 1K of memory, that the ROM with the test program in it is one page ($\frac{1}{4}$K), and the rest of the memory is empty. That comes out to 63K of NON-RAM and only 1K of (maybe defective) RAM in the memory map. If the SP could be pointed anywhere, the chances are 1 in 63 that the stack is in the RAM, and less than that for the stack to be in a working RAM.

We need to revise our hardware to include a small amount of RAM along with the ROM program and the test slot. This new hardware design is shown in Figure 20-13.

We also need a revised software design. The

RAM (stack) added to our hardware has been wired in at the top of memory. We designed the hardware so that the RAM (which is actually made from a couple of 7489 16 × 4 RAM chips) is located in the last 16 bytes of page FF. The SP counts backwards as numbers are PUSHed into the stack. It's fairly common practice in complete systems, to put the stack at the top of memory and the program at the bottom, to give both areas the largest range of expansion before they overlap. In this case, it's not necessary, but we've decided to do it the same way, even for this simple microprocessor-based board tester.

The stack pointer has been placed at line FF of page FF, which is the very top of the memory. As 16-bit words are PUSHed into the stack, the SP will back up in steps of two addresses apiece, to line FD, line FB, and so forth. This is shown in Table 20-4.

You can see that all we've done is to add one line at the beginning of the program. The line LXI SP, FF, FF ensures that the stack is in the RAM we added, located in a place where it can use the maximum amount of memory cells available in the RAM. You'll also notice if you inspect the machine language program carefully that all the addresses associated with labels (except START) are new numbers. All the Jump and Call instructions have to be rewritten for addresses that are the "old addresses plus three," because we added a three-byte instruction to the beginning of the program. You can see that modifying machine language programs, even by a single instruction, is clumsy. If we had already "burned" a ROM with the old program on it, the ROM

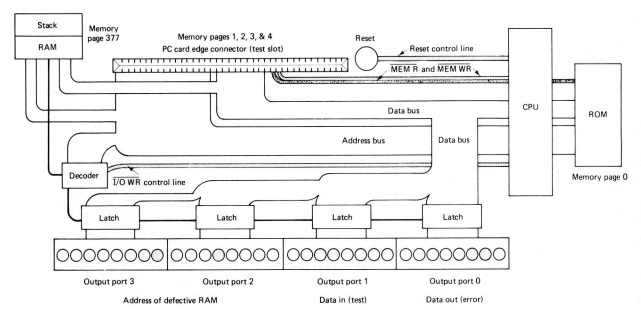

FIGURE 20-13 Hardware for a RAM troubleshooter.

TABLE 20-4

Revised machine language program

Assembly language (8080 code)	Address		Hex instruction	
	Line	Page	Opcode data	
START: LXI SP, FF, FF	00	00	31, FF, FF	added instruction
LXI H, 00, 01	03	00	21, 00, 01	
LOOP: MVI A, 00	06	00	3E	
CALL TEST	08	00	CD, 14, 00	
MVI A, AA	OB	00	3E, AA	
CALL TEST	OD	00	CD, 14, 00	
INX H	10	00	23	
JMP LOOP	11	00	C3, 06, 00	
TEST: MOV M, A	14	00	77	
MOV B, M	15	00	46	
CMP B	16	00	B8	
JNZ EOJ	17	00	C2, 1B, 00	
RET	1A	00	C9	
EOJ: OUT, 00	1B	00	D3, 00	
MOV A, B	1D	00	78	
OUT, 01	1E	00	D3, 01	
MOV A, H	20	00	7C	
OUT, 03	21	00	D3, 03	
MOV A, L	22	00	7D	
OUT, 02	23	00	D3, 02	
HLT	24	00	76	
	26	00		

would have to be thrown out (or completely reprogrammed if it is an EROM). There is no way we could use the old instructions, because the addresses they contain are no longer valid.

You can also see that we did not add a PUSH or POP to the subroutine; we know that register B is not used anywhere else in the program, and it's not necessary to preserve the value of B between the Call and the Return.

A last word on the problem we solved when we wrote this program. We designed the hardware and the ROM software to do a specific, inflexible task—to test RAM boards for opens and shorts. That is all the complete system does, once it's finished. We're not taking full advantage of the computer capabilities of the Z-80 or 8080 chip in this system.

In fact, the system isn't a general-purpose computer at all—it's a one-program computer. We call machines like this **dedicated machines.** The same job could have been done by a network of counters and logic gates, using all hardware. A RAM board tester unit could be made of gates to do exactly what our programmed microprocessor does, but it would be many times bigger and more expensive to build. If the layout of the RAM boards it was designed for is changed, the all-hardware tester would become instantly obsolete. It can't be adapted to a new design of the

RAM board without being totally rebuilt, and if we choose to change it to an I/O board tester—forget it!

With the microprocessor-based tester (it's not really a computer!), we can adapt the tester to a new RAM board design by pulling the ROM out of its socket, and replacing it with one that has a program for the new board design. Ultimately, the program written can test another, completely different type of board, such as an I/O board, provided that it plugs into the same socket and has its bus connections in the right places.

We can see that the microprocessor-plus-ROM combination called a dedicated machine is a much more powerful and flexible design than the random-logic board devised for only one kind of test. We can also see one other important thing: In this design, we've replaced hardware with software. Programming for logic design—using software to replace hardware—is the way to go in the design of "smart" test equipment, or anything else where logic circuits are needed.

20-4 INTERRUPTS

An interrupt is hardware-driven software. When hardware outside the computer can control what's happening inside (the software), we call it an interrupt. When the hardware signal becomes active and the special software that services the hardware signal begins to run, we say that the computer is doing an interrupt or servicing an interrupt. The name of the software that runs during the interrupt is the *interrupt service routine* (ISR).

There are three types of interrupt that we will discuss in this section of the chapter. An interrupt that uses the full hardware "handshaking" capabilities of the microprocessor is called a *vectored interrupt*. A shorthand version of this called a non-maskable interrupt is available on the Z-80 and many other microprocessors. It does not require as much hardware as the vectored interrupt, but does not have as much flexibility either (it's also not available on the 8080). The third type of interrupt is called a *polled interrupt*. It doesn't require any special hardware at all; it just uses the available I/O ports and is really more software than hardware. We'll begin our discussion of interrupts with the hardware used by each type of interrupt. After that, we'll discuss the ISRs and how they work hand-in-hand with the hardware to provide interrupt control of the computer.

20-4.1 Hardware

Vectored interrupts. There are two signals found on the 8080 (or Z-80) CPU called the *interrupt request* and the *interrupt acknowledge* (these signals are found on the control bus). Each manufacturer seems to have slightly different names for these signals, but in any case, they are an input that asks the computer CPU to begin an interrupt, and an output that the CPU uses to tell the outside world that it has accepted the request, and is beginning the interrupt procedure. When a pair of signals like this (a "request" and a "reply") are used in a digital system to indicate when a special state is being used, the two signals are called a *handshake*. We have already seen some other handshake signals in our discussion of microprocessor CPUs (Chapter 18).

When an interrupt request appears on the INT input of the CPU, five things happen:

1. The INT input becomes "active." If it is accepted, then . . .
2. The CPU completes the cycle it is in.
3. The CPU "replies" by making the INTA (interrupt acknowledged) output "active."
4. The CPU fetches any instruction it finds

on the data bus and executes it (the instruction should be a ReStarT).

5. After it completes the RST instruction, the CPU finds itself in the ISR. At the end of the ISR, it returns to the next step in the program that was running at the time of the interrupt.

Everything will "work" if the hardware attached to INT, INTA, and the data bus are designed to cooperate with the five steps above. For instance, when the INTA signal becomes active, some logic circuitry must be activated that will "gate" a RST opcode onto the data bus. When the INTA signal is active on the control bus, the memory and I/O control signals will be inactive, so the data bus will be floating. Whatever circuitry is used to gate the RST instruction onto the data bus will have no interference from anywhere else. Now, when the CPU does the RST instruction (executes it) there must be a subroutine, or at least a Jump to a subroutine (a **vector**), at the address the RST makes the program counter go to. Remember that the RST instruction is just like a CALL to a subroutine.

It is possible for the interrupt request (INT) signal to be *denied*. This is done by disabling interrupts. A flag in the CPU can be set or reset by software commands that will permit or block the INT input. The software command that enables the INT input to "get in" is called *enable interrupts* (EI). When this instruction is "run" during a program, it sets the interrupt flip-flop. From that moment on, the software can be interrupted.

The software command that disables the INT input is called *disable interrupts* (DI). When this instruction is run in a program, it resets the interrupt flip-flop, making the INT input go away. This is done when the program that is running is sensitive and cannot afford to be interrupted for unimportant reasons. For instance, suppose that a program is running that is calling the Fire Department because a fire alarm has been detected in the input network. We don't want a trivial interrupt, like the digital clock driver, to interrupt this program every time the clock ticks. The fire alarm program has priority. It outranks the clock driver program.

Figure 20-14 shows a circuit for an interrupt-driven input from a keyboard. The keyboard, represented by a block, is attached to an 8-bit input port. The D7 data bit is an active-LOW keypressed signal ($\overline{KP}$) and the remaining 7 bits are ASCII code from the key being pressed. The port to which the keyboard is attached is input port 6; the decoder only enables the buffer when

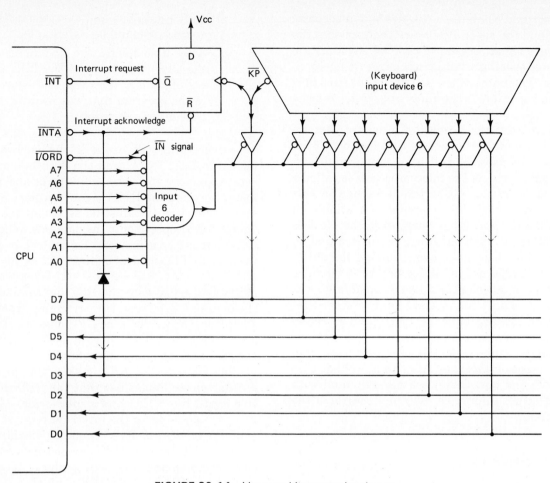

FIGURE 20-14 Vectored interrupt hardware.

the address on the address bus is a 6 and there is an active input signal on the control bus.

When any key on the keyboard is pressed, the $\overline{KP}$ signal becomes LOW. Its falling edge triggers the keypressed flip-flop, which sets. The output of the flip-flop makes the INT signal on the control bus become "active." For the Intel 8080, it is an active-HIGH signal; for the Zilog Z-80, it is active LOW (Figure 20-5 represents a Z-80).

As soon as the INTA becomes active (it is active LOW for both Intel and Zilog CPUs) it resets the flip-flop, and the INT request is turned off—since the CPU is already handling this interrupt, the request isn't needed anymore.

At the same time that it resets the flip-flop, the INTA signal also pulls down on data bus line D_3. The rest of the data bus lines float, and their terminating resistors pull up the unconnected lines to a logic 1 level. The diode is used to connect INTA to D_3 so that it doesn't conduct 1s to the D_3 line when INTA is inactive. That would fight with authentic data, and we don't want to do that.

Why do we want to pull down on the D_3 data

line when INTA becomes active? With all the other lines HIGH, this makes the number 11110111 appear on the data bus. This is the code for a RST 6 instruction. It will cause the same results as a call to a subroutine at line 30. Hopefully, the subroutine at line 30 will "pick up" a data number from input device 6 and do something with it, then return. We will look at this software in Section 20-4.2.

Nonmaskable interrupts. The Z-80 has an input called NMI (**nonmaskable interrupt**) which is like the INT (interrupt request), but is really an "interrupt demand." The NMI takes priority over any software that's running, even if the interrupt flip-flop has been disabled. That's what "nonmaskable" means. A "mask" covers something up and makes it go away. The INT input is "maskable" because a DI instruction can disable it. The NMI outranks everything else in the system.

An example of an NMI is the power failure indicator shown in Figure 20-15. If the power-supply voltage falls below a critical value, the de-

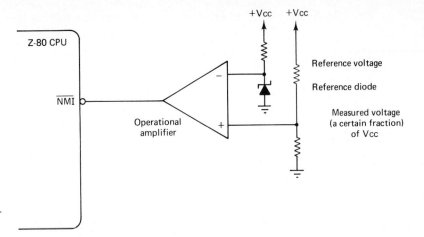

FIGURE 20-15 Nonmaskable interrupt if power fails.

tector triggers the NMI input of the CPU. Unlike the INT, the NMI doesn't have to fetch an instruction from the data bus to figure out where it's going. It just goes immediately to a predesigned location (in the Z-80, it's line 66). When it arrives, the CPU finds a program that switches off the connection to the power supply, and attaches the emergency batteries (we hope).

Why should this interrupt have priority over the fire alarm routine, for instance? Well, suppose that the fire burns through the power cables while the computer is calling up the fire department and reporting the fire. The report will never get finished, if the power fails, so the interrupt is interrupted by the NMI. Switching into the alternate voltage source probably takes a fraction of a second, and the small "hiccup" in the report will probably never be noticed at the fire station. The power failure interrupt should have a higher priority than the fire alarm interrupt, just as the fire alarm interrupt has priority over the digital clock interrupt that adds ticks to the clock every second.

Polled interrupts. The **polled interrupt** is a program. It uses conventional input port hardware instead of special signals on the control bus. In a polled interrupt, the computer program kills time until one bit of the input becomes active.

Since there's no special hardware, we didn't include a circuit here.

Reset. A **Reset** isn't an interrupt, but it is a hardware signal that can interrupt any software that's running. When the Reset is active, the program counter is automatically at memory line 0. Unlike the real interrupt, a Reset doesn't deposit the old value of the program counter in the stack. It's a case of "no deposit, no return."

You can think of a Reset as a Jump to line 0, but not a Call to line 0.

Reset is usually not used to get to a special program as much as it's used to get away from one. It's used when you want to get the computer back if it's lost in an endless loop or stuck at a Halt.

Since there's no way to get back to the program that was running at the time of the Reset (and you usually don't want to, anyway!), the normal thing that's put at the Reset's destination is the main operating system program for the computer. In our RAM troubleshooter earlier in this chapter, we used Reset to begin the RAM test. Since every test ended in a Halt, the Reset was used to restart the program when there was a new board to be tested. Of course, the program had to start at line 0.

20-4.2   Software (and ISRs)

Vectored interrupts. The software needed to support the hardware in Figure 20-14 must be located at line 30 (hex) on page 0 of memory. If the data from the keyboard have to be stored in memory and displayed on a video screen, and the memory pointer for the keyboard's memory space incremented, and so forth, the program to get a byte from keyboard 6 could be quite complicated. If there is a keyboard 7 which uses the RST 7 instruction, there won't be enough room for a complicated procedure like this between RST 6's destination and RST 7's destination. What's usually done to handle this is to put the ISR (in this case, the program that handles data from keyboard 6) in another part of the memory where there's lots of space. Let's call this routine KSR6 for Keyboard Service Routine 6), and suppose that our system's programmer has written it on line 00 of page 03. In this example, then, we won't find KSR6 at line 30. What we'll find instead is:

JP KSR6	30	00	C3
	31	00	00
	32	00	03

This Jump is called a *vector*. It's where the vectored interrupt gets its name.

We also have to remember that the INT is a maskable interrupt. The INT input is a "request" that can be denied if the program that's running at the time has higher priority. For instance, the fire alarm routine and the power failure routine we mentioned earlier would both contain a DI command at the beginning of their code. The keyboard in our example would be unable to get the computer's attention during these emergencies. At the end, there would be an EI command, to restore the system to interruptible status once the emergency has been handled.

Nonmaskable interrupts. The software for a nonmaskable interrupt would be exactly like that of the standard interrupt, except that the NMI, since it is reserved for emergency use, would always disable other interrupts when it is running. This is not done by hardware, so the ISR must change a little. Suppose that we want to vector to a place called ESR (Emergency Service Routine) on line 05 of page 07. Here are two ways we could do it:

Mnemonic	Line	page	contents
DI	66	00	F3
CALL ESR	67	00	CD
	68	00	05
	69	00	07
EI	6A	00	FB
RET	6B	00	C9

or

JP ESR	66	00	63
	67	00	05
	68	00	07
	.	.	.
ESR: DI	05	07	F3
	.	.	.

rest of routine ESR

	.	.	.
EI	85	07	FB
RET	86	07	C9

In these cases, the DI and EI commands could

either be a part of the ESR or of the vector. The second method is more commonly used.

EXAMPLE 20-3

In this example, the computer room is on fire, and the power cables have just burned through! Fortunately, the computer is equipped with fire and power-loss sensors capable of quick reaction, and a telephone dial-up system that can notify the authorities. To utilize these capabilities, the system is interrupt driven. Since the power loss was detected first, the computer is executing an emergency interrupt-service routine (ISR) to switch to auxiliary power. During this routine, the fire alarm detects the fire.

List the order in which the flowchart blocks are executed between the POWER ISR and its RETURN, assuming that the maskable interrupt for the FIRE interrupt-service routine arrives during the execution of block 2.

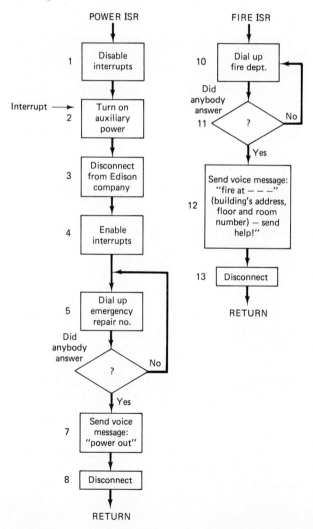

Solution: Although the interrupt request comes along during block 2, it is not accepted until line

4, because the computer is running higher-priority software (switching over to emergency power) and cannot afford to interrupt what it's doing. When interrupts are enabled again, the call to the fire department is made, then the call to the emergency repair number (which has a lower priority, and can be interrupted). The sequence is

START, 1, 2, 3, 4, 10, 11, 12, 13, 5, 6, 7, 8, END

Polled interrupts. Since a polled interrupt is all software, we present an example using hardware similar in purpose to the vectored interrupt hardware. Suppose that there is a keyboard attached to input port 6 in the standard I/O of our computer, and that its D7 data bit is its KP (active LOW) signal, while the remaining 7 bits are ASCII code for the key being pressed. Instead of using the KP signal to trigger the INT input, we use the software shown in Figure 20-16 to "wait" for the key to be pressed. Notice the use of a rotate instruction to see if the D7 bit is HIGH or LOW by testing the Carry flag. Sneaky, huh? Well, we told you those Rotate instructions were good for something.

In this program, the computer spins around in an input loop as long as the D7 bit is HIGH. As soon as D7 becomes LOW, which is the active condition for the KP signal, the computer proceeds to KSR6. Of course, when nobody's pressing on a key, the computer's just killing time—it can't do anything useful.

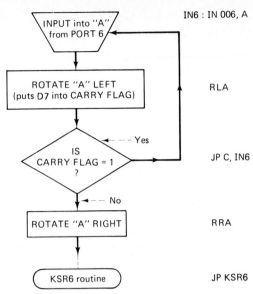

FIGURE 20-16   A polled interrupt

20-4.3   Summary

What are the advantages and disadvantages of various types of interrupts? The preceding paragraph should give you a hint: "Standard" (vectored) interrupts use the computer most efficiently, since the computer can run "useful" software between interrupts.

Polled interrupts use simpler hardware than vectored interrupts, but in between interrupts, they can't run "just any" software; the only software that they can run is the loop that keeps looking for an input bit to change.

QUESTIONS

20-1. Draw the symbol (flowchart block) for a:
  (a) Process
  (b) I/O operation
  (c) Decision
  (d) End-of-Routine

20-2. Sketch a flowchart with several blocks, using flow arrows to show what a loop is.

20-3. Describe how a decision block is used in a program to terminate a loop.

20-4. What are the five steps you must carry out to develop a computer program in machine language from the idea stage to the final program? (See Section 20-2.)

20-5. How can a subroutine call reduce the amount of code written to solve a problem? (You may use an example from this chapter if you wish to.)

20-6. Describe a procedure for testing a location of RAM to see if it does what it's supposed to do.

20-7. Describe a procedure for testing a 7400 quad-NAND gate chip instead of RAM. The chip will have to be connected to input ports and output ports.

20-8. Why is it necessary to worry about where the stack pointer is pointing in the RAM troubleshooter diagnostic program used in this chapter?

20-9. What is a dedicated machine?

20-10. Describe the differences among the following:
  (a) Vectored interrupt
  (b) Polled interrupt
  (c) Nonmaskable interrupt

20-11. What is "handshaking"?

20-12. Describe what happens during a maskable interrupt that is masked (disabled).

20-13. Is the hardware in Figure 20-14 used for a vectored, nonmaskable, or polled interrupt?

20-14. Describe what a RESET does, and how it can be used as a "super-high-priority" interrupt.

20-15. Draw a schematic showing how the keyboard in Figure 20-14 could be interfaced for use with a polled interrupt.

DRILL PROBLEMS

Follow the flowchart below and list the number of each block executed between the START of the program and its END.

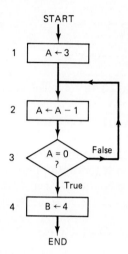

20-1. In what order are the blocks executed?
_____

20-2. What are the values of A and B at the END? A _____; B_____

20-3. Does this program fetch data from an input device? _____

20-4. Does this program deliver data to an output device? _____

20-5. Does this program contain constants? _____

20-6. Does this program contain variables? _____

20-7. Does this program contain a loop? _____

20-8. Does this program contain subroutines? _____

20-9. Does this program contain a conditional branch? _____

20-10. How many times is block 2 executed? _____

Follow the flowchart below and list the number of each block executed between the START of the program and its END. What is the value of B at the END?

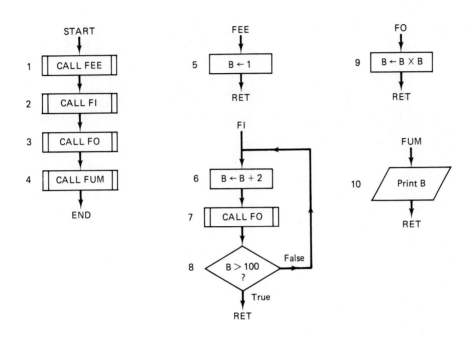

20-11. In what order are the blocks executed?
_____

20-12. What is the value of B at the END?
_____

20-13. Does this program fetch data from an input device? _____

20-14. Does this program deliver data to an output device? _____

20-15. Does this program contain constants?
_____

20-16. Does this program contain variables?
_____

20-17. Does this program contain a loop?
_____

20-18. Does this program contain subroutines?
_____

20-19. Does this program contain a conditional branch? _____

20-20. How many times is block 5 executed? ___

21
HIGH-LEVEL LANGUAGE PROGRAMMING

_____CHAPTER OBJECTIVES_____

By the time you finish this chapter, you will be able to:

1. Use a computer programmable in BASIC to emulate a machine language program.
2. Describe the features of the following medium-level and high-level software:
   a. An *operating system*
   b. A *monitor program*
   c. An *assembler* (and an *assembled language*)
   d. An *interpreter* (and an *interpreted language*)
   e. A *compiler* (and a *compiled language*)
3. Identify statements, and sequences of statements, in BASIC, that work like the machine language instructions discussed in the two preceding chapters.
4. Identify places where BASIC performs operations that don't exist at all in machine language.
5. Describe how an *algorithm* is used by a high-level language like BASIC to perform operations for which there is no hardware circuit (and thus no single opcode for the operation), such as a square-root calculation.
6. See how to access machine language through BASIC, which makes it possible to run mixed BASIC/*machine language* programs to get an optimum mix of simplicity and speed.

By the time you finish the end-of-chapter lab project (Example 21-4), you will be able to:

1. Use a BASIC program to POKE a machine-language program into memory and execute the machine language, on an IBM-PC where the machine language is 8086 code rather than the 8080/Z-80 code used for examples elsewhere in the chapter.
2. Perform a graphics hardcopy printout of any CRT screen produced by a BASIC program on the IBM-PC, using an EPSON printer. This printout will show both graphics and text exactly as it appears on the screen. In ''computerese,'' this program is a WYSIWYG (What You See Is What You Get).

21-1 THE VIRTUAL MACHINE

When you power-up a computer, there is a set of "rules" by which it behaves, or more appropriately, by which *you* must behave, if you are going to get the computer to do what you want. These rules are only partially defined by the microprocessor contained within the computer.

If, for example, you use a microprocessor trainer in the lab which is based on the 8085 or Z-80 processor, it will execute the two-byte instruction 3E,40 in exactly the same way no matter who makes it; it will load a hex 40 into register A in the CPU. That is built into the hardware of the 8085 CPU or the Z-80 CPU as the opcode for MVI A (Intel) or LD A (Zilog).

On the other hand, what do you do to make the program in memory execute? Do you press a key called "X," or "Go," or maybe "R" (for *run*)? This is not part of the microprocessor's design. It is part of the larger design of the microprocessor trainer you are using. This part of the computer's function could be done by hardware—a specific set of counting circuits could keep track of the beginning of the program and force a jump to that program when the "run" key is pressed—but that is not how usually it is done. Instead, software takes the place of this hardware, and decides when you are loading instructions into memory (and where), and when you want the computer to begin executing those instructions (and where to begin). The software itself is called the **operating system**. The rules you follow to make the computer work are called the **virtual machine**.

For instance, let's suppose that you do microprocessor lab experiments on a microprocessor trainer with an "X" (execute) key. A simplified set of rules might be "Press W, followed by an address, to write into memory starting at that address, but if you press X, followed by an address, you will execute the instructions of a program beginning at that address." Rules like this are defined by the **monitor program** fetched from ROM when the trainer is turned on. In larger systems, the operating system may be a **DOS (disk operating system)**, which uses diskette I/O and manages transfer of data to and from parts of the disk according to simple instructions like "copy b:*.*," which says to copy everything on the diskette in drive b to the disk you currently are using (drive A in most systems).

The most popular operating systems are CP/M, for older 8080/8085 and Z-80 machines; MS-DOS, for the more recent IBM-PC and PC-clone machines; and Applesoft BASIC, which is both a basic interpreter and disk operating system for Apple computers. The CP/M machines had Microsoft BASIC as a standard interpreter that came bundled* in with the hardware for the system, Apple computers came with Applesoft BASIC, and for the IBM-PC and clone machines, MS-DOS and MS-BASIC came "bundled" in with the package.

If you recognize the connection between "Applesoft" and "Microsoft," you'll see that all these machines have dialects of Microsoft BASIC as their primary high-level language.

The IBM-PC may be powered-up with an MS-DOS "system" disk in drive A. At power-up, a ROM program called a **bootstrap** checks to see if there is a disk in drive A. If there is, it attempts to load the MS-DOS (or another) operating system in from the disk. If there is *not* a disk in drive A within a reasonable time (about a minute), the ROM bootstrap jumps over to a ROM version of BASIC (without the fancy bells and whistles used to save and load programs onto disk), and the computer "wakes up" listening to you in BASIC. It now expects you to program it in BASIC.

In this case the "virtual machine" of the IBM-PC (without a disk in drive A) is a BASIC interpreter. With a DOS disk in the drive, the "virtual machine" would have a different set of rules—the software loaded in from the disk would have it listening to you in PC-DOS or MS-DOS—and would expect different instructions from you than the BASIC machine. The hardware, let us hasten to point out, is not a bit different; only the software has changed to make a different virtual machine.

In this chapter we look at the BASIC interpreter as a sort-of common denominator between micros of varying types. Although we won't examine the differing features among operating systems like CP/M, Applesoft, MS-DOS, or Unix, we will look at those BASIC instructions that they all have in common, and toward the end of the chapter, we will even see how to run machine language programs (8085/8086) from the BASIC interpreter. Although every microcomputer works internally in the machine language of its microprocessor, few interact with their users in this language. Machine language programming is awkward and slow, and errors are easy to make. In this chapter, we will find out what a high-level language is, and, through examples in the language called BASIC, we'll see how certain problems that are "messy" in machine language may be very easy in a high-level language.

---

*This means that you bought the software and hardware as a package deal.

BASIC (Beginner's All-purpose Symbolic Instruction Code) was chosen for this chapter's examples because more microcomputers "talk" to their users in BASIC than any other high-level language.

Many microcomputers are **turnkey systems**. This means that when the power is turned on (when you "turn the key") the computer "wakes up" communicating in BASIC. "How is this possible?", you ask; "Isn't the 'native' language of the computer the machine language of its microprocessor?" Yes and no; the internal language of the microprocessor is still "doing the work" but the external communications between the user's keyboard and the computer's video display are controlled by a program called the **operating system**. This program, on most microcomputers, is mapped into the memory permanently in ROM. As soon as power comes ON (a power-on reset circuit is usually used) this program is running. In some microcomputers, this operating system is a BASIC **interpreter**.

"What's an interpreter?", you wonder. A program that translates lines of alphanumeric code into commands of machine language may be called an *assembler*, an *interpreter*, or a *compiler*. The assembler, we saw in Chapter 20, is the program that carries out the fifth step in our five steps to machine language programming (Section 20-2). Assembly is a little like using a dictionary to translate from one language to another. There are as many instructions in machine language at the end as there were mnemonics in assembly language at the beginning. The assembly language program is a little like a skeleton. We fill out the program as we translate to machine language by replacing the mnemonics with suitable binary opcodes, and by replacing labels with real addresses that refer to places in the memory. When the procedure described above is done by a program, the program is called an assembler. The final result is a clone of the assembly language program, duplicated in machine language. It is this clone which the computer runs, because its instruction decoder doesn't understand the alphanumeric mnemonics and labels of assembly language. The clone of the assembly language program is stored in the memory and then executed by the computer. This machine language program is called the *object program* or *object code*. The original assembly language program is called the *source program* or *source code*.

The higher-level languages of this chapter are written in alphanumeric codes also. They are even further from the final machine language program level than the assembly language is. We might describe the compiler of a large mainframe computer like an assembler that translates programs written at (or near) the flowchart level directly down to machine language. Both interpreters and compilers do this, but there is one major difference between them.

The compiler constructs an object program from the original alphanumeric program, as the assembler does. Until the object program is completely compiled, the computer cannot execute it. The interpreter does not make its source program into an object program. It works with the source code in blocks called *statements*. Each statement in the BASIC language is identified; then the interpreter finds, and runs, a machine language subroutine that does the same thing.

This means that an **interpreted language** (BASIC is almost always interpreted, rather than compiled) must reidentify every line it runs, no matter how many times it has been run before. If a BASIC statement is repeated in a loop 50 times, it is interpreted over again as though it's a new statement every time.

A **compiled language** (FORTRAN, for instance, is similar to BASIC, but is almost always compiled) is completely transcribed into machine language before any statements ever run. There's no need to reidentify the statement when it's repeated in the loop, because the computer runs the program in machine language form. Because of this, high-level languages that use a compiler run fast and high-level languages that are interpreted run slow.

There is one important advantage of interpreted languages for microcomputers that gives them the edge over compiled languages. The object program of the compiled program is many times larger than the source code. When it's finished, a FORTRAN compiler, for instance, has its source code (the program in FORTRAN) and its object code (the final translation into machine code) both "living" in the computer's memory. The object code takes up a lot of memory space, even for a simple FORTRAN program . . . often more space than a microcomputer has!

The program being run by the interpreter, however, has only taken up enough memory space for its source code, and the program in object code never gets written. Although it is slow, the BASIC program to do a specific task will occupy far less memory space than its FORTRAN counterpart. In a microcomputer of limited capabilities, this is an important consideration. If a program exceeds available memory space, it must be stored on an auxiliary memory device, such as a tape or disk file, and the computer must run the program in blocks, taking each block from the tape or disk as it's needed. This slows down the execution of the compiled program to the speed of the tape or disk device, and the FORTRAN may actually be slower than the BASIC if a really large program is run on *virtual storage* (using the tape or disk to supplement RAM memory).

Now that we've established that BASIC is an interpreter language run on micros to conserve memory space, and that it's slow compared to compiled or assembled languages, what difference does that make to you? None whatever! (unless it turns up on an employer's aptitude test!!) If you can get to the computer in BASIC, you can troubleshoot it using that language as nicely as in machine code. Writing a diagnostic program will be easier, too.

There is a multitude of languages other than BASIC, and within the world of microcomputers, there are many dialects of BASIC. A program written in one dialect won't work on a computer that uses another. The interpreter will do what it can understand, then, when it finds a statement that's not part of its dictionary, it comes back with a "What's that?" called a **syntax error message**. At that moment, it goes on a sitdown strike, and refuses to go any further until you correct the statement it thinks is an error.

We'd like to show you how to program in BASIC in this chapter, but the question is: "Whose BASIC?" We can't show you, for example, how to program in "BASIC that works on everybody's computer" because no such BASIC exists. Instead, our goal in this chapter is to show you a few things about BASIC, which may vary from machine to machine. One goal will be to show you how the same operations done in machine language look in BASIC. This will be useful because, for the most part, assembly language and machine language are not as easily reached as BASIC on most microprocessors. A simple test can be programmed in BASIC very quickly and directly. A second goal of this chapter will be to see how the BASIC language does things more complicated than machine language allows; for instance, multiplication, which doesn't exist in the machine language of a Z-80, is part of BASIC on all Z-80 machines. A third point in seeing how BASIC (or any high-level language) works is to see how changes in a program are made. It's much easier to change your mind or insert an additional operation into the middle of a program in BASIC than in machine language.

21-2 THE BASIC LANGUAGE

The BASIC language is a machine-independent language. As far as possible, programs written in BASIC do not depend on how many registers are available in the microprocessor, what addressing modes the microprocessor permits, or what operations are possible in the ALU of the microprocessor. Any microprocessor equipped with a suitable BASIC interpreter should be able to execute the same source code by interpreting it in object code that uses the processor's capabilities. In fact, it shouldn't matter whether the program's running on a micro, mini, or mainframe.

That's not the way it works in reality. There are almost as many dialects of BASIC as there are manufacturers of micros. There is a small ray of hope, though. Microsoft BASIC (which is available for a large number of micros) has become the de facto standard version of BASIC in the microcomputer industry. The examples we use in the following sections of this chapter are developed from this version of BASIC.

The 8080 and Z-80 processors formed the basis of our understanding of machine language and microprocessor operation in previous chapters. Since Microsoft BASICs exist for machines that use the 8080 and Z-80 processors (the Interact and the TRS-80, for example), we'll use these dialects of BASIC in our examples, and later, we'll use the features of these BASICs to get to the machine language level of operation inside the virtual machine that we see through the eyes of BASIC.

21-2.1 BASIC Statements That Work Like Machine Language Instructions

Data-transfer instructions. We'll begin by looking at the Z-80 data-transfer instructions that we found at the beginning of Chapter 19. Each of these has a comparable operation in BASIC. To see how BASIC statements are used to do operations similar to the ones that micros do, we'll show a Z-80 operation, the infix description of what the operation does in the Z-80, and a BASIC instruction that does essentially the same thing. From this we'll see how some of our programs written in machine language could be written in BASIC, but more to the point, we'll get an appreciation of how BASIC works for some operations we're already familiar with. Here are some examples from Chapter 19.

Zilog code	Infix description	BASIC statement
LD H,B	H ← B	H = B
LD H,19	H ← 19	H = 19
LD L,136	L ← 136	L = 136
LD HL,5000	HL ← 5000	HL = 5000
LD A,(HL)	A ← (HL)	A = PEEK (HL)
LD (HL),A	(HL) ← A	POKE HL,A
LD A,(3000)	A ← (3000)	A = PEEK (3000)
LD (3000),A	(3000) ← A	POKE 3000,A

The Z-80 instructions use registers, like A, B, H, and L; register pairs (HL); constants, like 19, 136, and 5000; indirect addresses, like (HL); and absolute addresses, like (3000). The BASIC instructions look almost exactly like the infix notation, except for two major differences. One difference is cosmetic—the ← turns into a =. The other is more important. H, L, B, and HL are not registers or register pairs in the BASIC statements. They are variables, which are actually places in the memory used by the BASIC interpreter the same way the microprocessor uses registers. There are a great many more variables than the number of registers in the real microprocessor. You can name them after any letter of the alphabet (and on some BASICs, after any word of two, three, or more letters). There's no need to confine yourself to names that are acceptable Z-80 register names when you're writing in BASIC. The constants and addresses are exactly what they appear to be, although not all BASICs allow you access to the absolute memory with PEEK and POKE Instructions.

There's a reason for the use of the = sign instead of the ← sign. It's historical, rather than logical; at the time BASIC was developed, keypunch machines didn't exist that had a ← sign. BASIC's predecessor, FORTRAN, used the = sign instead of a ←, so Dr. Kemeny (the developer of BASIC) decided to do the same.

It's important to remember that sometimes in BASIC, the = means the same thing as a ←. In ordinary algebra, for instance, A = 25 is the same as 25 = A. This is not true of BASIC or FORTRAN. In BASIC, 25 = A means nothing, since 25 means the value of a number, not a memory address. The expression (25) ← A, although it's not legitimate in BASIC, means "put the value in A into memory location 25." The expression 25 ← A doesn't mean anything, even in infix notation.

Sometimes, the = means "is equal to" (as you would expect) when the = sign is used for comparisons. We'll see more about this when we see the BASIC instructions that do the same things as conditional JUMPs.

In 8080/Z-80 machine language, numbers were transferred between the CPU and PORTS. I/O ports send data in to the CPU (input ports) or out from the CPU (output ports). There are BASIC statements that are exactly the same as in the IN and OUT machine commands; here are some examples:

Zilog code	Infix description	BASIC statement
IN A,(7)	A ← port 30	A = INP (7)
OUT (7),A	port 7 ← A	OUT 7,A
IN B,(C)	B ← port (C)	B = INP (C)
OUT (C),B	port (C) ← B	OUT C,B

Of course, in the BASIC statements, A, B, and C are not really registers, they're variable names that "look" like registers. What the BASIC instructions do is the same as the action of the machine instructions. There are also instructions that do things similar to the Block Move and Block I/O instructions:

Zilog code	Infix description	BASIC statement
LDIR	memory ← memory block block	D$ = H$
INIR	memory ← block of block I/O ports	INPUT H$
OTIR	block of ← memory I/O ports block	PRINT H$

INPUT and PRINT aren't exactly the same as INIR and OTIR instructions; the input device is the computer keyboard (which is usually memory-mapped or treated as a single port rather than a block of I/O ports) and the PRINT output device is the video display (which is often memory-mapped instead of being true I/O).

Arithmetic instructions. In BASIC, the interpreter constructs a model of a virtual machine that works internally in BASIC and uses alphanumeric variables instead of registers and addresses. The ALU of this virtual machine is not limited by the arithmetic-logic capabilities in the microprocessor. The reverse is sometimes true; all the Boolean and arithmetic operations of a microprocessor may not be found in every version of BASIC (BASIC seldom has an XOR operator, although it's standard in microprocessors).

In the examples below, we'll look at the Boolean and arithmetic operations normally found in microprocessors, and at the BASIC way of expressing the same idea. In some of these examples, we'll see a combination of operations in the BASIC version of a Boolean operation:

Zilog code	Infix description	BASIC statement
Arithmetic Operations:		
ADD A,D	A ← A + D	A = A + D
SUB A,E	A ← A − E	A = A − E
ADD A,5	A ← A + 5	A = A + 5
SUB A,6	A ← A − 6	A = A − 6
Boolean operations:		
AND L	A ← A AND L	A = A AND L

(Bits of A are ANDed with corresponding bits of L.)

AND 1	A ← A AND 1	A = A AND 1

[A is ANDed with a binary 1. Only the 1 bit of the original number in A is left (D0). Bits D1 to D7 are 0. This is called **masking** bits D1-D7.]

OR H	A ← A OR H	A = A OR H

(Bits of A are ORed with corresponding bits of H.)

OR 7	A ← A OR 7	A = A OR 7

(A is ORed with 00000111. Bits D2-D0 are all 1s.)

The AND and OR operations are useful in machines like the 8080 which don't have Bit Test, Reset, and Set instructions. The AND instruction in the example above, for instance, can be used to *test* if D0 is on or off. If D0 is 0, the whole number will be zero. The condition (being zero) can be tested-for, using ordinary "test if zero" techniques. ANDing A with a number that contains only one 0 will *reset* the corresponding bit in A, while ORing A with a number that has only one 1 will *set* the corresponding bit in A.

Other Boolean examples:

Zilog code	Infix description	BASIC statement
XOR B	A ← A XOR B	A = (A OR (NOT B)) AND ((NOT A) OR B)

(Bits of A are XORed with corresponding bits of B.)

XOR A	A ← A XOR A	A = 0

(Used to zero the accumulator in machine language)

CP C	(Look-ahead subtraction)	T = A − C
CP 5	A − 5	T = A − 5

(T stands for "temporary" register. T is +, −, or 0.)

Compare and XOR instructions are useful in finding whether two bytes match, or if not, which one is larger. The XOR instruction is useful for "equality/inequality" testing—and also is used as the standard way of "zeroing" the accumulator. (No one would use XOR this way in BASIC, however; A = 0 is much easier.) Note that some BASICs don't have an XOR primitive (built-in function). The XOR must be implemented by its Boolean equivalent expression in AND, OR, and NOT.

The CP (compare) is used for testing whether two bytes match (the result sets the Zero flag) or if not, which is the larger byte. If the Sign flag indicates +, A is larger; if the sign flag indicates −, A is smaller than the other byte. Since there are no flags in BASIC, a temporary register T is set up in the BASIC instructions; T can then be tested to see if it is +, −, or 0.

There are also some **unary** operations in the 8080/Z-80 set. These are operations that have only one operand instead of two. (ADD and SUB are examples of **dyadic** operations—they have two operands, the addend and augend or subtrahend and minuend.) Examples of unary operations are:

Zilog code	Infix description	BASIC statement
Arithmetic:		
INC B	B ← B + 1	B = B + 1
DEC D	D ← D − 1	D = D − 1
DAA	(Binary BCD conversion)*	

(*BASIC is automatically displayed in decimal.)

Boolean:		
CPL	A ← $\overline{A}$	A = NOT (A)
NEG	A ← (−A)	A = −(A)

Notice that the INC and DEC operations use the standard ADD and SUB form in BASIC. (In BASIC, there is no difference.)

Look at the last two BASIC examples. We said that the = sign in BASIC didn't mean "is equal to" in these cases; the CPL and NEG should remove all doubt.

Control instructions. In machine language, the Jump, Call, and Return instructions transfer control of the computer from one area in "program space" to another. In the actual microprocessor, this takes place by altering the numbers in the program counter. The *virtual machine* created by the BASIC interpreter also has a program counter of sorts. Although the line numbers in a BASIC program are not really the same thing as addresses in a machine language program, they are used in a similar way by instructions called GOTO, GOSUB, and RETURN. If we accept, for a moment, the fiction that a line number in BASIC is the same as a line number in absolute address space, the following instructions match:

Zilog code	Infix description	BASIC statement
JMP 10	PC ← 10	GOTO 10
CALL 50	a) (SP) ← PC	GOSUB 50
	b) PC ← 50	
RET	PC ← (SP)	RETURN

Actually, the line numbers depicted in the GOTO and GOSUB statements do not work exactly like memory addresses, nor do they represent absolute memory locations. An example illustrates this point:

Absolute machine language	Z-80 Assembly	BASIC
00, 00 AF	START: XOR A	10 A = 0
00, 01 3C	LOOP: INC A	20 A = A + 1
00, 02 D3, 07	OUT (7),A	30 OUT 7, A
00, 04 C3, 01, 00	JMP LOOP	40 GOTO 20
	END	50 END

(The lines called END in both the assembly and BASIC programs are "noisewords" that identify the previous line as the last one in the program, but don't actually do anything themselves.)

Both of the examples above are programs that will count upwards from 0 and output each number counted to output port 7.

The machine language program (represented in octal) has one byte in each address. Where two-byte and three-byte instructions occur, they take up two and three memory locations, respectively. There are no gaps between numbers, for in the absolute memory of a computer there is something in every space. If we don't put an instruction into a space, leaving a gap, there'll be some code there anyway.

Contrast that with the BASIC program. The line numbers that identify places in the program (like line 20, so that the GOTO instruction can form a loop) start at 10, and jump to 20, 30, and 40 with total disregard for the numbers in between. Remember that these line numbers have meaning only in the virtual machine. They can follow any rules that the interpreter's designer wants them to. Why the gaps? Suppose that, as an afterthought, you realize you'd like the output to appear at port 3 as well as 7. In machine lan-

guage, there's no space to "sneak" a line into following the output to port 7. In BASIC, you would just type the line you wanted to add like this:

```
10  A = 0
20  A = A + 1
30  OUT 7,A
40  GOTO 20
50  END
```

```
35  OUT 3,A
```

and the virtual machine would slip it in between the other lines, so that if you made a LIST of the existing program in BASIC, you'd see:

```
LIST

10  A = 0
20  A = A + 1
30  OUT 7,A
35  OUT 3,A
40  GOTO 20
50  END
```

Notice that the line has been automatically spliced into the program in numerical order by the virtual machine. This is not a feature of the Z-80 hardware. It is only a part of the virtual machine defined by the BASIC interpreter, but it works the same way whether the BASIC is running on a TRS-80, an Apple, or a mainframe.

Conditional branch instructions. The conditional forms of Jump, Call, and Return have analogous instructions in BASIC. They all start with the word IF, and work like this:

Zilog code	Description	BASIC statement
JP NZ,5	Jump if Not Zero	IF A < > 0 THEN GOTO 5
JP Z,100	Jump if Zero	IF A = 0 THEN GOTO 100
JP NC,8	Jump if No Carry	IF A < 256 THEN GOTO 8
JP C,45	Jump if Carry	IF A > 255 THEN GOTO 45
JP P,30	Jump if Plus sign	IF A > 0 THEN GOTO 30
JP M,12	Jump if Minus sign	IF A < 0 THEN GOTO 12

(In case you aren't familiar with the >, <, and < > signs as they're used in BASIC, > means "is more than," < means "is less than," and < > means "is more or less than" (literally, "it is not equal to").) These comparative statements come from the "arithmetical" nature of BASIC. All relationships are expressed in terms of greater than (>), less than (<), or equal to (=), or some combination of them. Flags are not a part of the virtual machine of BASIC. Parity checking is not accessible to the BASIC user either, since

BASIC was designed as a high-level language for the user who does not want to bother with individual bits in each number. There could be a BASIC statement for "jump if parity odd" or "jump if parity even," but such statements would be highly complex, mathematical, and awkward.

Subroutines. The GOSUB (GOto SUBroutine) and RETURN instructions in BASIC are transparently obvious to machine language users. There's very little to say about how they work that you couldn't already infer from the way the Z-80 works. The only significant fact worth mentioning is that the stack used to save return addresses in BASIC is not as easy to get at as the stack in machine language. Few versions of BASIC contain POP and PUSH instructions, for instance, but without these, you can't "mess up" the stack as easily during a subroutine as in machine language (mainly, it's harder to make fatal mistakes this way!).

Addressing modes. All that fancy stuff about direct, indirect, absolute, and relative addresses for jumps, calls, and returns can be pretty much forgotten in BASIC. This way there are fewer things to confuse the programmer (at least, that's probably what the designer of the virtual machine had in mind). The only possible exception might be the BASIC command

ON N GOTO 5,10,15

which will go to different places when N equals 1, 2, and 3 (it will go to 5, 10, and 15, respectively). It's not exactly like a Z-80 indirect jump [JMP (HL), for instance], but it's something like it.

EXAMPLE 21-1

Convert the following 8085 program to BASIC (use Table 21-1).

```
        START:  IN 10d
                MVI C,1
                MVI B,0
        LOOP:   SUB C
                JM EXIT
                INR B
                INR C
                INR C
                JMP LOOP
        EXIT:   MOV A,B
                OUT 12d
                JMP START
```

Solution: We don't assume that the BASIC program is being run on the same computer hardware for which the 8085 machine language program will eventually be used. In other words, we don't assume that there are ports numbered 10 (decimal) and 12 (decimal), and that 10 is an input port while 12 is an output port, so we use the standard I/O assumed in the BASIC language's default conditions: input from the keyboard and output to the video screen.

START:	IN 10d	10	INPUT A
	MVI C,1	20	LET C = 1
	MVI B,0	30	LET B = 0
LOOP:	SUB C	40	LET A = A − C: LET CY = (A < 0)
	JM EXIT	50	IF A < 0 THEN GOTO 100
	INR B	60	LET B = B + 1
	INR C	70	LET C = C + 1
	INR C	80	LET C = C + 1
	JMP LOOP	90	GOTO 40
EXIT:	MOV A,B	100	LET A = B
	OUT 12d	110	PRINT A
	JMP START	120	GOTO 10

21-2.2 BASIC Statements That Do Things Machine Language Never Heard Of

The virtual machine created by the BASIC interpreter is capable of a great many mathematical (and other) operations that are not part of the Z-80 instruction set. Examples are

$$W = X * 25 \quad \text{(multiplication)} \quad W \leftarrow X \text{ times } 25$$
$$Y = Z / 12 \quad \text{(division)} \quad Y \leftarrow Z \text{ divided by } 12$$

Note the use of the * for "multiply," instead of x. Early microcomputers whose working language was BASIC didn't display uppercase and lowercase letters, like the "x" in "X x 25." They only displayed uppercase, which gives "X X 25." You can see that it's impossible to tell which "X" is an "X" and which means "times." On early keypunch machines, it was the same way. This led the designers of FORTRAN to adopt an alternate character found on the keyboard of a keypunch machine (a "*" is born!). BASIC followed FORTRAN, and maintained use of the same symbol, because in its early days, BASIC, too, was a punchcard-oriented language.

The BASIC virtual machines makes the use of multiplication and division possible, although they don't exist in the hardware of the Z-80 microprocessor, by the use of software modules called *algorithms*. Algorithms compute products (for multiplication) or quotients (for division) using nothing but ADD, SUB, and shift operations found in the Z-80. Each algorithm is a small subprogram in its own right, and the subject is

TABLE 21-1

Table for 8085 emulation in basic

M/L	A/L	Basic		M/L	A/L	Basic
00	NOP	REM		40	MOV B,B	LET B=B
01	LXI B,D16	LET B=(constant) (0→65535)		41	MOV B,C	LET B=C
02	STAX B	POKE B,A (B=0→65535)		42	MOV B,D	LET B=D
03	INX B	LET B=B+1 (B=0→65535)		43	MOV B,E	LET B=E
04	INR B	LET B=B+1 (B=0→255)		44	MOV B,H	LET B=H
05	DCR B	LET B=B−1 (B=0→255)		45	MOV B,L	LET B=L
06	MVI B,D8	LET B=(constant) (0→255)		46	MOV B,M	LET B=PEEK(H)
07	RLC	LET A=A*2 (A=0→255)		47	MOV B,A	LET B=A
08	−			48	MOV C,B	LET C=B
09	DAD B	LET H=H+B (H & B=0→65535)		49	MOV C,C	LET C=C
0A	LDAX B	LET A=PEEK(B) (B=0→65535)		4A	MOV C,D	LET C=D
0B	DCX B	LET B=B−1 (B=0→65535)		4B	MOV C,E	LET C=E
0C	INR C	LET C=C+1 (C=0→255)		4C	MOV C,H	LET C=H
0D	DCR C	LET C=C−1 (C=0→255)		4D	MOV C,L	LET C=L
0E	MVI C,D8	LET C=(constant) (0→255)		4E	MOV C,M	LET C=PEEK(H)
0F	RRC	LET A=A/2 (A=0→255)		4F	MOV C,A	LET C=A
10	−			50	MOV D,B	LET D=B
11	LXI D,D16	LET D=(constant) (0→65535)		51	MOV D,C	LET D=C
12	STAX D	POKE D,A (D=0→655345)		52	MOV D,D	LET D=D
13	INX D	LET D=D+1 (D=0→65535)		53	MOV D,E	LET D=E
14	INR D	LET D=D+1 (D=0→255)		54	MOV D,H	LET D=H
15	DCR D	LET D=D−1 (D=0→255)		55	MOV D,L	LET D=L
16	MVI D,D8	LET D=(constant) (0→255)		56	MOV D,M	LET D=PEEK(H)
17	RAL	LET A=A*2+CY (A=0→255)		57	MOV D,A	LET D=A
18	−			58	MOV E,B	LET E=B
19	DAD D	LET H=H+D (H & D=0→65535)		59	MOV E,C	LET E=C
1A	LDAX D	LET A=PEEK(D) (D=0→65535)		5A	MOV E,D	LET E=D
1B	DCX D	LET D=D−1 (D=0→65535)		5B	MOV E,E	LET E=E
1C	INR E	LET E=E+1 (E=0→255)		5C	MOV E,H	LET E=H
1D	DCR E	LET E=E−1 (E=0→255)		5D	MOV E,L	LET E=L
1E	MVI E,D8	LET E=(constant) (0→255)		5E	MOV E,M	LET E=PEEK(H)
1F	RAR	LET A=A/2+128*CY		5F	MOV E,A	LET E=A
20	RIM	LET A=PEEK(interrupt mask)		60	MOV H,B	LET H=B
21	LXI H,D16	LET H=(constant) (H=0→65535)		61	MOV H,C	LET H=C
22	SHLD Adr16	POKE (Adr),H		62	MOV H,D	LET H=D
23	INX H	LET H=H+1 (H=0→65535)		63	MOV H,E	LET H=E
24	INR H	LET H=H+1 (H=0→255)		64	MOV H,H	LET H=H
25	DCR H	LET H=H−1 (H=0→255)		65	MOV H,L	LET H=L
26	MVI H,D8	LET H=(constant) (0→255)		66	MOV H,M	LET H=PEEK(H)
27	DAA	no instruction		67	MOV H,A	LET H=A
28	−			68	MOV L,B	LET L=B
29	DAD H	LET H=H+H (H=0→65535)		69	MOV L,C	LET L=C
2A	LHLD Adr16	H=PEEK(Adr+1)#256+PEEK(Adr)		6A	MOV L,D	LET L=D
2B	DCX H	LET H=H+1 (H=0→65535)		6B	MOV L,E	LET L=E
2C	INR L	LET L=L+1 (L=0→255)		6C	MOV L,H	LET L=H
2D	DCR L	LET L=L−1 (L=0>255)		6D	MOV L,L	LET L=L
2E	MVI L,D8	LET L=(constant) (0→255)		6E	MOV L,M	LET L=PEEK(H)
2F	CMA	LET A=NOT(A)		6F	MOV L,A	LET L=A
30	SIM	POKE (interrupt mask),A		70	MOV M,B	POKE (H),B
31	LXI SP,D16	LET SP=(constant) (0→65535)		71	MOV M,C	POKE (H),C
32	STA Adr16	POKE (Adr),A		72	MOV M,D	POKE (H),D
33	INX SP	LET SP=SP+1 (SP=0→65535)		73	MOV M,E	POKE (H),E
34	INR M	POKE (H),(1+PEEK(H))		74	MOV M,H	POKE (H),H
35	DCR M	POKE (H),(PEEK(H)−1)		75	MOV M,L	POKE (H),L
36	MVI M,D8	POKE (H),(constant) (0→255)		76	HLT	STOP (OR END)
37	STC	LET CY=1		77	MOV M,A	POKE (H),A
38	−			78	MOV A,B	LET A=B
39	DAD SP	LET H=H+SP		79	MOV A,C	LET A=C
3A	LDA Adr16	LET A=PEEK(Adr)		7A	MOV A,D	LET A=D
3B	DCX SP	LET SP=SP−1		7B	MOV A,E	LET A=E
3C	INR A	LET A=A+1 (A=0→255)		7C	MOV A,H	LET A=H
3D	DCR A	LET A=A−1 (A=0→255)		7D	MOV A,L	LET A=L
3E	MVI A,D8	LET A=(constant) (0→255)		7E	MOV A,M	LET A=PEEK(H)
3F	CMC	LET CY=NOT(CY)		7F	MOV A,A	LET A=A

Note: D8 = Any 8-bit data word; D16 = Any 16-bit data word; Adr8 = an 8-bit I/O Device-code; Adr16 = a 16-bit memory address

TABLE 21-1

(Continued)

M/L	A/L	Basic	M/L	A/L	Basic
80	ADD B	LET A = A + B:LET CY = (A < 255)	C0	RNZ	IF (variable) < >0 THEN RETURN
81	ADD C	LET A = A + C:LET CY = (A > 255)	C1	POP B	LET B = STACK(SP):SP = SP + 1
82	ADD D	LET A = A + D:LET CY = (A > 255)	C2	JNZ Adr16	IF (variable) < >0 THEN GOTO (Adr)
83	ADD E	LET A = A + E:LET CY = (A > 255)	C3	JMPAdr16	GOTO (Adr)
84	ADD H	LET A = A + H:LET CY = (A > 255)	C4	CNZ Adr16	IF (variable) < >0 THEN GOSUB (Adr)
85	ADD L	LET A = A + L:LET CY = (A > 255)	C5	PUSH B	SP = SP − 1:LET STACK(SP) = B
86	ADD M	LET A = A + PEEK(H):CY = (A > 255)	C6	ADI D8	LET A = A + (constant):LET CY = (A > 255)
87	ADD A	LET A = A + A:LET CY = (A > 255)	C7	RST 0	GOSUB (fixed subroutine)
88	ADC B	LET A = A + B + CY:LET CY = (A > 255)	C8	RZ	IF (variable) = 0 THEN RETURN
89	ADC C	LET A = A + C + CY:LET CY = (A > 255)	C9	RET	RETURN
8A	ADC D	LET A = A + D + CY:LET CY = (A > 255)	CA	JZ Adr16	IF (variable) = 0 THEN GOTO (Adr)
8B	ADC E	LET A = A + E + CY:LET CY = (A > 255)	CB		
8C	ADC H	LET A = A + H + CY:LET CY = (A > 255)	CC	CZ Adr16	IF (variable) = 0 THEN GOSUB (Adr)
8D	ADC L	LET A = A + L + CY:LET CY = (A > 255)	CD	CALL Adr16	GOSUB (Adr)
8E	ADC M	LET A = A + PEEK(H) + CY:CY = (A > 255)	CE	ACI D8	LET A = A + (constant) + CY:CY = (A > 225)
8F	ADC A	LET A = A + A + CY:LET CY = (A > 255)	CF	RST 1	GOSUB (fixed subroutine)
90	SUB B	LET A = A − B:LET CY = (A < 0)	D0	RNC	IF CY = 0 THEN RETURN
91	SUB C	LET A = A − C:LET CY = (A < 0)	D1	POP D	LET D = STACK(SP):SP = SP + 1
92	SUB D	LET A = A − D:LET CY = (A < 0)	D2	JNC Adr16	IF CY = 0 THEN GOTO (Adr)
93	SUB E	LET A = A − E:LET CY = (A < 0)	D3	OUT Adr 8	OUT (Adr), (variable)
94	SUB H	LET A = A − H:LET CY = (A < 0)	D4	CNC Adr16	IF CY = 0 THEN GOSUB (Adr)
95	SUB L	LET A = A − L:LET CY = (A < 0)	D5	PUSH D	SP = SP − 1:LET STACK(SP) = D
96	SUB M	LET A = A − PEEK(H):LET CY = (A < 0)	D6	SUI D8	LET A = A − (constant):LET CY = (A < 0)
97	SUB A	LET A = A − A	D7	RST 2	GOSUB (fixed subroutine)
98	SBB B	LET A = A − B − CY:LET CY = (A < 0)	D8	RC	IF CY < >0 THEN RETURN
99	SBB C	LET A = A − C − CY:LET CY = (A < 0)	D9	—	
9A	SBB D	LET A = A − D − CY:LET CY = (A < 0)	DA	JC Adr16	IF CY < >0 THEN GOTO (Adr)
9B	SBB E	LET A = A − E − CY:LET CY = (A < 0)	DB	IN Adr8	LET A = INP(Adr)
9C	SBB H	LET A = A − H − CY:LET CY = (A < 0)	DC	CC Adr16	IF CY < >0 THEN GOSUB (Adr)
9D	SBB L	LET A = A − L − CY:LET CY = (A < 0)	DD	—	
9E	SBB M	LET A = A − PEEK(H) − CY:CY = (A < 0)	DE	SBI D8	LET A = A − (constant) − CY:CY = (A < 0)
9F	SBB A	LET A = A − A − CY:LET CY = (A < 0)	DF	RST 3	GOSUB (fixed subroutine)
A0	ANA B	LET A = A AND B	E0	RPO	no instruction
A1	ANA C	LET A = A AND C	E1	POP H	LET H = STACK(SP):SP = SP + 1
A2	ANA D	LET A = A AND D	E2	JPO Adr16	no instruction
A3	ANA E	LET A = A AND E	E3	XTHL	LET DTEMP = D:LET D = H:LET H = DTEMP
A4	ANA H	LET A = A AND H	E4	CPO Adr	no instruction
A5	ANA L	LET A = A AND L	E5	PUSH H	SP = SP − 1:LET STACK(SP) = H
A6	ANA M	LET A = A AND PEEK(H)	E6	ANI D8	LET A = A AND (constant)
A7	ANA A	LET A = A AND A	E7	RST 4	GOSUB (fixed subroutine)
A8	XRA B	LET A = A XOR B	E8	RPE	no instruction
A9	XRA C	LET A = A XOR C	E9	PCHL	GOTO (H) (won't work on most BASICs)
AA	XRA D	LET A = A XOR D	EA	JPE Adr16	no instruction
AB	XRA E	LET A = A XOR E	EB	XCHG	no instruction
AC	XRA H	LET A = A XOR H	EC	CPE Adr16	no instruction
AD	XRA L	LET A = A XOR L	ED	—	
AE	XRA M	LET A = A XOR PEEK(H)	EE	XRI D8	LET A = A XOR (constant)
AF	XRA A	LET A = A XOR A (A = 0)	EF	RST 5	GOSUB (fixed subroutine)
B0	ORA B	LET A = A OR B	F0	RP	IF (variable) > =0 THEN RETURN
B1	ORA C	LET A = A OR C	F1	POP PSW	LET A = STACK(SP):SP = SP + 1
B2	ORA D	LET A = A OR D	F2	JP Adr16	IF (variable) > =0 THEN GOTO (Adr)
B3	ORA E	LET A = A OR E	F3	DI	no instruction
B4	ORA H	LET A = A OR H	F4	CP Adr16	IF (variable) > =0 THEN GOSUB (Adr)
B5	ORA L	LET A = A OR L	F5	PUSH PSW	SP = SP − 1:LET STACK(SP) = A
B6	ORA M	LET A = A OR PEEK(H)	F6	ORI D8	LET A = A OR (constant)
B7	ORA A	LET A = A OR A	F7	RST 6	GOSUB (fixed subroutine)
B8	CMP B	LET TEST = A − B:LET CY = (A < 0)	F8	RM	IF (variable) < 0 THEN RETURN
B9	CMP C	LET TEST = A − C:LET CY = (A < 0)	F9	SPHL	LET SP = HL
BA	CMP D	LET TEST = A − D:LET CY = (A < 0)	FA	JM Adr16	IF (variable) < 0 THEN GOTO (Adr)
BB	CMP E	LET TEST = A − E:LET CY = (A < 0)	FB	EI	no instruction
BC	CMP H	LET TEST = A − H:LET CY = (A < 0)	FC	CM Adr16	IF (variable) < 0 THEN GOSUB (Adr)
BD	CMPL	LET TEST = A − L:LET CY = (A < 0)	FD	—	
BE	CMPM	LET TEST = A − PEEK(H):CY = (A < 0)	FE	CPI D8	LET TEMP = A − (constant)
BF	CMP A	LET TEST = A − A:LET CY = (A < 0)	FF	RST 7	GOSUB (fixed subroutine)

sufficiently important that it deserves a few words of discussion.

Both *dyadic* and *unary* operations exist in BASIC that have no counterparts in Z-80 machine code. One example is:

$$C = SQR (A)$$

where SQR (A) means "the square root of A." This is a unary operation of sufficient magnitude that its pencil-and-paper calculation is beyond most people. Even so, an algorithm exists that can do this computation using nothing but basic Z-80 ALU operations. In fact, there are several different algorithms that exist for doing a square root, each of which has different advantages. The smallest (in terms of number of machine language steps) that we've seen (although not the fastest in working time) is the one shown below. (Assume that the *radicand*—the number to be square-rooted—is already in register A when this subroutine is called.)

SQUARE ROOT ALGORITHM in 12 bytes

Z-80 Assembly		BASIC	
SQR:	LD B,1	1000	B = 1
	LD C,0	1010	C = 0
LOOP:	SUB A,B	1020	A = A − B
	RET M	1030	IF A < 0 THEN RETURN
	INC B	1040	B = B + 2
	INC B		
	INC C	1050	C = C + 1
	JMP LOOP	1060	GOTO 1020

This algorithm can be written in terms of BASIC instructions, shown at the right. It is often easier to try out an algorithm or test procedure in BASIC than in machine language. If you make a mistake, or think of something you want to add, it's much easier to slip in a BASIC line (or overwrite one if it's wrong) than to rewrite and reenter an entire machine language routine when one byte has to be added to the middle and all the addresses change.

You can use BASIC as a sort of testing ground for machine language diagnostic routines in this way. That's why we showed you, first, the instructions in BASIC that could be used to imitate the actions of a Z-80 running machine language.

Since most microcomputers "power up in BASIC," it's easiest to try out test routines in BASIC; then, when it's confirmed that they work, the routines can be rewritten in machine language.

What about BASIC statements that don't have matching machine language instructions? Should you use them?

If you're developing a BASIC program and you've got no plans to make it into machine language, the answer is "yes." You should go ahead and use all the power of a high-level language for mathematical, graphics, or string-handling operations.

On the other hand, if you're using the computer to perform simple control operations, as you would in a diagnostic program, you should use BASIC commands that are easily converted into machine language. Use the flashy stuff only if you know the machine language algorithms for each BASIC statement, and don't mind writing all the lines of machine code needed to convert the high-level statements into low-level code.

EXAMPLE 21-2

What does the following BASIC program segment do? List the line numbers in the order in which they will be executed.

```
10   LET A = 1
20   GOTO 70
30   LET A = A*A
40   PRINT A
50   END
60   LET A = 7
70   LET A = A + 5
80   GOTO 30
```

Solution: The list of line numbers is

$$10, 20, 70, 80, 30, 40, 50$$

At the end of the program run, the screen will contain the number 36 on one line, and a BASIC prompt like "OK" or "READY," indicating that the program is finished, on the next line. Line 60 will not be executed.

21-3 EXAMPLE BASIC PROGRAM

Now that we've gone and scared you off using high-level BASIC operations for developing machine language diagnostics, what's left? There's a whole world of interesting things that can be easily written using high-level statements. It's beyond the scope of this book, but if you pick up any one of a number of "using and programming BASIC" manuals on the shelves of your local computer store, you'll find many hours of interesting things to do while learning to use the language. Be prepared, though, to find things in these BASIC manuals that don't work in the dialect your computer uses. You'll need the programming manual for your own computer (with a description of how it handles its own "unique" commands) as well as the "how to program in BASIC" manuals. With a little practice, you'll get the knack of interpreting other people's BASIC into statements that work on your computer.

Some examples of BASIC statements that may be found in the virtual machine of a BASIC computer are given in Table 21-2.

There are a great many more functions possible in BASIC, and there are several in the table that will not be found in all BASICs. These are shown to give a feeling for what might be done in a fairly comprehensive BASIC program. As a final example of what a high-level language is like, we include a sample BASIC program (Table 21-3) to do a calculation of some values in an ac circuit. We'll "tear apart" the program to see how it works.

The first thing you'll probably notice is the large number of REM lines. These are lines with comments on them that are intended to explain what's happening to another programmer. REMarks are also useful to you if you want to look back at a program you wrote a long time ago, and need something to jog your memory to help you recall what you were doing at each part of the program. Putting REM statements in programs like this is called *documenting your code*. If the documentation accomplishes its task (makes the program easier to understand and the steps easier to identify), we say that it is a *well-documented program*.

The input statements on lines 30 to 60 contain output as well as input operations. Normally, when an input is done in BASIC, the virtual machine puts a ? on the video display and waits for you to enter a string of characters (usually a number). This ? is called a *prompt*. It is the machine's way of letting you know it's waiting for an input. In the inputs on lines 30 to 60, though, there are **literals** like R = and L =. These are strings of symbols that are actually printed on the screen in front of the ? sign. These add extra information to the prompt, letting you know which kind of number you should be entering at each input. In this program, several inputs are required. Instead of making you remember which one comes first, second, third, and fourth, the computer remembers for you. Remembering is one of the things that computers do better than people.

Literals are also used in the PRINT statements on lines 1000 to 1050. They are used for the same reason literals were used in the inputs; the program computes several different numbers and displays them. Without some sort of identification, it would be hard to remember which result was supposed to print first, second, and so forth. With the prompt, we let the computer "remember" for us.

Notice line 70. There is not just one state-

TABLE 21-2

Common basic statements

Basic statement	Mathematical function
Unary Operations	
C = SQR (A)	Computes the square root of A and puts it into C
Y = SIN (O)	Computes the trigonometric sine of O and puts it into Y
B = ABS (R)	Puts the absolute value of R into B (makes R positive)
D = LOG (N)	Computes the natural log of N and puts its value in D
T = INT (F)	Puts the integer part of F (the whole number part of the value in F) in T
P = EXP (Q)	Raises e to the Q power (e = 2.71828)
Dyadic Operations	
C = A ∧ B	Raises A to the B power (called exponentiation) and puts the result into C
W = X * Y	Multiplies X times Y and puts the product into W
U = T / V	Divides T over V and puts the quotient into U
PSET (X, Y)[a]	Set (turn ON) a graphics point at X and Y on the video display
PRESET (X, Y)[a]	Reset (turn OFF) a graphics point at X and Y on the video display
POINT (X, Y)[a]	Tests the state of a point at X and Y on the video display

[a]These are a combination of point-plotting and Boolean operations which combine the Cartesian coordinate system with a Boolean set, reset, or bit test function in the video RAM part of the memory.

TABLE 21-3

Sample basic program

```
10      REM PROGRAM TO FIND IMPEDANCE IN A
20      REM PARALLEL R, L, C AC CIRCUIT
30      INPUT "R="; R
40      INPUT "L="; L
50      INPUT "C="; C
60      INPUT "F="; F
70      I = 0 : J = 0
80      I$ = " INFINITE "
90      PI = 3.14159
100     W = 2 * PI * F
110     REM "W" IS OMEGA—THE ANGULAR VELOCITY OF
120     REM THE ROTATION OF THE AC GENERATOR
130     IF R=0 THEN J = 1 : GOTO 150
140     G = 1 / R
150     REM "G" IS CONDUCTANCE OF THE RESISTOR
160     XL = W * L
170     IF XL = 0 THEN J = 2 : GOTO 190
180     BL = 1 / XL
190     REM "BL" IS SUSCEPTANCE OF THE INDUCTOR (1/X)
200     BC = W * C
210     IF BC = 0 THEN I = 1 : GOTO 240
220     REM "BC" IS THE SUSCEPTANCE OF THE CAPACITOR
230     XC = 1 / BC
240     REM "XC" IS THE REACTANCE OF THE CAPACITOR
250     B = ABS ( BL − BC )
260     REM "B" IS TOTAL REACTANCE OF THE CIRCUIT
270     IF B = 0 THEN GOTO 290
280     X = 1 / B
290     REM "X" IS THE TOTAL SUSCEPTANCE OF THE CIRCUIT
300     Y = SQR ( G * G + B * B )
310     REM "Y" IS THE TOTAL ADMITTANCE OF THE CIRCUIT
320     IF J = 1 OR J = 2 THEN Z = 0 ELSE Z = 1 / Y
330     IF J = 2 THEN X = 0
340     REM "Z" IS THE TOTAL IMPEDANCE OF THE CIRCUIT
1000    PRINT "XL =";XL;"OHMS"
1010    PRINT "XC ="; : IF I = 1 THEN PRINT I$; ELSE PRINT XC;
1020    PRINT "OHMS"
1030    PRINT "X ="; X ;"OHMS"
1040    PRINT "Z (PARALLEL) ="; Z ;"OHMS"
1050    PRINT "INPUT NEXT SET OF PARAMETERS"
1060    GOTO 10
1070    END
```

ment on this line, there are two. The : symbol at the end of a statement lets the virtual machine know that there's another statement following it. Generally, only one statement is needed on a line, but occasionally, simple statements may be grouped on a single line. In this case, the variables I and J must be initialized at the beginning of the program (set to a starting value). Each initialization is a simple procedure, so we did all the initializing on a single line.

Line 80 is a *string variable*. Under certain circumstances (for example, if C = 0) the value of a certain number becomes uncomputably large. Our program must recognize when these circum-

stances occur and avoid the computation. In this case, we'd like to indicate that the results of calculating the number would be too large to express. The literal in line 80 is called a *string* when it's given a name of its own and isn't part of a PRINT or an INPUT. In this program, "I$" is used by a print statement later (on line 1010) and we could have written 'PRINT "INFINITE" ' instead of 'PRINT I$'. Operations that use string variables like I$ can connect them together, swap them, do internal manipulations within a string, or compare them. These sorts of operations are called *string manipulation* or *string handling*. Special programs called *word processors* use *string handling* routines to make writing, editing, business, and office typing easier and more convenient. (We are using one right now to write this book, although it's not a BASIC program.)

Line 90 defines a constant called PI (you can see that variables with names two letters long are permitted in this version of BASIC). Notice that this constant has a decimal point in it. It's not an integer, like many of the other numbers in this program. Such numbers are called *floating-point variables*. The input numbers may be floating-point variables also. All the numbers in this program are computed as floating-point numbers, whether they contain a decimal point or not. Integer calculation has to be specially marked by using a special type of variable name or defining the variable as an integer (DEFINT statement).

Line 100 is a *chain multiplication. Chain computation* is the placement of several mathematical operations in a single statement. Although it's impossible to do this in machine language, it's routinely done in BASIC.

Notice the IF statement on line 130. There are two statements on the line following the IF. Both of these statements will be done if the IF is TRUE, and neither will be done if the IF is FALSE. In an IF statement like this, all the statements on the same line with the IF are treated like a single routine to be done if the IF is TRUE and not done otherwise. Lines 170 and 210 are similar to 130 in this regard. The IF statements in lines 270 and 330 are simple IF statements that just do one thing. You'll notice that, unlike CONDITIONAL JUMP statements, these IF statements don't just do GOTOs when they're TRUE.

Line 320 is an "IF ... THEN ... ELSE" statement. It defines what to do if the IF is TRUE and also what to do if it is FALSE. This is even further from a machine language CONDITIONAL JUMP than the regular IF is, but it's a very powerful statement that can be used to handle a wide variety of conditional tests. Line 320 tests the possibility that Z is uncomputable by conventional methods. If this happens, the correct value is assigned to Z (Z = 0); otherwise, Z is computed by ordinary means.

Line 1060 is necessary to make this program a loop that repeats when new inputs are provided.

A PRINT statement like line 1000 can contain a mixture of literals and variables. In this line, both the words XL = and the value of XL are printed. The ; suppresses spacing between one field of characters and another (if a literal and a

variable are separated by a ; in a PRINT statement, they will be printed together).

Lines 1010 and 1020 can PRINT either of two ways. When I is equal to 1, the literal INFINITY is printed after XC = (line 1010) and before OHMS (line 1020). When I is not equal to 1, a variable value (XC) is printed according to what was calculated by line 230. You can see the use of the IF ... THEN ... ELSE statement in this example. You can also see how a "flag" (the variable called I) is passed from line 210 to this routine to handle the "uncomputable" case we described earlier. The flag is set when the number (XC) will be uncomputably large. It is reset at the beginning of the program, and will arrive at line 1010 as a 0 if the number (XC) is computable.

There are other flags in this program, used to pass information about the ultimate value of Z down to lines 320 and 330, where Z is computed. They also relate to whether Z is computable by normal means. We'll let you investigate how they work for yourself.

EXAMPLE 21-3

What does the following BASIC program segment do? List the line numbers in the order in which they will be executed.

```
10   FOR X = 1 TO 3
20   PRINT X
30   FOR Y = 1 TO 2
40   PRINT X + Y
50   NEXT Y
60   IF (X + Y) = 6 THEN PRINT "FINISHED"
70   NEXT X
80   END
```

Solution: The list of line numbers is

10, 20, 30, 40, 50, 40, 50, 60, 70,
20, 30, 40, 50, 40, 50, 60, 70,
20, 30, 40, 50, 40, 50, 60, 70, 80

and what prints on the screen is:

```
1
2
3
2
3
4
3
4
5
FINISHED
OK
```

To verify this, we follow the conditions of each of the variables as the program is running.

Line	X	Y	(X + Y)	Output
10	1	?	?	
20	1	?	?	1
30	1	1	2	
40	1	1	2	2
50	1	2	3	
40	1	2	3	3
50	1	3	4	
60	1	3	4	
70	2	3	5	
20	2	3	5	2
30	2	1	3	
40	2	1	3	3
50	2	2	4	
40	2	2	4	4
50	2	3	5	
60	2	3	5	
70	3	3	6	
20	3	3	6	3
30	3	1	4	
40	3	1	4	4
50	3	2	5	
40	3	2	5	5
50	3	3	6	
60	3	3	6	FINISHED
70	4	3	7	
80	4	3	7	

Comments:

Do not continue Y loop if Y > 2 —fall through to next X

Do not continue Y loop if Y > 2 —fall through to next X

Do not continue Y loop if Y > 2 —fall through to next X

Do not continue X loop if X > 3 —fall through to program end

21-4 ACCESSING MACHINE LANGUAGE THROUGH BASIC

One application of BASIC, we said, was to write small BASIC programs to test out whether an anticipated machine language program would work.

Once you're certain that the idea of the program works, you rewrite it into machine language. Now what do you do?

If the computer you're using has the POKE instruction, you'll POKE the program into memory addresses. Loading the program could be accomplished by a routine something like this:

Version I

```
10 N=65000
20 RESTORE
30 READ B
40 POKE N,B
50 N=N+1
60 IF N<=65011 THEN GOTO 30
70 END
100 DATA 33,0,60,1,232,3,175
110 DATA 119,35,16,251,201
```

Version II

```
10 RESTORE
20 FOR N=65000 TO 65011 STEP 1
30 READ B
40 POKE N,B
50 NEXT N
60 END
100 DATA 33,0,60,1,232,3,175
110 DATA 119,35,16,251,201
```

Notice the two versions of this routine. They both POKE the same bytes of code into the same area of memory, but the one below uses a short-cut called a FOR ... NEXT loop. The FOR and NEXT statements are used wherever a routine must be repeated a specific number of times. The number is included in the FOR as its index. The index is a counter that can be counted either up or down from an initial value to a limit. In the program on the left, the initial value is set by N = 65000, the counter is incremented by the step N = N + 1, and the limit is set by IF N <=65011 GOTO 30. The FOR loop on the right uses the statement FOR N = 65000 TO 65011 STEP 1 to count from 65000 to 65011. The NEXT N statement at the bottom of the routine does the same thing as the IF in the routine on the left. It says to loop back until the counter reaches (or over-shoots) the limit. The FOR ... NEXT loop is used in BASIC as a shorter way to do the same thing as the INITIALIZATION, INCREMENT, and IF statements in the "longer" BASIC routine. You can see in both program segments that the index (N) is used for more than just counting. In this POKE operation, N is the destination address of each byte of code taken from the data file by the READ statement. The numbers separated by commas in the data file are taken, one by one, in order, until all the data have been POKEd into memory addresses. Notice that there are 12 data numbers and 12 values of N when N is counted from 65000 to 65011. The loop will POKE 12

bytes of data into the memory—this is a 12-byte program to erase 1000 bytes of RAM, replacing the digits in the memory locations with zeros (ASCII nulls). By the way, we said the FOR loop could count either up or down, but the STEP size of 1 seems to count up only. How do we count DOWN? Use a step size of −1, of course! A negative step size will make the count become smaller as the counter is incremented (incidentally, in this case, the step size of 1 is optional; if no step size is included in the FOR loop, N will count up by ones anyway).

How do you RUN the program once it's poked into the memory? Some microcomputers have an instruction called CALL, others have a command called a USER CALL (USR). See the documentation for the particular BASIC on your computer to see how the machine code you POKE into the memory may be CALLed. Some machines do not permit a call to a machine language routine at all. Each one that does permit it seems to work in a different way. In the example above, the program segment is called by:

```
     ⋮
5000 POKE 16525,232:POKE 16526,253
5010 REM THIS IS THE STARTING ADDRESS OF THE
5020 REM MACHINE LANGUAGE ROUTINE (65000)
5030 X = USR (0)
5040 REM THIS STATEMENT WILL MAKE THE COMPUTER
5050 REM CALL THE MACHINE LANGUAGE ROUTINE THAT
5060 REM BEGINS AT MEMORY ADDRESS 65000
     ⋮
```

Of course, if the program is intended to do something that can be done directly in BASIC, there's no need to go to all this trouble, is there? If the BASIC interpreter is working, operations like RAM testing can be done just as easily using POKE and PEEK as the LD commands of machine language. After all that exposure to machine language in previous chapters, if this chapter has encouraged you to consider BASIC as an alternative option, then it's been a success.

EXAMPLE 21-4 BASIC PROGRAM TO LOAD 8086 CODE

Unlike the CP/M computers in this chapter's examples, the IBM-PC does not contain a Z-80 processor; it runs 8086 code instead. A program to load hexadecimal code instructions into the IBM-PC is a bit more complicated than the decimal Z-80 code loader at the end of this chapter, but it has the advantage of loading hex code, which is usually available directly from microprocessor manuals. This program drives an Epson printer directly from the graphics screen of an IBM-PC. If the DATA statements are altered, the hex code can be another program in 8086 code. Of course, if this program is loaded into a CP/M computer, it will only work if the DATA statements are machine language for a Z-80 microprocessor.

Enter the following program into an IBM-PC with an Epson printer. Save the program on a formatted disk in drive B.

Epson printer driver program from Graphics Screen 2

```
10 ' "B:EPSON.BAS"
1020 P$="":DEFINT A-Z:FOR N=0 TO 255
1030 READ PR$:IF PR$="END" THEN N=
     255:GOTO 1060
1040 PR=VAL("&H"+PR$)
1050 P$=P$+CHR$(PR)
1060 NEXT N
1070 ENTRY!=(PEEK(VARPTR(P$)+1))
     +(PEEK(VARPTR(P$)+2))*256
1080 IF ENTRY!>32768! THEN
     ENTRY%=ENTRY!-65536! ELSE
     ENTRY%=ENTRY!
1090 CALL ENTRY%
1095 LPRINT CHR$(27)"@"
1100 END
1110 DATA B0,1B, E8,68,00, B0,33, E8,63,00,
     B0,12, E8,5E,00, BA,00,00
1120 DATA B0,1B, E8,56,00, B0,4C, E8,51,00,
     B0,80, E8,4C,00
1130 DATA B0,02, E8,47,00, B9,00,00
1140 DATA 52, BB,08,00, D0,E7, B4,0D,
     CD,10, 0A,C0, 74,03, 80,CF,01
1150 DATA 42, FE,CB, 80,FB,06, 74,ED,
     80,FB,03, 74,E8, 80,FB,00, 75,E1,BA,C7,
     E8,1C,00, 5A, 41, 81,F9,80,02
1160 DATA 75,D0, B0,0D, E8,0F,00, B0,0A,
     E8,0A,00, 83,C2,08, 81,FA,90,01
1170 DATA 72,A6, CB, 52, B4,00, BA,00,00,
     CD,17, 5A, C3
1180 DATA END
```

On an IBM-PC with an Epson printer, load and run any program that develops a computer-graphics diagram in BASIC (SCREEN 2 graphics mode); then, while the picture is still on the screen, load and run this program using the command

RUN "B:EPSON.BAS"

(assuming that it has been saved on the disk in drive B:). Describe the results. Disassemble the code in hexadecimal (look up what the mnemonics are) and determine if there are any changes

you would make to improve the appearance of the pictures printed by the Epson printer. (*Note:* This program cannot be used to store more than 256 hex bytes of machine language.)

You could use this graphics program to make a picture for the "EPSON.BAS" screen-dump program to print onto paper:

```
10 '   "B:SHELL.BAS"
1000 CLS:SCREEN 2:DIM
     EKS(50,50),EX(50,50),WY(50,50)
1100 X0=120:Y0=100:M=0
1450 CLS:KEY OFF:'Printout Graphics
1500 FOR PHI=0 TO 3.2 STEP (.157/
     2):LASTEX=X0:LASTWY=Y0:N=0:M
     =M+1
1600 FOR THETA=1.57 TO 4.72 STEP (.157/
     2):N=N+1
1700 R=50*COS(THETA)*PHI + 50*SIN(PHI)
1800 X=R*COS(THETA)*1.2:Y
     =R*SIN(THETA)
1900 EX=X0-X*COS(PHI):WY=(Y0-Y)
2000 LINE (LASTEX,LASTWY)-(EX,WY)
2300 D=M-1:IF M=1 THEN D=M
2400 EX(N,M)=EX:WY(N,M)=WY
2500 LINE (EX(N,D),WY(N,D))-(EX,WY)
2800 LASTEX=EX:LASTWY=WY
2900 NEXT THETA
3000 NEXT PHI
```

The printout should look as shown in Figure LP 21-1.

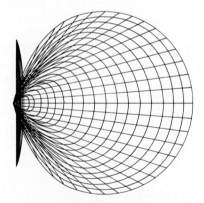

FIGURE LP21-1

Tip: Notice line 10 in both programs. It is a REMark statement using the shortcut form of the word REM: namely, an apostrophe ('). It's there to identify the program, but also so that you can reposition the BASIC cursor at the beginning of line 10 in the program LIST, and type SAVE over the numbers and apostrophe. Then all you have to do is press the ⟨ENTER⟩ key and the program will be saved onto disk B:

21-1. What is the main advantage of programming in a high-level language (like BASIC) that is interpreted?

21-2. What is the main advantage of programs written in a computer's assembly language?

21-3. What is the main advantage of programming in a high-level language (like FORTRAN) that is compiled?

21-4. Which executes in the least time, a compiled program or an interpreted one?

21-5. Which takes up the least memory space, a program that has been assembled from assembly language or a program that has been compiled from a high-level language?

21-6. What's a syntax error?

21-7. Write a brief explanation of the reason = takes the place of ← in BASIC and FORTRAN.

21-8. When names like B, C, H, and L are used in a BASIC program, does that mean that the microprocessor in the computer has registers called B, C, H, and L, as the 8080 has?

21-9. Several items listed below are available in both the microprocessor hardware and in the virtual machine you see when you write a program in BASIC. Some of the items are part of only the virtual machine (called "primitives" in programming jargon), and others are part of only the microprocessor hardware. For parts (a)–(f), identify whether the item is part of the virtual machine, the real microprocessor hardware, or both.
(a) Addition
(b) Multiplication
(c) Logical EXCLUSIVE OR
(d) Registers
(e) Variables
(f) Logical AND

21-10. Do line numbers in BASIC have to follow one another in consecutive, numbering order, without gaps (0,1,2,3,4,5,...) as addresses do in machine language?

21-11. Is the insertion of BASIC line numbers between other line numbers part of the virtual machine, or is it something built into the hardware of the microprocessor?

21-12. Can the programmer check the microprocessor's parity flag directly in BASIC?

21-13. Why do BASIC and FORTRAN use the * symbol for multiplication, instead of ×?

21-14. Define "algorithm" in your own words.

21-15. Explain why you might want to use BASIC to test out a diagnostic routine you intend to write later in machine code.

21-16. Look at the BASIC program for finding the impedance of a parallel *RLC* ac circuit. Is the number called PI stored in the computer as an integer or a floating-point number?

21-17. Is the impedance program an endless loop?

21-18. What is a prompt?

21-19. Write a BASIC statement that contains a chain computation.

21-20. Describe what a flag is in a high-level language.

21-21. Describe what a FOR ... NEXT loop does.

21-22. What BASIC statement permits you to enter bytes of machine code directly into RAM?

21-23. What are some BASIC statements you might use to call or run a machine language routine in RAM?

DRILL PROBLEMS

Questions 21-1 through 21-5 refer to the following BASIC program and flowchart:

The BASIC program and flowchart above are almost the same.

21-1. Which line in the program differs from the flowchart?_____

21-2. Which block in the flowchart differs from the program?_____

21-3. The_____(flowchart, program, both) will run endlessly.

```
10  LET A = 3
20  LET A = A - 1
30  IF NOT (A = 0) THEN GOTO 10
40  LET B = 4
50  END
```

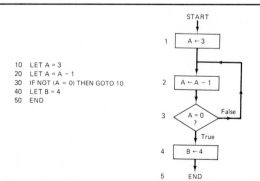

21-4. The_____(flowchart, program, both) contains a loop.

21-5. The_____(flowchart, program, both) will end with B = 4.

Questions 21-6 through 21-10 refer to the following BASIC program and flowchart:

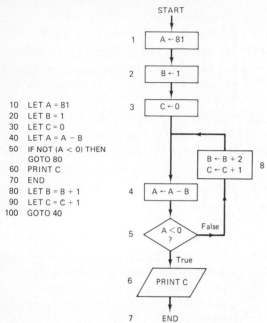

```
10   LET A = 81
20   LET B = 1
30   LET C = 0
40   LET A = A − B
50   IF NOT (A < 0) THEN
     GOTO 80
60   PRINT C
70   END
80   LET B = B + 1
90   LET C = C + 1
100  GOTO 40
```

The BASIC program and flowchart above are almost the same.

21-6. Which line in the program differs from the flowchart?_____

21-7. Which block in the flowchart differs from the program?_____

21-8. The_____(flowchart, program, both) will end with C = 9.

21-9. The_____(flowchart, program, both) contains a loop.

21-10. The_____(flowchart, program, both) will end with A = −10.

Questions 21-11 through 21-15 refer to the following BASIC program and flowchart:

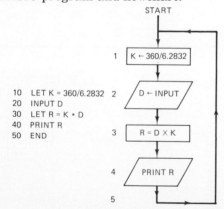

```
10   LET K = 360/6.2832
20   INPUT D
30   LET R = K * D
40   PRINT R
50   END
```

The BASIC program and flowchart above are almost the same.

21-11. Which line in the program differs from the flowchart?_____

21-12. Which block in the flowchart differs from the program?_____

21-13. The_____(flowchart, program, both) will run endlessly.

21-14. The_____(flowchart, program, both) contains a loop.

21-15. The_____(flowchart, program, both) will print one line only.

Questions 21-16 through 21-20 refer to the following BASIC program and flowchart below:

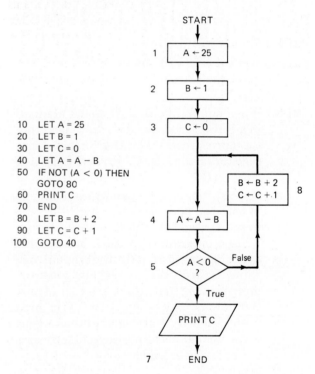

```
10   LET A = 25
20   LET B = 1
30   LET C = 0
40   LET A = A − B
50   IF NOT (A < 0) THEN
     GOTO 80
60   PRINT C
70   END
80   LET B = B + 2
90   LET C = C + 1
100  GOTO 40
```

The BASIC program and flowchart above are the same.

21-16. Does the program contain a conditional branch?_____

21-17. Does the program contain an unconditional branch?_____

21-18. The program_____(does, does not) contain subroutines.

21-19. The program_____(does, does not) contain a loop.

21-20. What is the final value of C?_____

22

TELECOMMUNICATIONS

CHAPTER OBJECTIVES

By the time you finish this chapter, you will be able to:

1. Explain why *serial data transmission* is used for long-distance data communication, rather than *parallel data transmission*.

2. Explain the function of the *UART* and *modem* blocks in a telephone data link.

3. Describe the FSK signaling used by a *modem* on the telephone lines.

4. Describe the RS-232 *interface* between the *UART* and the *modem* in a telephone data link.

5. Describe some of the characteristics of data transmission on long-distance lines.

6. Describe some of the commonly used *transmission media* employed in long-distance data transmission.

7. Describe some of the commonly used methods of *modulation* and *demodulation* employed in long-distance transmission.

8. Describe some reasons why optical (lightwave) methods are the wave of the future for data transmission.

9. Describe some of the major problems in long-distance data transmission.

10. Explain why optical technology may make electronics obsolete for computing and data transmission in the foreseeable future.

In previous chapters we looked at input and output devices that computers use to communicate with the outside world. In the examples we chose, the computer was part of a human–machine, document–machine, or machine–machine interface. When we described the machine–machine interface, we imagined that the computer was used for numerical control of some device like a "smart" drill or lathe. So far, we've never considered that the machine our computer outputs to or inputs from might be another computer. Why not? One computer's output can be another's input, and vice versa, provided that each one knows how to take turns with the other. In this chapter we'll look at the communication of data from one computer to another (data communication), and in particular, methods for transferring data from one computer to another over long distances (**telecommunication**).

Microprocessors are used in microcomputers, but can be made to imitate virtually any kind of digital network where extremely high speed is not an important consideration. We will often describe, in this chapter, how a microcomputer "talks" to another microcomputer over the telephone system or some other common carrier. Do not be misled by our choice of examples (in microcomputers) at each end of the data link. Microprocessors appear in all parts of the ESS (*electronic switching system*) between the source and destination of any long-distance communication. Many of our examples will use the telecommunications network as a glorified I/O device for a microcomputer (or two microcomputers). It's important to remember that that's not "all there is" to do with microprocessors in telecommunications, and we'll endeavor to show examples, from time to time, of the other "dedicated machine" aspects of microprocessor usage.

22-1 LONG-DISTANCE COMMUNICATION BETWEEN COMPUTERS

Suppose that there are two people with 8-bit computers in different cities. One person has a program that the other person would like to have. Instead of making a copy of the program on diskette or tape and sending it by mail, the people would like to have one computer "send" and the other "receive" 8-bit words between Chicago and Toronto.

How do they do it? If the computer in Chicago is sending and the computer in Toronto is receiving, the Chicago machine can be attached to an output port and the Toronto machine can be attached to an input port. Now all they need is a ribbon cable 500 miles long!

That's certainly not very practical. The cable would cost a fortune, and they'd have to get permission (or buy the right-of-way) to lay the cable. Someone has already done that. The telegraph company and telephone company have had conductors connected between Chicago and Toronto for over a century. The circuits are already there, and the cost of using them is not very high, compared to connecting their own circuits. Not only that, but in this case, they only intend to use this circuit one time. A *common carrier* is a communications network (like the telephone company) that is available to everybody, on a pay-for-time basis. The two people should connect their computers to one another via a common carrier.

22-1.1 A Parallel Communication System

In Figure 22-1, we see the most obvious way of carrying 8-bit data from the computer in Chicago to the one in Toronto. It seems simple enough, but there are several things wrong with it that are not evident in the diagram:

1. *Propagation delay.* There are eight separate telephone circuits. What are the chances that all of these circuits are equally long? Suppose that one path chosen to connect Chicago and Toronto is 100 miles longer than the others. Its data will arrive over 500 microseconds later than the other bits. That's not a long time to a human being, but it's "digital eternity" to a computer. A parallel output port can receive data hundreds of thousands of times a second, but if data are sent on this "cable" of unequal-length paths, if 2000 outputs are done per second, the bits will be out of phase when they arrive.

2. *Amplitude.* There's no guarantee that the logic pulses put into the telephone circuit (about 4 V peak to peak) will come out the other end exactly the same size. Nobody told Alexander, when he designed the system, that it would have to have unity gain (give or take less than 1 dB). "Can the guy on the other end hear me?" was the main concern. It wouldn't be unusual for the signals on the other end to be 12 dB higher or lower than the incoming signal. In one case, the voltage would swing about 1 V, peak to peak, and the logic gates wouldn't even know there was a signal present. In the other case, the voltage would swing 16 V, peak to peak, and turn all the 5-V logic in the receiver to "WAS GATES"(!) Not good for the well-being of TTL circuits.

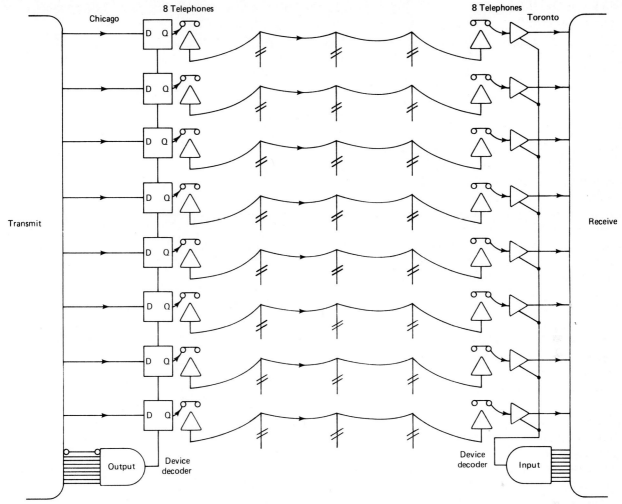

FIGURE 22-1 Eight-bit data communication—the hard way.

3. *Dc level.* Suppose there's a miracle, and a 4-V (p-p) wave input produces a 4-V (p-p) wave at every output. Is there any guarantee that the output doesn't have a LOW level of 18 V and a HIGH level of 22 V? No way! Nobody told Alexander, when he designed his system, that a wave that has the same shape, but is "centered" at a different dc level, makes any difference. "They both sound the same, don't they?" he might ask. True. But to the logic gates, 18 V is HIGH and 22 V is HIGH, and either one (if applied to a 5-V logic gate) is *goodbye.*

4. *Bandwidth.* While we're on the subject of what Alexander designed the system for, it was, you remember, designed for voice communications. The transmission-line characteristics and "repeater" amplifiers are designed to handle human voice frequencies (and not much more) from 300 to 3000 Hz. This is another limiting factor on the parallel output ports (which can handle *much* higher frequencies).

5. *Money.* Eight telephone calls to Toronto cost a bundle. Even if all the data are transmitted in a few seconds, the phone user will be charged for 3 minutes. If all the data could be sent on one circuit instead of eight, the data transmission would only cost an eighth as much. Even if transmission on one wire takes eight times as long as transmission on eight wires, money can be saved if the data are transmitted in less than (8 × 3) minutes.

22-1.2 Serial Data Transmission Using UARTs and Modems

Figure 22-2 shows how most computer "dial-up lines" work. In the diagram, the computer on the left is transmitting (Chicago) and the computer at the right is receiving (Toronto) by using the telephone system as its common carrier.

The CPU with information to send is using one of its output ports as a place where the data

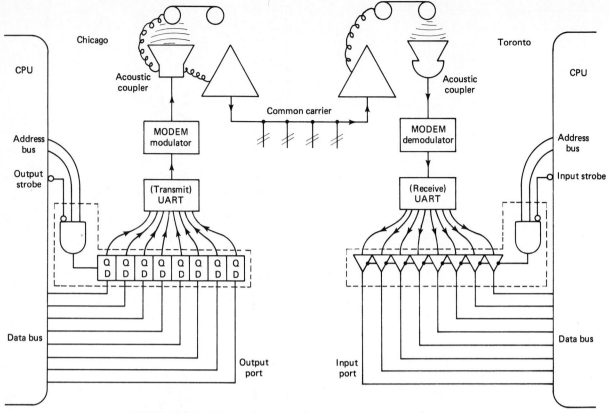

FIGURE 22-2 Eight-bit data communication—as it is really done.

are kept while they are being transmitted. In this example, 8-bit words are being sent from the (8-bit, parallel) output port, the parallel logic levels of the port's output port. The parallel logic levels of the port's output are *space-division multiplexed* (SDM), which means that you can tell what bit is being transmitted by where it is. There is a 1-wire, a 2-wire, a 4-wire, and 8-wire, and so forth. The location of the wire on the bus tells you what the binary place value of the bit is.

The UART block. The second block in this block diagram is used to convert the parallel data into serial data. The serial logic levels of the UART's output are *time-division multiplexed* (TDM), which means that you can tell what bit is being transmitted by when it is transmitted. Only one wire is used for the output, but at one moment, it is 1-time, then it becomes 2-time, then 4-time, then 8-time, and so forth. From the timing of the signal, you can tell what its binary place value is.

There is information on conversion of data from parallel to serial and serial to parallel form in Sections 11-5 and 11-6, but we'll briefly summarize how this UART block converts parallel data to serial data.

"UART" stands for universal asynchronous receiver-transmitter. When used in this sense, a receiver means a circuit that receives a serial data transmission and converts it into parallel form, while a transmitter converts parallel data to serial data before sending them.

In the circuits shown in Figure 22-3, the data are being transmitted. A parallel word is present at the input of each circuit, and each bit is separately "clocked" out of the circuit on a single serial output. The "clock" that coordinates the transmission of each bit is called a *baud rate generator*, and the frequency at which it clocks out bits at the output is called the *baud rate*. It measures the number of bits per second (also called bps) that are being transmitted.

In Figure 22-3(a), the input data word is clocked into an 8-bit latch in parallel (this is called a *broadside load*, because it resembles a "broadside" fired from a ship in the old pirate movies where all the cannons are fired at once). After the 8-bit word is transferred into the flip-flops of the 8-bit latch, the flip-flops are clocked for eight clock pulses as a shift register. At the end of eight clock pulses, the 8 bits latched into the register have been clocked out one at a time.

In Figure 22-3(b), the input data word is multiplexed onto a single output. We called this

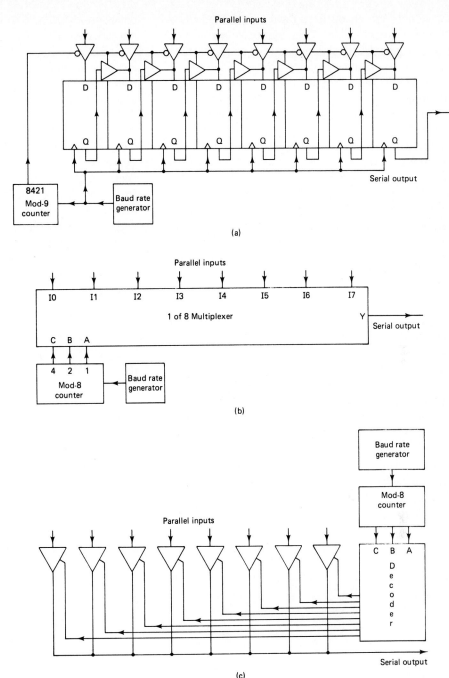

FIGURE 22-3 Transmitters (parallel-in, serial-out): (a) shift register; (b) multiplexer; (c) buffer.

process time-division multiplexing, after all, so it shouldn't be any surprise that a multiplexer is part of one design for parallel-to-serial conversion. In this design, a mod-8 counter is needed to select the sequence of inputs that will be passed to the output. If an up-counter is used, data will be transmitted with the least significant binary places first and the most significant binary places last—1-time, then 2-time, then 4-time, and so on—while a down-counter could be used to transmit bits starting with the most significant and ending with the least significant. Unlike the shift register design in Figure 22-3(a), this circuit has

the drawback that if its input data word changes in the middle of a transmission, that bits already transmitted will be from the first word, and the remaining half of the serial word transmitted will come from the second. With bits from the "front" half of one word and the "back" half of another, the serial word transmitted will be "garbage." There will have to be additional synchronization circuitry (not shown) to ensure that the input data change only when the counter is reset.

Figure 22-3(c) uses buffers to do the same thing as the multiplexer circuit. The buffers are enabled by a decoder driven by a mod-8 counter that selects which buffer's data is enabled onto the serial output wire. This design requires yet another block (the decoder) and still has the same drawback as the circuit in Figure 22-3(b). It can be "messed up" if its data change in mid-transmission.

Of the three designs, the first is the most commonly used for serial data transmission, and is virtually the only method used in UART design (for reasons which will become clear when we see receiver design).

It is the transmitter of the UART which converts the data from the output port from parallel data to serial data. In serial form, only one circuit is needed to carry the data instead of eight. A data sheet for the Motorola 6850 UART may be found in Appendix 4.

The modem block. The next block in the diagram (Figure 22-2) is a modulator. It is part of a two-way circuit called a **modem** (modulator–demodulator) that converts the dc logic levels (digital) from the serial output of the UART into a frequency-modulated audio output (analog). Since there are only two levels of voltage at the input, there are only two frequencies at the modem's output. The HIGH logic level produces a frequency called a *mark* and the LOW logic level produces a frequency called a *space*. The mark and the space are a 1 and a 0 encoded by *frequency shift keying* (**FSK**). The frequency shifts between the mark frequency and the space frequency give FSK its name.

Modulation and demodulation are done by circuits similar to those in an FM radio transmitter–receiver. Often, both are taking place at the same time, and the modem is transmitting data in one direction as it is receiving data in the other. This is called **full-duplex** communication. Yet if there is only one circuit for the voice conversation, how can signals be traveling in two directions, on one path, at the same time? It is done in the same way that two television sets can be attached to one antenna, yet receive two different channels at the same time. Each channel is carried at a different frequency (or in a different frequency *band*) from the other channel. The modem that placed the telephone call uses a frequency band called the **originate** frequencies, and the modem that answered the call uses a frequency band called the **answer** frequencies. An AT&T 103 standard modem uses answer frequencies of 1270 Hz for mark (or 1) and 1070 Hz for space (or 0) and originate frequencies of 2225 Hz for mark (or 1) and 2025 Hz for space (or 0).

Figure 22-4 shows the waveforms for the number 5 transmitted from the Chicago side of Figure 22-2 using AT&T 103 modem standard. (The differences between the frequencies have been exaggerated somewhat.) The parallel data on the output port are not shown, but the serial output bits of the UART are, and above them, the FSK output of the modem.

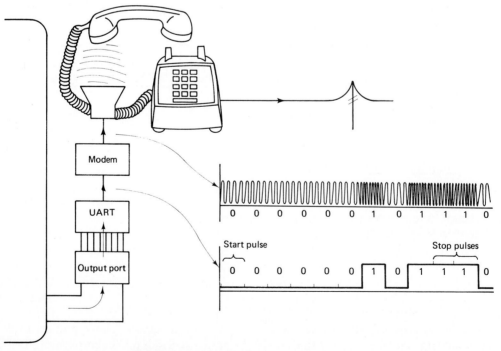

FIGURE 22-4 Modulation into FSK audio.

The acoustic coupler. The output of the modem is not attached directly to the telephone line as the parallel wires in Figure 22-1 were. Instead, the audio frequency output is converted from electrical energy to acoustic (sound) energy by an **acoustic coupler**, which is a speaker with suitable waveshaping circuity and amplification for the telephone mouthpiece. The telephone is now being used exactly as Alexander Graham Bell had intended. Sound arrives at the mouthpiece through the air, and leaves the earpiece of the receiving telephone through the air, and no "foreign" circuitry ever touches the unspoiled beauty of the phone company's equipment.

The common carrier. On both the sending and receiving end of the telephone system, the handset of the telephone fits into two rubber cups with a microphone in the cup that holds the earpiece and a speaker in the cup that holds the mouthpiece. The indirect nonelectrical coupling of the signal into the transmitting phone is reversed at the receiving end. The acoustic signal at the receiving phone's earpiece is converted by a microphone and suitable waveshaping and amplifying circuits into an FSK audio signal at the acoustic coupler. The FSK audio output is transferred to the modem, which demodulates it into serial dc logic levels. The serial logic output of the modem is transferred to a UART (the receiver is used here) that converts its serial pulses back into parallel 8-bit logic words. The parallel logic levels are then input through the input port of the receiving computer.

A moment's digression on the subject of receivers is called for here. The transmitter part of the UART was shown in three variations; there is only one primary way of doing the receiver's job. That is shown in Figure 22-5. Data that arrive serially are clocked into the receiver shift register by the baud rate clock. The oscillations of the baud rate clock must be synchronized with the clock in the transmitter on the other end of the telephone system. Not only must the receiver clock have the same frequency but they must be in phase as well. The receiver clock's pulses must start at exactly the time the transmitter clock's pulses would arrive (if a clock pulse were included in the transmission—which it is not). To do this without transmitting clock pulses side by side with the data (which would require another circuit), we use the same approach a TV transmission uses to keep the horizontal oscillator of your TV synchronized with the horizontal scan in the transmitted TV signal. Sync pulses are sent at the end of each line of scan in a TV waveform, and serial TTY audio transmission includes a similar type of sync information at the end of each 8-bit serial word that's transmitted. The stop pulses at the end of a "word" and the start pulse at the beginning of the next word comprise a sync pulse whose rising edge can be used to trigger the baud rate generator and get it back in step if it has drifted slightly since the last word was received. This means that the baud rate generator could drift as much as 10 percent in the 11 clock pulses of a word like the one in Figure 22-4, and the stop bits and start bit will still get the receiv-

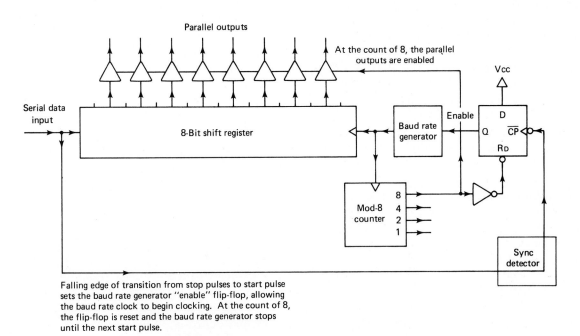

Falling edge of transition from stop pulses to start pulse sets the baud rate generator "enable" flip-flop, allowing the baud rate clock to begin clocking. At the count of 8, the flip-flop is reset and the baud rate generator stops until the next start pulse.

FIGURE 22-5 Receiver (serial-in, parallel-out).

er's baud rate clock back in step with the transmitter.

We have indicated this synchronization by a "magic" box in Figure 22-5 called "sync." Its function is similar to the detector for the sync pulses in a TV receiver.

22-1.3 Analysis of Reasons for Serial Transmission

Let's look back at the five objections we had to parallel data transmission as it was done in Figure 22-1, and see how the circuit in Figure 22-2 answers these objections:

1. *Propagation delay.* We now have only one telephone circuit. The bits are transmitted one after another, instead of all at once, so transmission is slower, but it's impossible for any bits to get ahead of any other bits. The wire that transmits the 1-bit has to be exactly as long as the wire that transmits the 2-bit, because they're exactly the same wire.

2. *Amplitude.* We don't have to worry too much about the amplitude of the received waveform, because it's frequency modulated rather than amplitude modulated. The only consideration in designing the transmitter-receiver combination is the "Can the guy on the other end hear me?" aspect. If the "guy" on the other end of the circuit is a microphone, the signal must be loud enough to operate the microphone reliably, but clipping and waveshaping circuits at the receiving end will accept any audible signal and make it acceptable for the modem.

The lack of a direct electrical connection to the phone system eliminates the need to do 5-V-to-48-V interfacing and permits direct use of 5-V logic where it would not be possible in a direct-coupled system. There are, however, direct-coupled modems that do not require an acoustic coupler. They use direct connection to telephone circuitry, and are more reliable than acoustic couplers because they are not affected by room noise. The hardware, however, does require level shifters, and must pass type approval by the FCC, which regulates all interstate communications in the United States. The equivalent Canadian agency (at the Toronto end of our line) must approve the receiving circuit hookup. The acoustic coupler, since it does not hookup directly to the communications network, does not require such approval, any more than a human voice does.

3. *Dc level.* As with our objections to amplitude shifts, the dc level shifts are unimportant to the circuit of Figure 22-2. The acoustic coupler

and use of frequency (rather than amplitude) to discriminate between a mark and a space input guarantee that "They both sound the same, don't they?" (despite different dc levels) is the only consideration that is needed for the FSK receiver.

4. *Bandwidth.* Frequency response of the phone company's voice lines is still 300 to 3000 Hz, and there's nothing we can do about that. There are, however, special lines that can be leased from the phone system, at added cost, that can handle 9000-, or even 19,000-baud transmissions. For most purposes, transmission at 300 baud (perhaps using 2.4 kHz and 1.2 kHz) will permit quite a bit of data to be transmitted at a reasonable cost (see below).

5. *Money.* We now have just one telephone circuit, which is actually transmitting data 11 times as slow as parallel transmission. We said before that there'd be a "break-even" point. If the data could be transmitted in (8×3) minutes, we'd pay no more than the eight phone calls (and actually pay less because the next 3 minutes is cheaper than the first). How much data can be transmitted in 24 minutes at 300 baud? We worked it out and found 432,000 bits, or 54,000 bytes, could be transmitted in that time.

22-1.4 The RS-232 Standard of Serial Data Transmission

RS-232 is a system of signaling used between digital systems. In the data communications area, RS-232 is the "de facto" standard way to connect data communications stuff to other data communications stuff. Any connection between one data communicating device and another, or between a data terminal and modem, probably uses the RS-232 connection to make the hookup. This type of connection will be discussed here.

22-1.5 RS-232 (RS-232A) and the DB-25 Connector

RS-232 is an interface for serial binary-encoded data. Usually, RS-232 is used to connect between a data terminal or a computer and a telecommunications network. The equipment used is going to depend on whether the network uses digital or analog lines. If analog, a modem is used to convert each "bit" of binary into a tone (or frequency) which can be transmitted down an analog line without deterioration. If the lines are digital, the digital format of RS-232 will be transformed into some *other* serial data transmission format (each manufacturer wants to have its own!) for use within the system. That doesn't mean that the RS-232 (which is serial digital data itself) can't be put onto a line directly. In some cases, RS-232 is used for short distances, usually under 100 meters.

The connector universally used to connect RS-232 devices together is the **DB-25 plug**. Figure 22-6 shows the layout of a DB-25 plug and socket, so-called because it has 25 pins. Don't worry! All 25 are seldom used. Usually, only eight or fewer are needed. The signals on these pins are used for four basic purposes:

Group	Purpose
I	Ground or common return
II	Data signals to Transmit and Receive bits
III	Control signals to show Busy/Ready status
IV	Timing signals for "synchronous" data (clock)

Since serial data are usually "asynchronous" (NOT-synchronous), group IV signals are seldom used.

The digital code most commonly transmitted on RS-232 serial data transmission is **ASCII**. This is the American Standard Code for Information Interchange, which we looked at in Chapter 7. It uses seven binary "bits" to represent each letter, number, punctuation mark, or control function found on a typewriter keyboard. In this case, each ASCII character is transmitted one bit at a time (serial data). Before the first bit of an ASCII character is transmitted, a **start bit** precedes it, and possibly a **check bit**, if transmission of odd- or even-parity code is desired. The 7-bit ASCII character is transmitted in seven separate data bits, then a **stop bit**, or possibly two stop bits, are transmitted after the data bits.

In Figure 22-7(a), a **bit stream,** or timing diagram, shows the data transmission of an ASCII character, the letter "A," using odd parity and one stop bit. The first bit transmitted is a start bit, which is a 0 in this example. It is followed by an 8-bit ASCII character. In the figure, **odd parity** (see Chapter 6) is used. That means that the check bit in each character is made 1 or 0, whatever is needed, so that the total bit count (per character) of 1s is 1, or 3, or 5, or 7 (an *odd* number). In this case the ASCII letter "A" contains two 1s, which is an even number; the check bit added is a 1, which brings the total bit count in the 8-bit ASCII character to 3, an odd number.

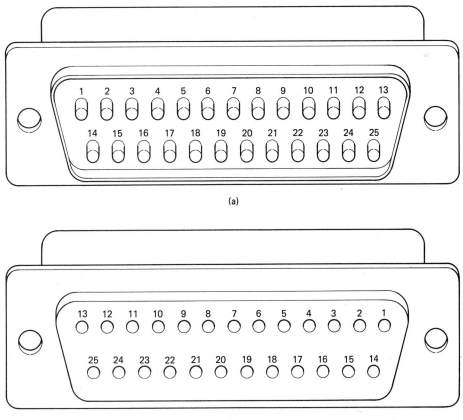

FIGURE 22-6 Layout of DB-25 connectors: (a) DB-25 (RS-232) plug; (b) DB-25 (RS-232) socket.

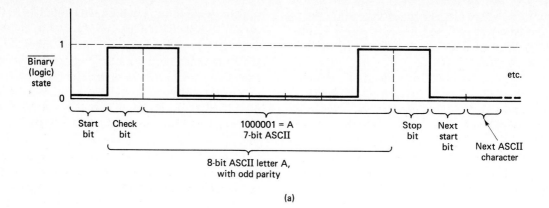

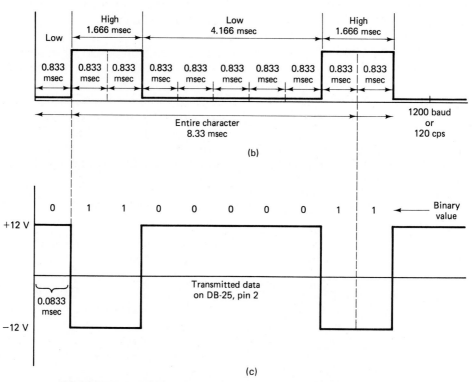

FIGURE 22-7 (a) Bit stream of ASCII letter A, using odd parity, one stop bit, 1200 baud; (b) timing of the transmitted bits; (c) bipolar, non-return-to-zero format RS-232 output.

The 8-bit ASCII character with odd parity is followed by a stop bit, which is a 1.

The reason for using parity is reliable transmission. It is a way to tell if the characters transmitted have lost or gained a wrong bit. If only odd-parity characters are transmitted, any character received with an even number of 1-bits is immediately *wrong*. The receiver can then request the transmitter to send that character again.

In Figure 22-7(b), we can see the timing of each bit transmitted. A pattern like this cannot be measured for its frequency, since the HIGH and LOW pulses that make up an "A" are not all

equal in length. Instead, the length of time each bit lasts—in this case, 0.833 ms—is measured, and a rate is calculated for the number of such bits that can be transmitted in a second. You can fit 1200 pulses of 0.833-ms duration into a full second. We say that the **data transmission rate** is 1200 **baud** (the units of data transmission speed). The actual voltage is not switching on and off (from HIGH to LOW) at anything like 1200 switchings per second in this figure, but that is the rate at which bits are being transmitted, and that is what a **baud rate** is. In the figure, although 10 bits are transmitted, there are only two positive pulses, so the pulse rate is much less than the bit (baud) rate. Very few characters have even half as many pulses as bits; a character with all "1s" would have no pulses or cycles at all because the voltage would always be ON.

Another way to measure the speed of data transmission is to look at how many characters we could send in a second at this speed. Since the character we see in Figure 22-7 has 10 binary digits (bits), including stop and start bits, a character is transmitted for every 10 bits. If 1200 bits are transmitted per second (that's 120 ten-bit groups), we are transmitting 120 **characters per second (cps),** as long as the bits in consecutive characters last 8.33 ms, like the ones in the figure. The actual transmission of data is *bipolar, non-return-to-zero* format (see Chapter 17). In Figure 22-7(c), we see *positive* 12 V used for binary 0, and *negative* 12 V = binary 1. The two polarities of 12 V are the "bipolar" voltages that represent the two binary states of logic. The "non-return-to-zero" part refers to the fact that only the positive and negative voltages are used for "legitimate" data. An open, nonconducting, or 0-V state on the data line would indicate that there were no data being transmitted.

Let's go back to the pins on the DB-25 plug and analyse what each is for. Sorted out by the four groups of functions their signals are used for:

I	II
Ground/Return	Data Xmit/Rcv
Pin No.	Pin No.
1, 7	2, 3, 14, 16*
III	IV
Control Signals	Clock
Pin No.	Pin No.
4, 5, 6, 8, 11,	15, 16*,
12, 13, 19, 20,	17, 18,
21, 22, 23, 25	24

*Depending on whether equipment uses EIA standard pin assignment or Bell 208A, this may be either a "secondary received data" pin or a clock pulse on a divided clock.

Pin 1 protective ground and pin 7 signal ground. These are the most important signals on the DB-25 plug. The signal ground/return pin (7) is *absolutely* necessary. Without it, none of the other signals could work. Whenever the other pins are made positive or negative, a current must flow out of them, through the wires, to the circuit at the far end of the connector, and back again, completing a current loop. This is the pin

to which currents from all the other pins return. They complete the current loops and are also at a zero-volt potential. Pin 1 is also at a zero-volt potential and may be attached to the outer shield of a shielded cable, used to prevent interference from stray external signals from reaching the conductors inside. This protective ground is optional and is sometimes strapped to pin 7.

Pin 2 transmitted data and pin 3 received data. These are the pins where the serial data come out (pin 2) and go in (pin 3) to the terminal. One end's output is the other end's input. The TD (transmitted data) at the near end is the RD (received data) at the far end, and vice versa.

For RS-232 data:

POSITIVE = binary 0
NEGATIVE = binary 1

Control and "handshake" signals. These are important signals that control the transfer of data, but are not data themselves:

Pin	Name	Symbol
4	Request to Send	RTS
5	Clear to Send	CTS
6	Data Set Ready	DSR
8	Data Carrier Detect	DCD
20	Data Terminal Ready	DTR
22	Ring Indicator	RI

Control signals are active (ON) when they are *positive.*

Some of the signals above "work together" in pairs; one "answers" a "question" posed by the other:

Request to Send (near end)	**Clear to Send** (far end)
+ Can I send to you?	+ Me? Sure, go ahead!
− Never mind . . .	− Me? No way! I'm busy.

Data Set Ready (interface equipment)	**Data Terminal Ready** (terminal)
+ I'm ready to send (receive).	+ I'm ready to receive (send).
− I'm busy right now—wait!	− I'm busy right now—wait!

The second pair of signals could use some explanation. A **data set** is usually a modem or some other interface between RS-232 and another communication link. A **data terminal** is a piece of station equipment we have already discussed. Suppose that your terminal has a keyboard and prints on paper, and it connects to a telephone modem through an RS-232 connection. Let's illustrate the "handshake" that goes on between the data set and the data terminal via these two signals. We can imagine it in the form of a dialogue something like this:

Checking Data Set Ready
Terminal: My human has just typed the letter "A" and I have digested it into RS-232 serial format. Before I send the serial bits to that modem over there, I better ask it if it's busy at the moment. "Excuse me—you, over there at the data set . . . ?"
Modem: Yes?
Terminal: I'd like to know, can you put this letter "A" onto the telephone line right now?

Data Set Ready—NEGATIVE
Modem: Sorry, right now, I'm receiving incoming bits from the distant end and can't send your letter "A".

Data Set Ready goes POSITIVE
Modem: There! I'm finished receiving those bits. It's a letter "Q" from the distant end. Now I can take your letter "A" and transmit it.

Checking Data Terminal Ready
Modem: Now I'm sending your "A". By the way, if you're able to handle it, I'd like to send you the letter "Q" that I just got from the distant end . . .

Data Terminal Ready—NEGATIVE
Terminal: Oh, too bad—my typewriter is in the middle of printing the letter "A" right now. It's the one my human just typed; I've got to do it or he'll pound on my keyboard. I don't have a buffer where I can store your letter "Q". Can you wait until I finish this "A"?
Modem: O.K. Let me know when you're ready, and I'll give you the "Q" to the print next.

Data Terminal Ready goes POSITIVE
Terminal: Hello? I just got done printing the "A". Do you have the "Q" for me now?
Modem: Sure. Here it is.

. . . And all of this is going on within a few hundredths of a second, as two bytes of binary data are being exchanged!

Clock signals (used only with synchronous data). You are not likely to encounter these unless synchronous transmission is used. Synchronous transmission does not depend on data bits to always be of a constant duration. A "clock" signal identifies when each new bit starts:

Pin	Name	Symbol
15	Transmitter Clock	TC
17	Receiver Clock	RC

A transmitter clock is an output the near end uses to tell the distant end when it is starting to send another bit on the TD lead. The receiver clock is an input that is the near end's way of finding out when the distant end is sending another bit on the RD. The near end's transmitter clock is the distant end's receiver clock, and vice versa.

Other signals. These signals are not used consistently, and probably will not appear on most interfaces. They are provided without further explanation, on the off chance that you encounter them in some equipment and need to know what they're called.

Pin	Name	Symbol
9	Positive Test Voltage (Testing only)	
10	Negative Test Voltage (Testing only)	
11	Equalizer mode (Bell modems only)	
12	Secondary Data Carrier Detect	(S)DCD
13	Secondary Clear to Send	(S)CTS
14	Secondary Transmitted Data	(S)TD
16*	Secondary Received Data (Bell modems: Divided Clock Transmitter)	(S)RD
18	Divided Clock Receiver (Bell modems only)	
19	Secondary Request to Send	(S)RTS
21	Signal Quality Detect	SQ
23	Data Rate Selector	
24	External Transmitter Clock	
25	Unassigned, sometimes used as a "Busy indicator."	

22-1.6 The "Null Modem"

A commonly used piece of equipment associated with RS-232 interfaces is called a **null modem** or **null modem cable.** The words "null" and "modem" imply that it is not a modem, and this is true. Most RS-232 equipment is documented as though the manufacturers expect you to attach it to the telephone system through a modem. A computer with an RS-232 interface might want to use a printing terminal with an RS-232 interface as a printer. As the manufacturers originally intended it to work, you would connect the computer to a telephone modem, call up the printer, and it would receive data through its telephone modem. But what if the printer is in the same room as the computer? It would seem wasteful to place a telephone call to yourself just to hook up the computer to the printer, and besides, you might not have two phones! A cable connecting the two RS-232 plugs together would seem to be the answer. Would a straight extension cord do the job?

A cable that connects pin 1 to pin 1, pin 2 to pin 2, and so forth, would not work. Both the near end and far end (the computer and the printer) would be sending signals out of pin 2 (TD) and waiting for signals to come in at pin 3 (RD). The telephone company takes care of this when they put the data that goes in the near end's mouthpiece through to the distant end's earpiece. The near end's transmitted data becomes the far end's received data automatically, because the phone company transfers the signal from the near end's transmitter to the far end's receiver, and vice versa. Without a phone company in the middle, the cable will have to take care of these reversals itself.

The null modem cable reverses, or swaps connections, for the "paired" signals we just described. Figure 22-8 shows one arrangement for a null modem. The computer's transmitted data (TD) output goes to the received data (RD) input of the printing terminal. The computer's RS-232 ground is connected to the ground on the printing terminal, directly. "Data Set Ready" and "Data Terminal Ready" are swapped, to eliminate the need to wait for data conversion at a telephone line (there is no telephone line or modem to wait for, now). That's basically all there is to a null modem.

The null modem is a handy way of connecting two pieces of station equipment via RS-232, without a modem and a telephone company. Certain companies will charge you $40 or more for a

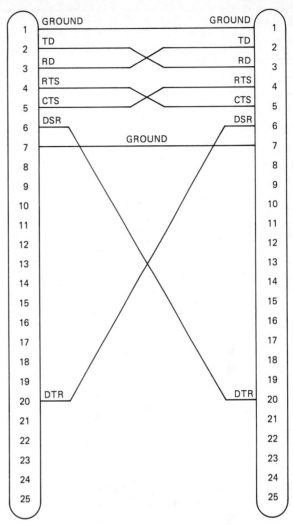

FIGURE 22-8 Null modem.

null modem, but all it contains are two $5 DB-25 plugs and a dozen-or-so wires. You can make one yourself for about a quarter of the catalog price, as long as you know what goes on in the RS-232 interface of your computer. As in other parts of high-tech, what you *don't* know can be very profitable—for somebody else.

A few words about the RS-449 interface. Another popular interface, similar in nature to RS-232, is RS-449. Instead of a 25-pin plug, a 37-pin connector is used, similar to the DB-25. RS-449 is designed for higher-speed transmission. Each signal pin has its own return line, instead of a common ground return, and the signal pairs (signal, return) are balanced lines, rather than a signal referenced to ground. This reduces noise and crosstalk, problems that we describe later in this chapter.

EXAMPLE 22-1

How long does it take to transmit a 5000-word text file via telephone if the transmission takes place at 2400 baud, using characters which are 7-bit ASCII with odd parity, with one stop bit? Assume that a text word is six characters—five characters plus a space before the next word.

Solution: The number of characters transmitted is

$$5000 \times 6 = 30,000$$

The number of bits per character is

$$10 = 7 \text{ (ASCII)} + 1 \text{ (parity)}$$
$$+ 1 \text{ (start)} + 1 \text{ (stop)}$$

The number of bits transmitted is

$$30,000 \times 10 = 300,000$$

At a rate of 2400 baud = 2400 bps, the time it takes to transmit 300,000 bits is

$$\frac{300,000}{2400} = 125$$

So, the entire text file is transmitted in 125 seconds, or 2 minutes, 5 seconds.

22-2 DC AND AC SIGNALING AND BANDWIDTH

We're going to look at some examples of dc and ac transmission of data on transmission lines (wires) to see the dc and ac characteristics of common carriers that might be used for digital data transmission.

22-2.1 DC Attenuation

For our example, we're going to go back to the nineteenth century and look at how a telegraph works [Figure 22-9(a) and (b)]. The transmitter is a *key*, which is just a switch, and the receiver is an electromagnet which makes its moving armature produce a "click" when it's magnetized (this is called a *sounder*). Between them is a length of wire intended to carry the signal long distances, and a return path for the signal represented by a ground symbol at the transmitter and receiver. (To reduce the cost of copper wire, the ground return path was actually a metal stake driven into the ground in early telegraph transmitters and receivers. The ground is a reasonable conductor—if the stake is driven several feet deep—there is less than 100 Ω resistance between stakes several miles apart.) The resistance of the copper wire gets larger as the wire gets longer. To transmit a long distance, the sounder coil had to be made of enough windings of wire to get a majority of the source voltage; it had to have more resistance than the transmission line.

It was not the invention of the key and sounder—or even of the Morse code—that made Samuel F. B. Morse a rich man. All of those were invented by his predecessors in early telegraphs going back to the 1820s. What made Morse's system practical was the *repeater*, shown in Figure 22-9(b). Consider what happens if you want to send a signal to a sounder coil wound with 100 Ω of wire. Suppose that the transmission line is No. 18 copper conductor, the same size as household lamp cord. That gauge of wire has about 20 Ω per thousand meters. By the time we sent the signal down 5 km of wire, only half the original voltage would be delivered to the sounder coil. We need either a larger sounder coil to go farther, or thicker wire. To go 10 km using the same sounder at the end of the line, we'd need to use No. 15 wire (twice the cross-sectional area). To go as far as Toronto is from Chicago (about 800 km), you would need No. 0000 gauge wire (which is about 1 cm thick). The wire would be as thick as a finger, and mass just a tad over 3 billion kilograms for the entire conductor (over $61 billion worth of copper pennies!). This is obviously out of the question, and a wire from New York to San Francisco would be five times as long and five times as thick in cross-sectional area, so it would cost 25 times as much—if there's that much copper in the world! Yet the Morse telegraph system had linked New York with San Francisco by 1869. How?

What Morse invented was a way to keep the same thickness of wire over longer and longer distances by "refreshing" the signal with a fresh battery every few kilometers. Morse's repeater is what we would call a *relay* today. The telegraph system was the first *digital data transmission* network (using pulse-width modulation—dots and dashes—to represent two logic states) and the repeater was the first *signal amplifier*.

You can see that Morse's idea was very simple. Instead of a sounder, the 100-Ω electromagnet at the end of 5 km of wire is an electromagnet that closes a key (a pair of switch contacts). When it receives its signal (which has lost half its voltage) from the first segment of line, it completes a circuit that puts current into a second 5-km wire from a fresh battery. At the end of the second span of wire, a relay electromagnetically closes

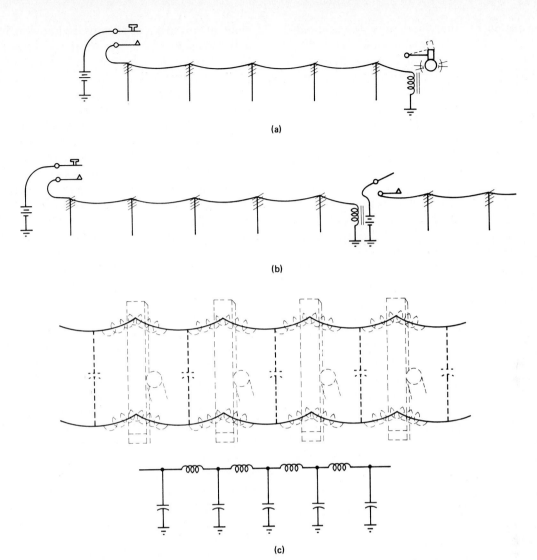

FIGURE 22-9 (a) Simple telegraph; (b) with repeaters; (c) equivalent circuit for parallel transmission line.

the contacts of another key after half the signal's voltage has been lost, and so on to the other end of the line.

Dc attenuation (loss of signal strength) is a characteristic of all data communication by wires. It is an automatic result of the fact that the longer a wire is, the more resistance it has. Whether it's Morse's telegraph or the telephone system, there must be amplifiers placed at intervals along the transmission line to repeat the signal with fresh power.

22-2.2 AC Attenuation

The ac characteristics of an open-wire transmission line are shown by Figure 22-9(c). The two wires going from pole to pole are shown with distributed inductance and capacitance. A pair of lines is used for the current loop in this system

instead of one wire and a stake driven into the ground. This avoids a certain "mixing" of signals that might occur if all voice circuits shared a common ground conductor. In the picture, you can see that, between the two parallel wires, a distributed capacitance between the lines appears as "dashed capacitors" in parallel between the lines. The lengths of wire themselves have inductance (represented as a "dashed coil" in series with the wires) that opposes changes in current through the wire. The collective inductive and capacitive characteristics of the circuit are shown below the picture of the wires on the poles as a schematic. The schematic is clearly an infinite number of "pi-section" filters of the type used in dc power supplies to filter ripple out of the dc output (the dc resistance is still there, but we've already discussed its effect). In a power supply, this is good. The low-pass filter gets rid of the "ripples" from a pulsating dc rectifier, and provides a steady, constant dc level. On a transmission line for digital data, or any kind of ac signal, this is bad news. The "ripple" is the information, and the longer the wire pair is, the more of a *low-pass filter* the circuit becomes. This means that there is more and more filtering of the ac signal as the line pair gets longer and longer.

Figure 22-10 shows the effect of this low-pass filtering on a digital square-wave transmission. At the top, a square wave is transmitted through a circuit with a negligible attenuation, because the RC and L/R times are very short compared to the wavelength. Traveling downward, we see the effects of increasing C and L. Both increases cause longer and longer time delays between the time the voltage starts to rise or fall and the time it gets there. By the fourth diagram down, the signal has been affected so much that its amount of rise and fall between waves is not enough for a digital gate to detect. The effect of differences in frequency is also shown. The pulses get closer and closer together as you go across the waveforms from left to right. This is equivalent to increasing the frequency. In the "short" transmission line at the top, the delay time only affects the very highest frequencies, but as the transmission line lengthens, serious attenuation of the signal happens at lower and lower frequencies. At the bottom, there's hardly anything left, even at the lower frequencies (pulses far apart). This is why we call the circuit characteristics of the open-wire line pair a low-pass filter. Only the low frequencies pass through; the higher frequencies are lost.

How much high-frequency attenuation is acceptable? What can we do about it? Each repeater in an analog system like the phone company's voice lines amplifies the signal to replace lost power, and can also boost the high-frequency attenuation. This process is not perfect, and the 300- to 3000-Hz bandwidth of ordinary voice lines is a result of the length of lines between repeaters and the characteristics of the amplifier at each repeater junction. To answer the question "How much attenuation is acceptable?" we must say, "as much as we can have without making spoken words impossible to understand." Actually, a 300- to 3000-Hz bandwidth provides voice communication that can be easily understood, which is less attenuation than the worst case. It is not, however, anywhere close to covering the entire frequency range of human hearing. To get larger bandwidth and handle higher frequencies, shorter segments of wire between repeaters would be needed. The voice communica-

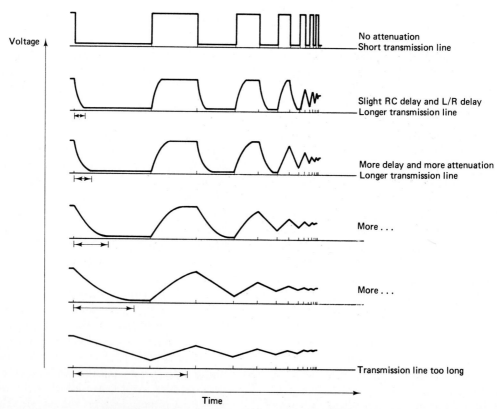

FIGURE 22-10   Effects of ac attenuation and distortion.

tion system was designed for the frequencies in the human voice. Leased lines that have better characteristics are more expensive because there are more repeaters, and the repeaters require higher-quality components. Circuits that permit data transmission at 9600 and 19,200 baud are not at all unusual, but you cannot transmit at these frequencies with ordinary voice lines.

22-3 TRANSMISSION MEDIA AND TECHNIQUES

When we used the word "transmitter" in describing part of a UART, you might have thought of a radio or television broadcast transmitter. That is certainly one kind of transmitter, although a "wireless" transmitter (as the British would call it) is hardly the only kind of transmitter that's possible. In fact, the telephone handset is also called a transmitter, when you're talking into the mouthpiece. In this section, we see some of the ways a transmitter can transmit.

22-3.1 Frequency-Division Multiplex

In an earlier section of this chapter we saw that space-division multiplexing (SDM) uses a separate wire for each signal being transmitted, and it is where the signal appears that tells you what it is. Time-division multiplexing (TDM) permits transmission of multibit numbers one bit at a time on the same channel. The time when a signal appears tells you what it is. Frequency-division multiplexing (FDM) is the technique used to separate channels of television or radio stations from one another. Radio transmitters use the same medium (the electromagnetic environment) to carry any different channels simultaneously.

There is absolutely no reason why this technique must be limited to "wireless" broadcast transmission. Electromagnetic broadcast allows you to use one antenna to transmit or receive many signals simultaneously. The total electromagnetic spectrum is divided into *bands* which are used by different types of transmitters, but there is really only one electromagnetic medium, and everyone shares it. If two channels want to transmit at the same frequency to someone in their vicinity, the receiver cannot separate the two channels of information. This problem can be solved by a combination of FDM and SDM. If wires or other conductors are used that confine the transmitter's signal into a "tight beam" so that it doesn't "slop over" into somebody else's

signal, the two channels of information can be separated even though both are transmitted at the same frequency. In fact, if FDM is used, a single wire can carry many hundreds of channels with different signals on them, and another wire nearby can carry more hundreds of channels (using the same frequencies, if need be) without any conflict. Transmission of multiple channels across a single physical link can be done by TDM or FDM.

The amount of information that can be transmitted on a physical link depends on the variety of frequencies that can be used without confusion. This variety depends on the *bandwidth* of the link. Each channel requires a band of its own, so that information can be carried on the signal without overlapping another channel that can't be filtered out at the receiver. The total range of frequencies that can be transmitted on the link will determine how many bands will fit between the lowest and highest frequencies the link can handle.

22-3.2 Modulation and Demodulation

In the preface to this book, we said that the original audience to whom this course was presented had already studied analog electronics before being introduced to digital electronics. It's not fair to assume that everybody who's reading this already understands modulation. We've already mentioned modems as modulators and demodulators, without any real attempt to explain what those terms mean. Since this is a digital electronics book, we feel justified in "copping out" a bit and not going into detail on modulator and demodulator circuits—which are not digital circuits anyway—but we do feel a brief description of the various methods of modulation is in order.

We are not going to try to answer the question "Why modulate?" We'll assume from past examples in this chapter that you understand the problems that happen if you just "dump" voice or digital information onto a wire or other medium that's already carrying a signal.

Modulation combines the low-frequency information you want to transmit with a higher-frequency *carrier* signal. The idea of frequency-division multiplexing should already be familiar to you as a way of carrying more than one signal on the same medium. Now there is still the problem of how you combine the low-frequency information with the carrier. We see, in the modem for telephone communication, a shift in the carrier frequency when the level of the low-frequency signal changes.

There are three parameters that are part of the signal produced by a carrier-wave oscillator. The wave produced by the transmitting circuit has amplitude, frequency, and phase. Any one of these three items can be changed to represent a change in the level of the low-frequency information signal.

If the low-frequency changes cause a corresponding change in the amplitude (the peak-to-peak voltage) of the transmitting circuit's waveform, we call this **amplitude modulation** (AM). In a sense, the changes in the dc voltage of the voice or digital signal cause a corresponding change in the ac voltage being transmitted. The receiver demodulates by receiving an ac signal at the carrier frequency and converting it to dc.

Frequency modulation which we already saw in an example) uses shifts in the dc voltage of the low-frequency signal to change the frequency that is transmitted. An FM (frequency modulation) modulator is a voltage-to-frequency converter. Frequency-shift keying (which we used in our modem) is just a specialized example of FM. To demodulate the FM signal, we do a frequency-to-voltage conversion, making variations in the frequency back into a changing dc level.

Phase-shift modulation (PSM) uses changes in the low-frequency audio or digital signal to shift the phase angle of the carrier wave. The amplitude and frequency of the transmitted wave do not change, but the phase angle shifts like the frequency shifts in FSK. A *phase-locked loop* (PLL) is one circuit that can be used to detect changes in the phase of the received signal and demodulate them by converting phase to a dc voltage.

In Figure 22-11, three diagrams show how a square-wave digital signal would affect the transmitted wave when the amplitude, frequency, and phase are modulated by the bits 1, 0, and 1 (transmitted serially).

Frequency modulation has supplanted amplitude modulation for most data transmissions because it is more noise-immune than AM. Phase-shift modulation is very similar to frequency modulation, but does not require as large a bandwidth to transmit the same digital information. This means that more channels of PSM can be transmitted on a band than would be possible with FM. At the time of this writing, FM is still more popular than PSM, but we don't think that can last forever, with one-chip PLL circuits being produced cheaply.

22-3.3 Transmission Media

We have already determined that the number of simultaneous channels that can be carried on a physical link (a *medium*) depends on how far it is between the lowest frequency and the highest frequency that can be used (the bandwidth of the link). We'll look at some media to see what methods are used, and how the method relates to the number of channels.

"Wireless" transmission. Radio broadcast (nonmicrowave) transmission (frequencies less than 1 GHz) spreads out like ripples on a pond. The number of frequencies possible is limited by the distance a transmitter can broadcast. Outside one transmitter's range, another can transmit at the same frequency. A receiver can't receive both transmissions, though, so only one channel can occupy each band. The number of messages that can be broadcast can be increased, though, if we transmit at higher frequencies, like . . .

Microwave (frequencies in the gigahertz range). The microwave broadcast is a directional transmission, like a beam of light. It is possible for a receiver to pick up microwave "beams" from different directions at the same frequency. This is a little bit like combining frequency-division multiplex with space-division multiplex. As long as the transmitter is in a line of sight, the receiver can pick up its signal, and the communication to the receiving station can contain as many channels as the microwave transmitter can fit. Changing the direction of the receiving antenna can allow you to receive a whole new set of channels on the same frequencies. Since the microwave "beam" is like a beam of light—not

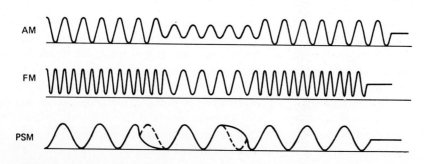

FIGURE 22-11 Three types of modulation.

"visible" beyond the horizon—the taller the tower is that carries the transmitting antenna, the farther the distance the transmission can carry. Microwave repeaters can relay the signal from one tower to another just as the repeaters on telephone and telegraph lines relay signals from one segment of transmission line to another. The number of repeater stations gets smaller as the towers get taller (and the signals can "see" farther and farther around the curve of Earth). Ultimately, we can relay microwave signals using an *extremely* tall tower, like a . . .

Satellite. If a satellite is placed in an orbit at an altitude of 37,000 km (23,000 miles), its orbital period is 24 hours. If placed in an orbit above the equator going in the same direction the earth is turning, the satellite will remain at the same point in the sky all the time, because the satellite will be in sync with Earth's rotation (such a satellite is called *geosynchronous*). It will then serve as an "orbital platform" from which microwaves can be relayed just like any other microwave tower (except that it's a couple of hundred thousand times taller). Because it's extremely high, this "tower" can "see" extremely far. With three satellites placed in the same circular orbit 120° apart, the whole earth can be covered, provided that one satellite can "talk" to another.

Early in the space program, passive satellites, which were nothing more than microwave "mirrors," were placed in *low earth orbit* (LEO). Radio signals were bounced off these satellites and picked up by ground stations as a way to "relay" information over long distances. The problem with this was that the satellite in LEO (450 miles up, or so) didn't stay in the sky very long. Like other microwave relay stations, the satellite was useless once it passed over the horizon. As many as 20 or 30 satellites would be needed in a LEO to keep one overhead all the time. Why didn't the users of these satellites use lower frequencies that could carry "over the horizon"? The answer is that a layer of the atmosphere called the *ionosphere* (or Kennelly–Heaviside layer) reflects low-frequency radio waves. High-frequency waves (like microwaves) penetrate the ionosphere. It's the reflection of the radio waves that enables them to "bounce" around the horizon. Unfortunately, the same reflection makes it hard for the radio waves to penetrate beyond the space between the ground and the ionosphere where they must go to reach the satellite.

Are the geosynchronous (24-hour orbit) satellites passive reflectors like the early ECHO I (which was just an aluminum-plated balloon)?

The answer is "no," and in fact, the transmission to the satellite (*uplink*) is not the same frequency used to receive from the satellite (*downlink*). There's a very good reason for this—in the form of a natural satellite (the moon)—because reflected signals from a passive reflector come back at the same frequency they're transmitted. With the moon out there, there'll always be reflected signals coming back at the transmitting frequency. To separate these from the signals desired from the geosynchronous satellite (the moon does not go around the earth in 24 hours!), the satellite repeater's downlink frequency is different from the uplink frequency, and it's arranged so that nobody ever uses an uplink frequency for a downlink.

Wires and other conductors. Frequencies like those used for microwave transmission have characteristics like light waves. At these frequencies, the electromagnetic vibrations can be shaped into narrow beams and focused by reflectors, for instance. Like light waves, also, microwave beams are blocked by buildings, trees, and don't transmit well when an airplane is flying through the beam. Unlike light waves, microwaves and other radio frequencies can be carried through electrical conductors. For the lowest electromagnetic frequencies, ordinary wires may be used. At higher frequencies, significant amounts of energy radiate from the wires between the transmitter and receiver. The loss of power from the signal is bad enough, but that lost power gets into "the airwaves" and interferes with radio receivers near the wires, which is even worse.

Special geometries (shapes) for the conductors help to confine the signal within the conductor. A *coaxial* conductor [see Figure 22-12(a)] uses a hollow outer conductor with a wire in the center, separated by an insulator. Conductors with coaxial geometry can transmit much higher frequencies than a wire pair. Coaxial conductors do not radiate much power, and in a cable with many coaxial tubes, there is very little "crosstalk" between conductors. The current flowing in the coaxial tube is mostly on the inner surface of the hollow conductor and on the outside of the center wire. The outer part of the hollow conductor acts as a Faraday shield, and prevents radiation of signal and penetration of that signal into other conductors in the cable.

For even higher frequencies, especially microwave frequencies, *waveguides* with a rectangular cross section are used [see Figure 22-12(b)]. With higher frequencies and wider bandwidths

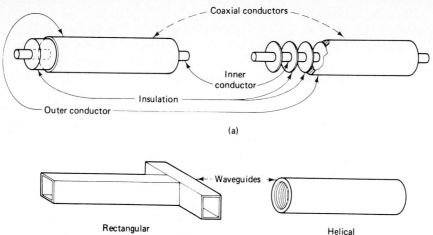

comes a higher capacity for carrying information. Circular waveguides are available. One type has a helical geometry, with a conductor in the form of a spiral (a helix) forming the inner surface of the tube. These waveguides look like pipes (in some cases, square pipes), and are put together by techniques more akin to plumbing than electrical wiring. Any sharp bends in the waveguide become sites of severe power loss, so waveguides must follow gentle curves, or be connected around corners with special couplers. Although these conductors are much more expensive than ordinary wires, they can carry far more information, because they have more bandwidth (at the higher frequency) and can handle more channels.

EXAMPLE 22-2

A satellite microwave link transmits multiplexed voice and data in digital form. One channel of voice is transmitted at 64,000 bps. (The analog signal is digitized into 8-bit serial binary code before transmission.)

(a) Is this sufficient to reproduce voice-quality sound, with a bandwidth of 300 to 3000 Hz?

(b) Is this sufficient to reproduce high-fidelity music, with a bandwidth of 20 to 20,000 Hz?

Solution: At 64,000 bps and 8 bits per sample, the number of samples transmitted per second is

$$\frac{64{,}000}{8} = 8000$$

To reproduce a cycle with a positive and a negative alternation, two samples are needed, minimum. This means that the rate at which this equipment can transmit, in cycles per second, is half of the samples-per-second rate, namely,

$$\frac{8000}{2} = 4000$$

So the maximum frequency audio that can be transmitted (4000 cycles per second, or 4000 Hz) is sufficient to reproduce voice-quality sound of 300 to 3000 Hz. Answer (a) is "Yes."

The sampling rate is *not* sufficient to reproduce high-fidelity sound with a bandwidth of 20 to 20,000 Hz. Answer (b) is "No."

Fiber-optic light pipe. At higher and higher frequencies, the transmission of electromagnetic (radio) waves looks more and more like the propagation of light. Why not use light as the transmitting medium? For broadcast in the open air, this would not be a very good idea. Clouds, small birds, rain, and dust in the air all get in the way of light beams. The same problems (to a smaller degree) occur in microwave links, and the best solution is to use coaxial cables and waveguides (which don't allow things to block the beam and can go around corners). Light can be transmitted through a sort of "waveguide" called a **light pipe.** Its working principle is total internal reflection, which is something that happens at the boundary between two dissimilar transparent materials. The speed of light in glass or silica is slower than its speed in air (about two-thirds as fast in silica as air). When light traveling through silica strikes air, if it hits the boundary between the silica and the air at a shallow angle, all the light that strikes the boundary is reflected back into the silica, and none of it ends up as a beam in the air.

This is a better reflector than the best metal mirror that can be made. A pure silver shiny-metal reflector reflects about 93% of the light that strikes it, and absorbs the remaining 7%. Total internal reflection is necessary to "trap" a beam of light in a light pipe (see Figure 22-13). Suppose that we have a silica fiber 0.1 mm in diameter and 1 km long. If the light loss at each reflection is 1 part per million, and the light strikes the silica-air boundary at a shallow angle so that it rebounds and strikes the other edge 1 mm down the fiber, there will be 1,000,000 reflections along 1 km of silica. At 1 part in a million loss per "bounce," there will be only 36% of the light left at the end of a kilometer. If 10 parts per million is lost at each bounce, only 45 millionths of the original light makes it to the far end of the fiber. Real mirrors lose 7%, at best. After 200 reflections (2 meters), less than half a millionth of the original light would remain. At the end of 1 km, nothing measurable would remain. Fiber light guides that use total internal reflection lose less than half the signal's power in 6 km. What loss there is happens because the silica isn't perfectly transparent, and has impurities in it. At the reflection, however, there must be 100% reflection—or at least 99.999999 ... with more nines than we'd like to count—otherwise practically nothing would arrive at the far end of the fiber. The silica used in optical fibers has exceptional transparency. To carry signals more than 1 km, the clarity of the silica must be better than (smoggy) city air.

Light frequencies are around 10^{15} Hz. This covers the range from infrared to visible light. Frequencies like this (10,000 times microwave frequencies) suggest a correspondingly higher capacity for multiple channels carried on a single fiber. Since the fibers are thin, and each thin fiber is quite flexible, a bundle of thousands of fibers can be made into a fiber-optic cable with flexibility as good as stranded-copper wire cables. One advantage of fiber-optic cables is that only one strand is needed for each "circuit," whereas copper-conductor cables need two wires for each line.

Optical (light) energy has the potential to be used for every logical and analog operation carried out in electrical circuits, and has many advantages.

We can conclude this discussion of fiber optics by comparing fiber optics (for carrying signals) to copper wire.

A. Places where silica fiber has an advantage over copper wire are:
    1. Abundant raw materials (sand) to make fiber
    2. Light weight
    3. Number of conductors needed per channel (one)
    4. Speed (Nothing faster than light—Einstein)
    5. Impervious to chemical corrosion
    6. Immune to electromagnetic interference
    7. Difficult to wiretap
    8. Signal cannot be shorted by water leakage
    9. No shock hazard
    10. Does not radiate electromagnetic energy
B. Places where copper wire has an advantage over optical fiber are:
    1. Interconnection (soldering, etc.) cost
    2. Fabrication cost
    3. Ductility/malleability/resistance to breakage

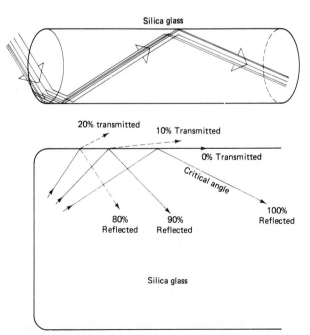

FIGURE 22-13 Fiber-optic light pipe.

22-4 PULSE-CODE MODULATION

The modem was a device used to transmit digital signals in analog form, permitting multiple channels to use the same line by frequency-division multiplex. It is possible that one day in the future, the telecommunications industry may use a different method of transmission. Multiple channels will be combined on a single line using time-division multiplexing and transmitted as serial digital data instead of analog. In this case, analog signals will need to be converted to digital form using pulse-code modulation (PCM). Digital signals will not need to be converted to anything at all. They will be "ready to go" as they are.

Instead of the digital users having to convert their data over to analog form for transmission, analog users will have to convert their data to digital form. To do this, they will need a device called a codec (coder/decoder). This is a D/A and A/D device that will sample the analog waveform at some moment; convert its momentary voltage level to a digital number, insert the number into the bit stream "when its time comes" (time-division multiplexing) and return to sample the analog waveform another time. This is the "cod" part of the codec. The "dec" part reverses the process, reconstructing the channel's analog waveform from the digital numbers that come along in the receiver's time slot.

This method's strong point is the fact that digital repeater can reconstruct the data pulses it received, "cleaning up" noisy pulses or pulses which have degenerated due to line distortions. Analog repeaters amplify the noise along with the information, since they must be able to reproduce all of the waveform, and can't "decide" what part of the waveform is "unimportant." Digitally transmitted signals, since they are repeatedly reconstructed into clean pulses, can be transmitted through a signal path that distorts them badly, yet still be recognized by the repeater and reincarnated as fresh, clean pulses. Another advantage of PCM transmission is that a higher frequency of transmission can be used than with analog transmission, because the distortion at the higher frequency doesn't matter very much.

PCM is likely to take over from analog methods because of its higher reliability and the fact that there is continuing reduction of price-per-gate in digital systems. Multiplexing and switching of digital signals is simpler, and uses simpler and less expensive circuitry, than for analog signals.

22-5 IMPERFECTIONS

In this, the last section of the chapter, we'll define the main problems that "mess up" digital telecommunications. Since we are interested in communicating digital data, we might conclude after looking at the telecommunications system that exists today, that it wasn't designed for us. In a very real sense, that is true. The telecom-

munications system was designed for voice communications originally. As you can see in the section just concluded, this is changing, and will change more in the future. Eventually, someday, modems will be obsolete and codecs will be necessary in "old-fashioned" analog communications. In the meanwhile, there are some problems that are inherent in any system, digital or analog, and some problems that plague analog systems more than digital ones. There are even a few problems that cause more trouble in digital systems than analog ones (not very many).

Since analog systems will be with us for years to come, we'll look at analog system design features that have to be "corrected" by clever engineering, as well as authentic transmission problems that afflict all transmissions, digital or otherwise.

22-5.1 Noise

Noise is a term that refers to any electrical signal you don't want, that "creeps in" on top of the information that you're transmitting and superimposes itself on your signal.

RFI. When a magnetic field moves across a conductor, it induces a voltage in the conductor. Stray magnetic fields cross transmission wires all the time, and add noise to the signal on the wire. As we already said, an analog system amplifies the noise with the signal at each repeater. If a certain amount of noise is picked up randomly on each mile of wire, the noise added to the signal will get larger as the signal travels further and further. Coaxial cable and waveguides are naturally shielded against this type of **RFI** (*radio frequency interference*) **noise,** and are a much better choice than open-wire pairs for long-distance transmission. Consequently, open-wire pairs are generally used only in the local area between the telephone user and the local office. Trunk lines will be coax or waveguides (and now, some fiber-optic lines with repeaters are being used). Since the silica used in fiber-optics light pipes is an electric insulator, it is totally immune to RFI.

Thermal noise. Another source of noise is the random motion (Brownian motion) of electrons in the wires themselves. This is called **shot noise** or **white noise.** There is also a similar type of noise caused by molecular and atomic vibrations within the semiconductor material. It is produced by thermal energy in the crystal lattice, and since it's in the amplifier material, it gets amplified. There's really nothing that can be done about it except to cool down the semiconductors (expensive) or tolerate a certain amount of it. White noise or thermal noise can be heard as a hissing sound in the background of all electronically produced audio.

Crosstalk. Signals on other lines or communications links that penetrate into your conductor and "walk" on your signal are called **crosstalk**. If you have a neighbor with an overpowered and badly adjusted CB that "gets in" on top of your TV signal and messes up your reception, that's an example of crosstalk in the broadcast area, although you might think of other things to call it.

The clicks, beeps, and occasional snatches of someone else's conversation that you sometimes hear on the telephone are crosstalk. Usually, the problem is caused by induction between one wire and another running alongside it, although a wire carrying multiple channels can have crosstalk if the channels are not effectively separated on the receiving end.

High-voltage spikes on lines caused by electromagnetic switching circuits or the ring voltage used to ring a phone are high-level signals that may penetrate into your circuit even though the lower-level voice signals do not.

Although the sources of noise on transmitted data may vary, the results are the same. As the noise accumulates on the signal waveform, it becomes harder and harder to tell the signal from the noise, and eventually the data are lost in the noise. Digital transmission of data using digital repeaters remedies this problem, since repeaters can clean up small amounts of noise on a digital signal, and do not have to keep amplifying the noise along with the signal as it is passed from repeater to repeater.

22-5.2 Distortion

When the waveform transmitted down a line changes or loses its waveshape due to characteristics of the transmission line itself, we call the results **distortion**. One example of distortion was already discussed earlier in this chapter, when we saw how an uncompensated transmission line acts as a low-pass filter (attenuation distortion). Coils called *loading coils* are normally added to the transmission line at intervals of every few kilometers to compensate for this effect, but they may fail, or not be perfectly matched to the impedance of the transmitting and receiving equipment. Changes in phase or amplitude of the transmitted signal can arise in the amplifier (re-

peater) circuits along the line. Poor electrical contact can give a "noisy" character to signals transmitted down a wire.

Signals that contain a number of frequencies (and all digital signals do) may travel down lines that do not conduct all frequencies at the same speed. If parts of the signal arrive at a different time than others—for example, if the mark frequencies of an FSK signal get ahead of the spaces—total garbage is received. Equalizing the analog telephone circuit for this type of distortion is more complicated than equalizing it for frequency response, and the solution is to either convert to digital transmission and let closely spaced repeaters on the line correct the problem every time the signals begin to drift out of step, or else slow down the data transmission rate on an analog link to a rate where the differences in delay times are negligible compared with the time used to transmit each bit.

Intermodulation distortion is caused by "mixing" of two frequencies in a FDM multiplexed line to form a third frequency. If this frequency happens to be at or close to the frequency of another channel, IM distortion (actually a form of crosstalk or noise) gets into the "innocent bystander" channel.

22-5.3 Line Failure

This is one of the things that happens to cause the computer to "hang up" on you in the middle of a transmission. If it happens, another circuit will be needed. This is basically a reliability problem. Temporary loss of transmission capability is called a **dropout**. Loss of signal that lasts until a repair crew fixes the fault are called **line failures**. Failures in the modem, terminal, software, or power are not, strictly speaking, line failures, but they have the same effect. The computer you're talking to "hangs up" unexpectedly.

Suppose that an airplane flies through the microwave beam of a relay tower. This will cause a dropout that may be interpreted by your terminal as a "hangup" signal. It isn't good for the signal or the pilot of the plane, but the pilot knows this, and isn't likely to do it often unless he's a dropout or has bad hangups.

Outages are the long-term nasties that happen when the earth-moving vehicle plays pick-up-the-cable, or Mother Nature decides to use 10 MV of lightning for a bell voltage. The best thing to do in these cases is have an alternative path available—that's the common carrier's problem—which is usually the case when the failure is on an intertoll trunk. For the most part, though, the outages happen at the local loops, and there's not a lot that can be done until the repair crew is finished.

22-5.4 Data Errors

As the frequency of data transmission (baud rate) goes up, the chance that a short glitch will cause the loss of a bit—or several bits—goes up, too. We mentioned earlier that one way to handle losses of the one-bit variety was **parity checking**. For more-than-one bit losses, methods that take advantage of redundancy of information are more valuable. As far back as the old telegraph days, important information like dollars-and-cents figures was repeated twice, whereas alphabetic information could be read easily enough even if the odd letter here and there was wrong.

For the most part, the telegraph technique was "ignore the errors, but repeat the important stuff." Special error-detecting codes have been devised. The general approach to handling an error is to have the receiver say "Eh? How's that again?" and cause the transmitter to retransmit the data block that contained the error.

22-5.5 Delays and Blocking

Data are transmitted and received after a delay called the **propagation delay**. This is the time it takes the signal to travel from the transmitter to the receiver. If half-duplex transmission is used, each end of the line "takes its turn" and then the other end transmits. The propagation delay determines how fast this "turnaround" can take place.

On telephone lines, the speed of transmission is less than the speed of light. In the case of satellite communications, transmission is at nearly the speed of light, but has a long way to go. To uplink and downlink to a satellite in geosynchronous orbit takes 270 ms (74,000-km round trip). To retransmit a block when an error has been detected, the satellite link is going to take a good deal longer than 270 ms to handle the "handshaking," and the receiver and transmitter will have to be able to "keep track" of what block of information was transmitted long after the transmission is over, so that if a "retransmit" request comes along after a long propagation delay, the transmitter will remember what block has to be retransmitted.

There aren't really any "magic" solutions to propagation delay time. The data transmitters and receivers will just have to be "smart" enough to work around the long delays involved in satellite communications and long cables.

22-6 TOMORROW IT WILL ALL BE OBSOLETE (OPTICAL COMPUTING)

It is entirely possible that electricity may become obsolete for computing and telecommunications in the foreseeable future. Optical switching elements (one type of optical transistor is known by the Star-Trekkish name of "transphasor") can switch on and off, latch data, and be used as logic gates, with signal delay times of 1 picosecond (pa). Josephson-junction switching, the only electrical switching technology with comparable speed, requires exotic refrigeration technology. The Josephson-junction switch requires cooling to superconducting temperatures. Although superconducting technology has undergone some astounding improvements in recent times, operation without any refrigeration at all may be difficult to achieve in superconducting systems. Optical switching requires no cooling.

We have already discussed fiber-optic conductors for the optical signal. Now let's look at the optical transistor, which, in combination with the laser as a power supply and silica fiber conductors, are the building blocks of an optical computer.

How do we use light to switch light? In a transistor an electrical signal is used to control a larger flow of electricity, and in a transphasor, light is used to control the flow of other light.

The key to this device is a **nonlinear optical conductor**. This material (actually, a semiconductor that's transparent to the light being used) passes light through itself at less speed than the speed of light in free space (empty space). The speed of light in the transphasor depends on the number of free electrons in the material (the more electrons, the slower the light goes). Light of the proper wavelength "excites" atoms in the semiconductor to break apart into hole–electron pairs. Although these holes and electrons eventually recombine, a semiconductor exposed to some level of light forms new holes and electrons to replace them. There is always a certain number of holes and electrons in the semiconductor, depending on how bright the light is. In turn, the speed with which light passes through the semiconductor changes as the brightness of the illumination changes (see Figure 22-14).

The second key to making an optical transistor is light-wave superposition, or **reinforce-**

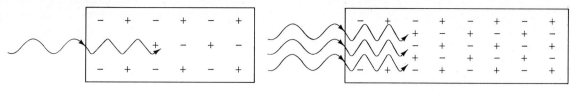

Lower brightness
Fewer free electrons
Faster speed of propagation

Higher brightness
More free electrons
Slower speed of propagation

FIGURE 22-14 Nonlinear optical medium.

ment (see Figure 22-15). In an optical device called a Fabry–Perot interferometer, light waves bounce back and forth between two mirrors through a light-conducting medium. The space between the mirrors is called a *cavity*, even if it is filled with semiconductor material. On every even-numbered bounce, light waves bouncing off the mirrors are traveling in the same direction as those that came in to the interferometer. Ordinarily, there are two outcomes:

1. If they are in phase, the waves will add up, and the brightness of light in the cavity will rise to a level much brighter than the light in the incoming beam.
2. If they are not in phase, the waves will cancel, and the brightness of light in the cavity will be less than the brightness of the incoming beam.

With ordinary optical materials, there is only one way to change the situation from condition 2 to condition 1. The light-carrying path has to be made longer or shorter, like an accordion. If the cavity between the mirrors is really an empty space, the mirrors can be moved closer together or farther apart. In the transphasor, there is no "accordion" action of the mirrors, but the brightness of the light has the same effect. Suppose that the distance between two mirrors is 5.66 wavelengths. This means that the light reaches the far end of the cavity in the time it takes 5.66 waves to enter the near end. Of course, 5.66 is an awkward number. If the light bounces back and forth between the mirrors, the even-numbered bounces (traveling in the same direction) will not be in phase, and **destructive interference** will occur. But suppose that the light is made brighter, and "pumps up" the semiconductor between the mirrors to contain a higher level of holes and electrons; then the light is slowed down by the extra electrons. If the light is bright enough, it would slow down transmission through the semiconductor until light reaches the far end of the semiconductor in the time it takes nearly *six* wavelengths of light to enter the near end. It is as though the mirror were moved farther apart. Now, 6 is a very convenient number. When the light bounces back and forth between the mirrors, the even-numbered bounces (traveling in the

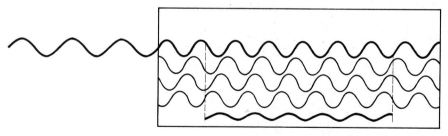

(a) Destructive interference

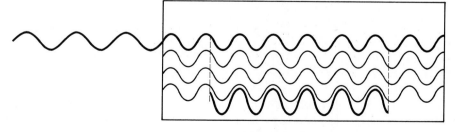

FIGURE 22-15 Effect of: (a) cancellation of even-numbered beams; (b) reinforcement of even-numbered beams.

(b) Constructive interference

same direction) will be in phase and **constructive interference** will occur. Suddenly the brightness builds up to very intense levels inside the cavity. As the path length between the mirrors approaches 6 wavelengths, the light will brighten as the waves are *almost* in phase, and the extra brightness (from reinforcement) will increase the electron count in the semiconductor until exactly 6 wavelengths fit between the mirrors. Once this happens, a slight increase or decrease in the brightness of the incoming light will have no effect, since most of the brightness in the cavity is from reinforcing, constructive interference of the light waves already bouncing around in the cavity. The Fabry-Perot interferometer is "locked on."

If the mirrors are made partially reflective, so that some light can enter the cavity from one side through mirror 1, and some of that light can get out the other side through mirror 2, the transphasor will pass a tiny part of the light through itself. When destructive interference exists within the cavity, almost no light will travel through the transphasor. When constructive interference exists within the cavity, the light that gets out through mirror 2 will be almost 100% of the incoming light (see Figure 22-16).

A light beam that is *almost bright enough* to cause constructive interference, but *not quite*, should cause relatively little transmitted light to leave the other end of the transphasor. By adding a tiny additional beam needed to push the transphasor "over the edge" into a "fit," the whole brightness of the two combined beams is transmitted through the transphasor. As a result, a weak **signal beam**, switching on and off, could

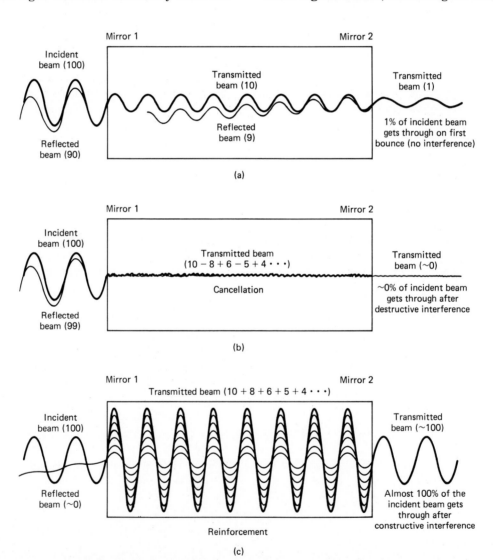

(a)

(b)

(c)

FIGURE 22-16 Switching of a transphasor: (a) before interference; (b) destructive interference; (c) constructive interference.

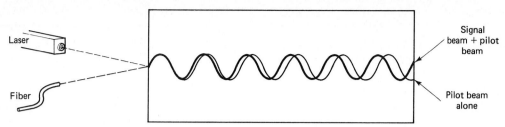

FIGURE 22-17   Effect of brightness on nonlinear medium.

be combined with a much stronger **pilot beam** (a steady beam from a laser) to switch the combined light of the two on and off at the far side of the transphasor (see Figure 22-17). This is the principle of amplification, which makes it possible to use a transphasor in a telecommunications link as a *repeater*, to reconstruct a strong signal from the weak output of a long fiber-optic cable. Most repeaters for optical signals presently convert the light into a (slow!) electrical signal, amplified by a (slow!) amplifier, and then convert it back into light for retransmission. Using a mixed optical/electrical repeater, we've put the bottleneck back onto the bottle, so to speak. With a transphasor, that bottleneck is removed.

Since the optical transistor (transphasor) has a tendency to lock–in on a good–fit condition, it can also be used as a latch. That means a memory element with a 1-pS (one picosecond) SET or RESET time can be made of the transphasor. For comparison, a 7475 TTL latch takes 75,000 pS (75000 picoseconds) for a signal to propagate through from its input to its output, and be latched.

Here is another spectacular possibility; there can be more than one good-fit condition. An interferometer whose mirrors are 100 wavelengths apart might be set up to lock-in on 100, 101, 102, and 103 wavelengths, latching at each with a slightly different level of brightness, but held in place by the same pilot beam in all cases. That would mean that the latch is not limited to binary arithmetic. One switch might hold a base-4 number or even a decimal number.

Finally, there is the matter of scale and complexity—if a decimal number is latched in a 7475 four-bit latch made of TTL devices, the actual number of *junctions* in the latch is several dozen. On the other hand, a transphasor contains just one layer of semiconductor, and is less complicated than *one* junction.

As if that isn't enough, *analog* functions can be performed with optical devices as well. Optical analog approaches seem to be well suited to problems such as picture recognition in artificial-in-

telligence systems, and the design of "neural net" computers patterned after the brain structure of living things.

It seems clear that if this technology is developed—and the team that developed it already has a contract from SDI (the Star Wars defense project)—it can make digital computing with electricity one with velocity control by buggy whip and computing by slide rule.

So don't be surprised if some day you find yourself going back to school to learn optics. A lot of electronics people who learned vacuum-tube theory had to go back to school to learn about transistors, and this kind of thing can certainly happen again.

22-7   CONCLUSION

The "digital invasion" is under way. Digital techniques are taking over from analog methods in almost every imaginable field of electronic communications. This book is dedicated to the technician who may someday service digital equipment in every guise from the "out-front" digital computer to the "hidden" VLSI chips in a fourth-generation television set. No textbook will ever keep you ahead of the game. The best you can hope for is to get the basic vocabulary and ideas, then keep abreast of the field by subscribing to trade magazines in your specialty. Learning—especially in the field of digital electronics—doesn't stop when you get your diploma. That's when the real learning, the OJT (on-the-job training) actually begins. Think of this text, and any course you take in digital electronics, as the "first-stage booster" you will need for a successful "launch" into the industry. *You* will have to supply the "sustainer engine" of your launch vehicle, and, unlike the ballistic missile from which this analogy is taken, you'll have very little time for "coasting flight." Good luck. Your adventure in the field of digital electronics promises to be as exciting as any astronaut's commitment to the space program.

22-1. What is a common carrier?

22-2. Which is more suitable for transmitting data from one 8-bit computer to another, parallel or serial data transmission?

22-3. Suppose that output signals from the data bus (outputs are placed on an 8080 data bus for 500 ns) of a computer are connected directly to the telephone system. The 500-ns pulses will not arrive at the other end of the telephone connection. Why not? What might happen to the logic devices in the data-bus buffer?

22-4. Describe what a UART does when it's transmitting, and what it does when it's receiving.

22-5. Describe what a modem does when it's transmitting, and what it does when it's receiving.

22-6. What is TDM, and how is it used in data communications?

22-7. Name three types of digital circuit that can be used for parallel-to-serial conversion.

22-8. What is a direct-connect modem? Do you think its 5-V logic (if it contains a 5-V logic) is directly connected to the telephone circuit?

22.9. Is level shifting and amplitude shifting important to a modem that uses an acoustic coupler?

22-10. What is attenuation? How does it affect pulses sent down a long transmission line?

22-11. What does a repeater do? How do repeaters make long-distance communication possible?

22-12. Briefly describe AM, FM, and PSM.

22-13. Why are satellites in a geosynchronous orbit better than those in low-earth orbit?

22-14. At various frequencies, different types of transmission lines are used to carry the signals. What types of conductors are used for transmitting electromagnetic waves at the following frequencies? (Choices below)
  (a) Audio (20 to 20,000 Hz)
  (b) Microwave (300 MHz to ? GHz)

  (c) Optical (400 to 800 THz)
    (1) Silica fibers
    (2) Open-wire pairs
    (3) Waveguide

22-15. Approximately what percent of light is reflected at each "bounce" as light travels down a fiber in a light pipe?

22-16. What does a codec do when it's transmitting?

22-17. What does a codec do when it's receiving?

22-18. PCM is used in transmissions from the Space Shuttle. Do the astronauts' voices travel through a codec before transmission?

22-19. Does a telemetry transmission (already digital data) from the Space Shuttle to a ground station travel through a codec? Is it transmitted in PCM?

22-20. Does an analog repeater amplify distortion and noise as it amplifies the signal? Does a digital (PCM) repeater?

22-21. Describe "crosstalk" briefly.

22-22. What are dropouts in telecommunications? How do they differ from line failures? What would you do to transmit data reliably where periodic dropouts in the data stream are expected?

22-23. What is the factor that causes propagation delays in satellite transmission of data?

22-24. What is meant by "answer" and "originate" with reference to a full-duplex modem?

22-25. How are signals carried in both directions on one telephone circuit when a modem is in full duplex mode?

22-26. Describe two reasons why optical computing may replace electronic computing in the future.

22-27. Explain why a transphasor would be a good device to use as a repeater in a fiber-optic data link.

22-28. What is a *null modem*?

22-29. What is the difference between a *data set* and a *data terminal*?

22-30. Describe two of the advantages of fiber-optics "light pipe" over copper wire.

The questions below are about the following data communications interface:

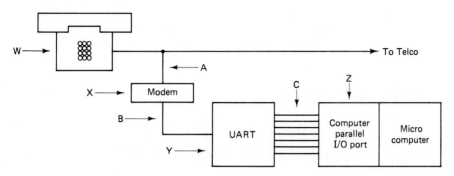

Identify the links that carry the following types of signals.

22-1. Parallel binary digital _____

22-2. Serial binary digital or RS-232 _____

22-3. FSK-encoded analog _____

Identify the block that:

22-4. Converts RS-232 or serial data into FSK-encoded audio. _____

22-5. Converts parallel data into serial or RS-232. _____

22-6. Picks up parallel data from input and places it onto the microprocessor's data bus when requested. _____

22-7. Converts serial data or RS-232 into parallel data. _____

22-8. Converts FSK encoded audio into RS-232 or serial data. _____

22-9. Latches parallel data from the microprocessor's data bus and delivers it to its outputs when requested._____

22-10. Permits connection to telco equipment when a direct-connect modem is not available. _____

22-11. Is identified in telecom jargon as the *data set*. _____

22-12. Is identified in telecom jargon as the *data terminal*. _____

The following 10 questions are about the RS-232 interface.

22-13. How many pins are there on a DB-25 plug or socket? _____

22-14. What line is *absolutely necessary* on any RS-232 link? _____

22-15. Which line carries data output? _____

22-16. Which line carries data input? _____

22-17. Name six control and "handshake" signals that control the transfer of data but are not data themselves.

22-18. Name the standard modem frequencies used as mark (logic 1) and space (logic 0) for *answer* and *originate* on a full-duplex transmission.

22-19. Name four electrical quantities that may be varied to modulate information onto carrier waves.

Appendix 1
OPCODES FOR 8080/8085 AND Z-80

ONE-BYTE OPCODES USED FOR 8080/8085 and Z-80

Single-byte instructions[a]

INR r inc r	00r100 binary	INX B inc bc	03 hex	POP B pop bc	C1 hex	RNZ ret nz	C0 hex	XCHG ex de,hl	EB hex
DCR r dec r	00r101 binary	INX D inc de	13 hex	POP D pop de	D1 hex	RZ ret z	C8 hex	XTHL ex (sp),hl	E3
		INX H inc hl	23 hex	POP H pop hl	E1 hex	RNC ret nc	D0 hex	SPHL ld sp,hl	F9 hex
MOV r,r' ld, r,r'	01rr' binary	INX SP inc sp	33 hex	POP PSW pop af	F1 hex	RC ret c	D8 hex	PCHL ip (hl)	E9 hex
						RPO ret po	E0 hex	HLT halt	76 hex
ADD r add a,r	10000r binary	DCX B dec bc	0B hex	PUSH B push bc	C5 hex	RPE ret pe	E8 hex	NOP nop	00 hex
ADC r adc a,r	10001r binary	DCX D dec de	1B hex	PUSH D push de	D5 hex	RP ret p	F0 hex	DI di	F3 hex
SUB r sub a,r	10010r binary	DCX H dec hl	2B hex	PUSH H push hl	E5 hex	RM ret m	F8 hex	EI ei	FB hex
SBB r sbc a,r	10011r binary	DCX SP dec sp	3B hex	PUSH PSW push af	F5 hex	RET ret	C9 hex		
ANA r and r	10100r binary							DAA daa	27 hex
XRA r xor r	10101r binary	DAD B add bc	09 hex	STAX B ld (bc),a	02 hex	RLC rlca	07 hex	CMA cpl	2F hex
ORA r or r	10110r binary	DAD D add de	19 hex	STAX D ld (de),a	12 hex	RRC rrca	0F hex	STC scf	37 hex
CMP r cp r	10111r binary	DAD H add hl	29 hex	LDAX B ld a,(bc)	0A hex	RAL rla	17 hex	CMC ccf	3F hex
		DAD SP add sp	39 hex	LDAX D ld a,(de)	1A hex	RAR rra	1F hex	RST n rst n	11n111 binary

[a]r = register code; B = 000; C = 001; D = 010; E = 011; H = 100; L = 101; M = 110; A = 111. Intel mnemonics are in capital letters; Zilog mnemonics are in lowercase letters.

Two-byte instructions[a]

ADI **	C6,**	IN pp	DB,pp	MVI r,**	00r110,**
add a,**	hex	in a,(pp)	hex	ld, r,**	binary
ACI **	CE,**	OUT pp	D3,pp		
adc a,**	hex	out (pp),a	hex		
SUI **	D6,**				
sub a,**	hex				
SBI **	DE,**				
sbc a,**	hex				
ANI **	E6,**				
and **	hex				
XRI **	EE,**				
xor **	hex				
ORI **	F6,**				
or **	hex				
CPI **	FE,**				
cp **	hex				

[a]** = any one-byte data number; pp = device code for a port.

Three-byte instructions[a]

JNZ adr	C2,LL,PP	CNZ adr	C4,LL,PP	LXI B,%	01,ln,pg
ip nz,adr	hex	call nz,adr	hex	ld bc,%	hex
JZ adr	CA,LL,PP	CZ adr	CC,LL,PP	LXI D,%	11,ln,pg
ip z,adr	hex	call z,adr	hex	ld de,%	hex
JNC adr	D2,LL,PP	CNC adr	D4,LL,PP	LXI H,%	21,ln,pg
ip nc,adr	hex	call nc,adr	hex	ld hl,%	hex
JC adr	DA,LL,PP	CC adr	DC,LL,PP	LXI SP,%	31,ln,pg
JPO adr	E2,LL,PP	CPO adr	E4,LL,PP		
JPE adr	EA,LL,PP	CPE adr	EC,LL,PP	STA adr	32,LL,PP
JP adr	F2,LL,PP	CP adr	F4,LL,PP	LDA adr	3A,LL,PP
JM adr	FA,LL,PP	CM adr	FC,LL,PP	SHLD adr	22,LL,PP
JMP adr	C3,LL,PP	CALL adr	CD,LL,PP	LHLD adr	2A,LL,PP

[a]adr = an address in memory; LL, PP = line, page of address; % = a 16-bit data word; ln, pg = line, page of 16-bit data.

OPCODES FOR Z-80 ONLY

Arithmetic/logic[a]

ADD A, (IX+d)	DD,86,dd	ADD A, (IY+d)	FD,86,dd
ADC A, (IX+d)	DD,8E,dd	ADC A, (IY+d)	FD,8E,dd
SUB A, (IX+d)	DD,96,dd	SUB A, (IY+d)	FD,96,dd
SBC A, (IX+d)	DD,9E,dd	SBC A, (IY+d)	FD,9E,dd
AND (IX+d)	DD,A6,dd	AND (IY+d)	FD,A6,dd
XOR (IX+d)	DD,AE,dd	XOR (IY+d)	FD,AE,dd
OR (IX+d)	DD,B6,dd	OR (IY+d)	FD,B6,dd
CP (IX+d)	DD,B6,dd	CP (IY+d)	FD,BE,dd
INC (IX+d)	DD,34,dd	INC (IY+d)	FD,34,dd
DEC (IX+d)	DD,35,dd	DEC (IY+d)	FD,35,dd
ADD IX,BC	DD,09	ADD IY,BD	FD,09
ADD IX,DE	DD,19	ADD IY,DE	FD,19
ADD IX,HL	DD,29	ADD IY,HL	FD,29
ADD IX,SP	DD,39	ADD IY,SP	FD,39
INC IX	DD,23	INC IY	FD,23
DEC IX	DD,2B	DEC IY	FD,2B
RLC (IX+d)	DD,CB,dd,06	RLC (IY+d)	FD,CB,dd,06
RRC (IX+d)	DD,CB,dd,0E	RRC (IY+d)	FD,CB,dd,0E
RL (IX+d)	DD,CB,dd,16	RL (IY+d)	FD,CB,dd,16
RR (IX+d)	DD,CB,dd,1E	RR (IY+d)	FD,CB,dd,1E
SLA (IX+d)	DD,CB,dd,26	SLA (IY+d)	FD,CB,dd,26
SRA (IX+d)	DD,CB,dd,2E	SRA (IY+d)	FD,CB,dd,2E
SRL (IX+d)	DD,CB,dd,3E	SRL (IY+d)	FD,CB,dd,3E
BIT b, (IX+d)	DD,CB,dd,01b110	BIT b, (IY+d)	FD,CB,dd,01b110
RES b, (IX+d)	DD,CB,dd,10b110	RES b, (IY+d)	FD,CB,dd,10b110
SET b, (IX+d)	DD,CB,dd,11b110	SET b, (IY+d)	FD,CB,dd,11b110
	hex		hex
	binary		binary
ADC HL,BC	ED,4A	SBC HL,BC	ED,42
ADC HL,DE	ED,5A	SBC HL,DE	ED,52
ADC HL,HL	ED,6A	SBC HL,HL	ED,62
ADC HL,SP	ED,7A	SBC HL,SP	ED,72
NEG	ED,44		
IM 0	ED,46		
IM 1	ED,56		
IM 2	ED,5E		
RLC r	CB,00000r	RL r	CB,00010r
RRC r	CB,00001r	RR r	CB,00011r
SLA r	CB,00100r		hex,binary
SRA r	CB,00101r	RRD	ED,67
	hex,binary	RLD	ED,6F
BIT b,r	CB,01br	RES b,r	CB,10br
SET b,r	CB,11br		hex,binary
	hex,binary		

[a] r = register code; B = 000; C = 001; D = 010; E = 011; H = 100; L = 101; M = (HL) = 110; A = 111; b = bit position; bit 0 (LSB) = 000; bit 1 = 001; bit 2 = 010; etc.; dd = any one-byte displacement added to *IX, or IY.*

Control Instructions

JR d	14,dd	DJNZ d	10,dd
JR Z,d	24,dd	JR NZ,d	20,dd
JR C,d	34,dd	JR NC,d	30,dd
JP (IX)	DD,E9	JP (IY)	FD,E9
RETN	ED,45	RETI	ED,4D

I/O group

IN r,(C)	ED,01r000	INIR	ED,B2
INI	ED,A2	INDR	ED,BA
IND	ED,AA		
OUT (C),r	ED,01r001	OTIR	ED,B3
OUTI	ED,A3	OTDR	ED,BB
OUTD	ED,AB		

Appendix 2
IEC/IEEE LOGIC SYMBOLS

The IEC (International Electrotechnical Commission) and the IEEE (Institute of Electrical and Electronics Engineers) have developed a new set of symbols to represent logical devices. In part, the new symbols use rectangular shapes to replace the familiar distinctive shapes by which AND, OR, NOT, and so on, are represented. For instance, the two forms of the AND gate are

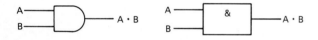

while the two forms of the OR gate are

If this were all that the new notation accomplished, it would be of very little use. When large arrays of gate symbols are reduced to a small size, the older distinctive shapes still retain their recognizable identity as AND, OR, and so on, while the IEC/IEEE symbols lose their identity, since the small & and ≥1 symbols within the rectangles become unreadable long before the rectangles lose their identity as rectangles. For single gates and large systems of AND, OR, and NOT (in random-logic arrays), where a lot of gates have to be shown in a small space, the older system of distinctive symbols has many advantages over the IEC notation. The majority of engineering

and computer-hardware specialists we have spoken to *despise* this new notation. However, you may notice that I prefaced this paragraph with the phrase, "If this were all that the new notation accomplished . . . "; it is not. The main reason for this new notation is to represent the functional block diagram of a logic system and to show how each input affects each output without having to show how the internal logic does it.

DEPENDENCY NOTATION

It is obvious from the diagrams of AND and OR gates just presented that the shape of the symbols is not significant. The **general qualifying symbols** inside the outline tell you what logical operation combines the effects of the inputs to produce an output. These are shown in Table A2-1. (We have already seen, for example, the symbols for AND and OR.) Special features of an input or output, such as inversion, active LOW or active HIGH, rising- or falling-edge triggered, open-collector or open-emitter, and so on, are indicated by **special input–output qualifying symbols** used to join the line to the block. This is shown in Table A2-2. In some cases, symbols used inside the outline indicate characteristics for which there were no previous symbolic representation. For example, a rising-edge triggered clock that causes a counter to count up by 1, or another rising-edge clock that causes a shift register to shift left, are indicated by the same symbol inside

Symbol	Description
▷ or ◁	Buffered or power-driver output
&	AND logic (output active if *all* inputs are asserted)
≥1	OR logic (output active if *any* inputs are asserted)
=	Logic identity (output is active if all inputs are HIGH or all LOW)
=1	XOR logic [output is active if one (and only one) input is asserted]
ALU	Arithmetic-logic unit
COMP	Digital comparator
CPG	Look-ahead carry generator
CTR DIVm	Counter (frequency divider) with modulus m
CTRm	Counter with modulus 2^m
DMUX or DX	Demultiplexer
FIFO	First-in, first-out memory
I=0	Powers-up in RESET state
I=1	Powers-up in SET state
2k	An even number of inputs are asserted (even parity)
2k+1	An odd number of inputs are asserted (odd parity)
MUX	Multiplexer
P−Q	Subtracter
RAM	Random access memory (read/write memory)
RCTRm	Ripple counter with modulus 2^m
ROM	Read-only memory
SRGm	Shift register with m stages
X/Y	Code converter (e.g., BCD to seven-segment)
⊓	Retriggerable monostable multivibrator
¹⊓	Nonretriggerable monostable multivibrator (one-shot)
⊓!G⊓	Astable multivibrator that is triggered to start it
⊓G!⊓	Astable multivibrator that is triggered to stop it
⊓G⊓	Astable multivibrator
⊓	Schmitt triggered input or element with hysteresis
π	Multiplier
Σ	Adder
ϕ	Complex function output

the outline of traditional logic. The new notation makes it possible to tell what the rising edge is try. A group of bus lines labeled from 0 to 15 would indicate a 16-bit data bus in exactly the same way on the new symbols as it did in the past. These are shown in Table A2-3.

Dependency is related to what inputs (or expected to do, and that was never a standardized part of the counter or shift-register diagram before. In other cases, the symbols used to represent groupings (like a data bus) are the same ones that are presently in use in the microcomputer industry outputs) are *affecting* and what inputs (or outputs) are *affected by* other signals. Certain letters are used to indicate what type of dependency is involved (see Table A2-4). The input or output doing the affecting is labeled with one of these letters, followed by an identifying number. The other inputs or outputs affected by that input or output are labeled with the same number.

For example, a gated R-S flip-flop would appear like this:

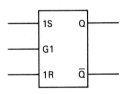

This is the simplest case, called **G dependency**, in which the signals S and R, labeled with the number 1, are dependent on input G (the gate or clock input) in an AND relationship. It is also called **AND dependency**. If inputs G AND S are high, the flip-flop will be set (input S functions); if inputs G AND R are high, the flip-flop will be reset (input R functions). This indicates that if the G input is LOW, it is ANDed with affected inputs/outputs; this overrides the state of the affected inputs/outputs and makes them LOW.

Notice that the *affecting* input is labeled with a letter followed by a number, while the *affected* inputs are labeled with numbers first. This will generally be the case in IEC symbols. The

—d	Inverted input. External 0 produces internal 1.
p—	Inverted output. Internal 1 produces external 0.
—⌐|	Active-LOW input. Same as —d in positive logic.
⌐—	Active-LOW input (Left-to-right signal flow).
⌐—	Active-LOW output. Same as p— in positive logic.
—⌐	Active LOW output (left-to-right signal flow).
—◄—	Signal flow from right to left.
—◄►—	Bidirectional signal flow.
—◁	Falling-edge triggered.
—▷	Falling-edge triggered.
—▷	Rising-edge triggered.
—├ ┤—	Nonlogic connection. (Defined inside symbol.)
—⊓⌐	Analog input on a primarily digital device.
—#⌐	Digital input on a primarily analog device.
—┬—	Internal connection (noninverting).
—⊙—	Internal connection (inverting).
—▷—	Internal connection (item on right is edge-triggered by the signal entering at the left).
—⌐	Internal input not connected to an external terminal (comes from another part of the internal circuit.
⌐—	Internal output not connected to an external terminal (goes to another part of the internal circuit).

letters following the numbers in the affected inputs will not always be there; in this case, the set (S) and reset (R) letters follow the number 1 to denote the special effect of enabling each input. In other cases, the inputs may be labeled with letters separately from the dependency notation, as follows:

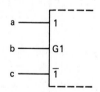

In this case, the inputs of a two-channel multiplexer are shown; input a is selected if input b is HIGH, and input c is selected if input b is LOW.

If two or more inputs (or outputs) have the same *letter and number*, they have an OR relationship to one another:

In this case, input c will be enabled if either a or b goes HIGH.

This is not the same thing as OR dependency. Input c is still gated by any G input in an AND-logic relationship. If input c AND any G input are HIGH, the output will be high.

OR dependency is identified by the letter V. It is also called **V dependency**, and indicates that if the V input is HIGH, it is ORed with affected inputs/outputs; this overrides the state of the affected inputs/outputs and makes them HIGH.

In both G and V dependency, the inputs/outputs are enabled in their normal states when one level is present, and disabled when the other level is present. As with AND, OR, NAND, and NOR gates, the disabled outputs remain stuck HIGH or LOW (depending on the kind of logic), and the disabled inputs become *don't care* inputs. You may recall that all the basic gates introduced in Chapters 2 and 3 may be used to enable or disable signals in this way, *except for XOR and XNOR gates*. They become *controlled inverters* instead.

N dependency, also called **exclusive-OR dependency**, is identified by the letter *N*, which

TABLE A2-3

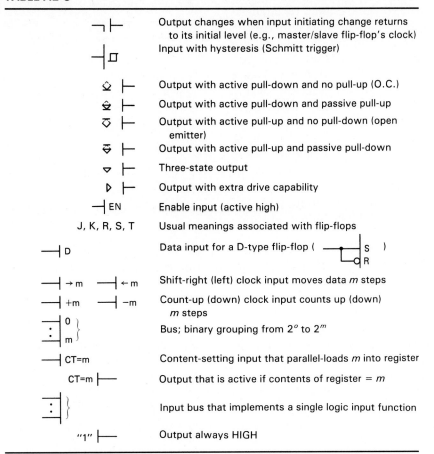

Symbol	Description
¬⊢	Output changes when input initiating change returns to its initial level (e.g., master/slave flip-flop's clock)
⊣□	Input with hysteresis (Schmitt trigger)
◇⊢	Output with active pull-down and no pull-up (O.C.)
◈⊢	Output with active pull-down and passive pull-up
▽⊢	Output with active pull-up and no pull-down (open emitter)
▽⊢	Output with active pull-up and passive pull-down
▽⊢	Three-state output
▷⊢	Output with extra drive capability
⊣EN	Enable input (active high)
J, K, R, S, T	Usual meanings associated with flip-flops
⊣D	Data input for a D-type flip-flop
⊣→m ⊣←m	Shift-right (left) clock input moves data m steps
⊣+m ⊣−m	Count-up (down) clock input counts up (down) m steps
⊣0...m	Bus; binary grouping from 2^0 to 2^m
⊣CT=m	Content-setting input that parallel-loads m into register
CT=m⊢	Output that is active if contents of register $= m$
⊣...}	Input bus that implements a single logic input function
"1"⊢	Output always HIGH

TABLE A2-4

G	AND (disabled by LOW)
V	OR (disabled by HIGH)
N	XOR (controlled inverter)
Z	Interconnection
X	Transmission
C	Control
S, R	Set and Reset
EN	Enable
M	Mode
A	Address

indicates that if the N input is HIGH, it *inverts* affected inputs/outputs; if the N input is LOW, the affected inputs/outputs are *not inverted*. An example of this is the CARRY output of the adder/subtracter described in Chapter 6.

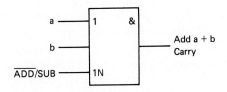

Z dependency, also called **interconnection dependency**, indicates that an internal connection exists between the Z input or output and affected inputs and outputs with the same number.

X dependency, also called **transmission dependency**, indicates that a bidirectional connection is enabled between one bidirectional input/output line and another. The affected input/output lines have the same number as the X control line affecting them. For example, a bidirectional 8-bit bus buffer enabled by input d would look like this:

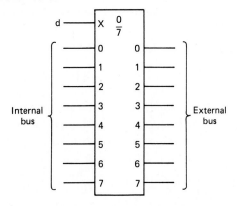

Note that the suffix (number) after the X handles all of the buffer lines from 0 through 7.

C dependency, also known as **control dependency**, indicates that inputs affected by C are enabled when C is HIGH, and disabled from having any effect otherwise. This is different from a G dependency only because C is generally used to enable inputs whose effect is latched, while G enables inputs whose effect is generally transient, unless the input is identified by an S or an R, or some other input identifier that indicates latched results explicitly.

S dependency (set) and **R dependency** (reset) have exactly the same meanings as the set and reset functions of a flip-flop defined in Chapter 10.

EN dependency, also known as **enable dependency**, distinguishes (in dependency notation) between an input marked EN (which affects *all* outputs) and an input marked EN1, EN2, and so on, which affects only inputs and outputs containing the same number as the EN input.

M dependency, also called **mode dependency**, indicates that the input marked M may depend on the mode in which the device is operating. For example, in the 74LS190, the clock input may count up (as in a 74LS90) when the count-direction pin (5) is LOW but will count down when the count-direction pin's level changes to HIGH. It is the same clock input (on pin 14) in either case, but it has a different effect based on the mode (upcount or downcount) selected by another pin. The up/down pin will be labeled with an M identifier (maybe several) and the clock pin will show dependency on the up/down pin.

At this point, an examination of the IEU symbol for the 74LS190 up/down counter seems appropriate. We will take the symbol apart to see what its contents indicate. There will be a few things that we haven't already seen in this symbol, but most of its parts should be recognizable if we check back into the definitions already presented. One of the new things we see in the symbol is a **common-control block**. Its function should become apparent as we work our way through the description of what is going on in the 74LS190.

First, we identify the paddle-shaped section at the top of Figure A2-1. You might picture it as a "funnel" that you can pour signals into, and it feeds them into the lower portion of the diagram. This is the common control block just mentioned. It receives signals that will affect all of the rest of the symbol that it "funnels into." In the lower portion of the figure we see the four flip-flops of

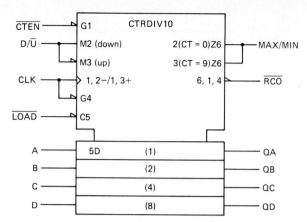

FIGURE A2-1 IEU logic symbol for a 74LS190 IC.

the decade counter. Although they have separate inputs (A, B, C, D) and outputs (QA, QB, QC, QD), they are all subject to the actions of the signals entering the common-control block. There is an exception to this: the 5D shown in the (1) block (which applies to all four blocks despite its appearance in only one). It indicates that inputs A, B, C, and D are enabled to latch data by the C5 (LOAD) input in the common-control block.

Let's look at the inputs to the common-control block. Input G1 (CTEN) is an active-LOW count-enable input, affecting the CLK input functions and the RCO (ripple-carry output). All inputs and outputs showing 1 in their label are enabled when this input is active. Without this input, this chip cannot count, although the chip can still be loaded through the parallel latches as a parallel register. How do we know this? The clock input (CLK) contains labels $(1, 2-/1, 3+)$ which show that it depends on G1. The latch inputs, labeled with 5D as mentioned before, are D latches that do not show a 1 in their labels, so they are affected only by input C5.

The M2 and M3 mode inputs are the D/U (down-up) count mode selector pin. They are actually the same external connection, with mode M2 selected when D/U is HIGH (downcounting) and mode M3 selected when D/U is LOW (upcounting). All inputs or outputs showing 2 or 3 in their label are affected by this mode selection. We see this in the clock input labeling: $(1, 2-/1, 3+)$. The clock is downcounting if mode 2 exists $(2-)$ and upcounting if mode 3 exists $(3+)$. Although it is not shown, it is assumed that the clock counts up or down one count at a time. For any other situation, the number of counts upward or downward would follow the + or − sign.

The clock input (CLK) is positive-edge (rising-edge) triggered, as we can see from the symbol, but it also has an effect when it is LOW, as indicated by G4. The only input or output affected is the ripple-carry output (RCO), which contains a 4 in its label (6, 1, 4) and also requires that G1 (count enable) and Z6 (an internal connection to the MAX/MIN output) be active at the same time that G4's external connection is LOW.

The parallel load input (LOAD) is active LOW. It is indicated by C5. Since it causes data to be latched, it is a C rather than a G or an EN. The label at the inputs of the four latches (5D) indicates that they are affected by this input.

Now, let's look at the common-control block's outputs. The MAX/MIN output indicates overflow or underflow as the counter completes its modulus. Its labels ($2(CT = 0)Z6$, $3(CT = 9)Z6$) indicate that in mode 2 the output is active if the count = 0. If the counter is counting down, the next count will go below zero (and need to borrow a one from the next decimal place), so the MIN (minimum) condition is in effect now. The labels also indicate that in mode 3 the output is active if the count = 9. If the counter is counting up, 9 is its last count, and the next count will "roll over" to zero (and need to carry a one to the next decimal place), so the MAX (maximum) condition is in effect now. The joined output lines indicate that either condition affects the same pin on the IC. Also notice that both conditions are delivered to internal output Z6. This internal output is used by the RCO line.

Output RCO is active LOW, as indicated by its appearance. Its label indicates that it is affected by 6, 1, and 4. That is, when Z6 (MAX/MIN) is active AND when G1 (CTEN) is active, AND when G4 (CLK) is LOW. These all happen during the half-cycle at the end of the counter's count. RCO is usually attached to the CLK input of another counter, sharing a common connection to D/U (so that both counters are going in the same direction). As this counter counts up past 9, the RCO condition ends, and a rising edge advances the next counter (thus carrying a 1 to the next decimal place). As this counter counts down below 0, the RCO condition ends, and a rising edge downcounts the next counter (thus borrowing a 1 from the next decimal place).

Now, we can look at the rest of the symbol. The outputs QA, QB, QC, and QD are not labeled, but because they receive commands funneled in from the common-control block, they are affected by everything, unless otherwise indicated. The inputs A, B, C, and D show this; they are affected (enabled) only by 5D, and are explicitly labeled that way.

Now that we've looked at the 74LS190, hopefully the mysterious stuff in these symbols

has become a bit more meaningful. There is one remaining type of dependency that appears in computers and memory devices. That is **A dependency**, also called **address dependency**.

Of course, it is possible to represent a memory device by showing every latch in the thing, but nobody would do so if it could be avoided. A dependency works out a way to avoid this by representing one cell of the memory, and letting the address bus indicate how many total cells like it there are.

Let's use as an example the 7489 RAM described in Chapter 14. It has sixteen 4-bit latches, selected by four address lines. There is a write-enable input and a chip-enable input. In Figure A2-2 we can see that the CE (chip enable) and

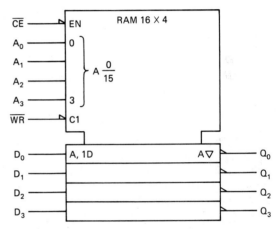

FIGURE A2-2 IEU logic symbol for a 7489 RAM.

WE (write enable) are in the common-control block, so they affect all the cells of the RAM array. The address bus A_0 through A_3 is responsible for selecting which 4-bit latch is being accessed. The numbers 0 and 15 indicate how many latches there are in the "A" dependency label attached to the bus.

In the section below the common-control block, we see a typical 4-bit latch (one cell of the memory). Inside the topmost latch of the cell, and representing the situation in all four latches, is the label A, 1D at the input, and A (with a triangle for three-state control) at the output. Beginning with the input, A, 1D means that latch A (any one from 0 to 15) has D-type inputs dependent on (enabled by) common-control input C1. The output of the latch is identified by A indicating address dependency. Only the latch outputs accessed by the address bus are enabled. We can also see the triangle that indicates the outputs are Tri-Stated or of high impedance when they are not selected. The output is active-LOW, indicating that data are inverted compared to the input, which is the case with the 7489.

Whether there are 16 latches, or 16 thousand, the symbol will not be very different. Only the address bus and the larger number in the dependency symbol will change.

This concludes our explanation of dependency and IEC symbols. We expect that their application to complex integrated circuits will increase with time. The more complicated the circuits are, the more useful these symbols become. Although this description has been only a brief overview, we hope your will find it useful in interpreting such symbols where they occur in computer schematics and digital systems. More detailed information can be obtained from the Institute of Electrical and Electronics Engineers as IEEE Standard 91-1984.

Appendix 3
KARNAUGH MAPPING

You already know how to derive a sum-of-products expression from a truth table. If you spot the appropriate relationships between the terms and factors in the *not simplified* expression, you can probably use the Boolean identities in Chapter 4 to simplify the expression. This takes a good deal of effort and some skill at algebra manipulations. However, some people are more adept with pictures than they are with equations, and there is a useful way to take advantage of this ability. Boolean simplification can be done using a graphic/pictorial type of solution instead of math. This simplification method is called **mapping.**

KARNAUGH MAPPING

We are going to reorganize the information in a truth table into a grid of numbers called a **Karnaugh map.** A Karnaugh map is a two-dimensional grid of numbers, which contains the 1s and 0s from the output portion of the truth table. It is also called a **Veitch diagram.** The first thing that we will need to know is how to organize the truth table's data into a Karnaugh map; the second thing we need to know is how to use it to simplify (or minimize) a Boolean expression for the logic system being analyzed.

SETTING UP THE MAP

To set up the Karnaugh map, look at the truth table and determine how many variables there are. Divide up the variables into two groups (if possible, they should be equal-sized groups, but if there are three or five variables, for example, you would split the variables up into two and one, or three and two) and mark the letters of the variables on the top and side of the map. In general, the map should be an array of spaces organized as 2^n rows by 2^m columns, where n is the number of variables marked alongside the map, and m is the number of variables marked above the map. To show how this works, we will use the "Alice problem" from Chapter 4. If you don't remember the Alice problem, it was one where an output sets off an alarm when Alice (the daughter) is with one of both of her boyfriends and is not chaperoned by her mother (Betty). We will begin by setting up the truth table for the "Alice alarm" like this:

Alice	Betty	Chuck	Dave		
→A	↘B	C↙	D↩	Alarm	
Inputs:	0 0	0 0		0	*Outputs:*
(0 = people outside	0 0	0 1		0	(0 = no
the cabin)	0 0	1 0		0	need for
	0 0	1 1		0	alarm)
(1 = people inside	0 1	0 0		1	
the cabin)	0 1	0 1		1	(1 = turn
	0 1	1 0		1	on alarm;
	0 1	1 1		0	chaperone
	1 0	0 0		1	needed)
	1 0	0 1		1	
	1 0	1 0		1	
	1 0	1 1		0	
	1 1	0 0		0	
	1 1	0 1		0	
	1 1	1 0		0	
	1 1	1 1		0	

Since there are four variables, the easiest way to split them up is two and two. We will assign variables A and B to the side of the map, and C and D to the top. Since $n = 2$ and $m = 2$, the map will be four rows by four columns, making it a square.

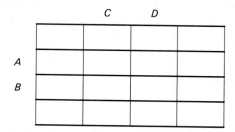

The combinations of 1 and 0 states of each pair of variables may take on four values; these are shown alongside the rows and above the columns:

		C 00	01	D 11	10
	00				
A	01				
B	11				
	10				

Notice that we didn't arrange the binary numbers in counting sequence along the rows and columns. Instead, they are arranged to count up in *Gray code* from left to right and top to bottom. This is important and will be explained shortly. Now, in addition to the pairs of numbers, we convert the pairs of numbers to pairs of letters representing the AB row or CD column state, placing them at the other end of each row and column:

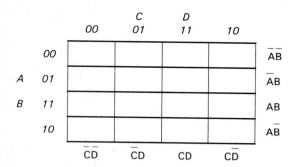

Into each space of the Karnaugh map, we put the output of the truth table for the line that has the combination of ABCD variables indicated by the row and column. To see what the input combinations should be, combine the row and column algebraic expressions as shown:

		C 00	01	D 11	10	
	00	$\overline{A}\,\overline{B}\,\overline{C}\,\overline{D}$	$\overline{A}\,\overline{B}\,\overline{C}D$	$\overline{A}\,\overline{B}CD$	$\overline{A}\,\overline{B}C\overline{D}$	$\overline{A}\,\overline{B}$
A	01	$\overline{A}B\overline{C}\,\overline{D}$	$\overline{A}B\overline{C}D$	$\overline{A}BCD$	$\overline{A}BC\overline{D}$	$\overline{A}B$
B	11	$AB\overline{C}\,\overline{D}$	$AB\overline{C}D$	$ABCD$	$ABC\overline{D}$	AB
	10	$A\overline{B}\,\overline{C}\,\overline{D}$	$A\overline{B}\,\overline{C}D$	$A\overline{B}CD$	$A\overline{B}C\overline{D}$	$A\overline{B}$
		$\overline{C}\,\overline{D}$	$\overline{C}D$	CD	$C\overline{D}$	

This means that the input lines of the truth table which "belong" to each space in the Karnaugh map are

		C 00	01	D 11	10	
	00	0000	0001	0011	0010	$\overline{A}\,\overline{B}$
A	01	0100	0101	0111	0110	$\overline{A}B$
B	11	1100	1101	1111	1110	AB
	10	1000	1001	1011	1010	$A\overline{B}$
		$\overline{C}\,\overline{D}$	$\overline{C}D$	CD	$C\overline{D}$	

And the outputs that belong to each space in the Karnaugh map are the outputs for the lines on the truth table that have the inputs shown above. When we replace the inputs above with their outputs on the Alice table, we get

		C 00	01	D 11	10	
	00	0	0	0	0	$\overline{A}\,\overline{B}$
A	01	1	1	0	1	$\overline{A}B$
B	11	0	0	0	0	AB
	10	0	1	1	1	$A\overline{B}$
		$\overline{C}\,\overline{D}$	$\overline{C}D$	CD	$C\overline{D}$	

Now that we have constructed the Alice map, what do we do with it?

SUBCUBES AND SIMPLIFICATION

Let's begin by defining a structure that is part of a Karnaugh map and consists of a bunch of 1s clustered together in the map. Wherever a block of adjacent 1s is bunched together, and there are 2^n (2, 4, 8, etc.) of them in the bunch, the bunch is called a **subcube.** Here are some two-cell subcubes in the Karnaugh map of the Alice problem:

0	0	0	0	$\bar{A}\bar{B}$
$\boxed{1 \quad 1}$		0	1	$\bar{A}B$
0	0	0	0	AB
0	1	1	1	$A\bar{B}$

Columns: C D over: $\bar{C}\bar{D}$ $\bar{C}D$ CD $C\bar{D}$

2.

0	0	0	0	$\bar{A}\bar{B}$
1	1	0	1	$\bar{A}B$
0	0	0	0	AB
0	$\boxed{1 \quad 1}$		1	$A\bar{B}$

Columns: $\bar{C}\bar{D}$ $\bar{C}D$ CD $C\bar{D}$

3.

0	0	0	0	$\bar{A}\bar{B}$
1	1	0	1	$\bar{A}B$
0	0	0	0	AB
0	1	$\boxed{1 \quad 1}$		$A\bar{B}$

Columns: $\bar{C}\bar{D}$ $\bar{C}D$ CD $C\bar{D}$

4.

0	0	0	0	$\bar{A}\bar{B}$
$\boxed{1}$	1	0	$\boxed{1}$	$\bar{A}B$
0	0	0	0	AB
0	1	1	1	$A\bar{B}$

Columns: $\bar{C}\bar{D}$ $\bar{C}D$ CD $C\bar{D}$

The first three subcubes are obvious. They are formed of pairs of 1s that are obviously adjacent. But what about the fourth subcube? It certainly doesn't *look* like the cells are adjacent. In fact, they appear to be as far apart as they could get!

KARNAUGH'S BAGEL

The Karnaugh map is not really a square grid of cells, as it appears in the diagram. Since Gray code is used, and we remember that Gray code goes with a type of codewheel, it is not surprising that the rows of cells, like Gray code, actually wrap around from the extreme right side of the map to the extreme left side of the map. In a way, the Karnaugh map is like a flat map of the globe.

When you go off the map on the right-hand edge, you come back on the left (because the world is round). Is the Karnaugh map round, too? It is, but not in the way you might suppose. When you reach the end of a row, it's true that you come back to the other end of the map, so that the cells in the fourth diagram are really adjacent, despite appearances, but the Karnaugh map wraps around from bottom to top, too. Each column of cells is attached from its bottom to the top cell, and wraps around top to bottom. What kind of shape would make this possible?

You can see that the *torus* (the shape of a doughnut or bagel) can have a map printed on its outside that works just like a Karnaugh map (see Figure A3-1). Cells at the top wrap around to the bottom, and cells around the equator go all the way around the bagel. The Karnaugh map is like a flat-map projection of the bagel, so let's call it

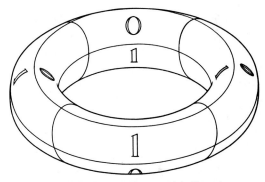

FIGURE A3-1   Karnaugh's Bagel

Karnaugh's bagel, to help you visualize what's adjacent to what.

Problem: In the example shown below, are the four minterm cells in the Karnaugh map adjacent? Do they form a subcube?

1	0	0	1	$\bar{A}\bar{B}$
0	0	0	0	$\bar{A}B$
0	0	0	0	AB
1	0	0	1	$A\bar{B}$

Columns: C D over: $\bar{C}\bar{D}$ $\bar{C}D$ CD $C\bar{D}$

Solution: If you wrap the Karnaugh map around a bagel, you will see that all four 1s in the map actually meet. They form a 4-cube, in rows $\bar{A}\bar{B}$ and $A\bar{B}$, and columns $\bar{C}\bar{D}$ and $C\bar{D}$. Since the rows have the same B factor ($\bar{B}$) and the columns have the same D factor ($\bar{D}$), the expression for the subcube is $\bar{B}\bar{D}$. We will explain how the K-map helps us develop this expression shortly. If you would like to check that it is correct, develop a truth table for $X = \overline{B}\overline{D}$ and fit it into a K-map. Happy mapping!

USING THE SUBCUBES

Why are these subcubes important? Because they provide a way to get a simplified Boolean (minimized) expression from the original minterms, without math. Look at the variables beside the rows and beneath the columns of the map. We arranged them using Gray code, so that *only one* of the variables would change for any two adjacent cells. This is important (as we said before) because it makes it possible to reduce the complexity of a minterm sum-of-products expression in the following way:

Look at the cells in the subcube. Identify what variables they have in common (for two-cell subcubes like the ones in the example, they will all be identical, except for one variable). Throw out the variable that has two different states in the subcube, and write the minterm expression by writing down what's left. This is the expression for the subcube. If the areas covered by subcubes cover all the 1s in the map, the map will simplify considerably compared to the expressions from the minterm sum of the truth table (unsimplified). Let's do this for the Alice map developed earlier. We will select the following subcubes (note that it is OK to use the same 1 in two overlapping subcubes):

We have covered all the 1s in the map with four subcubes. These are subcubes w, x, y, and z. How do we simplify them?

Subcube w, the subcube with adjacent cells that don't *look* adjacent, covers two 1s on the second row of the K-map. This is the $\overline{A}B$ row, in columns $\overline{C}\overline{D}$ and $C\overline{D}$. Both cells have $\overline{A}B\overline{D}$ in common with C varying, so we throw out C, and the expression for subcube w is $\overline{A}B\overline{D}$. Subcube x covers two 1s in row $\overline{A}B$, in columns $\overline{C}\overline{D}$ and $\overline{C}D$. Since the only item that varies between the the

two cells is D, we drop the D factor, and the expression for subcube x is $\overline{A}B\overline{C}$. Subcube y covers cells in row $A\overline{B}$ with columns $\overline{C}D$ and CD. Dropping the C factor, which varies, gives $A\overline{B}D$ for subcube y. Finally, subcube z has 1s in row $A\overline{B}$ and columns CD and $C\overline{D}$. Since D is the only variable that is different between these cells, we drop the D factor, and the expression for subcube z is $A\overline{B}C$.

We combine the expressions for the subcubes in the same way that we combine minterms from a truth table. Since they all describe different ways to get a 1 at the output, either one OR another, OR a third will produce an output, and we "OR together" all the subcube expressions to get the overall expression:

$$\text{alarm} = w + x + y + z$$
$$= \overline{A}B\overline{D} + \overline{A}B\overline{C} + A\overline{B}D + A\overline{B}C$$

We can factor out the common factor $\overline{A}B$ in the first two terms, and the factor $A\overline{B}$ in the second two, and we are able to simplify the overall expression a little more, to

$$\text{alarm} = \overline{A}B(\overline{D} + \overline{C}) + A\overline{B}(D + C)$$

which, we know from Chapter 4, is as far as we can go with this expression. Look back at Chapter 4. Getting this simplification just using Boolean identities took a lot of steps of algebraic simplification. In this case, using subcubes, the final result was really only one step away from the expression for the subcubes. With nonoverlapping subcubes, the expression is already as simple as it can get, and you wouldn't need to do any algebraic manipulation, not even the factoring we did here.

"ORPHAN" CELLS AND MAPPING

Occasionally, there are one or more cells in a Karnaugh map that are not part of a subcube. An example is shown below:

Notice that the cell in the lower-right corner is not part of a subcube, although all the other cells are part of a 4-cube in the middle of the map. Simplifying the 4-cube is a relatively simple matter. We note that the two columns the subcube is in ($\overline{C}D$ and CD) share a D, while C changes, and that the two rows the subcube is in ($\overline{A}B$ and AB) share a B, while A is different in each row. When we "AND together" the row and column expressions that are left, the expression for the 4-cube is BD. Now, what do we do about the lonesome 1 in the corner?

If we decide to consider the isolated 1 as a subcube of one cell, no simplification is possible, but its formula is the formula for the cell it occupies, namely $A\overline{B}C\overline{D}$. As with the subcubes of the Alice problem, we sum the terms from each subcube (treating the orphan cell as a subcube of 1). We get

$$BD + A\overline{B}C\overline{D}$$

If a map is *all* orphan 1s, with nothing but isolated 1s surrounded by 0s everywhere, no simpli-

The CAPS LOCK key on an ASCII keyboard is supposed to change the lowercase alphabetic letters in the keyboard into capital letters. It is not a shift key, and is not supposed to change any of the other characters on the keyboard. Design a circuit that can identify the lowercase alphabetic characters on the keyboard *only*, and produce an output HIGH state when any one of these characters is detected.

The K-map of the ASCII lowercase characters will have 1s in cells 1100001 through 1111010 (a through z). Here is one way to separate the 7 bits of ASCII code, for mapping:

row bits G, F, E	column bits D, C, B, A
(eight rows)	(16 columns)

We'll begin by making a K-map for the ASCII lowercase characters; most rows are 0s, since most ASCII characters aren't lowercase letters. In the diagram below we've skipped the all-0 rows and shown the two rows that contain 1s:

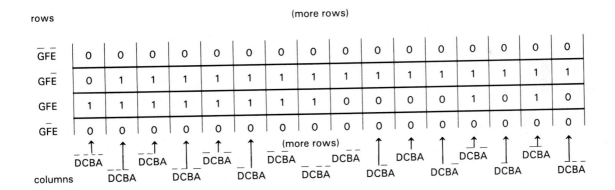

fication of the resulting minterm sum is possible by Karnaugh mapping.

KARNAUGH KNOWLEDGE

The basics of simplification by K-mapping have now been covered. Since all the examples we've looked at have been done for four variables, in closing, let's look at a problem with a larger number of variables and see how it's done. For example, let's consider the following problem:

Looking for subcubes, we find a row of fifteen 1s on row $GF\overline{E}$. This can be thought of as two overlapping subcubes of eight 1s in a row, or as a row where all 16 columns are represented, *except for the first column*. This second interpretation will turn out to be more useful, as we shall see. On row GFE, we find a row of eight cells containing 1s (a straight line is OK as a subcube; all that's important is that it has 2^n adjacent 1s). On row GFE we also see two isolated 1s, which appear to be orphan cells, but they can be combined with 1s above them, which are adjacent, to make 2-cubes.

Here are the subcubes on the map:

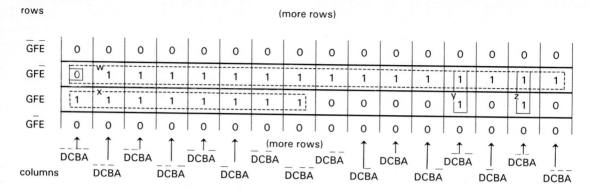

We have marked out subcubes w, x, y, and z on the map. Subcube w is a bit funny, since one of its 1s is really a 0. We're going to handle that very soon, but for now, let's pretend the 0 isn't there. The expression for subcube w, since it covers all the cells on the row, is the expression for the row itself, namely $GF\overline{E}$. Subcube x covers cells on row GFE that are in columns containing $\overline{D}$. You can verify that all the other letters in these eight columns vary, but they all contain $\overline{D}$. There are also no other columns on this row that contain $\overline{D}$. The expression for subcube x is the expression for the row, GFE, and the common factor, $\overline{D}$, found in all eight columns. We put these together and get $x = GFE\overline{D}$. Subcubes y and z are 2-cubes on rows $GF\overline{E}$ and GFE. Both rows have GF in common, and the expressions for the subcubes are simply GF ANDed with their column expressions. The result of this is: $y = GFD\overline{C}B\overline{A}$ and $z = GFD\overline{C}\overline{B}A$.

So now we put together the expressions for the subcubes, and get

lowercase
$$= w + x + y + z$$
$$= GF\overline{E} + GFE\overline{D} + GFD\overline{C}B\overline{A} + GFD\overline{C}\overline{B}A$$

But wait! We forgot something! What about the 0 in subcube w? The expression for the column on row $GF\overline{E}$ with the 0 in it is $\overline{D}\,\overline{C}\,\overline{B}\,\overline{A}$. Since the inverse of this is $\overline{\overline{D}\,\overline{C}\,\overline{B}\,\overline{A}}$ (= d + c + b + a), the expression for that part of subcube w which *doesn't* include the 0 is

$$GF\overline{E} \cdot (D + C + B + A)$$

so the overall expression is

$$\text{lowercase} = GF\overline{E} \cdot (D + C + B + A)$$
$$+ GFE\overline{D} + GFD\overline{C}B\overline{A} + GFD\overline{C}\overline{B}A$$

You might have noticed that all the terms in this expression contain the factors GF, so one step of factoring (like that done in the Alice problem) may be done to further simplify the Boolean expression. Factoring out the GF factor, we get

$$\text{lowercase} = GF \cdot (\overline{E} \cdot (D + C + B + A)$$
$$+ E\overline{D} + D\overline{C}B\overline{A} + D\overline{C}\overline{B}A)$$

Further internal factoring may be possible, but it won't reduce the expression much further, and we can probably consider this expression complete. A logic diagram could now be constructed to implement this expression in AND, OR, and NOT gates. Most designers would probably implement this design with a ROM or a PLA (PAL) instead.

Clearly, mapping can't do everything, but in identifying what may be done to simplify a Boolean expression—or in working from a truth table to get a logic diagram directly—it is a useful tool and does much to reduce the number of steps required, compared to examining unsimplified minterms from a truth table and using Boolean algebra to reduce them.

APPENDIX 4
DATA SHEETS

National Semiconductor

DM54LS154/DM74LS154 4-Line to 16-Line Decoders/Demultiplexers

General Description

Each of these 4-line-to-16-line decoders utilizes TTL circuitry to decode four binary-coded inputs into one of sixteen mutually exclusive outputs when both the strobe inputs, G1 and G2, are low. The demultiplexing function is performed by using the 4 input lines to address the output line, passing data from one of the strobe inputs with the other strobe input low. When either strobe input is high, all outputs are high. These demultiplexers are ideally suited for implementing high-performance memory decoders. All inputs are buffered and input clamping diodes are provided to minimize transmission-line effects and thereby simplify system design.

Features

■ Decodes 4 binary-coded inputs into one of 16 mutually exclusive outputs
■ Performs the demultiplexing function by distributing data from one input line to any one of 16 outputs

■ Input clamping diodes simplify system design
■ High fan-out, low-impedance, totem-pole outputs
■ Typical propagation delay
   3 levels of logic 23 ns
   Strobe 19 ns
■ Typical power dissipation 45 mW

Absolute Maximum Ratings (Note 1)

Supply Voltage	7V
Input Voltage	7V
Storage Temperature Range	−65°C to 150°C

Note 1: The "Absolute Maximum Ratings" are those values beyond which the safety of the device cannot be guaranteed. The device should not be operated at these limits. The parametric values defined in the "Electrical Characteristics" table are not guaranteed at the absolute maximum ratings. The "Recommended Operating Conditions" table will define the conditions for actual device operation.

Connection and Logic Diagrams

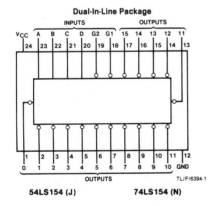

Dual-In-Line Package

54LS154 (J)   74LS154 (N)

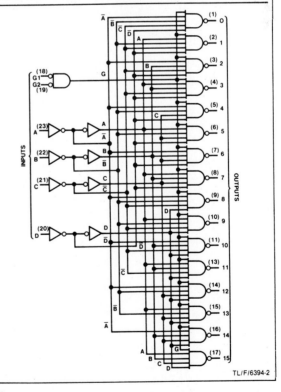

TL/F/6394-1
TL/F/6394-2

National Semiconductor

DM54LS153/DM74LS153 Dual 4-Line to 1-Line Data Selectors/Multiplexers

General Description

Each of these data selectors/multiplexers contains inverters and drivers to supply fully complementary, on-chip, binary decoding data selection to the AND-OR-invert gates. Separate strobe inputs are provided for each of the two four-line sections.

Features

■ Permits multiplexing from N lines to 1 line
■ Performs at parallel-to-serial conversion
■ Strobe (enable) line provided for cascading (N lines to n lines)
■ High fan-out, low impedance, totem pole outputs

■ Typical average propagation delay times
  From data 14 ns
  From strobe 19 ns
  From select 22 ns
■ Typical power dissipation 31 mW

Absolute Maximum Ratings (Note 1)

Supply Voltage	7V
Input Voltage	7V
Storage Temperature Range	−65°C to 150°C

Note 1: The "Absolute Maximum Ratings" are those values beyond which the safety of the device cannot be guaranteed. The device should not be operated at these limits. The parametric values defined in the "Electrical Characteristics" table are not guaranteed at the absolute maximum ratings. The "Recommended Operating Conditions" table will define the conditions for actual device operation.

Connection Diagram

Dual-In-Line Package

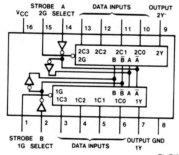

54LS153 (J)   74LS153 (N)

TL/F/6393-1

Function Table

Select Inputs		Data Inputs				Strobe	Output
B	A	C0	C1	C2	C3	G	Y
X	X	X	X	X	X	H	L
L	L	L	X	X	X	L	L
L	L	H	X	X	X	L	H
L	H	X	L	X	X	L	L
L	H	X	H	X	X	L	H
H	L	X	X	L	X	L	L
H	L	X	X	H	X	L	H
H	H	X	X	X	L	L	L
H	H	X	X	X	H	L	H

Select inputs A and B are common to both sections.
H = High Level, L = Low Level, X = Don't Care

Logic Diagram

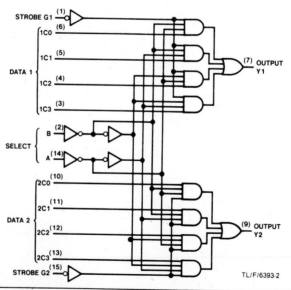

TL/F/6393-2

Reprinted with permission of National Semiconductor Corp.

National Semiconductor

DM54LS165/DM74LS165 8-Bit Parallel In/Serial Output Shift Registers

General Description

These are 8-bit serial shift registers which shift the data in the direction of Q_A toward Q_H when clocked. Parallel-in access is made available by eight individual direct data inputs, which are enabled by a low level at the shift/load input. These registers also feature gated clock inputs and complementary outputs from the eighth bit.

Clocking is accomplished through a 2-input NOR gate, permitting one input to be used as a clock-inhibit function. Holding either of the clock inputs high inhibits clocking, and holding either clock input low with the load input high enables the other clock input. The clock-inhibit input should be changed to the high level only while the clock input is high. Parallel loading is inhibited as long as the load input is high. Data at the parallel inputs are loaded directly into the register on a high-to-low transition of the shift/load input, regardless of the logic levels on the clock, clock inhibit, or serial inputs.

Features

- Complementary outputs
- Direct overriding (data) inputs
- Gated clock inputs
- Parallel-to-serial data conversion
- Typical frequency 35 MHz
- Typical power dissipation 105 mW

Absolute Maximum Ratings (Note 1)

Supply Voltage	7V
Input Voltage	7V
Storage Temperature Range	−65°C to 150°C

Note 1: The "Absolute Maximum Ratings" are those values beyond which the safety of the device cannot be guaranteed. The device should not be operated at these limits. The parametric values defined in the "Electrical Characteristics" table are not guaranteed at the absolute maximum ratings. The "Recommended Operating Conditions" table will define the conditions for actual device operation.

Connection Diagram

Dual-In-Line Package

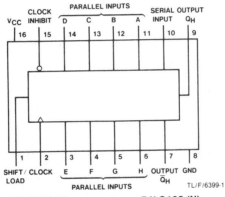

54LS165 (J)   **74LS165 (N)**

TL/F/6399-1

Logic Diagram

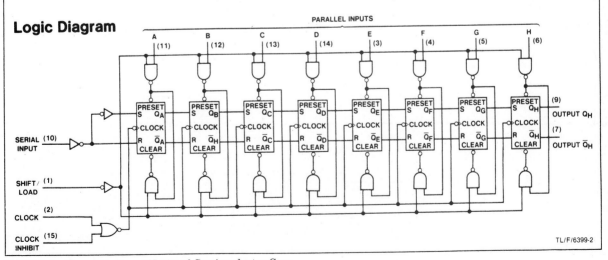

TL/F/6399-2

DM54LS164/DM74LS164 8-Bit Serial In/Parallel Out Shift Registers

General Description

These 8-bit shift registers feature gated serial inputs and an asynchronous clear. A low logic level at either input inhibits entry of the new data, and resets the first flip-flop to the low level at the next clock pulse, thus providing complete control over incoming data. A high logic level on either input enables the other input, which will then determine the state of the first flip-flop. Data at the serial inputs may be changed while the clock is high or low, but only information meeting the setup and hold time requirements will be entered. Clocking occurs on the low-to-high level transition of the clock input. All inputs are diode-clamped to minimize transmission-line effects.

Features

- Gated (enable/disable) serial inputs
- Fully buffered clock and serial inputs
- Asynchronous clear
- Typical clock frequency 36 MHz
- Typical power dissipation 80 mW

Absolute Maximum Ratings (Note 1)

Supply Voltage	7V
Input Voltage	7V
Storage Temperature Range	−65°C to 150°C

Note 1: The "Absolute Maximum Ratings" are those values beyond which the safety of the device cannot be guaranteed. The device should not be operated at these limits. The parametric values defined in the "Electrical Characteristics" table are not guaranteed at the absolute maximum ratings. The "Recommended Operating Conditions" table will define the conditions for actual device operation.

Connection Diagram

Dual-In-Line Package

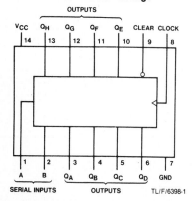

54LS164 (J)   74LS164 (N)

Function Table

Inputs				Outputs			
Clear	Clock	A	B	Q_A	Q_B	...	Q_H
L	X	X	X	L	L	...	L
H	L	X	X	Q_{A0}	Q_{B0}	...	Q_{H0}
H	↑	H	H	H	Q_{An}	...	Q_{Gn}
H	↑	L	X	L	Q_{An}	...	Q_{Gn}
H	↑	X	L	L	Q_{An}	...	Q_{Gn}

H = High Level (steady state), L = Low Level (steady state)

X = Don't Care (any input, including transitions)

↑ = Transition from low to high level

Q_{A0}, Q_{B0}, Q_{H0} = The level of Q_A, Q_B, or Q_H, respectively, before the indicated steady-state input conditions were established.

Q_{An}, Q_{Gn} = The level of Q_A or Q_G before the most recent ↑ transition of the clock; indicates a one-bit shift.

Logic Diagram

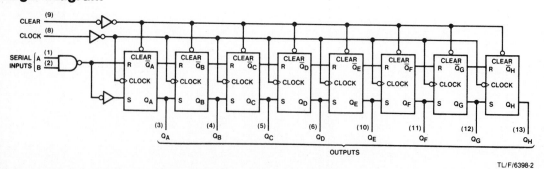

TL/F/6398-2

Reprinted with permission of National Semiconductor Corp.

National Semiconductor

DM54LS90/DM74LS90, DM54LS92/DM74LS92, DM54LS93/DM74LS93 Decade, Divide by 12, and Binary Counters

General Description

Each of these monolithic counters contains four master-slave flip-flops and additional gating to provide a divide-by-two counter and a three-stage binary counter for which the count cycle length is divide-by-five for the LS90, divide-by-six for the LS92, and divide-by-eight for the LS93.

All of these counters have a gated zero reset and the LS90 also has gated set-to-nine inputs for use in BCD nine's complement applications.

To use their maximum count length (decade, divide-by-twelve, or four bit binary), the B input is connected to the Q_A output. The input count pulses are applied to input A and the outputs are as described in the appropriate truth table. A symmetrical divide-by-ten count can be obtained from the LS90 counters by connecting the Q_D output to the A input and applying the input count to the B input which gives a divide-by-ten square wave at output Q_A.

Features

- Typical power dissipation 45 mW
- Count frequency 42 MHz

Absolute Maximum Ratings (Note 1)

Supply Voltage	7V
Input Voltage (Reset)	7V
Input Voltage (A or B)	5.5V
Storage Temperature Range	−65°C to 150°C

Note 1: The "Absolute Maximum Ratings" are those values beyond which the safety of the device cannot be guaranteed. The device should not be operated at these limits. The parametric values defined in the "Electrical Characteristics" table are not guaranteed at the absolute maximum ratings. The "Recommended Operating Conditions" table will define the conditions for actual device operation.

Connection Diagrams (Dual-In-Line Packages)

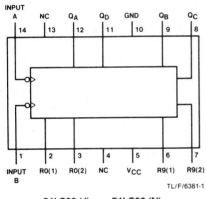

TL/F/6381-1

54LS90 (J)   74LS90 (N)

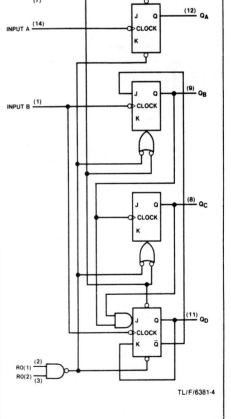

LS90

TL/F/6381-4

LS90 BCD COUNT SEQUENCE (See Note A)

Count	Output			
	Q_D	Q_C	Q_B	Q_A
0	L	L	L	L
1	L	L	L	H
2	L	L	H	L
3	L	L	H	H
4	L	H	L	L
5	L	H	L	H
6	L	H	H	L
7	L	H	H	H
8	H	L	L	L
9	H	L	L	H

LS90 BI-QUINARY (5-2) (See Note B)

Count	Output			
	Q_A	Q_D	Q_C	Q_B
0	L	L	L	L
1	L	L	L	H
2	L	L	H	L
3	L	L	H	H
4	L	H	L	L
5	H	L	L	L
6	H	L	L	H
7	H	L	H	L
8	H	L	H	H
9	H	H	L	L

Reprinted with permission of National Semiconductor Corp.

National Semiconductor

DM54LS192/DM74LS192, DM54LS193/DM74LS193
Synchronous 4-Bit Up/Down Counters with Dual Clock

General Description

These circuits are synchronous up/down counters; the LS192 circuit is a BCD counter and the LS193 is a 4-bit binary counter. Synchronous operation is provided by having all flip-flops clocked simultaneously, so that the outputs change together when so instructed by the steering logic. This mode of operation eliminates the output counting spikes normally associated with asynchronous (ripple-clock) counters.

The outputs of the four master-slave flip-flops are triggered by a low-to-high level transition of either count (clock) input. The direction of counting is determined by which count input is pulsed, while the other count input is held high.

The counters are fully programmable; that is, each output may be preset to either level by entering the desired data at the inputs while the load input is low. The output will change independently of the count pulses. This feature allows the counters to be used as modulo-N dividers by simply modifying the count length with the preset inputs.

A clear input has been provided which, when taken to a high level, forces all outputs to the low level; independent of the count and load inputs. The clear, count, and load inputs are buffered to lower the drive requirements of clock drivers, etc., required for long words.

These counters were designed to be cascaded without the need for external circuitry. Both borrow and carry outputs are available to cascade both the up and down counting

functions. The borrow output produces a pulse equal in width to the count down input when the counter underflows. Similarly, the carry output produces a pulse equal in width to the count down input when an overflow condition exists. The counters can then be easily cascaded by feeding the borrow and carry outputs to the count down and count up inputs respectively of the succeeding counter.

Features

- Fully independent clear input
- Synchronous operation
- Cascading circuitry provided internally
- Individual preset each flip-flop
- Typical count frequency 32 MHz
- Typical power dissipation 95 mW

Absolute Maximum Ratings (Note 1)

Supply Voltage	7V
Input Voltage	7V
Storage Temperature Range	−65°C to 150°C

Note 1: The "Absolute Maximum Ratings" are those values beyond which the safety of the device cannot be guaranteed. The device should not be operated at these limits. The parametric values defined in the "Electrical Characteristics" table are not guaranteed at the absolute maximum ratings. The "Recommended Operating Conditions" table will define the conditions for actual device operation.

Connection Diagram

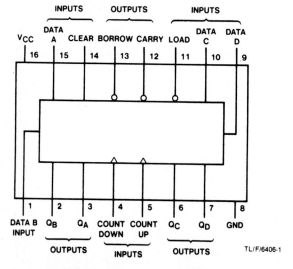

Dual-In-Line Package

Note: Low input to load sets $Q_A = A$, $Q_B = B$, $Q_C = C$, and $Q_D = D$.

54LS192 (J)	74LS192 (N)
54LS193 (J)	74LS193 (N)

TL/F/6406-1

Reprinted with permission of National Semiconductor Corp.

Logic Diagrams (Continued)

LS193

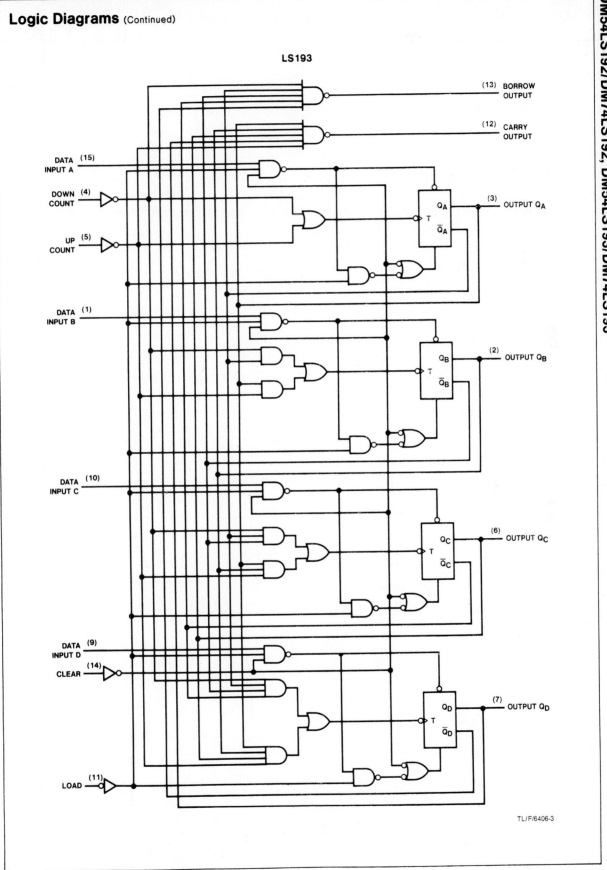

(13) BORROW OUTPUT

(12) CARRY OUTPUT

DATA INPUT A (15)

DOWN COUNT (4)

UP COUNT (5)

(3) OUTPUT Q$_A$

DATA INPUT B (1)

(2) OUTPUT Q$_B$

DATA INPUT C (10)

(6) OUTPUT Q$_C$

DATA INPUT D (9)

CLEAR (14)

(7) OUTPUT Q$_D$

LOAD (11)

TL/F/6406-3

Reprinted with permission of National Semiconductor Corp.

TMS4256, TMS4257
262,144-BIT DYNAMIC RANDOM-ACCESS MEMORIES

MAY 1983 — REVISED JANUARY 1984

- **262,144 X 1 Organization**
- **Single +5-V Supply (10% Tolerance)**
- **JEDEC Standardized Pin Out**
- **Upward Pin Compatible with TMS4164 (64K Dynamic RAM)**
- **Performance Ranges:**

DEVICE	ACCESS TIME ROW ADDRESS (MAX)	ACCESS TIME COLUMN ADDRESS (MAX)	READ OR WRITE CYCLE (MIN)
TMS4256-10 TMS4257-10	100 ns	50 ns	200 ns
TMS4256-12 TMS4257-12	120 ns	60 ns	230 ns
TMS4256-15 TMS4257-15	150 ns	75 ns	260 ns
TMS4256-20 TMS4257-20	200 ns	100 ns	330 ns

TMS4256, TMS4257 . . . JL OR NL PACKAGE
(TOP VIEW)

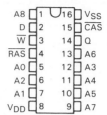

PIN NOMENCLATURE	
A0-A8	Address Inputs
$\overline{CAS}$	Column Address Strobe
D	Data-In
Q	Data-Out
$\overline{RAS}$	Row Address Strobe
$\overline{W}$	Write Enable
V_{DD}	+5-V Supply
V_{SS}	Ground

Dynamic RAM and Memory Support Devices

functional block diagram

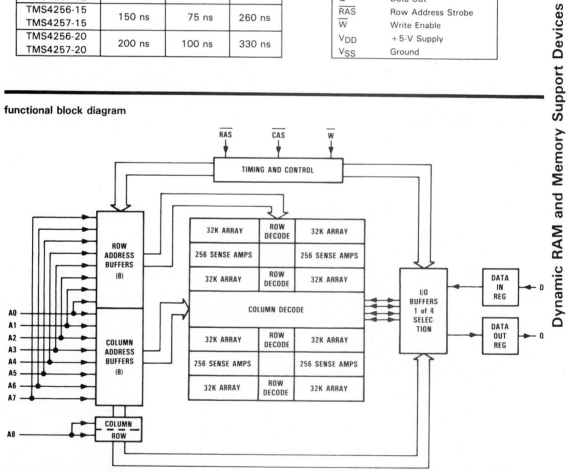

Copyright © 1984 by Texas Instruments Incorporated

TEXAS INSTRUMENTS

POST OFFICE BOX 225012 • DALLAS, TEXAS 75265

Used with permission of Texas Instruments, Incorporated.

National Semiconductor

DM54LS373/DM74LS373, DM54LS374/DM74LS374
TRI-STATE® Octal D-Type Transparent Latches and Edge-Triggered Flip-Flops

General Description

These 8-bit registers feature totem-pole TRI-STATE outputs designed specifically for driving highly-capacitive or relatively low-impedance loads. The high-impedance state and increased high-logic-level drive provide these registers with the capability of being connected directly to and driving the bus lines in a bus-organized system without need for interface or pull-up components. They are particularly attractive for implementing buffer registers, I/O ports, bidirectional bus drivers, and working registers.

The eight latches of the DM54/74LS373 are transparent D-type latches meaning that while the enable (G) is high the Q outputs will follow the data (D) inputs. When the enable is taken low the output will be latched at the level of the data that was set up.

(Continued next page)

Features

- Choice of 8 Latches or 8 D-Type Flip-Flops in a Single Package
- TRI-STATE Bus-Driving Outputs
- Full Parallel-Access for Loading
- Buffered Control Inputs
- Clock/Enable Input Has Hysteresis to Improve Noise Rejection
- P-N-P Inputs Reduce D-C Loading on Data Lines

Absolute Maximum Ratings (Note 1)

Supply Voltage	7V
Input Voltage	7V
Storage Temperature Range	−65°C to 150°C

Note 1: The "Absolute Maximum Ratings" are those values beyond which the safety of the device cannot be guaranteed. The device should not be operated at these limits. The parametric values defined in the "Electrical Characteristics" table are not guaranteed at the absolute maximum ratings. The "Recommended Operating Conditions" table will define the conditions for actual device operation.

Connection Diagrams

Dual-In-Line Package

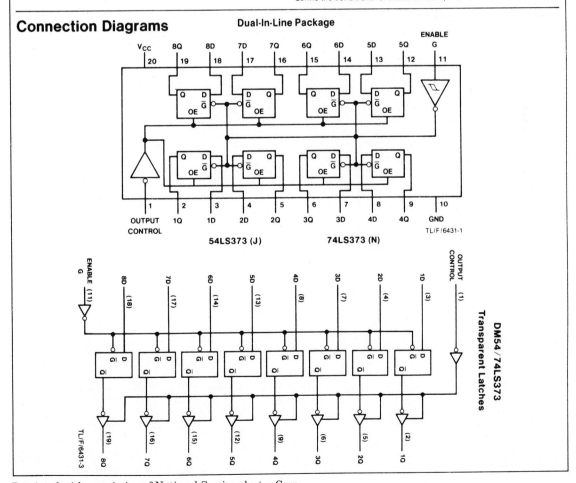

TL/F/6431-1

54LS373 (J)   74LS373 (N)

DM54/74LS373 Transparent Latches

TL/F/6431-3

Reprinted with permission of National Semiconductor Corp.

8155H/8156H/8155H-2/8156H-2
2048-BIT STATIC HMOS RAM
WITH I/O PORTS AND TIMER

- **Single +5V Power Supply with 10% Voltage Margins**
- **30% Lower Power Consumption than the 8155 and 8156**
- **100% Compatible with 8155 and 8156**
- **256 Word x 8 Bits**
- **Completely Static Operation**
- **Internal Address Latch**
- **2 Programmable 8-Bit I/O Ports**

- **1 Programmable 6-Bit I/O Port**
- **Programmable 14-Bit Binary Counter/ Timer**
- **Compatible with 8085AH, 8085A and 8088 CPU**
- **Multiplexed Address and Data Bus**
- **Available in EXPRESS**
  - **Standard Temperature Range**
  - **Extended Temperature Range**

The Intel® 8155H and 8156H are RAM and I/O chips implemented in N-Channel, depletion load, silicon gate technology (HMOS), to be used in the 8085AH and 8088 microprocessor systems. The RAM portion is designed with 2048 static cells organized as 256 x 8. They have a maximum access time of 400 ns to permit use with no wait states in 8085AH CPU. The 8155H-2 and 8156H-2 have maximum access times of 330 ns for use with the 8085AH-2 and the 5 MHz 8088 CPU.

The I/O portion consists of three general purpose I/O ports. One of the three ports can be programmed to be status pins, thus allowing the other two ports to operate in handshake mode.

A 14-bit programmable counter/timer is also included on chip to provide either a square wave or terminal count pulse for the CPU system depending on timer mode.

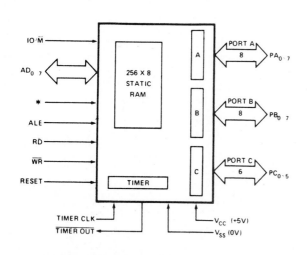

*8155H/8155H-2 = $\overline{CE}$, 8156H/8156H-2 = CE

Figure 1. Block Diagram

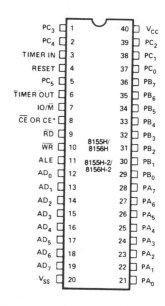

Figure 2. Pin Configuration

Courtesy Intel Corporation, copyright 1989.

8086/8086-2/8086-4
16-BIT HMOS MICROPROCESSOR

- **Direct Addressing Capability to 1 MByte of Memory**

- **Assembly Language Compatible with 8080/8085**

- **14 Word, By 16-Bit Register Set with Symmetrical Operations**

- **24 Operand Addressing Modes**

- **Bit, Byte, Word, and Block Operations**

- **8-and 16-Bit Signed and Unsigned Arithmetic in Binary or Decimal Including Multiply and Divide**

- **5 MHz Clock Rate (8 MHz for 8086-2) (4 MHz for 8086-4)**

- **MULTIBUS™ System Compatible Interface**

The Intel® 8086 is a new generation, high performance microprocessor implemented in N-channel, depletion load, silicon gate technology (HMOS), and packaged in a 40-pin CerDIP package. The processor has attributes of both 8- and 16-bit microprocessors. It addresses memory as a sequence of 8-bit bytes, but has a 16-bit wide physical path to memory for high performance.

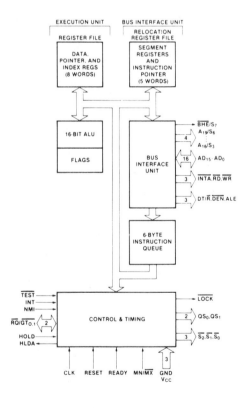

Figure 1. 8086 CPU Functional Block Diagram

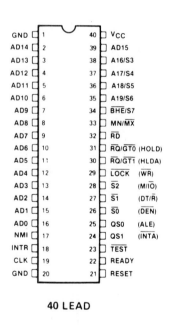

40 LEAD

Figure 2. 8086 Pin Diagram

Courtesy Intel Corporation, copyright 1989.

MOTOROLA

ASYNCHRONOUS COMMUNICATIONS INTERFACE ADAPTER (ACIA)

The MC6850 Asynchronous Communications Interface Adapter provides the data formatting and control to interface serial asynchronous data communications information to bus organized systems such as the MC6800 Microprocessing Unit.

The bus interface of the MC6850 includes select, enable, read/write, interrupt and bus interface logic to allow data transfer over an 8-bit bi-directional data bus. The parallel data of the bus system is serially transmitted and received by the asynchronous data interface, with proper formatting and error checking. The functional configuration of the ACIA is programmed via the data bus during system initialization. A programmable Control Register provides variable word lengths, clock division ratios, transmit control, receive control, and interrupt control. For peripheral or modem operation three control lines are provided. These lines allow the ACIA to interface directly with the MC6860L 0-600 bps digital modem.

- Eight and Nine-Bit Transmission
- Optional Even and Odd Parity
- Parity, Overrun and Framing Error Checking
- Programmable Control Register
- Optional ÷1, ÷16, and ÷64 Clock Modes
- Up to 500 kbps Transmission
- False Start Bit Deletion
- Peripheral/Modem Control Functions
- Double Buffered
- One or Two Stop Bit Operation

MOS

(N-CHANNEL, SILICON-GATE)

ASYNCHRONOUS COMMUNICATIONS INTERFACE ADAPTER

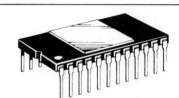

L SUFFIX
CERAMIC PACKAGE
CASE 716

NOT SHOWN: **P SUFFIX**
PLASTIC PACKAGE
CASE 709

MC6850 ASYNCHRONOUS COMMUNICATIONS INTERFACE ADAPTER BLOCK DIAGRAM

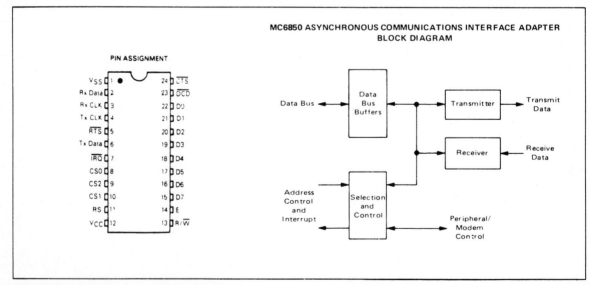

PIN ASSIGNMENT

V$_{SS}$	1	24	$\overline{CTS}$
Rx Data	2	23	$\overline{DCD}$
Rx CLK	3	22	D0
Tx CLK	4	21	D1
$\overline{RTS}$	5	20	D2
Tx Data	6	19	D3
$\overline{IRQ}$	7	18	D4
CS0	8	17	D5
CS2	9	16	D6
CS1	10	15	D7
RS	11	14	E
V$_{CC}$	12	13	R/$\overline{W}$

ANSWERS TO SELECTED DRILL PROBLEMS

CHAPTER 1

1. 1
2. 0
3. 1
4. 0
5. open
6. closed
7. open
8. closed
9. open
10. George Cayley Boole
11. series
12. AND
13. $Z = A \cdot B$
14. input
15. output
17. OR
18. AND
19. base
20. base, collector, and emitter
21. NPN
22. does not
23. PNP
24. bipolar

25. bipolar
26. gate, drain, source, and substrate
27. gate
28. field effect transistor
29. negative
30. positive

CHAPTER 2

2. Four
3. Figure 2-1
4. A and B are the inputs in Figure 2-1
6. 3
7. Figure 2-2
8. Z is the output in Figure 2-2
9. 10
10. 6
11. zero
12. 1011
13. 0100
14. 8
15. 16

16. logic multiplication matches algebra multiplication, but only for binary numbers with one digit
17. AND
18. ·
19. +
20. OR
21. Because, in everyday language, we use AND for "plus" (addition). AND logic is much closer to multiplication than addition, so the results of AND logic don't sound like we expect
22. Figure 2-8
23. A
24. In Figure 2-8, the output is either $\overline{A}$ or Z, depending on where you look
25. 0
26. FALSE
27. $\overline{O} = 1$
28. $\overline{1} = O$

29. NOT
30. No effect

CHAPTER 3

2. Four
3. 3
4. Figure 3-1
5. NOT, AND
7. 1
8. Figure 3-2
9. NOT OR
10. NAND
11. NOR
12. $\overline{A \bullet B} = \overline{A} + \overline{B}$; $\overline{A+B} = \overline{A} \bullet \overline{B}$
13. "Break the bar, and change the sign"
14. one, output
15. two, inputs
16. Figure 3-3 (a)
17. Figure 3-3 (c)
18. NAND and NOR
20. Figure 3-6 (left)
22. 2
23. Figure 3-6 (right)
24. 1
25. 3
26. 3
27. 2
28. 1110
29. 1000
30. 0110

Chapter 4

1. $Q = \overline{D} \cdot \overline{C} \cdot B \cdot A$
2. $Q = \overline{D} \cdot C \cdot \overline{B} \cdot A$
3. $\overline{Q} = D \cdot \overline{C} \cdot \overline{B} \cdot \overline{A}$
4. $\overline{Q} = D \cdot \overline{C} \cdot B \cdot A$
5. $Q = D \cdot C \cdot \overline{B} \cdot \overline{A}$
6. $Q = D + C + B + A$
7. $Q = D + \overline{C} + B + \overline{A}$
8. $Q = \overline{D} + C + B + A$
9. $Q = \overline{D} + C + \overline{B} + \overline{A}$
10. $Q = \overline{D} + \overline{C} + B + \overline{A}$
16. $Q = D \cdot C(B \oplus A)$ ($\oplus$ is an exclusive OR symbol)
17. $Q = \overline{C} \cdot B \cdot A$
18. $Q = D \cdot \overline{C} \cdot B$
19. $Q = \overline{D} \cdot B \cdot A$
20. $Q = D \cdot \overline{C} \cdot \overline{B} \cdot A$
21. $Q = \overline{D} \cdot C$
22. $Q = D \cdot C \cdot B \cdot A$
23. $Q = \overline{D} \cdot C (B + A)$

24. $Q = C (B + A)$
25. $Q = 1$
26. $Q = 0$
27. $Q = 1$
28. $Q = D \cdot \overline{C}$
29. $Q = D \cdot C \cdot A$
30. $Q = \overline{D} \cdot C \cdot A$

CHAPTER 5

1. 9
2. 128
3. 80
4. 13
5. 0101
6. 01111000
7. 01000100
8. 11001000
17. many, one
18. one, many
19. inputs, output
20. outputs, input
21. 6
22. 3
23. 5
24. 0
25. e
26. 4
27. 5
28. 3
29. 0
30. b

CHAPTER 6

1. carry 0, and sum 0
2. carry 0, and sum 1
3. carry 1, and sum 0
4. carry 1, and sum 0
5. carry 1, and sum 1
6. 111
7. 1100
8. 1001
9. 10100
10. 11000010
11. A or B
12. B or A
13. D
14. C
15. borrow 0, and difference 1
16. borrow 0, and difference 0
17. borrow 1, and difference 1
18. borrow 1, and difference 1

19. borrow 1, and difference 0
20. 1000
21. 1010
22. 1110, borrow 1
23. 00100011
24. 11111111, borrow 1
25. A
26. B
27. D
28. C
29. 00000101 (-5)
30. 00000011 (-3)
31. 00000001 (-1)
32. 10000000 (-128)
33. 101000001
34. 001000011
35. 100110000
36. 000110001
37. 001000001
38. 101000011
39. 000110000
40. 100110001

CHAPTER 7

1. 303
2. 666
3. 166
4. 323
5. 011000011 (leading zeroes optional)
6. 100110
7. 011111111 (leading zeroes optional)
8. 001110110 (leading zeroes optional)
9. AB
10. 4F
11. 3D
12. 12
13. 10100001
14. 00111101
15. 11001011
16. 00010110
17. 0010, 0101
18. 0111, 0110
19. 0001, 1001, 1001, 0011
20. 0010, 0000, 0010, 0000
21. 1993
22. 25
23. 99
24. 3210
25. E
26. 6
27. 11-9

28. 0-9

	BCDIC	EBCDIC
29.	001001	11111001
30.	011001	11101001
31.	101001	11011001
32.	111001	11001001

	HEX	BINARY
		(A leading 0 may
		be added to these)
33.	50	1010000
34.	41	1000001
35.	30	0110000
36.	20	0100000

37. 37
38. −91
39. −121
40. −8
41. 10.1
42. 11000011.11
43. 0.011
44. 10110100.000001
45. 26.26
46. 64.64
47. 12.12
48. 99.99
49. 65,535
50. 1,000

CHAPTER 8

1. 1010
2. 1000
4. 22.5°
5. 200K ohms
6. 2.0 mV
7. 2%
8. 5
9. 5
10. 50
11. 50
12. 10%
13. 1%
14. track C, the 4-bit
15. 16
16. 4
17. 1 in 16
18. 22.5°
19. track B, the 2-bit
20. 8
21. 3
22. 1 in 8
23. 45°
24. 5%
25. .4 volts
26. 11

27. 10
28. 5% or less
29. 10
30. no

CHAPTER 9

1. saturated
2. cutoff
3. saturated
4. cutoff
5. pulled-up
6. HIGH
7. larger
8. cutoff
9. saturated
10. cutoff
11. saturated
12. pulled-down
13. LOW
14. smaller
15. NOT
16. pull-down, pull-up
17. no The transistors should not have circles around them in IC's
18. saturated
19. cutoff
20. pulled-up
21. HIGH
22. cutoff
23. saturated
24. pulled-down
25. LOW
26. more
27. complementary symmetry
28. NOT
29. no
30. superconductor

CHAPTER 10

1. high
2. high
3. low
4. low
5. set
6. high
7. high
8. low
9. low
10. reset
11. can't tell without X
12. can't tell without X

13. can't tell without Y
14. can't tell without Y
15. latched
16. high
17. high
18. low
19. low
20. latched set
21. low
22. low
23. high
24. high
25. reset
26. low
27. low
28. high
29. high
30. set
31. can't tell without X
32. can't tell without X
33. can't tell without Y
34. can't tell without Y
35. latched
36. low
37. low
38. high
39. high
40. latched reset
41. low
42. high
43. reset
44. low
45. high
46. set
47. can't tell without old Q
48. can't tell without old $\overline{Q}$
49. latched
50. can't tell without old Q
51. can't tell without old $\overline{Q}$
52. latched
53. synchronous
54. high
55. high
56. level
57. high
58. low
59. set
60. high
61. low
62. reset
63. can't tell without old $\overline{Q}$
64. can't tell without old Q
65. toggle
66. can't tell without old Q
67. can't tell without old $\overline{Q}$
68. latch

69. synchronous
70. asynchronous
71. low
72. low
73. high
74. edge
75. no, yes

CHAPTER 11

1. right
2. one
3. edge
4. falling
5. clears
6. three
7. FIFO
8. serial
9. parallel
10. D or RS
11. parallel
12. three
13. either
14. falling
15. clears
16. one
17. latch
18. parallel
19. parallel
20. D or RS
21. Johnson
22. will not
23. 1
24. 1's
25. six
26. LOW
27. ring
28. Z
29. W
30. X

CHAPTER 12

1. binary
2. asynchronous
3. down
4. 8
5. 2MHz
6. 1MHz
7. 500KHz
9. high
10. does
11. binary
12. synchronous

13. up
14. 8
15. 4MHz
16. 2MHz
17. 1MHz
19. does not
20. 1.92MHz
21. 240KHz
22. 24KHz
23. 1.5KHz
24. 60 Hz
25. 256,000
26. asynchronous
27. up
28. 5
29. 3MHz

CHAPTER 13

1. 3
2. 5v
3. 4v
4. 8, 16, 24
5. 0 to 7
6. 1
7. 3
8. 7
9. 142,857
10. no
11. 6
12. 1
13. 4
14. up-down
15. 255
16. 0 to 255
17. 1
18. 255
19. yes
20. no
21. 7
22. 7
23. 1
24. 4
25. 4
26. 0 to 15
27. 1
28. 4
29. yes
30. yes

CHAPTER 14

1. 1K
2. no

3. destructive
4. large
5. AND
6. 1
7. 1
8. 8
9. 8
10. 8
11. 4K
12. yes
13. common
14. high
15. low
16. low
17. static RAM
18. 4
19. 2
20. 4
21. 1M
22. yes
23. common
24. high
25. low
26. low
27. dynamic RAM
28. 4
29. 2
30. 1

CHAPTER 15

1. 2
2. 4
3. data
4. address
5. 1
6. 4
7. control
8. hardware
9. software
10. 5, 3
11. CPU
12. mem
13. I/O
14. data
15. address
16. control
17. data
18. data
19. CPU
20. CPU
21. 4K
22. 2
23. 4

24. 4
25. low
26. 1M
27. 4
28. 2
29. 1
30. no decoder needed

CHAPTERS 16 AND 17

1. out
2. both
3. out
4. both
5. out
6. in
7. in
8. out
9. in
10. in
11. out
12. out
13. in
14. in
15. both
16. out
17. in
18. out
19. out
20. out
21. out
22. in
23. out
24. both
25. in
26. out
27. out
28. in
29. out
30. out
31. in
32. in
33. in
34. in/both
35. out
36. out
37. out
38. out
39. out
40. in
41. out
42. out
43. out
44. in

45. out
46. in
47. both
48. out
49. in
50. in
51. doc-m
52. doc-m
53. doc-m
54. doc-m
55. doc-m
56. hum-m
57. hum-m
58. hum-m
59. hum-m
60. m-m
61. m-m
62. doc-m
63. m-m
64. hum-m
65. doc-m
66. doc-m
67. doc-m
68. hum-m
69. hum-m
70. hum-m
71. hum-m
72. hum-m
73. hum-m
74. doc-m
75. doc-m
76. hum-m
77. doc-m
78. doc-m
79. hum-m
80. doc-m
81. m-m
82. m-m
83. m-m
84. doc-m
85. hum-m
86. doc-m
87. hum-m
88. m-m
89. doc-m
90. hum-m
91. doc-m
92. doc-m
93. hum-m
94. doc-m
95. doc-m
96. m-m
97. m-m
98. doc-m
99. hum-m
100. m-m

CHAPTER 18

1. 22
2. 10
3. yes
4. yes
5. yes
6. yes
7. yes
8. yes
9. yes
10. yes
11. yes
12. yes
13-21. Clock, Reset, Read, Wait, Write, Interrupt Request, DMA Request, Interrupt Granted, DMA Granted
22-25. Program Counter, Stack Pointer, Flags, Accumulator
26-29. Program Segment, Data Segment, Stack Segment, Extra Segment
30. 16

CHAPTER 19

1. A, 00001000; B, 00001000
2. A, 01000000; B, 01000000
3. A, 01001000; B, 00001000
4. A, 01001001; B, 00001000
5. A, 00111000; B, 00001000
6. A, 00110111; B, 00001000
7. A, 00000000; B, 00001000
8. A, 01001000; B, 00001000
9. A, 01001000; B, 00001000
10. A, 01000000; B, 00001000
11. 0 (positive), 0 (not < 0)
12. 3000H
13. 1003H
14. 3000H
15. 3000H
16. 1003H
17. 3000H
18. 1003H
19. 1003H
20. 3000H
21. AL, 00001000; BL, 00001000
22. AL, 01000000; BL, 01000000
23. AL, 01001000; BL, 00001000
24. AL, 01001000; BL, 00001000
25. AL, 00111000; BL, 00001000
26. AL, 00110111; BL, 00001000
27. AL, 00000000; BL, 00001000

28. AL, 01001000; BL, 00001000
29. AL, 01001000; BL, 00001000
30. AL, 01000000; BL, 00000111

CHAPTER 20

1. START, 1, 2, 3, 2, 3, 2, 3, 4, END
2. 0, 4
3. NO
4. NO
5. YES
6. YES
7. YES
8. NO
9. YES
10. 3
11. START, 1, 5, 2, 6, 7, 9, 8, 6, 7, 9, 8, 3, 9, 4, 10, END
12. 14641
13. NO
14. YES
15. YES
16. YES
17. YES
18. YES
19. YES
20. 1

CHAPTER 21

1. 30
2. 3
3. program
4. both
5. flowchart
6. 80
7. 8
8. flowchart
9. both
10. program
11. 50
12. 5
13. flowchart
14. flowchart
15. program
16. yes
17. yes
18. does not
19. does
20. 5

CHAPTER 22

1. C
2. B

3. A
4. X
5. Y
6. Z
7. Y
8. Y
9. Z
10. W
11. X
12. Y or Z
13. 25
14. ground
15. TD
16. RD
17. RTS (Ready to Send), CTS (Clear to Send), DSR (Dataset Ready), DCD (Data Carrier Detect), DTR (Data Terminal Ready), RI (Ring Indicator)
18. Answer Frequencies:
    Mark = 1270 Hz
    Space = 1070 Hz
    Originate Frequencies:
    Mark = 2225 Hz
    Space = 2025 Hz
19. Amplitude, Frequency, Phase, Pulse-width

GLOSSARY

A/D (*see* Analog-to-digital converter)

Absolute addressing A memory addressing mode wherein the address affected by the machine language instruction is a part of the instruction itself.

Accumulator A register in a CPU which is used by the ALU to store the results of arithmetic-logic operations.

ACIA (asynchronous communications interface adapter; *see* UART)

Active An input or output is active when it is being used.

Active HIGH An input or output that is high when it is being used.

Active LOW An input or output that is low when it is being used.

Active pull-down A switching element that is used to switch the output to ground (Vss) (positive logic).

Active pull-up A switching element that is used to switch the output to Vcc (positive logic).

Actuator An output device that can be used to switch on (and off) a substantial amount of drive power (usually for an electromechanical device).

Addend A number that will be added to another number.

Address A number (usually binary) that enables the transfer of data into or out of one device (like a memory cell) selected from among many.

Address bus A group of lines (conductors) that carry the individual bits of an address.

Algorithm A procedure that yields a computation unavailable as a single instruction, by using a combination of instructions.

Alphameric code (*see* Alphanumeric code)

Alphanumeric code (alphameric code) A binary number code that uses the codes to represent letters, numbers, and punctuation marks, rather than their actual binary value. The codes are symbols for something else.

ALU (*see* Arithmetic-logic unit)

Amplitude modulation Superimposing input information onto a waveform by varying its peak value as the input changes.

Analog A quantity capable of varying continuously [taking on all possible values between its limits (an infinite number)] is analog.

Analog comparator A two-input device whose output is digital, which changes states when one of its inputs becomes larger than the other.

Analog-to-digital converter (A/D) A circuit that takes any continuously variable input and converts its value into a binary code.

AND circuit (AND gate) A circuit whose output is active if all of its inputs are active.

AND gate (*see* AND circuit)

Answer frequency The frequency band a modem receiving a call uses when it is using full-duplex carrier modulation (1270 Hz = mark; 1070 Hz = space).

Arithmetic-logic unit (ALU) The part of a computer's CPU that does arithmetic and logical operations. An adder/subtracter can be an ALU.

ASCII (American Standard Code for Information Interchange) A seven-level alphanumeric code comprising 32 control characters, an uppercase and lowercase alphabet, numerals, and 34 special characters.

Assembler A program for translating assembly-language source code into machine-language object code.

Assembly language A symbolic language to represent machine-code instructions in a format easier for people to understand.

Asserted (*see* Active)

Associative law [A · (B · C) = (A · B) · C = A · B · C] Law which states that when everything is ANDed together, it doesn't matter what was ANDed first and what was ANDed later. (Rule also applies to OR logic.)

Associative principle (*see* Associative law)

Astable multivibrator (square-wave oscillator) A circuit that generates a recurrent square wave by switching back and forth between two unstable states.

Asynchronous counter (ripple counter) A counter in which each stage is the clock for the next stage.

Asynchronous logic Logic circuits whose output responds to their inputs without requiring a clock pulse.

Audio recording A tape recording made on a recorder designed to operate in its linear region. The magnetic material on the tape is never driven by a signal large enough to saturate it magnetically.

Augend A number that will have another number added to it.

Backlash (*see* Hysteresis voltage)

Band printer (*see* Chain printer)

Bandwidth On a frequency spectrum, the area between the lowest and highest frequencies a circuit will respond to (or will produce).

Barcode (*see* OCR)

Base The value a counting system uses powers of to represent each place in a number, or else, simply a number raised to a power.

Base address An address in memory from which a displacement, or relative address, is measured.

BASIC (Beginners' All-purpose Symbolic Instruction Code) A high-level scientific/engineering language, similar to FORTRAN. It is usually interpreted rather than compiled.

Baud rate The number of signaling elements per second transmitted as data. For a binary system, this is the same as *bits per second*.

Baudot code A five-level alphanumeric code that uses two shift characters to reuse the 30 remaining characters twice, as letters and figures. Also called Murray code.

BCD code (*see* Binary-coded decimal)

BCD full-adder A circuit that adds BCD numbers by adjusting the results of a binary adder. Software does this with a DAA instruction. (*see* Decimal Adjust)

BCD interchange code (*see* BCDIC)

BCDIC (binary-coded decimal interchange code) A six-level alphanumeric code which provides alphabetic (caps), numeric, and 28 special characters.

Belt printer (*see* Chain printer)

Bidirectional Capable of carrying signals in two directions. Wires are bidirectional; gates, generally, are not. A common I/O bus used for input and output alternately must be buffered by bidirectional buffers.

Binary code Representation of numbers in which the only two digits used are 0 and 1, and each position's value is twice that of its right-hand neighbor (with the rightmost place having a value of 1).

Binary counter A counter that counts in natural binary code.

Binary flash converter An A/D converter that uses D/A ladders, analog feedback, and comparators to digitize input voltage. Its output is natural binary code.

Binary fractions Numbers to the right of the "decimal point" in a binary number have values of $\frac{1}{2}$, $\frac{1}{4}$, and so forth, each half the place value of its left-hand neighbor. The sum of all the 1s in these places is a fraction less than 1.

Binary numbering system (*see* Binary code)

Binary-coded decimal (BCD) code A code in which the first ten 4-bit (hexadecimal) codes are used for the decimal digits, and each nibble represents 1 decimal place. Codes "A" through "F" are never used.

Binary-to-decimal decoder A circuit with four inputs and 10 outputs that activates one output suitable to the BCD number put into its inputs.

Binary-to-Gray decoder A circuit that converts a binary number into a Gray-code number with an equal number of bits.

Biphase (*see* Manchester standard)

Bipolar transistor A solid-state switching element whose output current is controlled by its input current.

Biquinary counter (symmetrical divide-by-10) A modulus-10 counter, whose most-significant bit is high for five counts, and low for five counts. The code it produces is not BCD.

Bistable multivibrator (*see* Flip-flop)

Bit A binary digit. A 1 or a 0.

Bit stream (*see* Timing diagram)

Bit switches An input that enters one binary digit for each switch.

Block transfer A data transfer that moves more than one byte in a computer with a single opcode.

Boilerplate (slang) The software needed for setup and configuration at the beginning of a program used, for instance, to initialize a PIA.

Boolean expression A symbolic representation of the input states required to get an output in a logic system.

Boolean identity A basic theory of logic, such as NOT ON = OFF.

Boolean simplification Finding a reduced, but equivalent logic expression for a logic system.

Bootstrap program A machine-code routine (usually in ROM) that is activated at power-up, and used to load the operating system or monitor.

Borrow An output (or input) of a binary subtracter circuit that becomes active when a large digit is subtracted from a smaller one.

Breadboard A temporary solderless hookup of the components of a circuit for testing purposes. (Components were originally clipped together with Fahnestock clips on little wooden blocks. The resulting assembly of wooden blocks reminded someone of a breadboard, hence the name.)

Broadside load (*see* Parallel load)

Bubble memory A serial magnetic memory device with no moving parts.

Buffer 1. A gate whose output is logically the same as its input. The output may be disabled by making a second input, the *enable*, inactive. If the output has more drive power than the input, it is called a *driver*. 2. A block of memory used as a temporary storage area for data.

Bus-oriented system A system designed with a common set of conductors used to synchronize the transfer of data to all parts of the system.

C-MOS (complementary-symmetry MOSFET logic) A series of logic devices which have active pull-ups and active pull-downs. The switching elements used for the pull-ups are P-channel MOSFETs and the pull-downs are N-channel MOSFETs.

Call An instruction that saves the program counter on the stack before placing a new address on the program counter. Used to go to subroutines.

Carry An output (or input) of a binary adder circuit that becomes active when a sum is larger than 1. The output when a sum exceeds the size of the accumulator intended to receive it.

Carry flag A flip-flop where the carry output of an adder goes when the sum exceeds the size of the accumulator. Often also used for the borrow output of a subtracter when the result is less than zero.

CAS (*see* Column-address strobe)

Cathode ray tube (CRT) The "picture tube" of an oscilloscope or television set. It uses an electron beam, steered by electric or magnetic fields, to trace patterns on its screen. ("Cathode rays" = electrons.)

CD-ROM (*see* Laser disk)

Central processing unit (CPU) The portion of a computer that contains the ALU and circuitry to fetch, decode, and execute instructions.

Chain printer (band or belt printer) A high-speed impact printer using a horizontal moving-font type chain struck by print hammers.

Character generator ROM A memory with patterns burned-in to each address that can be sent serially to produce a dot-matrix pattern for a CRT raster display. Each address contains one row of a character's matrix. Also used for matrix printers.

Check bit A parity bit that makes the total number of 1s in a coded character even (for even-parity code) or odd (for odd-parity code).

Clear (reset-direct) An asynchronous reset input on a flip-flop.

Clock A gating pulse or wavetrain that enables part of a circuit. Often, this wavetrain triggers a circuit as it rises or falls.

Code wheel (*see* Disk encoder)

Coincident-current technique In core memory, a technique that allows control of many locations with few wires. Each location is at the intersection of two currents, each too small to have any effect, but collectively strong enough to change the state of the location (AND logic).

Column-address strobe (CAS) A signal that becomes active when one-half of the address (the most-significant bits) are sent to a dynamic RAM integrated circuit. The other half is sent through the same pins when a signal called RAS (row-address strobe) is active.

Common bus A conductor or group of conductors that are shared by a large number of devices. Each device "takes its turn" using the bus.

Common carrier A vendor of telecommunication services, available to the public.

Common I/O In memory devices, a scheme whereby the same pins on an integrated circuit are alternately used for data-out and data-in.

Commutative law [A + B + C = A + C + B = B + C + A, etc.] Law which states that the order of terms in a logic sum, or factors in a logic product, can be rearranged without affecting the overall logic of the expression.

Commutative principle (*see* Commutative law)

Comparator In digital logic a circuit that compares two multidigit binary-code inputs, and delivers an output indicating if they are equal or unequal. (*see also* Analog Comparator)

Compiled language A high-level language (e.g., FORTRAN) that is the source code for a compiler.

Compiler A program that translates high-level source code into machine-language object code.

Complement (invert) In a binary word, all 1s are changed to 0s, and all 0s into 1s.

Computer A programmable calculating machine capable of carrying out a sequence of instructions.

Computer card (*see* Punchcard)

Condition codes register (*see* Flag byte)

Conditional branch An instruction that puts a new value into the program counter if a logical condition (usually about a flag) is true.

Constants Numbers used in a computer program which maintain the same value throughout the course of the program run. Usually, these are written in the program as numbers, not represented symbolically by names.

Constraint A condition to avoid. The unstable output of a flip-flop when inputs for SET and RESET are both active at the same time.

Control bus A group of signals used to synchronize actions in a computer system, which are not data or address information.

Control line An input that affects the way in which a logic device processes other signals.

Control word A binary pattern sent to a programmable device to configure its mode of operation.

Controlled inverter A gate that passes a signal through itself without inversion, until a control line becomes active. Then the gate inverts the signal passing through itself.

Controller (*see* Actuator)

Conversion time The time it takes an A/D device to complete the translation of an analog input level into binary code. Usually, this depends on the number of bits of binary output.

Core memory A random-access magnetic memory device with no moving parts; it accesses each location through a grid of current-carrying wires.

Core plane A two-dimensional grid of wires with a magnetic ferrite toroid at every intersection. It allows access to one bit at a time.

Counter-type A/D converter A circuit that compares the analog output of a counter-driven D/A to the input, stopping the counter when they match.

CPU (*see* Central processing unit)

Cross-coupled A circuit with feedback wherein the output of one section is the input for the other, and vice versa. Usually a flip-flop.

Crossover slivers (*see* Glitch)

CRT (*see* Cathode ray tube)

D/A (*see* Digital-to-analog converter)

D latch (*see* D-type flip-flop)

D-type flip-flop A synchronous circuit with one data (D) input that copies the level of the input at its output when the clock is triggered.

Daisywheel An impact printer mechanism with a rotating typefont wheel. Characters are on separate "petals" struck by a print hammer.

Data Binary information that is the operand of an instruction or program, rather than the codes for the instructions.

Data bus A bidirectional multiline path that transfers the contents of memory, ports, or CPU registers back and forth in a computer.

Data direction register A register in a PIA that controls which pins of the port are inputs, and which are outputs.

Data set A modem or other interface adapter between a digital system and common-carrier telecommunications equipment.

Data terminal A source or destination of data. An instrument that allows entry and reception of data into a network.

Data transfer instructions Instructions that copy binary information from a source to a destination in a computer system.

DB-25 plug A connector commonly used with the RS-232 interface. It has 25 pins or sockets in two rows arranged in a trapezoidal profile.

Debouncer A circuit or software that converts a closure of metallic contacts (which may "bounce" open and closed several times before they settle into solid contact) into a single transition.

Decade counter A modulus-10 counter that counts in BCD code.

Decimal adjust (BCD adjust) A process that adds the number six to any nibble where a number greater than 10 has resulted from BCD addition.

Decimal-to-binary encoder A circuit with 10 inputs and four outputs, that produces appropriate binary code when any input is activated.

Decoder A device whose output is active for only one pattern of binary digits at its inputs.

Decoder matrix A circuit with one group of inputs and many outputs. It has one active output for each code pattern presented to its inputs.

Decoupling capacitor A capacitor attached to the power leads of an IC to suppress switching noise coupled to the power supply from the IC.

Dedicated machine A computer that always runs the same program.

DeMorgan's theorem "Break the bar and change the sign." An inverted Boolean expression is equivalent to one where the inversion bar over the whole expression has been broken between the parts of the expression, and the AND signs changed to ORs, or ORs changed to ANDs.

Demultiplexer (DEMUX) A circuit that selects which of several outputs an input will be routed to. An address applied to the select lines determines which of the outputs will receive the input.

DEMUX (*see* Demultiplexer)

Designation number A binary number made by listing the outputs of a logic system's truth table, in binary counting order (e.g., 0110 = XOR).

Device code An address for an I/O device that selects which of the I/O system devices should be activated.

Device decoder A circuit which identifies that a specific I/O device has been selected, and activates it.

Device-select pulse (DSP) The strobe, clock, or enable pulse used to activate an I/O device when its device code is received.

Diagnostic A program for troubleshooting.

Difference An output of a subtracter circuit that shows what the minuend-minus-subtrahend or minuend-minus-subtrahend-minus-borrow is.

Differentiator An analog circuit is proportional to the rate at which its input signal is changing.

Digital A system having a discrete number of output levels (usually two), and a finite number of possible states, is digital.

Digital clock A timekeeping instrument with numeric displays and no moving parts.

Digital voltmeter A measuring instrument that indicates the level of voltage at its input on numeric displays.

Digital-to-analog converter (D/A) A circuit that converts its binary code input into a voltage (or some other continuously variable quantity).

Digitizer (*see* Analog-to-digital converter)

Diode logic (DL) Logic circuits that use diodes as their only active switching elements. Only AND and OR are possible.

Diode matrix encoder (*see* Encoder matrix)

Diode-transistor logic (DTL) Logic circuits that combine diodes and bipolar transistors as their active switching elements.

DIP (*see* Dual-in-line package)

Direct addressing (*see* Absolute addressing)

Direct memory access (DMA) A peripheral device that reads or writes data directly into memory by generating its own address and control signals, while the computer is put into a "hold" condition, is using DMA.

Disable (inhibit) To put a logic device into a state where it blocks or stops a signal that would otherwise go through it.

Discrete circuit A circuit that is assembled from electronic devices, each of which is in its own separate package.

Disk encoder A wheel or disk marked with sectors of binary code so that the code can be read off the wheel by sensors as it turns.

Disk operating system (DOS) A software system of programs or modules that controls the use of peripheral (I/O) devices in a computer system, especially the disk drives.

Disk pack A multiple-platter assembly of magnetic disks. It is part of a hard-disk system, and may be removable.

Diskette (*see* Magnetic disk)

Displacement In computer addressing, this is a number that indicates how far away the memory location is from some reference point location. Relative jump instructions and instructions that use indexed addressing contain a displacement from a location, rather than the whole address.

Display A digital output device that provides a visible indicator of what's sent to it.

Distributive law [A · (B + C) = A · B + A · C] Law which states that a variable can be ANDed onto a group of other variables in the expression by ANDing it with each variable separately. (Rule also applies to OR logic.)

Distributive principle (*see* Distributive law)

Divider chain A cascade of flip-flops in which each one's output is the clock signal for the next. The last stage's output frequency is the input frequency divided by 2^n, where n is the number of flip-flops.

Divider String (*see* Divider chain)

DL (*see* Diode logic)

DMA (*see* Direct memory access)

DOS (*see* Disk operating system)

Dot-matrix display A digital output device that provides a visible indicator of what's sent to it, in the form of a symbol composed of dots.

Dot-matrix printer A printer that forms characters on paper as an array of dots.

Double inversion Any digital signal that is inverted an even number of times has the same logic as if it hadn't been inverted at all.

Double precision A binary number with twice as many bits as the standard word size of the system it occupies.

Down-counter A binary counter that counts in reverse, decreasing the number it contains with each clock pulse.

DRAM (*see* Dynamic RAM)

Driver (*see* Buffer)

Drum memory A rotating magnetic memory storage device that has been largely replaced by disk drives in modern systems.

Drum printer A high-speed impact printer using a vertical moving-font rotating print drum struck by print hammers.

DSP (*see* Device-select pulse)

DTL (*see* Diode-transistor logic)

Dual-in-line package (DIP) A carrier for an integrated-circuit chip. It has multiple-pin connectors arranged along the two long sides of a rectangular plastic or ceramic holder that has the chip embedded inside.

Dual-slope A/D converter A digitizer that uses an up-down counter and an up-down ramp wave to reduce the conversion time of samples taken from a variable input after the initial sample is digitized.

Dyadic An operation that requires two operands. Addition is a dyadic operation, since it must have an addend and an augend. So is subtraction.

Dynamic RAM (DRAM) A random-access memory device that uses a MOSFET's interelectrode capacitance as a storage element. Due to leakage, data in DRAM must be periodically rewritten or they will fade away.

EBCDIC (expanded BCD interchange code) An eight-level alphanumeric code comprising control codes, an uppercase and lowercase alphabet, numerals, and special characters.

ECL (*see* Emitter-coupled logic)

Edge detector A circuit whose output is a short pulse produced when there is a change in the state of the input.

Electromechanical device A device containing parts that move when driven by an electrical signal.

Emitter-coupled logic (ECL) Logic devices containing bipolar transistors, which are faster than other bipolar logic but consume more electrical power to operate.

Enable To put a logic device into a state where it allows a signal to go through it. (Sometimes called *transparent*)

Encoder A device whose outputs become active with an appropriate binary pattern whenever its input becomes active.

Encoder matrix A multi-input circuit which, when one input becomes active, delivers an appropriate code pattern to its outputs.

Equality gate (*see* XNOR)

Even parity An arrangement in which the bit count of every coded character always contains an even number of 1s.

Excitation table A table that describes what inputs would be needed to get certain changes at the outputs to occur.

Exclusive-NOR (*see* XNOR)

Exclusive-OR (*see* XOR)

Expanded BCD interchange code (*see* EBCDIC)

Exponent The part of a scientific-notation or floating-point number which is used to relocate the *mantissa's* decimal point (decimal example: in 1.234×10^{23}, the exponent is 23). The exponent determines the number of decimal (or binary) places to shift the point.

Factoring A method of simplification that involves extracting a common factor from all terms in which it occurs, so that the factor appears only once, multiplied by all the terms.

Factors Things with "·" signs between them.

Falling-edge triggered (trailing-edge triggered) A device whose inputs are enabled when a one-to-zero transition of the clock occurs.

False A logical input or output which is not operating in the condition in which it was intended to be operating.

Fan-in The amount of current required to drive the input of a logic device. Whichever condition (1 or 0) requires the most current defines the fan-in of the device. The fan-in for an elementary input is called a *unit load*.

Fanout The number of inputs of similar devices that can be driven by a device's output. Fanout is usually measured in *unit loads*.

Feedback The return (input) of information about a system's output that the system uses to modify that output.

Fiber optics (light pipe) A conductor for light that allows it to be carried around corners.

Field-effect transistor A solid-state switching element whose output current is controlled by its input voltage.

Flag byte (condition codes register) A group of flip-flops (a register) which holds the status of a mathematical result. Examples are the zero flag, sign flag, and carry flag.

Flash converter (A/D network) An A/D converter that uses a voltage divider and comparators to digitize input voltage. Its output is not in natural binary code, and is usually of fairly low resolution.

Flatbed plotter (*see* Plotter)

Flip-flop The solid-state logic equivalent of a latching switch, which can be set or reset into either of two stable states. The inputs are enabled by a clock input, generally by a transition of its level.

Floating high A disconnected input that registers as a 1 to the internal logic of its logic device.

Floating low A disconnected input that registers as a 0 to the internal logic of its logic device.

Floating-point number The binary-number version of scientific notation (decimal example 1.234×10^{23}), which has a binary *mantissa* and a binary *exponent* stored as separate parts of one number.

Floating-point variable (*see* Floating-point number)

Floppy disk (*see* Magnetic disk)

Flowchart A block diagram of software. A flowchart bears the same relation to a machine language program that a schematic diagram does to an actual electronic circuit.

Forward bias The polarity that makes a diode or other semiconductor junction conduct current, namely, + to P-type and − to N-type material.

FPLA (*see* Programmable logic array)

Frequency How often a repeated action occurs per second. The units of frequency measurement are called hertz.

Frequency modulation Superimposing input information onto a waveform by varying its frequency as the input changes.

Frequency-division multiplex (FDM) Transmission of multiple channels of information on a single conductor by modulating each channel onto a separate carrier frequency.

Frequency-shift keying (FSK) A variant form of frequency modulation in which there are only two frequencies used, representing the two levels of logic.

Frozen An output is frozen when it stays HIGH or LOW without any response to the input. A disabled gate has a frozen output.

Frozen high (*see* Frozen)

Frozen low (*see* Frozen)

FSK (*see* Frequency shift keying)

Full duplex Communication on one telecommunication link in two directions at the same time.

Full-adder A binary adder circuit with three inputs that adds addend + augend + carry to form a sum.

Full-adder/subtracter A circuit that can be used to

add with carry, or subtract with borrow, depending on the state of a control line.

Full-subtracter A binary subtracter circuit with three inputs, that subtracts minuend − subtrahend − borrow to form a difference.

Fusible link A resistive element on an integrated circuit that is melted or vaporized by a pulse of current to record a binary digit.

Game controller An analog input device, such as a potentiometer, which is used for proportional control of things on a video screen.

Gated (*see* Enable)

Glitch An output that becomes active when it's not supposed to.

Graphic plotter (*see* Plotter)

Gray code A binary code that doesn't follow the positional notation of true binary code. Only one bit changes from any Gray number to the next.

Gray-to-binary decoder A circuit that converts a Gray number into a binary code number with an equal number of bits.

Half-adder A binary adder circuit with two inputs that adds addend + augend to form a sum.

Half-adder/subtracter A circuit that can be used to add (without carry) or subtract (without borrow) depending on the state of a control line.

Half-duplex Communication on one telecommunication link in two directions, one at a time.

Half-subtracter A binary subtracter circuit with two inputs that subtracts minuend − subtrahend.

Half-carry flag A flip-flop that is set whenever the result of a binary addition results in a nibble greater than 10.

Handshaking The return of a signal indicating that another signal has been received.

Hard disk (*see* Magnetic disk)

Hardcopy Computer output printed or plotted on paper.

Hardware Actual circuit components, wiring, and connections.

Hexadecimal byte code A two-digit hexadecimal number for each byte, having a value between 00 and FF.

Hexadecimal code A base-16 number code which uses the letters A, B, C, D, E, and F as one-digit numbers for 10 through 15.

High On or positive (in positive-logic). Same as TRUE or 1.

High-impedance state An open or nonconducting condition in which the terminal (usually an output) does not pull up or down on a bus line.

High-level language A language that uses machine-independent statements which are more similar to programming at the flowchart level than at the machine-language level. (*see* Macro)

High-Z state (*see* High-impedance state)

Hollerith card (*see* Punchcard)

Hollerith code A 12-level binary code used on punchcards to represent alphanumeric characters. Holes on the punchcard are 1s, unpunched = 0s.

Horizontal sync A pulse in the video signal used to trigger horizontal retrace at the beginning of each scan line.

Hysteresis A memory or partial memory effect, in which a state triggered by a certain signal level remains even after the signal level is lowered below the triggering threshold.

Hysteresis loop A graph of response versus signal level that shows when a response remains after the signal is reduced below trigger level.

Hysteresis voltage In a Schmitt trigger, the difference between the upper trip point voltage and the lower trip point voltage.

I/O device (input/output device) A device that converts some action in the outside world into digital code for a computer, or vice versa.

I/O mapped (true I/O) An addressing scheme in which addresses exist for the I/O ports which are the same as memory addresses, but the computer system can tell them apart because a different control signal is active.

IIL (*see* Integrated injection logic)

Immediate addressing A computer addressing mode in which the source data is contained in the instruction itself, not in memory or a register.

Impact printer A printing mechanism that transfers ink to paper by using a metal hammer to press an inked ribbon against the paper. The pressure allows carbon copies to be made as the original is being printed.

Impedance Opposition to the passage of electrical current. A circuit that does not allow *any* current flow is called an *open circuit*.

Implementation in random logic (*see* Random logic)

Inactive An input or output is inactive when it is not being used.

Indexed addressing A computer addressing mode; the address is the sum of an operand in the instruction, and a number in an index register.

Inequality gate (*see* XOR)

Infix notation 1. A way of writing assembly language in which addresses are written inside parentheses to indicate that they are not data [example: MOV A, 5 puts a 5 into A, while MOV A, (5) gets the number in *memory location* 5 and puts it into A].
2. A method of representing mathematical operations in which the operation code is between the operands [example: in A + B = . . . the operation code (+) is between A and B. ADD A, B is an example of *prefix* notation and A (enter) B (add) is *postfix*, also called "reverse Polish" notation].

Inhibit (*see* Disable)

Inhibit wire A line used in three-dimensional core memory to disable a core plane.

Initialization Setting up the starting values needed in a program before the actual computing can be done. This includes PIA setup.

Ink-jet printer A nonimpact printer that transfers ink onto the paper by squirting it in droplets through a nozzle, steering the droplets by electrostatic deflection.

Input The part of a digital system where signals that control the system enter.

Input data bus In as separate I/O system, the pins where data enter the system are an input data bus. They are attached to the inputs of memory latches, and are on separate pins from the latch outputs.

Input device In a computer system, the devices that convert actions in the outside world into digital code for the computer are input devices.

Input port An interface that allows data from an input device to reach the computer's data bus *only* when it is selected (by an address).

Instruction decoder The section of an I/O system that enables or strobes the input and output ports when each one's address appears on the address and control buses.

Instruction pointer (*see* Program counter)

Instruction set The binary codes used by the manufacturer of a computer for its operations. The operations made available by the manufacturer are also called the instruction set of the computer.

Integrated circuit A circuit in which all the components are in a single package, usually assembled on the surface of a single piece of semiconductor material. If they are in separate packages, they are called *discrete components*.

Integrated injection logic (IIL) A bipolar family of logic having a multicollector transistor (the injector) to pull up its inputs, and an open-collector type of output configuration.

Inter-record gap (IRG) A space between data blocks on magnetic tape. Blocks (records) on multi-track tape are separated by spaces to allow starting and stopping time on reel-to-reel machines.

Interblock gap (*see* Inter-record gap)

Interconnects The conductors used to join components together, either in discrete circuits or integrated circuits.

Interpreted language A high-level computer language in which each statement is identified, then used to call an appropriate machine-language subroutine. The source code is never converted into an object program.

Interpreter A program that identifies statements in a high-level language, then calls an appropriate machine-language subroutine for each.

Interrupt A hardware-driven subroutine call.

Interrupt acknowledge (interrupt granted) A response to "interrupt-request" which is output from the central processor when it can accept the request signal, and begin an interrupt.

Interrupt flag A flip-flop or latch in the computer's flag byte that indicates the enabled or disabled status of an interrupt input.

Interrupt request A hardware signal that becomes active when an external device wants the computer to execute an interrupt subroutine.

Interrupt service routine (ISR) A subroutine that is called when a hardware signal (interrupt request) is received and acknowledged.

Inverted input An input whose 1s become 0s, and vice versa, before affecting the internal logic of a device. In positive logic, this is the same as an active-LOW input.

Inverted output An output whose 1s become 0s, and vice versa, before the internal state of the device logic reaches the outside.

Inverter (*see* NOT circuit)

IRG (*see* Inter-record gap)

ISR (*see* Interrupt service routine)

I²L (*see* Integrated injection logic)

J-K flip-flop A flip-flop that is set if it is clocked and its set (J) input is active, provided that it is *not already set*—and that is reset if its reset (K) input is active, provided that it is *not already reset*.

Johnson counter (shift counter) A shift register with feedback, set up so that it will fill with 1s when clocked, until filled, then will fill with 0s until filled, then begins the whole cycle over.

Josephson junction A superconducting device with latching capability which can be set and reset in a picosecond or less.

Joystick A controller for a computer which usually contains two potentiometers and is used to digitize the x-y position of the handle in two-dimensional movements.

Jump A computer command that jams a new number into the program counter and forces processing to resume with the next instruction at a different memory location than the counter would normally go to.

Kansas City standard A tape recording standard of FSK, 300-baud modulation, in which a mark = 8 cycles of 2400 Hz and a space = 4 cycles of 1200 Hz.

Keyboard (keypad) An array of momentary-contact switches used to enter numeric or alphanumeric information into a computer.

Keypressed signal A signal that becomes active whenever any key on a keyboard is pressed. It helps identify when data output from the keyboard is legitimate, and sometimes, whether it has had time to stabilize.

Label A symbolic representation of an address in assembly language; it is used to identify a line of code for future reference.

Ladder network (ladder-type D/A converter) A network of resistors used to deliver an analog voltage level to an output. The level is proportional to the binary code applied to its inputs.

Lamp driver A logic output device with enough voltage and current handling ability to light a lamp.

Laser disk A high-density storage medium for digital code. This has been read-only technology in the past, but now includes the read/write magneto-thermo-optical disk.

Latching switch A switch that "remembers" to stay on (or off) after you let go of its control handle, as opposed to a momentary-contact switch, which goes on when you push its button, off when you let go of the button.

LCD (*see* Liquid-crystal display)

Leading-edge triggered (*see* Rising-edge triggered)

LED (*see* Light-emitting diode)

LED driver (*see* Lamp driver)

Level-triggered latch (transparent latch) A latch that responds to its data input (Q = D) as long as its clock/enable is active, and then latches its last state when the clock/enable becomes inactive.

Light-emitting diode (LED) A semiconductor that emits photons of light when its electrons release energy as they fall into holes.

Light pen A computer input device used to draw on a CRT screen.

Light pipe (*see* Fiber optics)

Liquid crystal display (LCD) An electro-optical display device that works by rotating polarized light when its electrodes are activated.

Literals (*see* Strings)

Location (*see* Memory cell)

Logic addition OR logic. The "+" symbol is used for OR because its truth table is similar to that for binary addition.

Logic multiplication AND logic. The "·" symbol is used for AND because its truth table is identical to that for binary multiplication.

Logic probe A test instrument that uses lights to display the TRUE and FALSE conditions of a test point. Additional indications vary from model to model. The probe is generally a static-test instrument.

Logical address An address mapped into a smaller number of bits than the real memory requires. It is combined with a number in a register to form the *physical address*.

Loop A software structure that allows the computer to "recycle" a group of instructions instead of requiring them to be written over and over in the program code for repeated use.

Low Off, or not positive (in positive logic). Same as False or 0.

Low-power Schottky (LS TTL) A bipolar family of logic in which low-voltage Schottky diodes speed up performance and reduce power needs.

LS TTL (*see* Low-power Schottky)

Machine code The binary codes used internally when a computer runs a program. (*see also* Instruction set)

Machine language (*see* Machine code)

Macro A pre-defined sequence of machine-language instructions treated as a single opcode mnemonic in assembly language. One macro in source code generates many lines of object code.

Magnetic disk A random-access mass-memory storage medium used to hold computer programs and data separately from the computer memory.

Magnetic tape A serial-access mass-memory storage medium used to hold computer programs and data separately from the computer memory.

Main program The calling program for a subroutine call. The RETurn command at the end of the subroutine comes back to this program.

Manchester standard (phase-encoded recording) Also called "biphase," this is a magnetic-recording standard that uses the rising edge of a wave to encode a 0, and the falling edge to encode a 1.

Mantissa The part of a scientific-notation or floating-point number which is multiplied by the *exponent* (decimal example: in 1.234×10^{23}, the mantissa is 1.234). The mantissa determines the precision of the number.

Mark A logic 1 in FSK encoding or telecommunication transmission.

Masked ROM A read-only memory chip which uses factory-placed wiring on the surface of the integrated circuit for 1s and 0s.

Master/slave A two-part design (flip-flops) which allows the inputs to affect the flip-flop while its lock is inactive, but the output doesn't "get out" until the clock becomes active. Avoids *racing*.

Matrix printer (*see* Dot-matrix printer)

Maxterm A Boolean expression describing the inputs required to get a 0 at the output, usually taken from a line of a truth-table.

Maxterm sum (*see* Product-of-sums expression)

Mean time between failures (MTBF) The average number of times a system can function properly without a "mistake."

Meatball (slang) The circle used at an input or output in positive-logic to indicate inversion or active-LOW status.

Memory A latching device that contains words (latches of one or more bits) of which one is selected at a time by an address, for reading or writing.

Memory cell A single latch of one or more bits, located inside a memory chip with a lot of others having different addresses. Sometimes a cell is called a *location*.

Memory mapped An addressing scheme in which some memory addresses are used for the I/O ports; each port is wired into a place that would otherwise have been a memory location.

Memory refresh (*see* Refresh)

Microprocessor A single-chip device that contains all or most of the parts of a computer's CPU.

Minterm A Boolean expression describing the inputs

required to get a 1 at the output, usually taken from a line of a truth table.

Minterm sum (*see* Sum-of-products expression)

Minuend In subtraction, the number you take-away-from is the minuend.

Mnemonic A symbolic (assembly-language) representation for an opcode.

Mode A state of operation where the inputs of a digital device have certain effects that they may not have in other modes.

Modem A modulator/demodulator that converts logic levels to FSK (similar to FM) modulation, and vice versa.

Modulation Superimposing information onto a carrier sine wave by varying one of its parameters, such as its amplitude, frequency, phase angle, pulse width, or timing (pulse position).

Modulus A number that described how many different states a synchronous digital system will have as it is clocked.

Monitor program A program that handles input, output, and operation of a small computer system at a low level—a higher-level program for a larger system is called an *operating system*.

Monostable multivibrator (one-shot) A device with one stable state that can be triggered into its unstable state, but returns to its stable state after a short while.

Morse code A pulse-width code developed in the 1850s for the telegraph system. It was the first digital serial data transmission code to become commercially successful.

MOSFET (*see* Field-effect transistor)

MTBF (*see* Mean time between failures)

Multiplexed readout (multiplexed display) A device that has multiple digits, can only display one at a time, but displays each digit in turn so rapidly that all the digits appear to be simultaneously present.

Multiplexer (MUX) A circuit that selects which of several inputs will be routed to its output. An address applied to the select lines determines which of the inputs will go to the output. (*see also* TDM bus)

Murray code (*see* Baudot code)

MUX (*see* Multiplexer)

NMOS (N-channel MOSFET logic) A family of logic devices in which the active pull-down element and the passive pull-up are N-channel MOSFETs.

N.C. (*see* Normally closed)

N.O. (*see* Normally open)

NAND circuit (NAND gate) A gate whose output is low when both (or all) of its inputs are high.

NAND gate (*see* NAND circuit)

NAND implementation Construction of a logic circuit using all NAND gates instead of AND, OR, and NOT gates in random logic. Sum-of-products expressions convert into NAND implementation especially well.

NAND latch A pair of cross-coupled NAND gates

which "remembers" the last state delivered through its S and R inputs.

Negative logic Logic in which the active or high level is the voltage from the negative terminal of the power supply.

Noise Unwanted voltage pulses or waveforms that get into electronics.

Noninverting logic (*see* Buffer)

Nonmaskable interrupt A hardware signal that forces a subroutine call, and which cannot be disabled.

NOR circuit (NOR gate) A circuit whose output is low if either (or any) of its inputs is high.

NOR gate (*see* NOR circuit)

NOR implementation Construction of a logic circuit using all NOR gates instead of AND, OR, and NOT gates in random logic. Product-of-sums expressions convert into NOR implementation especially well.

Normally closed (N.C.) A momentary-contact switch whose contacts open when it is operated.

Normally open (N.O.) A momentary-contact switch whose contacts close when it is operated.

NOT circuit (NOT gate) A circuit whose output is low when its input is high, and vice versa.

NOT gate (*see* NOT circuit)

NRZ (non-return-to-zero) A magnetic tape recording standard in which 1s and 0s are represented by the two opposite polarities of magnetism.

NRZI (modified non-return-to-zero) A magnetic tape recording standard in which a polarity transition is a 1 and no transition represents a 0.

Null modem A connector that makes one device's data output go to another device's data input, and vice versa. Additional signal reversals and feedbacks are used according to the complexity of the interface.

O.C. TTL (*see* Open-collector TTL)

Object code The machine language produced when an assembly program or high-level language program is translated into a machine-language program by an assembler or a compiler.

Object program A program compiled or assembled into machine code by a compiler or assembler. (*see also* Object code)

OCR (optical character recognition) Reading of characters printed on paper by optical sensors. Includes barcode and alphanumeric printout.

Octal byte code A code that uses three octal characters to represent each 8-bit byte. Codes range from 000 to 377.

Octal code A code that represents every 3-bit group of binary with a single numeral between 0 and 7.

Odd parity A system of error checking used in data transmission that adds to each transmitted code group a check bit whose value is 1 or 0, according to what is needed to make the sum of its 1s equal an odd number.

Off (*see* Inactive)

On (*see* Active)

Ones'-complement code (*see* Signed numbers)

Opcode A binary number used to activate a circuit or perform an operation in a computer's central processor. In the expression $2 + 2 = 4$, the 2s are *operands*, and the $+$ is an *operation*. Any binary code for the "$+$" would be an *operation code*, or opcode.

Open A circuit that does not conduct electrical current.

Open-collector TTL A variant of the basic TTL design in which there is no pull-up resistor or transistor in the output section.

Optocoupler A device that passes a signal from a light-emitting diode to a phototransistor without electrical connection, preventing "noise" from the output device from returning to the input.

OR circuit (OR gate) A circuit whose output is active if any one of its inputs is active.

OR gate (*see* OR circuit)

Originate frequency The frequency band a modem placing a call uses when it is using full-duplex carrier modulation (2225 Hz = mark; 2025 Hz = space).

Oscilloscope An electronic display instrument which functions as a very fast graph-plotting voltmeter. Its display shows voltage versus time.

Output The part of a digital system, controlled by its internal logic, where signals come out and are used to control external devices.

Output data bus In as separate I/O system, the pins where data leave the system are an output data bus. They are attached to the outputs of memory latches, and are on separate pins from the latch outputs.

Output device In a computer system, the devices that convert digital code from the computer into actions in the outside world are input devices.

Output port An interface that allows data from the computer's data bus to reach an output device *only* when it is selected (by an address).

Overflow flag A flip-flop in the flag byte that indicates when the results of a mathematical operation are larger than the word size of the register where it was done. It is *not* set by borrow or negative results.

PMOS (P-channel MOSFET logic) A family of logic devices in which the active pull-down element and the passive pull-up are P-channel MOSFETs.

Paddle (*see* Game controller)

PAL (programmable array logic) *see* Programmable logic array)

Parallel A circuit which is wired so that current from one terminal of the power supply can reach the other terminal through several different paths.

Parallel data Data in which all the bits are available in different places at the same time.

Parallel full-adder A circuit employing a half-adder and several full-adder gates, which adds two multidigit binary numbers in a single operation, not requiring multiple clock pulses.

Parallel full-adder/subtracter A circuit that can add or subtract two multidigit binary numbers, depending on whether a control signal is in the "add" or "subtract" mode. The circuit performs either addition or subtraction in a single operation, not requiring multiple clockpulses.

Parallel full-subtracter A circuit that can subtract one multidigit binary number from another in a single operation, not requiring multiple clock pulses.

Parallel I/O port A port in which all bits are transferred at once.

Parallel load A transfer operation which puts new bits into all the flip-flops of a multibit register (or counter) at one time.

Parallel register A group of flip-flops whose bits may all be accessed (reached for reading or writing) at one time (both input and output).

Parallel-in, parallel-out register (*see* Parallel register)

Parallel-in, serial-out register (PISO) A register whose flip-flops are all loaded (given an input) at once, but whose output data are clocked out a single bit at a time. A shift-register.

Parallel-to-serial converter (*see* Parallel-in serial-out register)

Parity checker (*see* Parity detector)

Parity detector A circuit that can examine the bits of a binary word, and deliver an output that indicates whether the sum of all the 1s in the word is an *even* or an *odd* number.

Parity flag A flip-flop in the flag byte which contains the output of a parity detector (see above) that examines a word in the accumulator.

Parity generator A circuit whose output adds a 1 or zero to a binary word in order to make the whole combination have odd or even parity.

Passive pull-down A resistor used to ensure that an output is low unless there is an active element pulling up on it (trying to make it high).

Passive pull-up A resistor used to ensure that an output is high unless there is an active element pulling down on it (trying to make it low).

PC (personal computer) A computer based on a microprocessor, and inexpensive enough (with its memory and peripherals) that anyone who can afford to purchase a used car can afford to purchase one of these.

PC (printed circuit) **board** or **card** A sheet of insulating plastic containing foil conductors on its surface(s) that are used to wire-together electrical components into a useful circuit. It replaces hand wiring.

PC (program counter) A counter inside the central processing unit of a computer, used to keep track of where the next instruction in memory will be fetched from.

Peripheral (*see* I/O device)

Persistence A tendency for the human eye to continue to see something that is no longer there for a fraction of a second after it is gone.

Phase inverter (*see* Phase splitter)

Phase splitter A circuit involving a transistor with a load resistor in its collector current path and its emitter current path. The signals at the emitter and collector are inverted with respect to one another, but approximately the same size (same peak-to-peak voltage).

Phase-shift modulation Superimposing information on a sine-wave carrier by varying the phase angle of the wave, compared to a fixed-frequency reference wave.

Photocell A device that converts light energy striking it into eletrical energy. It is usually used as a sensor in input devices.

Phototransistor A transistor amplifier whose base–emitter junction is a photocell, and which uses the base current generated by light to switch on the collector current path as a low-resistance conductor.

Physical address The address formed by combining a *logical address* and a number in a register. It is larger than the logical address found inside a 16-bit computer instruction.

PIA (peripheral interface adapter) A parallel port device that is programmable; that is, its inputs and outputs are defined by software.

PISO (*see* Parallel-in, serial-out register)

PLA (*see* Programmable logic array)

Plotter (graphic plotter) A graphic hardcopy device for turning computer data into pictures (drawings).

Polled interrupt A software program that "polls" (examines) a number of input bits, and executes subroutines according to which inputs are active.

Pop Fetch a number into a register pair from *stack memory*.

POR (*see* Power-on reset)

Positional notation The system of identifying the different positions to the left of a decimal point as ones, tens, hundreds and so on, which is extended to other number bases as powers of 2 (binary), 8 (octal), or 16 (hexadecimal). True binary code is positional; Gray code is not.

Positive logic Logic in which the active or high level is the voltage from the positive terminal of the power supply.

Postfix notation (*see* Infix notation, definition 2)

Power-on reset (POR) An asynchronous input used to reset a flip-flop, counter, or register when power is turned on. At power-up, the program counter of an 8080 or Z-80 computer running CP/M is reset to zero.

Prefix notation (*see* Infix notation, definition 2)

Preset (set-direct) An asynchronous input on a flip-flop that sets it to a Q = 1 state when it is activated.

Product-of-sums expression (maxterm sum) An expression derived from the maxterms of a truth-table, when the maxterms are DeMorganized to given an expression for output = 1. It is of the form S1 · S2 · S3 · . . . , where each S item is a sum of the form I1 + I2 + I3 + · · · (and the Is are input states).

Program A list of instructions, either in machine code or some higher-level language, used to make a computer perform a useful task.

Program counter (*see* PC)

Programmable counter A counter or timer that can be loaded with a number by a parallel load and will trigger an output when the counter has counted down to zero (or up from zero to the number loaded).

Programmable logic array (PLA) A circuit that can emulate the behavior of the truth table. It contains AND-OR and NOT gates attached to all inputs, which can be disconnected from the inputs by fusible links.

Programmable timer (*see* Programmable counter)

PROM (programmable read-only memory) It contains fusible links that can be burned out (opened) to insert 1s into bits of each memory cell.

Propagation delay The time required for an active signal at an input to produce a noticeable effect at the output of a logic device. This may vary, depending on whether the signal is rising or falling.

Pull-down resistor (*see* Passive pull-down) Usually an external resistor.

Pull-up resistor (*see* Passive pull-up) Usually an external resistor.

Pulse-code modulation Information transmitted as serial binary code, after being digitized by an A/D converter and converted from parallel to serial data.

Pulse-width modulation Superimposing input information onto a square waveform by varying its pulsewidth as the input changes.

Punchcard (computer card) Various media that store alphanumeric data as 12-level code (*see* Hollerith code) or six-level code (*see* BCDIC) in the form of rectangular or circular holes punched in a cardboard sheet.

Punched paper tape A medium that contains binary code as circular holes to represent alphanumeric characters. Holes represent 1s, unpunched = 0s.

Push Store a number from a register pair into *stack memory*.

Q output Letter usually used to designate the non-inverted output of a latch or flip-flop. The inverted output is called $\overline{Q}$.

R-2R ladder network (*see* Ladder network)

r/p head(s) Electromagnets used in magnetic tape and disk systems to record magnetic fields onto the surface, and read back magnetic fields as the magnetic surface moves by under the head.

Racing A condition where a sequential logic circuit does more things during one clock pulse than it is supposed to do.

RAM (*see* Random access memory)

Ramp wave (*see* Sawtooth wave)

Random access memory (RAM) A memory device containing latches of one or more bits (cells) that are accessed (enabled to record data or play it back) by an address decoder when an address is delivered to the device. Since address decoding takes the same amount of time regardless of location, the locations could be accessed serially (in numeric order) or randomly (in any old order) without any difference in access time.

Random logic A circuit composed of AND, OR, and NOT gates used to implement some Boolean expression exactly as it is written.

Random scan (*see* Vector graphics)

RAS (*see* Row-address strobe)

Raster graphics (raster scan) Display by using a CRT screen to produce a dot-matrix image of any picture, with the dots arranged on horizontal lines called scan lines. An ordinary television picture is a raster scan.

RB (return-to-bias) A magnetic tape recording standard in which 1s are recorded in one polarity and the rest of the tape is magnetized in the opposite polarity.

Read Input, or playback of information from a memory device.

Read-only memory (ROM) A memory device with burned-in or wired-in 1s and 0s that cannot be altered (overwritten), only read (played back).

Readout (*see* Display)

Rearrange In Boolean simplification, items that are commutative can be rearranged into any order on the line, for convenient grouping.

Receiver (*see* Serial-in, parallel-out register)

Record A unit of information recorded onto tape or disk. A disk record contains an address and a number of bytes of data, varying between a hundred and several thousand. A tape record usually does not contain an address, but on multitrack tape, may be identified by inter-record gaps.

Redundancy In Boolean simplification, the rule that any term in a sum, or factor in a product, that appears twice only needs to be written once.

Refresh (memory refresh) A rewriting operation done in DRAM to prevent loss of data due to charge leakage. (*see also* DRAM)

Register A latch (usually holding more than one bit). If input and ouput to the latch are parallel, it is a *parallel register*. If input and output to the latch are serial, it is a *shift register*.

Register pair(s) Groupings of two registers to form a larger word than the basic register size of the processor. (*see also* Double precision)

Register set The registers found in a central processing unit.

Relative addressing An addressing mode that uses an offset or displacement from the present program-counter value to identify a location elsewhere in memory.

Relay An electromechanical device that closes a pair of metallic contacts when an electromagnet is energized (magnetized by electrical current).

Reset To turn off a latch's Q output (Q = 0).

Resolution A measure of the smallest step digitized (by an A/D) or the smallest variation in the output level (of a D/A), compared with its total range of performance, or measured in absolute units such as voltage.

Restart A one-byte subroutine call used by 8080/8085 and Z-80 processors to direct a *vectored interrupt* to the location of its vectoring command.

Return A control instruction usually found at the end of a subroutine. It pops the program counter value stored on the stack (at the time the subroutine was called) back into the program counter.

Reverse bias The polarity that makes a diode or other semiconductor junction stop conducting, namely, − to P-type and + to N-type material.

Rigid disk (*see* Magnetic disk)

Ring Counter (circulating bit register) A shift register with feedback, set up so that, when clocked, it will shift a single 1 at the beginning flip-flop step by step to the end of the register, then transfer it back to the beginning, and start the entire cycle over.

Ripple counter (*see* Asynchronous counter)

Rising-edge triggered (leading-edge triggered) An input or inputs that are enabled by the zero-to-one transition of the clock.

ROM (*see* Read-only memory)

Rotary switch A switch with a mechanical wiper (contact) that is rotated to touch each of several contacts arranged in a circle. It is used to select different channels in older or less-expensive TV tuners.

Rotate An instruction that circulates bits by using a computer's accumulator as a ring counter.

Row-address strobe (RAS) A signal that becomes active when one-half of the address (the least-significant bits) is sent to a dynamic RAM integrated circuit. The other half is sent through the same pins when a signal called CAS (column-address strobe) is active.

RS-232 standard A serial data transmission interface standard for computer peripherals. It uses a specific plug-and-socket pin arrangement for a variety of signals related to telephone data transmission.

RZ (return-to-zero) A magnetic tape recording standard in which 1s are represented by one magnetic polarity, 0s by the opposite polarity, and the spaces between 1s and 0s are unmagnetized (no polarity, or zero). The magnetic field level always returns to zero between bits, hence the name.

Saturation recording Recording data onto magnetic tape or disk by overdriving the r/p head and magnetizing nearly 100% of the magnetic domains on the tape in the direction desired.

Sawtooth wave Also called a serrasoid or ramp, it has this shape:

Schematic A diagram that symbolically represents the components and connections of an electronic or logic circuit. It is to hardware what a flowchart is to software.

Schmitt trigger TTL Family of bipolar logic devices with inputs that recognize a 1 only if the input voltage goes higher than 2 V, and only recognize a 0 if the input voltage goes below 0.8 V.

Schottky TTL Family of bipolar logic devices that contains low-voltage Schottky junctions instead of ordinary P-N junctions. They switch faster than ordinary TTL. (*see also* Low-power Schottky)

Sectors Wedge-shaped areas on the surface of a disk encoder or magnetic disk. The round surface is cut into sectors that resemble pieces of pie.

Segment register(s) In the Intel 8086 family of 16-bit processors, a segment address and logical address are combined (with the segment address offset) to form the physical address (which is what goes onto the address bus). The segment registers hold the segment addresses for various types of segments (stack segment, code segment, data segment).

Segmented display A readout made of rectangular or line-shaped parts (segments) that can individually be turned on or off to form characters.

Select inputs (address inputs) The inputs of a multiplexer (or demultiplexer) that select which channel is enabled at the inputs (or outputs).

Self-clocking code A technique for recording on tape which contains both bits of data and clock pulses on the same track, alternately.

Sense wire In a core memory, the wire through the core which develops as induced-EMF voltage pulse when a toroid's magnetic field is collapsed.

Separate I/O In semiconductor memory, a design that has separate pins for the input data bus and output data bus.

Sequential access memory (*see* Shift register)

Serial data Data in which all the bits are transmitted on a single conductor at different times.

Serial full-adder An adder circuit that adds the bits of a multidigit binary number separately, in the same adder, at separate times. One clock pulse is required for each bit of the sum.

Serial I/O port An input or output port at which bits are received or transmitted one at a time.

Serial-in, parallel-out register (SIPO) A shift register into which data are clocked during several clock cycles, but once the register is filled, all bits are available at separate outputs at the same time.

Serial-in, serial-out register (*see* Shift register)

Series A circuit in which current from one power supply terminal must pass through all the components before reaching the other terminal.

Serrasoid wave (*see* Sawtooth wave)

Set To turn on a latch's Q output (Q = 1).

Shaft encoder A cylinder with strips of binary or Gray code marked on its surface in various places so that its position can be read by sensors.

Shift counter (*see* Johnson counter)

Shift register A cascade of flip-flops with data outputs and inputs connected so that data transfers down the chain a distance of one flip-flop with every clock pulse.

Shorted Connected together without electrical resistance.

Sign bit The most significant bit of a signed number. It does not have any significance as a binary value; it is a "−" sign if equal to 1.

Sign flag A flip-flop in the flag byte that holds a copy of the sign bit.

Signed numbers Numbers whose bits are (positive) true binary if the most significant bit (sign bit) is a 0 (+ sign) and (negative) twos'-complement if the most-significant bit (sign bit) is a 1 (− sign).

Signed twos'-complement numbers (*see* Signed numbers)

Simplex Transmission in only one direction. Commercial radio and TV broadcast are simplex (one-way communication).

Simplification (*see* Boolean simplification)

Single-slope A/D converter A digitizer that uses an up-counter and an up-slope ramp wave to digitize an input level. The counter counts until the ramp wave equals the input. Since the time is proportional to the level to which the ramp wave has risen, the count is proportional to the input level, and the number in the counter is the digitized value of the input. The counter and ramp must be reset to zero before another sample can be digitized.

Sink load A load (usually the input of another gate) that requires more drive current when driven by a logic gate's output that is LOW than it does when driven by a HIGH output.

SIPO (*see* Serial-in, parallel-out register)

Software The 1s and 0s (codes) stored in a binary system, especially when a sequence of these codes is used to activate parts of a CPU (a program).

Source code The high-level or symbolic language statements or mnemonics used in a source program, which will be translated into machine code by an assembler or compiler.

Source load A load (usually the input of another gate) that requires more drive current when driven by a logic gate's output that is HIGH than it does when driven by a LOW output.

Source program A program written in source code. (*see also* Source code)

Space A logic 0 in FSK encoding or telecommunication transmission.

Space-division multiplex (SDM) Transmission of multiple channels of information at one time on separate conductors.

Split-load phase inverter (*see* Phase splitter)

Stack pointer The up-down counter in a central processor that keeps track of the current "top-of-stack" location in stack memory.

Standard TTL (*see* Transistor-transistor logic)

Start bit A binary 0 level at the beginning of an asynchronously transmitted serial data character, added to keep the receiving equipment in synchronization with the transmitting equipment.

State table A table of last-state/next-state conditions for a circuit's outputs and inputs. The information it contains is similar to an STD.

State transition diagram (STD) A logic-flow diagram showing the states of a circuit's outputs at each timing state (clock pulse) in its normal sequence of operation.

Statement An instruction in a high-level language, often not equivalent to any single operation in machine code.

Static sensitivity A problem high-impedance inputs have because they cannot leak-off electrostatic charges fast enough to prevent buildup of voltage to dangerous levels.

Steering network A pair of feedback lines in a J-K flip-flop that makes the enabling of the J or K input depend on which output is high.

Stop bit(s) A binary 1 level (sometimes two 1s) at the end of an asynchronously transmitted serial data character, added to keep the receiving equipment in synchronization with the transmitting equipment.

Strings Groups of binary bytes intended as ASCII or some other alphanumeric code. They are usually groups of letters that spell some word or phrase, and are therefore not to be interpreted as numeric data.

Strobe (*see* Clock)

Structured programming Design of software so as to take maximum advantage of subroutining, loops, and relocatable code.

Stuck high (*see* Frozen high) Stuck low (*see* Frozen low)

Subroutine A segment of code, ending in a RETurn command, that is used several times at different ponts in the main program. An algorithm for multiplication, in a computer without hardware multiplication available, would be a good thing to write as a subroutine. The CALL to the subroutine would become, for all practical purposes, the opcode for "multiply."

Subtracter A circuit whose output is the difference between two binary numbers, and if a borrow to a higher digit is not found, the difference becomes a twos'-complement negative number.

Subtrahend A number being subtracted from another number (the minuend).

Successive-approximation A/D converter A counter-type A/D converter that uses an unusual counter; the count varies the most-significant bit, then the next, and so on down to the least-significant bit, and finds a digital approximation to the input in a number of clock pulses equal to the number of bits in the result.

Successive-approximation register The counter in a successive approximation A/D converter (see description above).

Sum The output of a half-adder or full-adder that indicates the result of addition if the result is a one-digit number. It indicates the least-significant digit of the result if the result is a two-digit number, and the other digit (the carry) is added to the next column of numbers (if there is one).

Sum-of-products expression (minterm sum) An expression derived from the minterms of a truth table, combined to given an expression for output = 1. It is of the form $P1 + P2 + P3 + \cdots$, where each P item is a product of the form $I1 \cdot I2 \cdot I3 \ldots$ (and the Is are input states).

Summing amplifier An analog op-amp circuit with several inputs, whose output is the sum of the inputs, each multiplied by the gain factor appropriate to its input resistor (R_i) and the feedback resistor (R_o).

Switch diagram A logic circuit schematic diagram in which all the active elements in the circuit are mechanical switches.

Switching amplifier An amplifier that digitizes its input into pulse-width modulation before amplifying it. As a result, the amplifying device (the *gain element*) does not have to be linear, or even proportional to the input. If the gain element can also be driven into saturation or cut off at both ends of the digital input signal (class D operation), its heat dissipation (wasted power) can be reduced to near-zero as well.

Switching power supply (*see* Switching regulator)

Switching regulator (switching power supply) A voltage-regulating power-output device that can operate directly from line voltage. It uses a digital chopper device to gate current to a capacitor, keeping the charge at a nearly constant level. Because the current-conducting chopper element is operating as a class D amplifier, its heat dissipation is very low, and the power supply can be very small compared to conventional designs for the same wattage.

Symbol (symbolic diagram) A shape or figure representing a logic gate. The traditional logic symbols distinguish the function of the gate according to its distinctive shape. The IEC logic symbols distinguish different gate types according to a character written inside the gate.

Synchronous counter A counter designed so that as it counts, all the bits that are going to change, change at the same time.

Synchronous logic Logic that is triggered to respond to its inputs when a signal at a clock input, usually a rising edge or falling edge, changes.

T-type flip-flop A flip-flop whose only input is its clock, and whose only response is to toggle at the rising or falling edge of the clock.

Tablet A graphics input device on which you draw; sensors under its surface convey the position of the drawing stylus to the computer.

Tapemark A code or foil reflector at the beginning or end of a reel of multitrack reel-to-reel digital tape.

TDM bus A bus or cable that carries separate information to many destinations from many sources, by doing one transfer at a time.

Telecommunications Long-distance transfer of information.

Terminal (*see* Data terminal)

Terms Things with "+" signs between them.

Three-state TTL [also called Tri-State (trade name)] A family of logic devices whose output may pull up on a bus line when HIGH, pull down on a bus line when LOW, or go into a high-impedance state when the output is disabled.

Threshold current The minimum amount of current required to trigger a reversal of magnetic field polarization in a core-memory toroid.

Throughput The rate at which input data are processed through to the output by a computer.

Time-base generator An oscillator or multivibrator that provides clock pulses to a digital circuit, or a sweep waveform to an analog circuit like an oscilloscope.

Time-division multiplex (TDM) Transmission of multiple channels of information on one conductor at separate times.

Timing diagram A graph of voltage versus time, like an oscilloscope trace, but having only TRUE and FALSE levels, and used to display the relative time relationship of all input and output signals active in some digital operation.

Timing states The intervals between clock pulses, or during each cycle of the clock.

Toggle A change in the outputs of a flip-flop (each output is the complement of its previous state), following a clock pulse.

Tracking A/D converter (*see* Dual-slope A/D converter)

Tracks On a code wheel or magnetic disk, zones or bands that circle the center of the disk at evenly spaced radial distances from the center.

Trailing-end triggered (*see* Falling-edge triggered)

Transfer circuit A circuit in which data are transferred from one flip-flop's outputs to another flip-flop's inputs when a clock pulse occurs.

Transistor–transistor logic (TTL) A large number of families of logic devices based on the bipolar junction transistor, and characterized by multiple inputs attached to the emitters of a multiemitter transistor.

Transition A change in logic state from zero to one or one to zero, sometimes called an "edge," as in "rising edge" or "falling edge."

Transmitter A device for transmission of serial data when the original data are parallel. (*see also* Parallel-in, serial-out register)

Transparent latch (*see* Level-triggered latch)

Transphasor An optical amplifier that controls light and is switched on and off by light.

Tri-State TTL (*see* Three-state TTL)

Trigger An input that sets a sequence of actions in motion, like a clock or the trigger input of a one-shot, which starts a timed pulse.

True A logical input or output operating in the condition in which it was intended to be operating.

True binary code A binary code in which the value of the number is the sum of all the place values containing 1s. The place values are increasing powers of 2 to the left of the radix point (the "decimal point" in a non-decimal-number base). This code is for positive numbers.

Truth table A grid of binary numbers that represents the state of the output of a digital system, for every possible pattern of states at the inputs of that system. (Answers the question: For each possible input, what is the output doing?)

TTL (*see* transistor–transistor logic)

Turnkey system A computer that is "ready to go" for the application it was designed for, as soon as power is turned on.

Twos'-complement code (*see* Signed numbers)

T²L (*see* Transistor–transistor logic)

UART (universal asynchronous receiver/transmitter) A serial-to-parallel and parallel-to-serial converter in one interface, usually intended for transmission and reception of serial data via telephone lines, in connection with the parallel input and output ports of a computer.

Unary Having only one operand. A mathematical operation is unary if it doesn't combine its target number with anything else. A square root is an example of a unary operation; so is a sine or cosine.

Unconditional branch The contents of the program counter is replaced by a number in the instruction or an address formed by the instruction.

Unit load The fan-in of one elementary input of a logic device.

Unit record (*see* Punchcard)

Universal asynchronous receiver/transmitter (*see* UART)

Universal gate (universal logic) A NAND or NOR gate, or any logic gate from which all other types of logic can be done by a combination of that gate. If the universal gate can be made into AND, OR, and NOT circuits, it can be made into any other digital circuit.

Universal logic (*see* Universal gate)

Unsigned numbers Numbers in which none of the bits is used for a sign bit and every number is intrinsically assumed to be larger than zero.

Up counter A counter that (when clocked) only counts numbers that increase in value, until it reaches the end of its modulus and starts over.

Up-down counter A counter that can either count increasing values when clocked, or decreasing values. It has the same modulus either way.

UV-PROM A read-only memory device that stores 1s as charged regions that can be discharged or erased only by exposing the silicon surface to ionizing radiation [ultraviolet (UV)], which discharges the 1s via the photoelectric effect.

Variables Quantities that change in value as a computer program is running. They are usually represented by symbolic names (alphabetic).

V_{CC} The positive voltage terminal of a power supply used for digital ICs. Technically, this is the collector supply for a bipolar transistor amplifier, but it has come to be used as *any* positive supply, usually $+5$ V.

VCO (*see* Voltage-controlled oscillator)

Vector graphics (vector scan or random scan) Producing a picture on a CRT picture tube by steering the electron beam around the face of the tube in the outline of the object being drawn.

Vectored interrupt A hardware-driven subroutine call that goes to an address in memory where there is a jump to another address (the *vector*).

V_{EE} The negative terminal of the power supply. Often, a voltage more negative than the ground reference in a circuit with split supply.

Vertical sync The pulse that returns the electron beam to the top of the screen in a raster-scan CRT display, so that it can repeat or interlace a frame.

Video RAM The memory device that holds the codes for an alphanumeric CRD display screen. These characters are circulated through the video system so that it can redraw them with every frame repetition.

V_{IH} The lowest input voltage that is still accepted as a 1.

V_{IL} The highest input voltage that is still accepted as a 0.

Virtual machine The rules by which the computer makes its inputs and outputs behave when it is interacting with its user. To the user, the way the inputs and outputs work *is* the machine, hence the name.

V_{OH} The lowest 1 level expected at a functioning output.

V_{OL} The highest 0 level expected at a functioning output.

Voltage divider A series of resistors used to develop various reference voltages for a flash converter.

Voltage-controlled oscillator An oscillator whose output frequency is proportional to the voltage applied to its input terminal.

V_{SS} The negative or ground terminal of a power supply used for digital ICs. Technically, this is the source supply for a MOSFET amplifier, but it has come to be used for the negative terminal of *any* positive supply.

Wait state A condition in which the processor holds all active signals on address, data, and control buses for an extra clock cycle (or more) to allow slow memory or a slow peripheral to respond to the bus signals.

Winchester disk (*see* Magnetic disk)

Wire-AND (*see* Wire-OR)

Wire-OR If open-collector outputs of several gates are tied together to a common bus line, a zero from the first gate, OR the second, OR the third, will deliver a zero to the common bus line. Since the bus line can be controlled by gate A OR by B OR by C, and so on, this is called wire-OR logic. Since the signals are active LOW, it can also be called wire-AND.

Wiring diagram Different from a schematic in that the actual connections are shown, indicating how to make the connections (for example, on a breadboard setup), rather than represented symbolically.

Word A word is however many bits a computer puts into its accumulator. For a 16-bit microprocessor, it is 16 bits; for an 8-bit processor, a word is 8 bits, and so forth.

Word length The size of a *word*, in bits (see above).

Word size (*see* Word length)

WORM (*see* Laser disk)

Write Output, or recording of information into a memory device.

x-wire A column wire in a 2D or 3D core memory.

X-Y drive A mechanical interface for a flatbed plotter.

Xerographic printer A laser-driven electrostatic printer.

XNOR (exclusive-NOR) **circuit** (exclusive-NOR gate) A gate whose output is HIGH if its two inputs are in the same state, LOW otherwise.

XOR (exclusive-OR) **circuit** (exclusive-OR gate) A gate whose output is HIGH if its two inputs are in different states, LOW otherwise.

y-wire A row wire in a 2D or 3D core memory.

z-axis modulation Using some information signal to vary the brightness of an electron beam on a CRT display.

Zero flag The flip-flop that is set if all the bits resulting from some mathematical operation in a central processor are zero.

INDEX

B

C

Phase-shift modulation, 450
Phase splitter, 125
Photocell detector, 107, 339
Physical address, 361
PIA (parallel interface adapter), 278
Piezo-optic effect, 254
Pilot beam, 459
Pixel, 311
PLA (programmable logic array), 248
Plotter, 334
PMOS, 137
Polled interrupt, 407, 409, 411
POP instruction, 372, 383, 403
Positional notation, 52
Positive logic, 119
Power-on reset (POR) circuit, 207, 208
Preset, 207
Process, 396
Program, 263, 349, 394
Program counter, 350, 354, 377
Programmable binary up/down counter (74193), 218
Programmable up-down counter, 217
PROM (field-programmable ROM), 246
Proof by perfect induction, 13
Propagation delay, 145, 456
Pull-down resistor, 121
Pull-up resistor, 119
Pulse-width modulation, 229
Punch mechanism, 335–36
Punchcard, 306
Punched paper tape, 309, 336
PUSH instruction, 372, 383, 403
Pushbutton, 2

Q

Q output, 161, 244, 255
Qualifying symbols, 466

R

Racing, 170, 188
RAM (random-access memory), 239
RAM test diagnostic, 398–406
Ramp generator, 228
Random logic, 56, 65
Random scan, 323
Random-access memory (RAM), 239
RAS (row-address select) input, 255
Raster scan, 323, 325
RB (return-to-bias) tape recording, 337
R dependency, 470
Read (RD) control signal, 351
Read-in, 235
Reading, 235, 237, 241
Read-only memory (ROM), 239, 245
Read-out, 236, 237

Receiver, 197
Record (tape), 301
Redundancy, 40, 43
Reel-to-reel tape drives, 300
Refresh, 256
Register, 11, 354
Register pair, 371
Reinforcement, 456
Relative jump, 378
Relay, 338
Repeater, 446
Reset, 162, 351, 409
Resolution, 106, 112, 226
RET (return) instruction, 379, 384, 402
Reverse bias, 119
RFI (radio-frequency interference), 273, 454
Rigid disk, 303
R (reset) input, 162, 242
Ring counter, 194
Ripple counter, 204
Rise time, 145
ROM (read-only memory), 239, 245, 249, 294, 415
Rotary switch, 62
Rotate and shift instructions, 380–81
Row-address strobe (RAS), 255
R/P (record/playback) head, 300, 302, 337
RS-232, 440–45
RS-449, 445
R-S master/slave flip-flop, 172
RST (restart) instruction, 379, 409
RTS (request to send), 443
RZ (return-to-zero) tape recording, 337

S

SAR (successive-approximation register), 227
Saturated, 127
Saturation recording, 300, 337
Scan (keyboard), 290–93
Schmitt trigger, 133
Schmitt trigger TTL, 133
Schottky TTL, 132
SCR motor controller, 341
S dependency, 470
SDM (space-division multiplex), 436
Sector, 302
Segmented display, 318
Segment registers, 360
Segments (memory), 361
Semiconductor memory, 239
Sense wire, 236
Sensor, 312
Sequential-access memory, 239
Serial data, 193, 435
Serial full adder, 191, 192
Serial port, 282
Serial-to-parallel converter, 197
Series, 3
Serrasoid (sawtooth) wave, 225, 230

Unsigned binary, 90
Up-counter, 204
Uplink, 451
Upper trip point, 133
UV-PROM, 247

V

V (IH), 147
V (IL), 147
V (OH), 147
V (OL), 147
VCO (voltage-controlled oscillator), 229
V dependency, 468
Vector, 410
Vectored interrupt, 407, 409
Vector graphics, 325
Veitch diagram, (see Karnaugh map)
Vertical sync, 326
Video digitizer, 311
Video divider chain, 324, 325
Video interface, 325
Video RAM, 324, 325
Virtual machine, 415
Virtual storage, 417
Voice recognition, 314
Volatile, 242
Voltage-controlled oscillator (VCO), 229

W

Wait, 352
Winchester disk, 302

Wire matrix printer, 331
Wire-AND, 131
Wire-OR, 131, 140, 269
Wiring diagram for a Boolean expression, 47, 48
Word, 239
Word length, 90
WR input, 243
Write (WR) control signal, 351, 352
Writing, 235, 240

X

X decoder matrix, **236, 242**
X dependency, 469
Xerographic printer, 332
XNOR gate, 27, 30, 32
XOR gate, 26, 30, 32
X wire, 235, 242
X-Y drive, 334

Y

Y decoder matrix, 236, 242
Y wire, 235, 242

Z

Z dependency, 469
Zero flag, 356
Zilog Z-80, 353, 462, 464
Zilog Z-8000, 358